Oxford Revision Guides

AS & A Level

BIOLOGY

Through Diagrams

W R Pickering

OXFORD
UNIVERSITY PRESS

Great Clarendon Street, Oxford OX2 6DP

Oxford University Press is a department of the University of Oxford.
It furthers the University's objective of excellence in research, scholarship,
and education by publishing worldwide in

Oxford New York

Auckland Bangkok Buenos Aires Cape Town Chennai
Dar es Salaam Delhi Hong Kong Istanbul Karachi Kolkata
Kuala Lumpur Madrid Melbourne Mexico City Mumbai Nairobi
São Paulo Shanghai Taipei Tokyo Toronto

First published 2009

British Library Cataloguing in Publication Data

Data available

ISBN-13: 9780199180912

10 9

Typeseting, artwork and design by Steve Evans Design and Illustration

Printed in Great Britain by Bell and Bain Ltd, Glasgow

CONTENTS

Specification structures

	AQA	Method of assessment
Unit 1	**Biology and disease:** pathogens; lifestyle; digestive system; action of enzymes; structure of molecules; cells and membranes; gas exchange; blood and circulation	Examination paper (60 raw marks = 100 UMS). Composed of 5-7 short answer structured questions, plus 2 longer questions (a short comprehension and a structured question requiring continuous prose). 1 h 15 min AS 33% A 16.7% **Available in January and June**
Unit 2	**Variety of living organisms:** variation; DNA and proteins; DNA and diversity; variation in molecules and cells; the cell cycle; cells; tissues and organs; adaptation and evolution; classification; biodiversity	Examination paper (85 raw marks = 140 UMS). Composed of 7-9 short answer structured questions plus 2 longer questions (one data handling and 1 assessing analysis and evaluation). 1 h 45 min AS 46.7% A 23.3% **Available in January and June**
Unit 3	**Practical skills:**	Either BIO3T, Centre Marked Route T (50 raw marks/60 UMS). Practical Skills Assessment (PSA – 6 raw marks). Investigative Skills Assignment (ISA – 44 raw marks). Or BIO3X, Externally Marked Route X (50 raw marks/60 UMS). Practical Skills Verification (PSV – teacher verification). Externally Marked Practical Assignment (EMPA – 50 raw marks). AS 20% A 10% **Available in June only**
Unit 4	**Populations and environment:** factors affecting populations; ATP and energy; respiration; photosynthesis; ecosystems; cycling of nutrients; isolation and natural selection	Examination paper (75 raw marks / 100 UMS). 6 – 9 short answer questions plus 2 longer questions involving continuous prose and *How Science Works*. 1 h 30 min A 16.7% **Available in January (from 2010) and June**
Unit 5	**Control in cells and in organisms:** stimuli and responses; chemical and electrical coordination; homeostasis and negative feedback; oestrus cycle; DNA determines protein structure; gene technology	Examination paper (100 raw marks / 140 UMS). 8 – 10 short answer questions plus 2 longer questions (a data-handling question and a synoptic essay - choice of 1 out of 2). 2 h 15 min A 23.3% **Available in June only**
Unit 6	**Practical skills:**	Either BIO6T, Centre Marked Route T (50 raw marks/60 UMS). Practical Skills Assessment (PSA – 6 raw marks). Investigative Skills Assignment (ISA – 44 raw marks). Or BIO6X, Externally Marked Route X (50 raw marks/60 UMS). Practical Skills Verification (PSV – teacher verification). Externally Marked Practical Assignment (EMPA – 50 raw marks). A 10% **Available in June only**

WHAT ARE
Structured questions?

This type of question is broken up into smaller parts. There is usually only a single mark, or a few marks, for each part. Some parts will ask you to

- recall a fact – the name of a biological molecule, for example
- define a biological term – ecological succession, for example
- obtain information from graphs or tables of data
- draw labelled diagrams, or add labels/annotations to a diagram.

Other parts might

- lead you to the explanation of a complex problem by asking you to provide a series of answers in sequence – a calculation of probability, for example.

	OCR	Method of assessment
F211	**Cells:** cell structure; cell membranes; cell division; cell diversity and cell organisation **Exchange and transport:** exchange surfaces and breathing; transport in animals; transport in plants	Candidates answer all questions. 1 h written paper 60 marks. AS 30% A 15% **Available in January and June**
F212	**Molecules, diversity, food and health:** biological molecules; food and health; biodiversity and evolution	Candidates answer all questions. 1 h 45 min written paper 100 marks. AS 50% A 30% **Available in January and June**
F213	**Practical skills:**	Candidates complete three tasks set by OCR. Tasks are marked by the centre using mark schemes provided by OCR. Coursework 40 marks. AS 20% A 10% **Available in June only**
F214	**Communication, homeostasis and energy:** communication; homeostasis and excretion; photosynthesis; respiration	This unit contains some synoptic assessment and stretch and challenge questions. 1 h written paper 60 marks. A 15% **Available in January and June**
F215	**Control, genomes and environment:** cellular control; gene technology; ecosystems; responding to the environment	This unit contains some synoptic assessment and stretch and challenge questions. 1 h 45 min written paper 100 marks. A 25% **Available in June only**
F216	**Practical skills:**	Candidates complete three tasks set by OCR. Tasks are marked by the centre using mark schemes provided by OCR. Work is moderated by OCR. This unit is syntopic. Coursework 40 marks. A 10% **Available in June only**

Comprehension questions?
In this type of question you will be given a passage about a biological topic and then asked a series of questions to test your understanding of the topic and the scientific principles in it. The actual content of the passage may be biological material that you're not familiar with.

'Essay' questions?
These questions will test your ability to recall information and to organise it into the form of written prose. Some credit will be given for 'style' (good organisation of material) as well as for the biological content.

Synoptic questions?
These questions are only used for A2 examination papers. They will test your ability to deal with information from different parts of your course, and to use biological skills in contexts which might be unfamiliar to you. These questions may be structured in style, but will be longer than those used in the AS papers. 20% of the A level marks are allocated to synoptic questions.

	WJEC	Method of assessment
BY1	**Basic biochemistry and organisation:** biological molecules; cell structure and organisation; cell membranes and transport; enzymes; nucleic acids	20% 1 h 30 min written paper 70 marks (120UM). Outline of paper structure – short and longer structured questions, choice of 1 from 2 essays. **Available in January and June**
BY2	**Biodiversity and physiology of body systems:** evolutionary history and classification; adaptations: gas exchange; transport; reproduction; nutrition and parasitism	20% 1 h 30 min written paper 70 marks (120UM). Outline of paper structure - short and longer structured questions, choice of 1 from 2 essays. **Available in January and June**
BY3	**Practical skills:**	10% Internal assessment 44 marks (60UM). Experimental work set in centre, completed by candidates over 3 month period. Marked by board plus, low power plan microscope drawing. **Available in June only**
BY4	**Metabolism, microbiology and homeostasis:** importance of ATP; respiration; photosynthesis; microbiology; populations; homeostasis; nervous system; plant coordination	20% 1 h 45 min written paper 80 marks (120UM). Outline of paper structure - short and longer structured questions, choice of 1 from 2 essays. Small % synoptic marks. **Available in January and June**
BY5	**Environment, genetics and evolution:** genetic code and cell function; reproduction in Humans and plants; inheritance; variation and evolution; applications of reproduction and genetics; energy and ecosystems; human effects	20% 1 h 45 min written paper 80 marks (120UM). Outline of paper structure – short and longer structured questions, choice of 1 from 2 essays. Small % synoptic marks. **Available in June only**
BY6	**Practical skills:**	Experimental work set in centre, completed by candidates over 3 month period. Marked by board plus one microscope drawing and calibration. **Available in June only**

	CCEA	Method of assessment
AS 1	**Molecules and cells:** molecules: enzymes; DNA technology; viruses; cells, cell physiology and cell division; tissues and organs	Written examination, externally assessed. 1 h 30 min AS 40% A 20% **Available in January and Summer**
AS 2	**Organisms and biodiversity:** transport and exchange; adaptations of organisms; biodiversity	Written examination, externally assessed. 1 h 30 min AS 40% A 20% **Available in January and Summer**
AS 3	**Practical skills:**	Internal practical assessment. AS 20% A 10% **Summer only**
A2 1	**Physiology and ecosystems:** homeostasis and excretion: endocrine control and osmoregulation; immunity; coordination and control; ecosystems	Written examination, externally assessed. 2 h A2 40% A 20% **January and Summer**
A2 2	**Biochemistry, genetics and evolutionary trends:** respiration; Photosynthesis; DNA as the genetic code; gene technology; patterns of inheritance; mechanism of change	Written examination, externally assessed. 2 h A2 40% A 20% **Summer only**
A2 3	**Practical skills:**	Internal practical assessment. A2 20% A 10% **Summer only**

Pathways

The following pathways identify the ***main sections*** in this book that relate to course units for each Examination Board. Note that

- you will not necessarily need all of the material given in any one section
- there may be material in other sections that you need to know
- you should identify the relevant material by referring to the specification you are following
- if you own the book you could highlight all of the relevant information.

Units 1, 2 and 3 are AS units and Units 4, 5 and 6 are the A2 units. All units make up the full 'A' level.

	AQA	Book sections
Unit 1	**Biology and disease:** pathogens lifestyle digestive system action of enzymes structure of molecules cells and membranes gas exchange blood and circulation	1, 2, 3-21, 31-38, 40-56
Unit 2	**Variety of living organisms:** variation DNA and proteins DNA and diversity variation in molecules and cells the cell cycle cells, tissues and organs adaptation and evolution classification biodiversity	18, 19, 20, 22-26, 39, 57-62, 64-72, 83-85, 101-106, 119, 120
Unit 3	**Practical skills:**	200-208
Unit 4	**Populations and environment:** factors affecting populations ATP and energy respiration photosynthesis ecosystems cycling of nutrients isolation and natural selection	67, 68, 69, 70, 73-80, 82, 87, 89-93, 95-100, 111, 122, 137-164
Unit 5	**Control in cells and in organisms:** stimuli and responses chemical and electrical coordination homeostasis and negative feedback oestrus cycle DNA determines protein structure gene technology	166-184, 188-203
Unit 6	**Practical skills:**	200-208

	OCR	Book sections
F211	**Cells:** cell structure; cell membranes; cell division; cell diversity and cell organisation **Exchange and transport:** Exchange surfaces and breathing; transport in animals; transport in plants	12-20, 24-26, 30, 31, 34, 35 42-45, 49-52, 60-62, 117-121
F212	**Molecules, diversity, food and health:** biological molecules food and health biodiversity and evolution	1, 2, 3-11, 22, 23, 64, 78-80, 101-105, 106-110, 124-133
F213	**Practical skills:**	200-208

	OCR	Book sections
F214	**Communication, homeostasis and energy:** communication homeostasis and excretion photosynthesis respiration	115, 122, 145-147, 144-154, 149, 170, 174, 176-181, 189-202
F215	**Control, genomes and environment:** cellular control gene technology ecosystems responding to the environment	69, 70, 73-82, 86, 88, 91-100 137-142, 149, 155-156, 162, 166-169, 182-187
F216	**Practical skills:**	200-208

	WJEC	Book sections
BY1	**Basic biochemistry and organisation:** biological molecules cell structure and organisation cell membranes and transport enzymes nucleic acids	4-15, 18-24, 27-31, 34-37, 69, 71, 72, 83, 87
BY2	**Biodiversity and physiology of body systems:** Evolutionary history and classification Adaptations: gas exchange; transport; reproduction; nutrition and parasitism	2, 32, 38, 42-44, 47, 49-52, 57, 62, 78-80, 101-105, 112, 115-119, 121
BY3	**Practical skills:**	200-208
BY4	**Metabolism, microbiology and homeostasis:** importance of ATP respiration photosynthesis microbiology populations homeostasis nervous system plant co-ordination	122, 134, 135, 136, 138-142, 144-146, 151-154, 158-160, 162, 165, 166, 168-170, 177-181, 190, 195-198, 200, 201
BY5	**Environment, genetics and evolution** genetic code and cell function reproduction in Humans and plants inheritance variation and evolution applications of reproduction and genetics energy and ecosystems human effects	64-67, 69, 70, 73-79, 81, 82, 84, 85, 88-90, 92, 93, 95, 96, 97, 107, 111, 137, 149, 155-157, 163, 164, 202
BY6	**Practical skills**	200-208

	CCEA	Book sections
AS 1	**Molecules and cells:** molecules enzymes DNA technology viruses cells, cell physiology and cell division tissues and organs	3-36, 71, 83, 84, 87, 96, 98-100, 115, 132
AS 2	**Organisms and biodiversity:** transport and exchange adaptations of organisms biodiversity	38, 39, 42-44, 48-52, 57-62, 64-66, 101-105, 108-111, 116, 117, 137
AS 3	**Practical skills:**	200-208
A2 1	**Physiology and ecosystems:** homeostasis and excretion endocrine control and osmoregulation immunity coordination and control ecosystems	53-56, 107, 137- 142, 149, 155-157, 163, 166-173, 175-181, 183, 188, 190, 195-200
A2 2	**Biochemistry, genetics and evolutionary trends:** respiration photosynthesis DNA as the genetic code gene technology patterns of inheritance mechanism of change	68-70, 73-82, 89-95, 144-147, 150-154, 212
A2 3	**Practical skills:**	200-208

There is no one method of revising which works for everyone. It is therefore important to discover the approach that suits you best. The following rules may serve as general guidelines.

GIVE YOURSELF PLENTY OF TIME

Leaving everything until the last minute reduces your chances of success. Work will become more stressful, which will reduce your concentration. There are very few people who can revise everything 'the night before' and still do well in an examination the next day.

PLAN YOUR REVISION TIMETABLE

You need to plan your revision timetable some weeks before the examination and make sure that your time is shared suitably between all your subjects.
Once you have done this, follow it – don't be side-tracked. Stick your timetable somewhere prominent where you will keep seeing it – or better still put several around your home!

RELAX

Concentrated revision is very hard work. It is as important to give yourself time to relax as it is to work. Build some leisure time into your revision timetable.

GIVE YOURSELF A BREAK

When you are working, work for about an hour and then take a short tea or coffee break for 15 to 20 minutes. Then go back to another productive revision period.

Four steps to success in examinations

You may already have read about the brain. If you have you might remember that an important function of the cerebral hemispheres is to act as an ***integration centre*** – input of information is compared with previous experience and an appropriate action is taken. As you prepare for an examination you will be inputting factual material and skills and you will be hoping that, when you're faced with examination papers, you will be able to make the appropriate responses! The effort you make in ***revision*** and your willingness to ***listen to advice on techniques*** will greatly affect your likelihood of success, as outlined here.

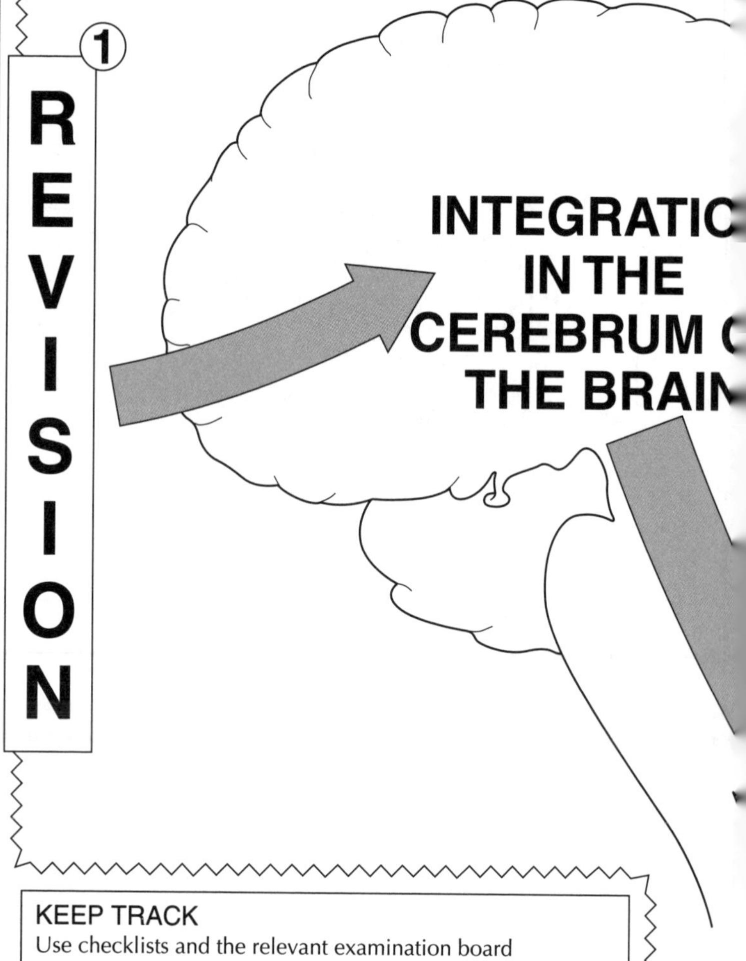

KEEP TRACK

Use checklists and the relevant examination board specification to keep track of your progress. Mark off topics you have revised and feel confident with. Concentrate your revision on things you are less happy with.

MAKE SHORT NOTES, USE COLOURS

Revision is often more effective when you do something active rather than simply reading material. As you read through your notes and textbooks make brief notes on key ideas. If this book is your own property you could highlight the parts of pages that are relevant to the specification you are following. Concentrate on understanding the ideas rather than just memorising the facts.

PRACTISE ANSWERING QUESTIONS

As you finish each topic, try answering some questions. There are some in this book to help you (see pages xxvii and xxxii). You should also use questions from past papers. At first you may need to refer to notes or textbooks. As you gain confidence you will be able to attempt questions unaided, just as you will in the exam.

2 KNOW WHAT TO DO

LEARN THE KEY WORDS

Name: the answer is usually a technical term (mitochondrion, for example) consisting of no more than a few words. ***State*** is very similar, although the answer may be a phrase or sentence. ***Name*** and ***state*** don't need anything added, i.e. there's no need for explanation.

Define: the answer is a formal meaning of a particular term i.e. 'what is it?'. ***What is meant by...?*** is often used instead of ***define***.

List: You need to write down a number of points (each may only be a single word) with no need for explanation.

Describe: your answer will simply say what is happening in a situation shown in the question, e.g. 'the temperature increased by 25 °C' – *there is no need for explanation.*

Suggest: you will need to use your knowledge and understanding of biological topics to explain an effect that may be new to you. You might use a principle of enzyme action to suggest what's happening in an industrial process, for example. There may be more than one acceptable answer to this type of question.

Explain: the answer will be in extended prose, i.e. in the form of complete sentences. You will need to use your knowledge and understanding of biological topics to write more about a statement that has been made in the question or earlier in your answer. Questions like this often ask you to ***state and explain ....***

Calculate: a numerical answer is to be obtained, usually from data given in the question. Remember to

- give your answer to the correct number of significant figures, usually two or three
- give the correct unit
- show your working.

3 IN THE EXAMINATION

- Check that you have the correct question paper! There are many options in some specifications, so make sure that you have the paper you were expecting.
- Read through the whole paper before beginning. Select the questions you are most comfortable with – there is no rule which says you must answer the questions in the order they are printed!
- Read the question carefully – identify the key word (why not underline it?).
- Don't give up if you can't answer part of a question. The next part may be easier and may provide a clue to what you should do in the part you find difficult.
- Check the number of marks allocated to each section of the question.
- Read data from tables and graphs carefully. Take note of column headings, labels on axes, scales, and units used.
- Keep an eye on the clock – perhaps check your timing after you've finished 50% of the paper.
- Use any 'left over' time wisely. Don't just sit there and gaze around the room. Check that you haven't missed out any sections (or whole questions! Many students forget to look at the back page of an exam paper!). Repeat calculations to make sure that you haven't made an arithmetical error.

4 SUCCESS!

Practical assessment (units 3 and 6)

Your practical skills will be assessed at both AS and A level. Make sure you know how your practical skills are going to be assessed.
You may be assessed by:

- **Coursework, set by your teacher**
- **Practical examination, set by the examination board**

The method of assessment will depend on the specification you are following and the choice of your school/college. You may be required to take a combination of different types of assessment.

PRACTISING THE SKILLS

Whichever assessment type is used, you need to learn and practise the skills during your course.

Specific skills

You will learn specific skills associated with particular topics as a natural part of your learning during the course; for example, accurate weighing when investigating water uptake by plant tissue. Make sure that you have hands-on experience of all the apparatus that is used. You need to have a good theoretical background of the topics on your course so that you can

- devise a sensible hypothesis
- identify all variables in an experiment (see p. 201)
- control variables
- choose suitable magnitudes for variables
- select and use apparatus correctly and safely
- tackle analysis confidently
- make judgements about the outcome.

Designing experiments and making hypotheses

Remember that you can only gain marks for what you write, so take nothing for granted. Be thorough. ***A description that is too long is better than one that leaves out important detail.***

Remember to

- use your knowledge of AS and A2 level biology to support your reasoning
- give quantitative reasoning wherever possible
- draw clear labelled diagrams of apparatus
- provide full details of measurements made, equipment used, and experimental procedures
- be prepared to state the obvious.

A good test of a sufficiently detailed account is to ask yourself whether it would be possible to do the experiment you describe without needing any further infomation.

The four basic skill areas are:

Implementing – carrying out the practical work
Recording and communicating – describing method and presenting results
Analysing – using your biological knowledge to explain what the results mean
Evaluating – explaining how limitations in techniques or design could affect results, collections and conclusions drawn.

GENERAL SKILLS

The general skills you need to practise are:

- the accurate reporting of experimental procedures
- presentation of data in tables (possibly using spreadsheets) (see p. 202)
- graph drawing (possibly using IT software) (see p. 203)
- analysis of graphical and other data
- critical evaluation of experiments.

When ***analysing data*** remember to:

- use a large gradient triangle in graph analysis to improve accuracy
- set out your working so that it can be followed easily
- ensure that any quantitative result is quoted to an accuracy that is consistent with your data and analysis methods
- include a unit for any result you obtain.

When ***evaluating data*** make sure that you:

- identify any systematic errors in the experiment
- suggest whether your results are valid
- suggest alternative conclusions that could be drawn
- suggest alternative approaches that might have improved the validity of the experiment.

Answering the question

There are many different styles of examination question. Some will simply expect you to recall facts, others will expect you to apply your knowledge or to demonstrate your ability to use certain skills.
In this section you will encounter a number of different question types and receive guidance on how to go about answering them. The different types are identified as

1. factual recall
2. experimental design and practical technique
3. data analysis
4. genetics and statistical analysis
5. prose and comprehension
6. essay-style and free response.

Whichever question style you are working on, it is important that you remember the following guidelines:

1. **read the question** – in particular look out for 'instructional' words such as 'explain' or 'comment on'.
2. **check the marks** – the number allocated for each section of a question will give you a good idea of how much you need to write. A rough guide is '1 appropriate point = 1 mark'.

Factual recall often involves **filling in gaps** and **very short (often one word) answers**, for example:

Example 1

Identify the structures labelled A, B, C, D, and E. *(5 marks)*

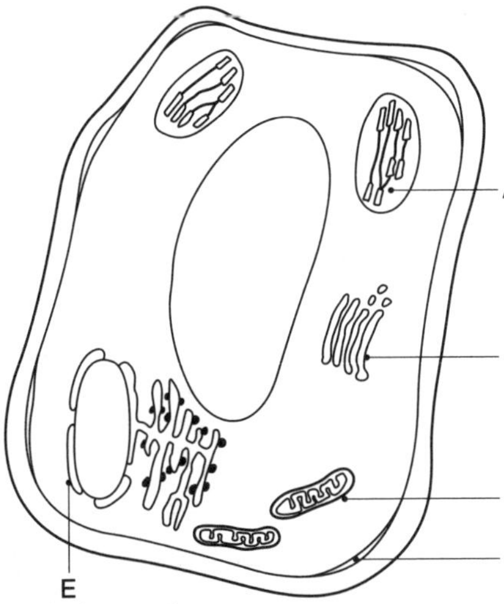

There's little alternative but to **know your work** – there's not much scope for alternative answers here!

Example 2

Read through the following description of DNA and protein synthesis and then write on the dotted lines the most appropriate word or words to complete the passage:

The DNA molecule is made up of four types of ...nucleotide... molecule, each of which contains ..deoxyribose.. sugar, a ..nitrogen-containing/organic.. base, and a phosphate group. Within the DNA molecule the bases are held together in pairs by ..hydrogen.. bonds – for example, guanine is always paired with ..cytosine.. .
DNA controls protein synthesis by the formation of a template called ..messenger RNA.. This molecule is single stranded and, compared with DNA, the sugar is ..ribose.. and the base ..uracil.. replaces the base ..thymine.. This template is made in the nucleus by the process of ..transcription.. and then passes out into the cytoplasm where it becomes attached to organelles called ..ribosomes.. A third type of nucleic acid called ..transfer RNA.. brings amino acids to these organelles where they are lined up on the template according to ..complementary.. base-pairing rules. The amino acids are joined by covalent links called ..peptide bonds.. to form a polypeptide molecule.

(14 marks)

Always **read through the whole question before writing in any answers** to make certain that the completed section makes good sense.

Examiners give some leeway – for example either 'nitrogen-containing' or 'organic' base would be acceptable – but note that they ask for the **most appropriate** term. For example, 'deoxyribose' is better than 'pentose', and 'thymine' is not the same molecule as 'thiamine'!

Other gap fillers might involve **completion of a table**:

Example 3

The table below refers to three components of human blood. If the statement is correct for the component place a tick (✔) in the appropriate box and if the statement is incorrect place a cross (✘) in the appropriate box.

Function	Red blood cell	Thrombocyte (platelet)	Plasma
Transports oxygen	✔	✘	✔
Transports hormones	✘	✘	✔
Contains enzymes involved in blood clotting	✘	✔	✔
Carries out phagocytosis	✘	✘	✘
Transports carbon dioxide	✔	✘	✔

(5 marks)

Don't forget to write in the tick or cross – a blank space would be regarded as incorrect, and so would writing 'yes' or 'no'. Many candidates don't obey this simple instruction and so lose straightforward marks.

Experimental/practical questions can also be thought of as recall, but often recall of a **general process** rather than of a particular example:

Example 4

Describe how a sample of ribosomes could be obtained from animal cells.

The technique of differential centrifugation relies on differences in mass/density of small structures to separate them from a mixture. So:

1. Break open the animal cells (homogeniser or by osmosis) to produce a mixture of parts (called an homogenate).
2. Keep mixture cool and at neutral pH, to avoid damage to organelles.
3. Spin in an ultracentrifuge.
4. Discard nuclei (first fraction) and mitochondria (second fraction) but keep pellet containing ribosomes.
5. Test for ribosomes by checking that this pellet can carry out protein synthesis.

(4 marks)

Note that by writing this in a list form you will (a) show a clear sequence of steps and (b) convince yourself that you've followed the rule of 'one sensible point = 1 mark'. The fifth point written here is an extra one – well done if you thought of it!

Many **experimental/practical questions** test whether you understand the **principles** of experimental design rather than your **recall** of particular procedures. Remember that the principle is:

> an experiment involves measuring the effect of a **manipulated (independent) variable** on the value of a **responding (dependent) variable** with all other variables **fixed** or **controlled**.

For example:

Example 5

The diagram below shows the apparatus used to compare the carbon dioxide production of different strains of yeast.

The yeast population to be investigated is suspended in sucrose solution in tube A. Nitrogen gas is bubbled through the apparatus during the experiment to ensure that respiration of the yeast is anaerobic.

Tube C contains hydrogencarbonate indicator solution through which carbon dioxide has been bubbled. This allows the colour in the tube to develop and this tube is then used as a standard. Hydrogencarbonate indicator is red when neutral, purple in alkaline and yellow in acid conditions.

The time taken for the colour to develop in tube B to match the colour in tube C is recorded.

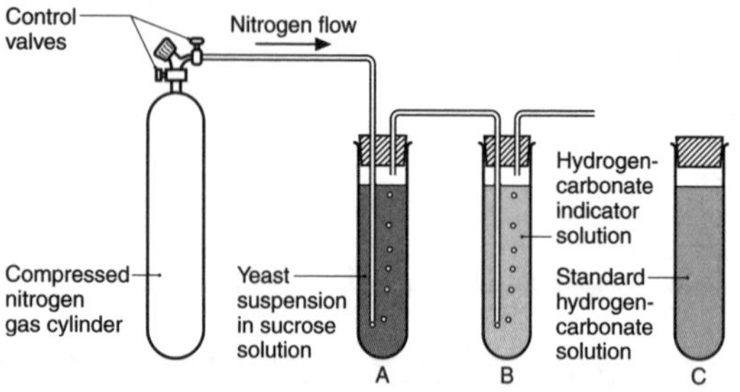

(a) (i) State how yeast suspension in tube A can be maintained at a constant temperature during the experiment.

Constant temperature/thermostatically controlled water bath.

(1 mark)

(ii) State the colour change which would occur in the hydrogencarbonate indicator solution in tube B during the course of the experiment.

From red (neutral) to yellow (acidic).

(1 mark)

(iii) Suggest why matching the colour by eye might not be a reliable method of determining the end-point.

Too subjective/could be colour-blind.
Reference solution may change colour.

(1 mark)

(b) Describe, giving experimental details, how you would use the apparatus to compare the carbon dioxide production in two strains of yeast.

CO_2 is responding variable – how would you measure it? The yeast is the manipulated variable. What are the fixed variables and how would you fix them?

(5 marks)

(c) An experiment was carried out to investigate the effect of light intensity on the rate of photosynthesis of an aquatic plant, using the apparatus shown in the diagram below.

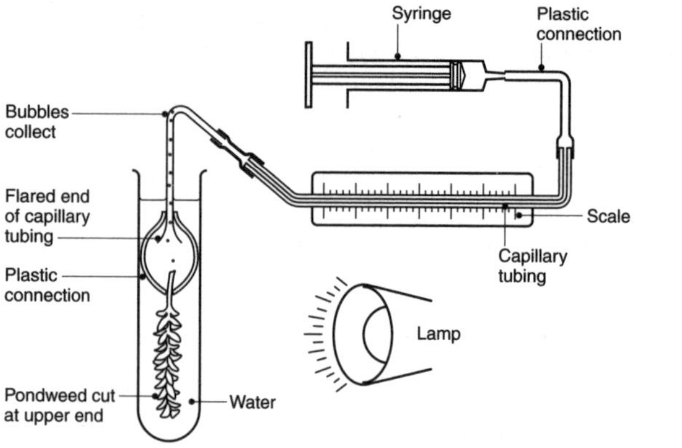

Read!
N.B. you *could* use **wavelength** of light.

State *two* environmental conditions, *other* than light intensity, which would need to be controlled. For each condition, describe how control could be achieved.

i.e. name the **apparatus** you would use.

2 conditions + 2 methods. *(4 marks)*

Data for analysis may be presented to you in the form of a **graph**:

Example 6

An experiment was carried out with cells of parsnip tissue to investigate the effect of temperature on the absorption of sodium ions. Cubes of the tissue were bathed in a sodium chloride solution of known concentration, and changes in the concentration of sodium ions in the solution were measured over a period of six hours. The solutions were aerated continuously, and the experiment was carried out at 2 °C and at 20 °C.

The results are shown in the graph below:

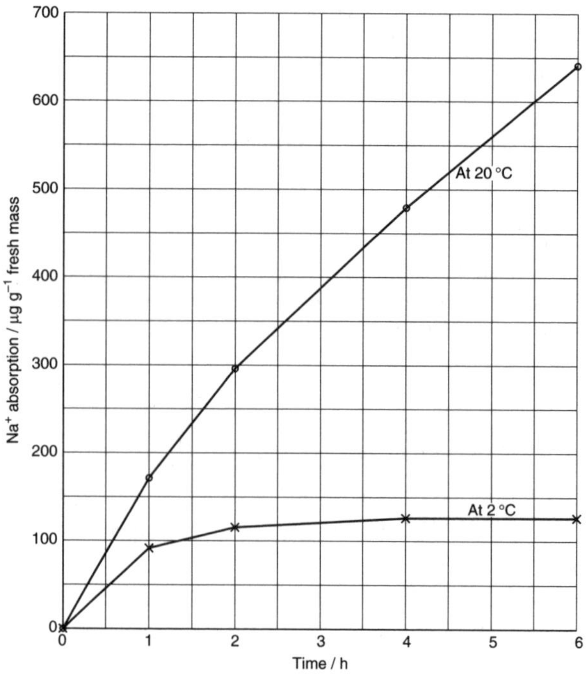

(i) Calculate the mean rate of absorption of sodium ions at 20 °C between 2 and 6 hours. Show your working.

Forget this and lose marks!!

Abs. at 6 h = 640
Abs. at 2 h = 295

Mean rate between 2 and 6 = $\frac{640 - 295}{4}$

Number of hours

= 86.25 $\mu g\ g^{-1} h^{-1}$

1 mark for number

1 mark for units

(3 marks)

(ii) Compare the rates of absorption of sodium ions at 2 °C and at 20 °C during this experiment.

You must present **both** rates.

i.e. between 0 and 6 h

Be quantitative!

At 2 °C, max. rate of absorption (135 units) is achieved by 4 h and there's very little increase from 2 to 4 h but at 20 °C, rate is higher (480 units at 4 h, for example) and continues to increase for the whole experimental period.

Good word for comparison

N.B. **No** request to explain these observations here

(3 marks)

(iii) Suggest an explanation for the differences in the rates of absorption of sodium ions at the two temperatures.

i.e. try to relate these results to your biological/physical background knowledge.

Higher rate at 20 °C suggests
- ions move more quickly at higher temperature
- active transport involved: higher temperature releases more energy by respiration.

Some uptake at 2 °C: suggests some uptake without active transport (diffusion).

Plateau at 2 °C suggests diffusion uptake system can be saturated (max. diffusion rate achieved by 4 h).

(3 marks)

Data analysis could also involve dealing with data presented in a ***table***:

Example 7

The numbers of stomata on each surface of an iris leaf and an oat leaf were estimated by counting the stomata in several 4 mm^2 areas of epidermis. From these measurements it was possible to calculate the mean number of stomata per cm^2. These are presented in the table below:

Species	Mean number of stomata per cm^2	
	Upper epidermis	Lower epidermis
Iris	3550	1850
Oat	5100	6300

The rate of water loss from these leaves was also measured. The results of these measurements are presented in the table below:

Species	Rate of water loss / arbitrary units
Iris, lower epidermis	1.5
Iris, upper epidermis	2.2
Oat, lower epidermis	4.0
Oat, upper epidermis	5.0

Explain the relationship between the rate of water loss and the number of stomata.

Don't just **state** the relationship.

Stomata regulate water loss, so more stomata offer greater risk of water loss: oat total (11 400 stomata per square centimetre – 9.0 units water loss) is greater than iris total (5400 stomata per square centimetre – 3.7 units water loss).

Examiners reward you for being **quantitative.**

(2 marks)

i.e. **name** them

State two other structural features of a leaf, apart from the number of stomata, which might influence the rate of water vapour loss. In each case give a reason for your answer.

Don't forget – half of the marks will be for your reasons.

Feature 1

Reason

Feature 2 Leaf rolling

Reason Traps humid atmosphere close to leaf surface to reduce water potential gradient.

(4 marks)

Genetics problems often involve **simple statistics**. Don't be afraid of this – you will only be asked to put numbers into a formula and interpret the result. **You won't be asked to recall the formula or to explain anything about it.**

Example 8

Wild type individuals of the fruit fly *Drosophila* have red eyes and straw-coloured bodies. A recessive allele of a single gene in *Drosophila* causes grey eye (g), and a recessive allele of a different gene causes black body (b).

A student carried out an investigation to test the hypothesis that the genes causing grey eye and black body show autosomal linkage. When she crossed pure-breeding wild type flies with pure breeding flies having grey eye and black body, the F_1 flies all showed the wild type phenotype for both features. On crossing the F_1 flies among themselves, the student obtained the following results for the F_2 generation.

Eye	Body	Number of flies observed in F_2 generation (O)
Wild	Wild	312
Wild	Black	64
Grey	Wild	52
Grey	Black	107
Total		535

(a) Using appropriate symbols, write down the genotypes of the F_1 flies and the grey-eyed, black-bodied F_2 flies.

F_1 flies *Gg Bb*

GB from one parent, gb from the other.

Grey-eyed, black-bodied F_2 flies *gg bb*

Must be homozygous to express recessive characteristics.

(2 marks)

(b) In order to generate expected numbers of F_2 flies for use in a χ^2 test, the student used the *null hypothesis* that the genes concerned were *not* linked.

i.e. genes are segregating independently.

(i) State the ratio of F_2 flies expected using the null hypothesis.

Red, straw — 9 : 3 : 3 : 1 — Grey, black

Red, black — Grey, straw

(1 mark)

(ii) Complete the table below to give the numbers of F_2 flies expected (*E*) using the null hypothesis, and the differences between observed and expected numbers (*O* – *E*).

Eye	Body	Number of flies observed (*O*)	Number of flies expected (*E*)	*O* – *E*
Wild	Wild	312	301	312 – 301 = 11
Wild	Black	64	100	64 – 100 = –36
Grey	Wild	52	100	52 – 100 = –48
Grey	Black	107	33	107 – 33 = 74

= Total (i.e. 535) multiplied by fraction e.g. 535 × $\frac{9}{16}$ = 301 red, straw flies

Nearest whole numbers

(2 marks)

(c) (i) Use the formula

$$\chi^2 = \Sigma \frac{(O-E)^2}{E}$$

You will **not need to recall any formulae** – simply put numbers in and understand **what the answer means.**

to calculate the value of χ^2. Show your working.

Note this

Add two columns to table:

$(O-E)^2$	$\frac{(O-E)^2}{E}$
121	0.40
1296	12.96
2304	23.04
5476	165.94

χ^2 = 202.34

(2 marks)

(ii) How many degrees of freedom does this test involve? Explain your answer.

3: number of d.o.f.
= (number of possible classes (4, in this case) – 1) i.e. 4 – 1 = 3.

(2 marks)

(iii) For this number of degrees of freedom, χ^2 values corresponding to important values of P are as follows.

Value of P	0.99	0.95	0.05	0.01	0.001
Value of χ^2	0.115	0.352	7.815	11.34	16.27

What conclusions can be drawn concerning linkage of the g and b alleles? Explain your answer.

χ^2 = 193.21
From this table, P must be less than 0.001, i.e. probability that genes are not linked is <0.001. Thus it is highly probable (more than 99.9% probable) that the g and b alleles are linked.

(3 marks)
(Total 12 marks)

If you are **confident** and **careful** you can **score full marks, very quickly** on this type of question.

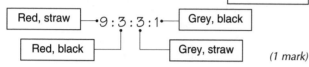

Prose passages are used as **tests of comprehension** and to **build on factual recall**:

Example 9

Read the passage, and answer the questions which follow it.

Haematologists in Britain have been investigating the use of a blood substitute. The substitute might be used in blood transfusions and might allow the use of blood donations that have passed their 'use-by' date.

The substitute is effectively a haemoglobin solution – red cell membranes have been removed so that there are no problems with incompatible blood transfusions. The product has been developed in the United States and has overcome the problem that pure haemoglobin tends to break into two molecules and to be rapidly lost from the body. Scientists at the US-based company have developed a way of locking the two sub-units together so that the haemoglobin remains an efficient oxygen-deliverer and permits heat treatment to destroy viruses.

The artificial blood was first developed for use in accident, shock, and injury cases where oxygen delivery and fluid volume were required, but a new and exciting possibility is that it may be useful in the treatment of strokes. In tests on animals the haemoglobin solution has been able to by-pass blood clots and to reach the parts of the brain being oxygen-starved by the clot. The effects of stroke damage to the brain, including paralysis and partial loss of speech, have been shown to be reduced if the solution is infused soon after the stroke has occurred.

Even if the trials of the blood substitute prove successful there will still be a need for blood donors because the haemoglobin solution is derived from human blood. The benefits, however, are that there will be less waste of out-dated blood and artificial blood can be stored for longer periods.

Triggers recall

Ability to use 'given information for explanation

Understanding the passage

Hint: it is often worth reading the questions before the passage – you may then know what to look out for!

Note three sections to question… and 6 marks. Remember that 6 = 3 × 2!

(a) Donated blood has a 'use-by' date because red blood cells have a shorter life span than most other cells in the body. Give one reason, connected with their structure, why red blood cells have such a short life span.

Red cells have no nucleus – no replacement of mRNA for protein synthesis so limited cell repair.

(1 mark)

(b) Explain why the blood substitute can 'reach the parts of the brain being oxygen-starved by the clot'.

Has no cells so less viscous/can pass obstructions more easily.

(2 marks)

(c) Apart from its use in stroke victims, give *two* advantages of using the blood substitute rather than natural whole blood.

(i) Reduce waste of 'out-of-date' blood.

(ii) No cross-matching for blood type is necessary.

(2 marks)

(d) Explain the role of haemoglobin in the loading, transport, and unloading of oxygen.

- Uptake where pO_2 is high/co-operative binding.
- 'Plateau' on association curve means O_2 retained as oxyhaemoglobin at pO_2 arteries.
- Steepness of curve suggests almost complete unloading for small change in pO_2 at respiring tissues.
- Uptake/release can be modified by local conditions of temperature and pH (Bohr effect).

(6 marks)

Why not use a simple **diagram** of oxygen–haemoglobin association curve?

Essay style (**free response**) **questions** may gain from 10–25 marks. The key to scoring highly is to **break down the question into smaller units**.

Example 10

Write an essay on water pollution.

The plan for your essay might look like this.

WATER POLLUTION

- Definition: 'addition of the products of human activities, usually to levels which are harmful to natural systems'
- Freshwater
  - sewage
  - fertiliser run-off from fields
    - (sewage and fertiliser run-off) both → eutrophication and BOD (Biological Oxygen Demand)
  - heat (from power station outflow) → establishment of new species e.g. piranha
  - heavy metals e.g. Pb and Hg from industry/mining/? fishermen
- marine
  - Usually think of oil e.g. Exxon Valdez in Alaska → oiling of birds and marine mammals

If you have a framework something like this it is not difficult to add your own examples and to gain marks quickly.

Conclusion? Humans have the potential to exploit the environment – we must try to ensure that we do so in a responsible manner so that it is not made unsuitable for other organisms.

- It is much easier to gain 2 + (4 × 4) + 2 marks for definition + 4 paragraphs + conclusion than 1 × 20 or 2 × 10 marks!
- Each section should contain some hard biological fact – you get no credit for general 'waffle', and only in the conclusion do you gain from opinion.
- Writing in **continuous prose** means leaving out lists **unless you specifically introduce them,** e.g. 'Sewage contains a number of components: organic waste, water, discarded nappies, dust, and detergents'.
- You can include diagrams and flow charts, but 'style' requires you to introduce them, e.g. 'The following flow diagram illustrates the principle of feedback control of blood sugar concentration'.

High scores on essays usually mean high scores overall – they are good indicators of performance.

Self assessment questions

1. The following table contains descriptions of a number of organelles. Identify the organelles from their descriptions.

Description	Name of organelle
A system of membranes that packages proteins in a cell	A
Contains the genetic material of a cell: surrounded by double membrane	B
Usually rod-shaped, and surrounded by a double membrane; inner membrane folded to increase its surface area	C
Approximately spherical, and often responsible for a 'rough' appearance of cell membranes; sometimes isolated attached to messenger RNA	D
Disc-shaped structure – surrounded by a double membrane and containing a series of grana	E

2. Complete the following table to compare the features of a eukaryotic and a prokaryotic cell. Use a + if the feature is present and a – if it is absent.

Feature	Prokaryotic cell	Eukaryotic cell
Nuclear envelope		
Cell surface membrane		
DNA		
Mesosome		
Mitochondria		
Ribosomes		
Microtubules		

3. The diagram shows a typical plant cell.
   (a) Identify the structures labelled **A–F**.
   (b) Suggest a label for the scale line – give your answer to the nearest 5μ.

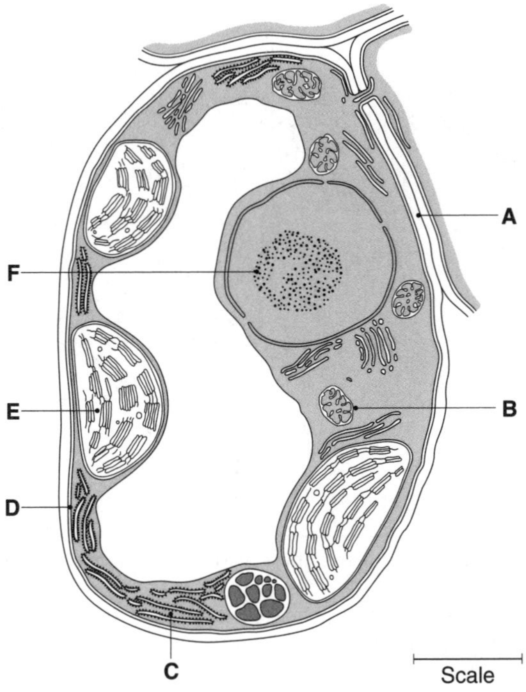

4. The diagram shows a section of cell membrane. Identify the components labelled **A–F**.
   What is the length of the line XY?

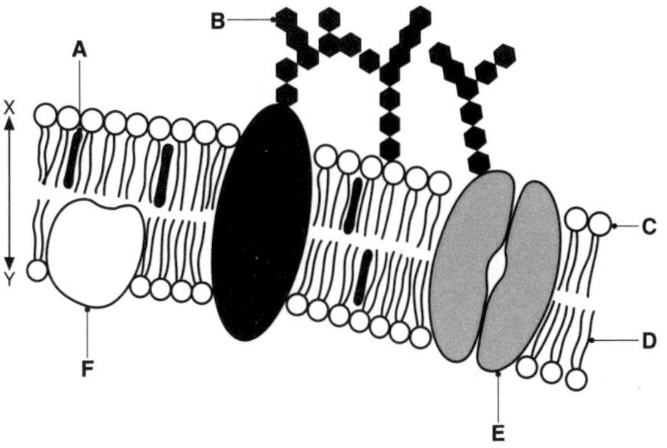

5. The diagram shows some of the structures present in an animal cell:

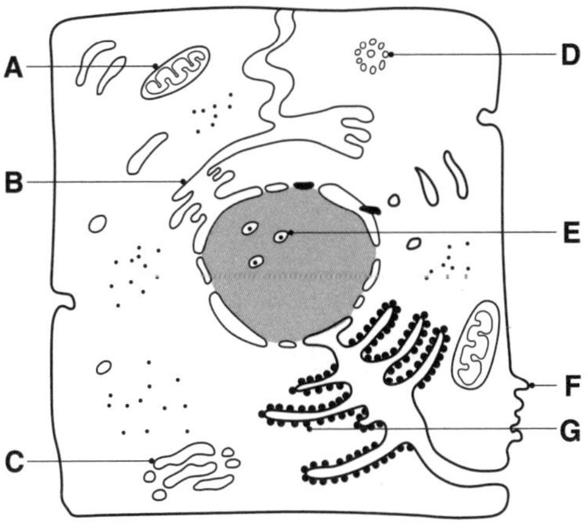

Which of these structures is responsible for
(a) manufacture of lipids and steroids
(b) release of energy
(c) manufacture of hormones and digestive enzymes
(d) production of spindle fibres in cell division
(e) endo- and exocytosis?

6. The following diagrams show some molecules found in cells.
   (a) Match the diagrams to the labels supplied:

   triglyceride; α-glucose; β-glucose; amino acid; purine; steroid; ribose

A

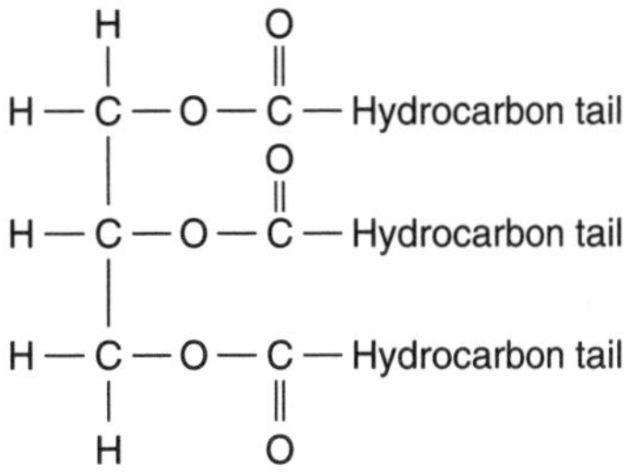

B

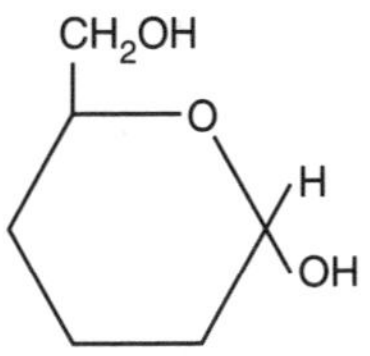

C

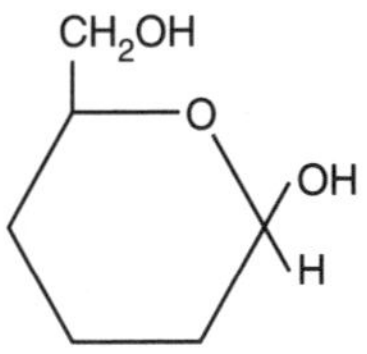

D

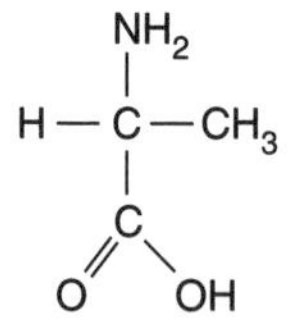

E

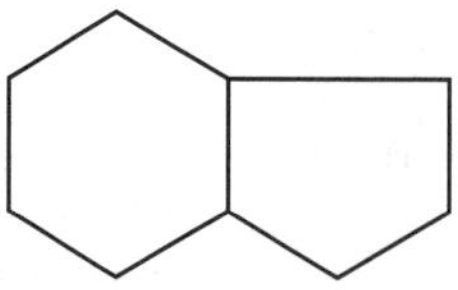

F

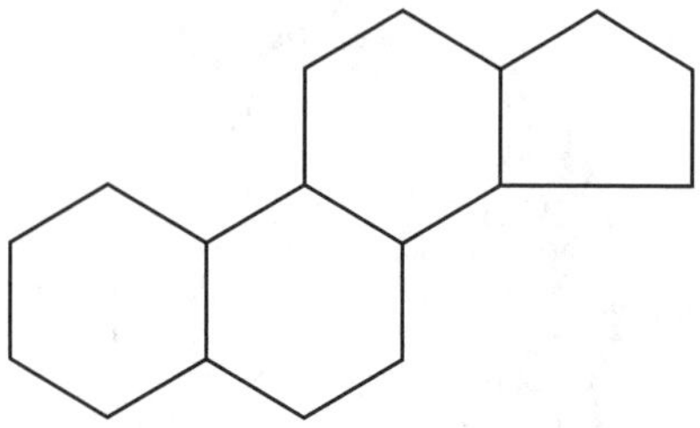

G

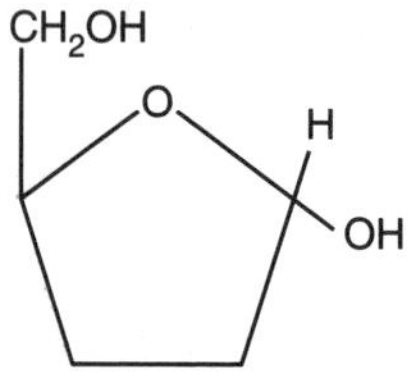

(b) Which of these molecules would be found in
(i) glycogen
(ii) insulin
(iii) DNA
(iv) cellulose
(v) amylase?

7. The diagram represents one possible mechanism for enzyme action.
Match the letters to the following labels:

catalase
oxygen
hydrogen peroxide
water
E–S complex

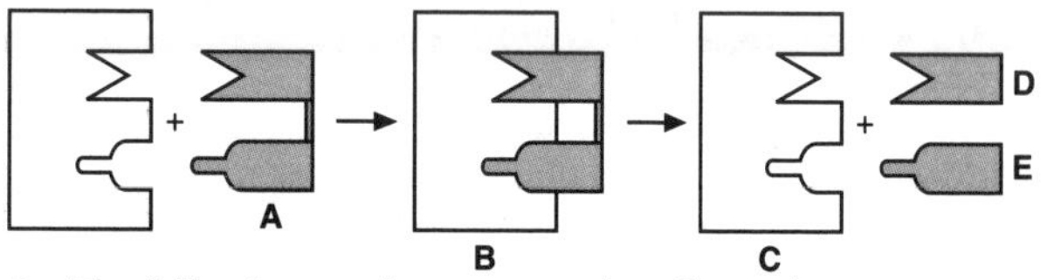

8. The following graph represents the effects of some compounds on the action of an enzyme:

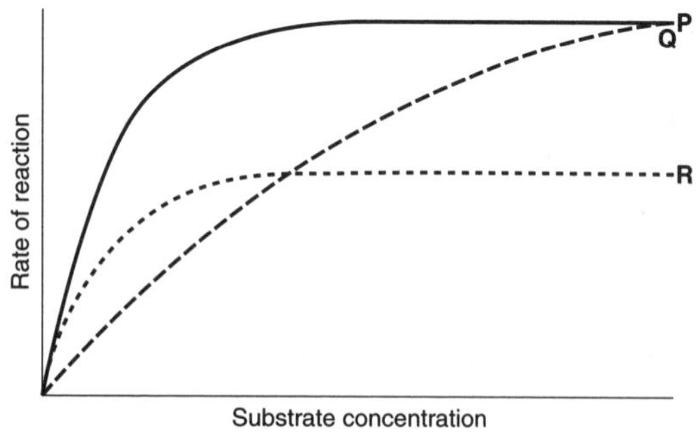

(a) Which of the following correctly identifies the three lines on the graph?

	Uninhibited enzyme	Enzyme + competitive inhibitor	Enzyme + non-competitive inhibitor
A	**P**	**Q**	**R**
B	**P**	**R**	**Q**
C	**Q**	**R**	**P**
D	**R**	**P**	**Q**
E	**R**	**Q**	**P**

(b) Name **(i)** an enzyme, its normal substrate, and its non-competitive inhibitor
(ii) an enzyme, its normal substrate, and its competitive inhibitor.

9. The following diagram represents a section of an important biological molecule:

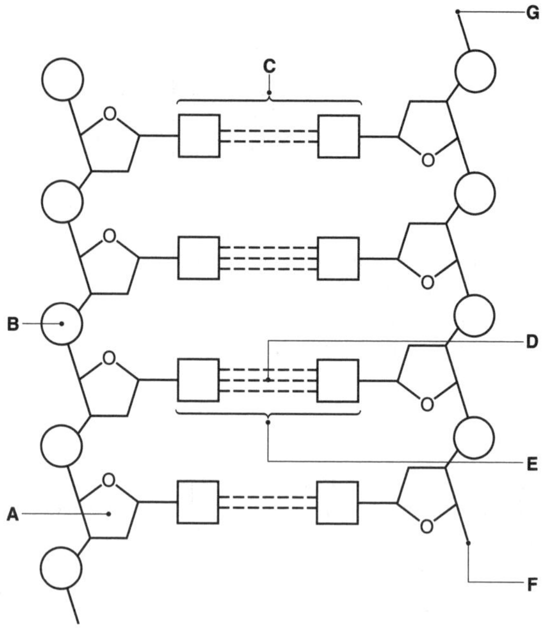

(a) Identify the molecule.
(b) Match the following labels to the letters on the diagram: A–T base pair; C–G base pair; hydrogen bond; 5′ end; 3′ end; deoxyribose; phosphate/pentose

10. The diagram represents a structure normally found on the underside of leaves. Match the letters on the diagram to appropriate labels from the list below:

chloroplast; epidermal cell; nucleus; site of K^+ pump; stomatal pore; mitochondrion; guard cell; cellulose cell wall

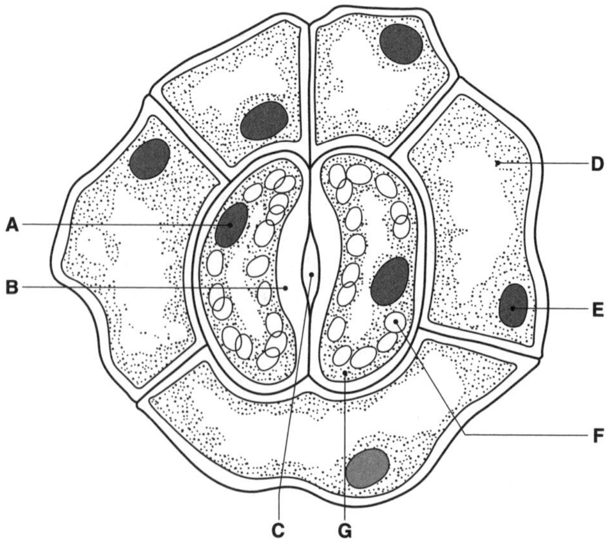

Explain why
(a) these structures are typically found on the *underside* of the leaf
(b) these structures may be sunk into pits in Xerophytes
(c) the pores usually open if internal leaf concentration of carbon dioxide falls
(d) the pores usually close as light intensity falls.

11. The following diagram represents cells from a plant tissue.

(a) Choose appropriate labels from the following list to match with the letters on the diagram:

sieve plate; companion cell; sieve tube; cytoplasmic strand; plasmodesma; mitochondrion; lignin

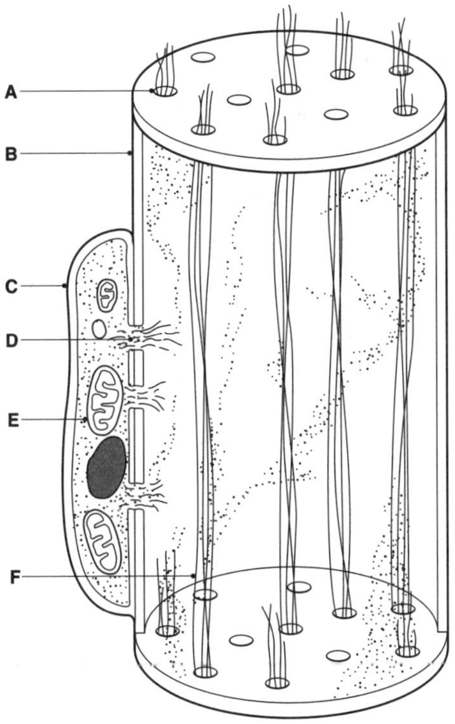

(b) Name the tissue of which these cells are a part. What is the function of this tissue?

(c) Choose an example from the human alimentary canal to illustrate the sequence ORGANELLE–CELL–TISSUE–ORGAN–SYSTEM.

12. The diagram shows the results of an experiment on the digestion of carbohydrates:

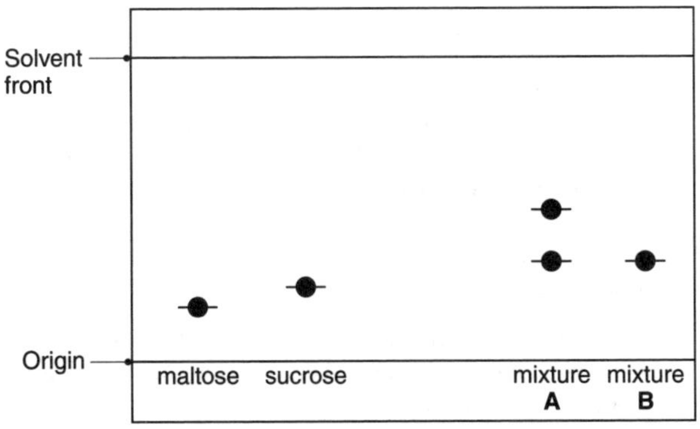

(a) Name the technique which would have produced this result.

(b) Name the two components of the mixture **A**.

(c) Calculate the R_f value for fructose in this solvent.

(d) State the site of secretion of the enzymes responsible for these digestive reactions.

(e) What is the general name given to reactions which 'break down large molecules to smaller ones by the addition of water'?

13. The graph shows the results of an experiment in which two groups of young rats were fed on a basic diet which could be supplemented with milk.

(a) Calculate the mean rate of growth of group A rats between 0 and 20 days. Show your working.

(b) Suggest why group B rats were able to grow from 0 to 10 days, even in the absence of milk.

(c) At day 30, calculate group A body mass as a percentage of group B body mass. Show your working.

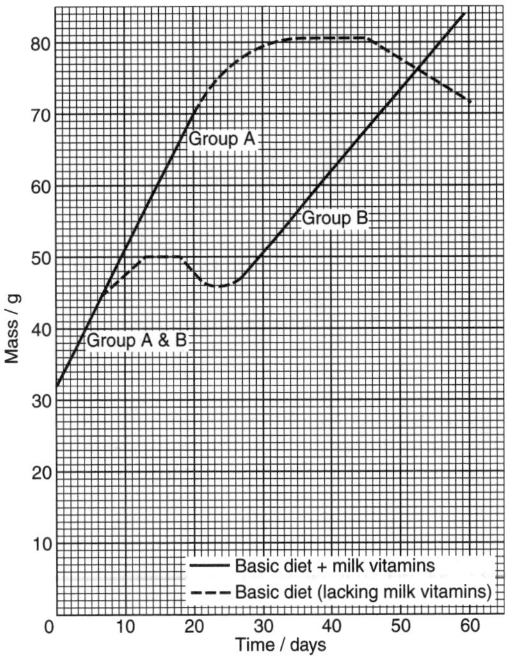

14. This diagram represents two possibilities for the flow of water over the gills of a fish:

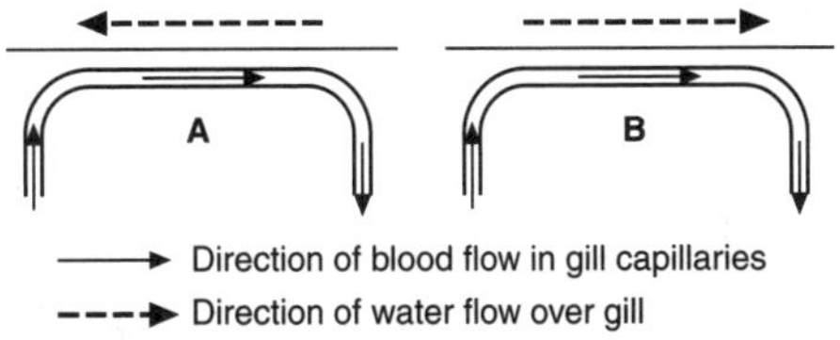

(a) Which of the two possibilities is the one you would expect to find in a living fish? Explain why this is the more efficient method for gas exchange.

(b) List three features of gas exchange surfaces which contribute to their efficiency.

15. The diagram represents the breathing movements of a human. Match the letters on the diagram to the appropriate labels from the following list:

inspiration; expiration; trachea; contraction of external intercostal muscles; relaxation of diaphragm; contraction of diaphragm; air exhaled down pressure gradient; relaxation of external intercostal muscles; air inhaled down pressure gradient

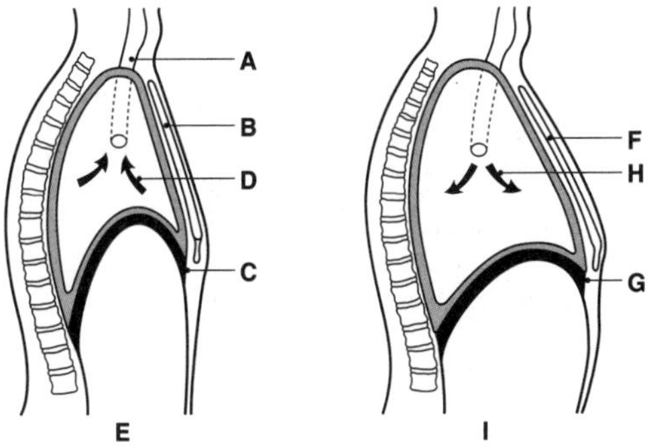

16. Choose words from the following list to complete the passage about blood and its functions. You can use each word once, more than once, or not at all.

white cells; glucose; urea; bone marrow; albumin; hydrogencarbonate; sodium; haemopoiesis; calcium; platelets; plasma; serum; solute potential; red blood cells; stem cells; homeostasis

Blood consists of a liquid called ...... in which are suspended several types of 'cell'. These include ...... (or erythrocytes), ... (including neutrophils), and ...... , which are really fragments of cells and are involved in blood clotting. All of these cells are produced from ...... by a process called ......, which occurs in the ....

Water is the main component of blood and may carry several dissolved ions including ...... (the most abundant cation), ... (another factor involved in blood clotting), and ...... (mainly formed from the solution of carbon dioxide in water). There are also plasma proteins present, including fibrinogen and ...... – as well as having individual specific functions these plasma proteins also affect the physical properties of the blood such as its viscosity and .... Blood with cells and fibrinogen removed is called ......

The blood is the major transport system of the body. For example, ... is transported from the liver to the kidney for excretion and ...... is distributed to the cells as a source of energy.

17. The following diagram shows some properties of different parts of the mammalian circulatory system.

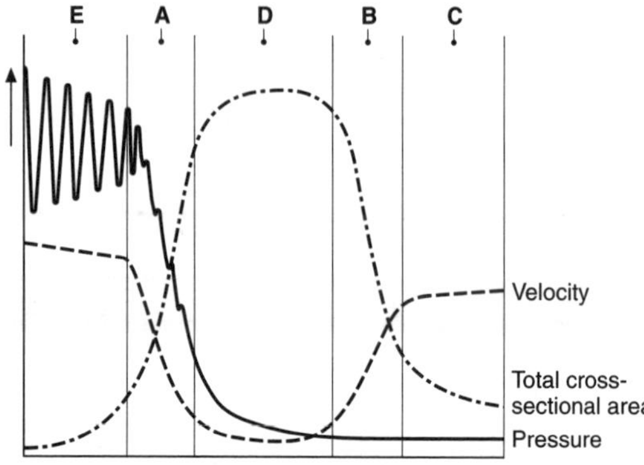

(a) Match the following structures with letters on the diagram:

arteriole; capillary; vein; artery; venule

(b) Use the letters to identify the structure in which
- **(i)** the pulse can be felt most strongly
- **(ii)** valves are most likely to be present
- **(iii)** the walls contain most elastic and muscular tissue
- **(iv)** exchange of soluble substances is most likely to take place.

18. The diagram shows a section through the mammalian heart:

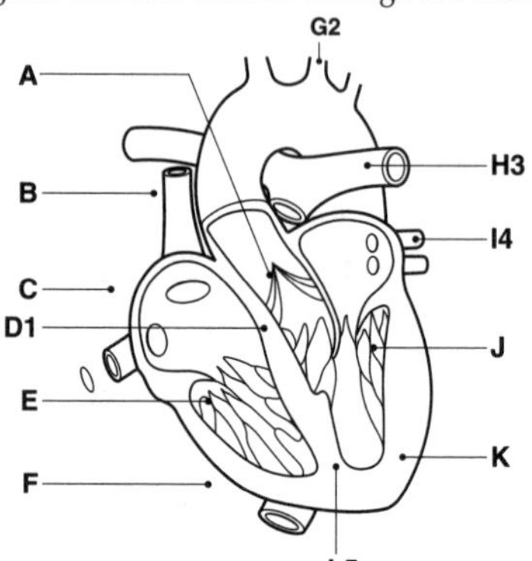

(a) Match the *letters* on the diagram with appropriate labels from the following list:

pulmonary artery; superior vena cava; wall of right atrium; pulmonary vein; chordae tendinae; tricuspid valve; aortic valve; wall of right ventricle; interventricular septum; carotid artery; wall of left ventricle; A–V node

(b) Use the *numbers* on the diagram to identify
- **(i)** the location of the bundle of His
- **(ii)** the site of receptors sensitive to blood pressure changes
- **(iii)** vessels carrying deoxygenated blood to the lungs
- **(iv)** a structure sensitive to electrical impulses arriving from the atria
- **(v)** vessels transporting oxygenated blood.

19. The graph represents the changes in pressure measured in the left side of the heart during the cardiac cycle:

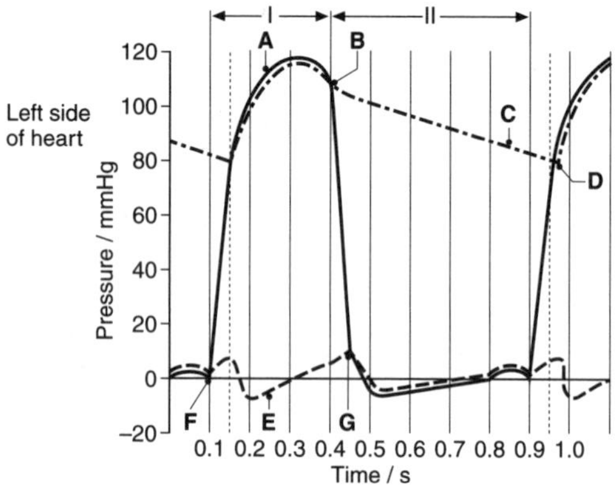

(a) Match the letters on the diagram with the appropriate label from the following list:

opening of aortic valve; pressure change in left ventricle; pressure change in left atrium; bicuspid valve opens; bicuspid valve closes; pressure changes in aorta; aortic valve closes

(b) If a doctor measured the blood pressure of this patient what would the result be?

(c) Calculate the pulse (rate of heartbeat) of this individual. Show your working.

(d) The peak systolic pressure in the left ventricle is approximately five times that in the right ventricle. Calculate the peak systolic pressure in the right ventricle. Explain why this value is different from the value measured in the left ventricle.

20. Use words from the following list to complete the passage about cell division:

pole; cell wall; interphase; cytokinesis; chromatids; nucleolus; nucleus; telophase; metaphase; centromere; spindle

During the cell cycle DNA replication takes place during ...... – sometimes mistakenly called 'resting phase'. At the beginning of prophase the chromosomes shorten and thicken so that they become visible – they are seen to consist of two identical ..... joined at the .... The ...... and the nuclear membrane are broken down, and a ..... develops in the cell. During ...... the chromosomes line up at the equator, and each one becomes attached to the ...... by its .... During anaphase one chromatid from each chromosome is pulled towards each ..... of the cell and during the final phase two new cells are formed as a result of the 'pinching' of the cytoplasm, called ....

21. Use words from the following list to identify the lettered structures on the diagram:

spindle organiser; chromatid; centromere; nuclear membrane; spindle fibre; chromosome; cell wall

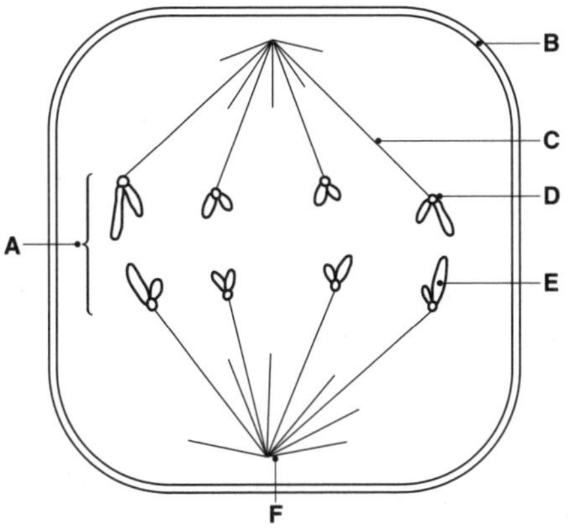

22. Use words from the following list to identify the lettered structures on the diagram:

anticodon; hydrogen bonds; unpaired folds; amino acid attachment site

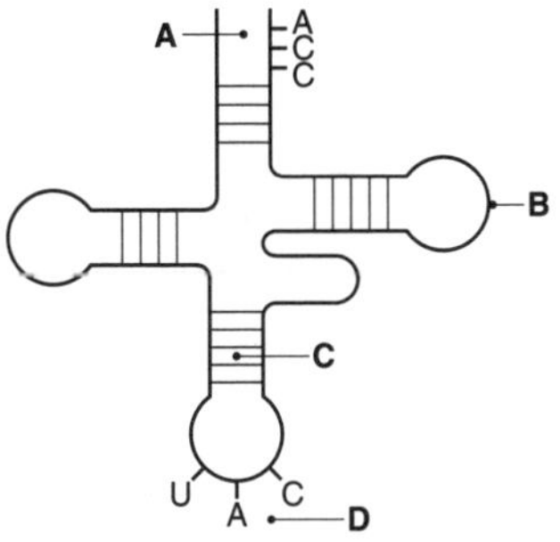

23. (a) The diagram shows the DNA content of a cell during meiosis. Use words from the following list to identify the lettered stages on the diagram:

DNA replication; separation of homologous chromosomes; separation of chromatids; haploid; cytokinesis

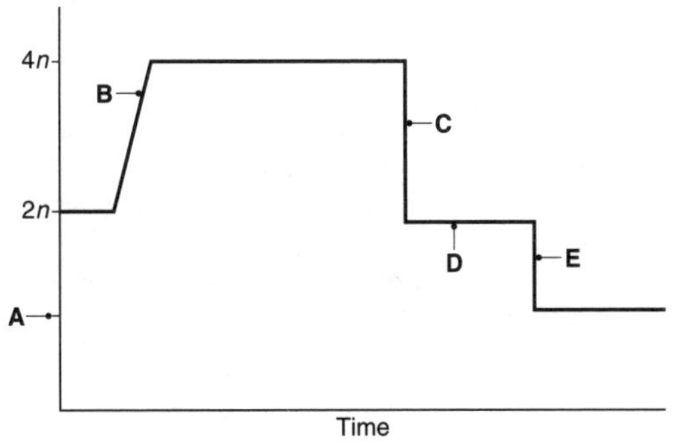

(b) Give two reasons why this process is important in the life cycle of living organisms.

24. Use words from the list to complete the passage which follows:

polyploidy; aneuploidy; X; Y; gene; crossing over; fusion; Down's; mutation; assortment

A ...... is any change in the structure or amount of DNA in an organism. Changes at a single locus on a chromosome are called ...... mutations – examples include cystic fibrosis and sickle cell anaemia. The loss or gain of a whole chromosome is called ...... – important examples include ...... syndrome (an extra 21st chromosome) and Klinefelter's syndrome (an extra ...... chromosome in males). The presence of additional whole sets of chromosomes, or ...... , has many important plant examples, including modern wheat.
All of these changes in DNA may be 'reshuffled' by free ...... during gamete formation, ... during meiosis, and random ...... during zygote formation.

25. The following flow diagram represents the stages involved in the formation of a section of recombinant DNA.
Match labels from the following list to complete the flow diagram:

DNA polymerase; plasmid; plasmid with 'sticky ends'; DNA ligase; restriction endonuclease; recombinant DNA; reverse transcriptase

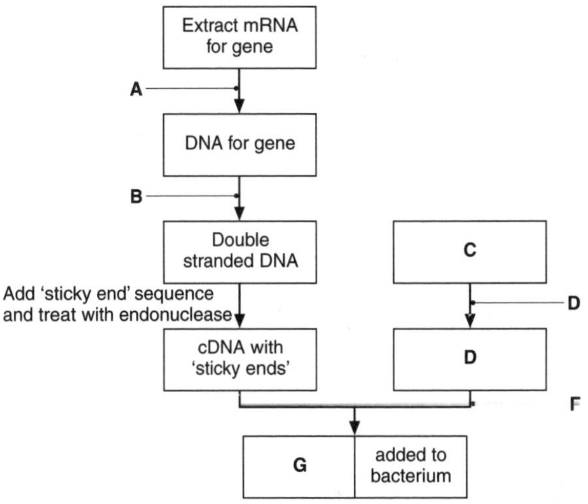

26. The diagram shows a single mitochondrion:

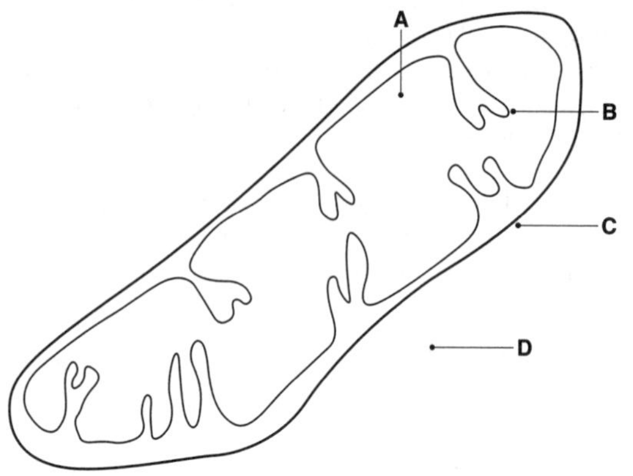

(a) Match the labels in the following list to the letters on the diagram:

glycolysis; TCA cycle; electron transport; pyruvate transport

(b) In the absence of molecular oxygen the following reactions occur. Match labels in the list to the letters on the flow chart:

NAD; ATP; ADP; alcohol; carbon dioxide; lactate; $NADH_2$

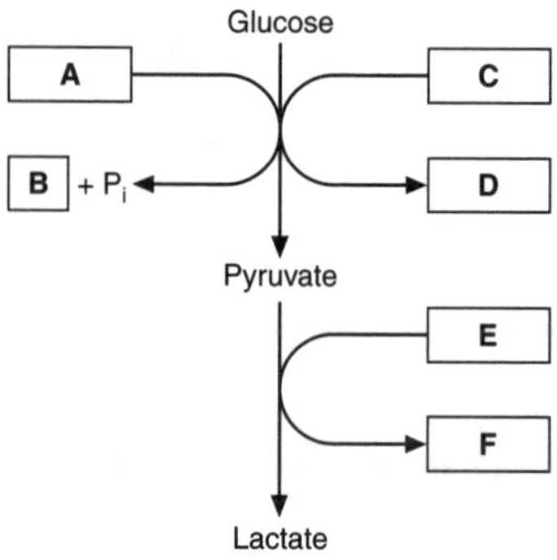

(c) What is the name given to the volume of oxygen required to reoxidise the lactate following anaerobic respiration?

27. The graph represents the effect of light intensity on the rate of photosynthesis of a culture of algae:

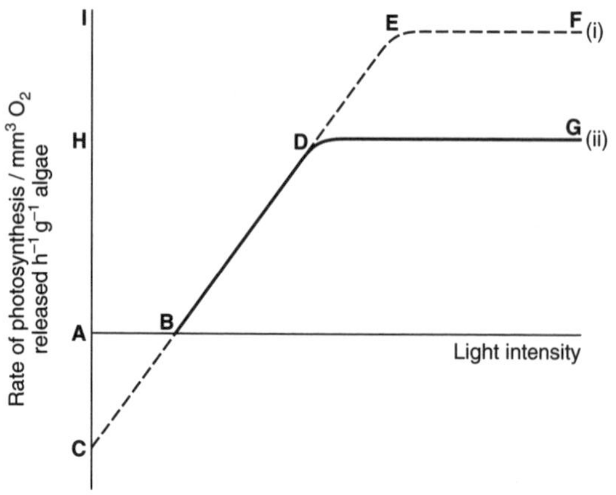

(a) Which region of the graph (use the letters to give your answer) corresponds to

(i) the rate of respiration at zero light intensity

(ii) a period when rate of photosynthesis is directly proportional to light intensity

(iii) the effect of raising the value of a second limiting factor, temperature

(iv) the maximum rate of photosynthesis possible at this light intensity and temperature?

(b) Similar results were obtained by plants in a woodland as light intensity increased during the day. What name is given to the point marked **B** on this graph?

28. The diagram represents a possible system for the light-dependent stages of photosynthesis.

(a) Match labels from the list to the letters on the diagram. A label may be used more than once.

light energy; photosystem I; photosystem II; excited electrons; ATP synthesis; protons; water; $NADPH_2$

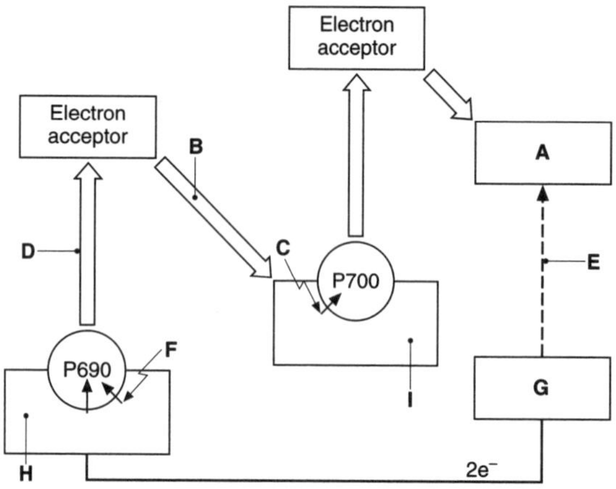

(b) These reactions occur in the structure drawn below:

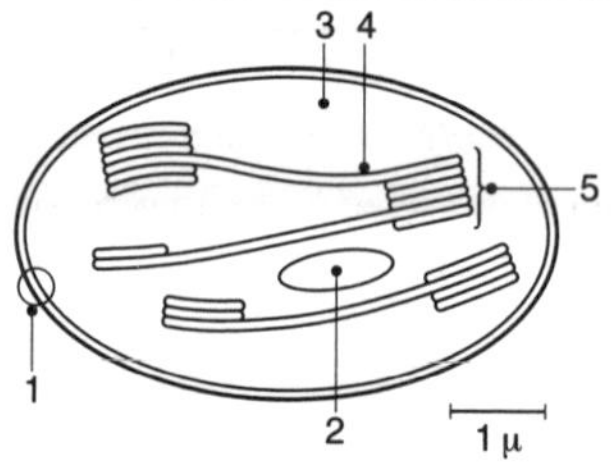

(i) Name this structure.

(ii) Which of the numbered regions carries out the biochemical processes above?

(iii) Use the scale to estimate the length of this structure.

(iv) Name an enzyme found in region 3, and state its function.

29. Complete the following table describing the effects of some plant growth substances. Use + to indicate YES and – to indicate NO.

Effect	Auxins	Gibberellins	Abscisic acid	Ethene
Stimulate cell elongation				
Stimulate root formation in cuttings				
Stimulate fruit ripening				
Inhibit development of lateral buds				
Stimulate breaking of dormancy in seeds				
Stimulate leaf fall in deciduous trees				

30. The following is a list of some important ecological terms. Match the terms to the list of definitions:

TERMS
A Biomass
B Community
C Abiotic
D Population
E Quadrat
F Niche
G Transect
H Ecosystem

DEFINITIONS
1 a piece of apparatus used to sample an area of the environment
2 the role of an organism in its environment
3 all of the living organisms within a defined area
4 the amount of organic matter present per unit area
5 the total number of any one species present within a defined area
6 non-living factors, such as temperature, which affect the distribution of living organisms
7 living organisms together with their non-living environment
8 a means of sampling a change in the living or non-living environment across a defined area.

31. Choose the most appropriate terms from the following list to complete the passage below:

climax; herbivore; population; ecology; habitat; ecosystem; succession; evolution; community

...... is the study of the interactions of living organisms with each other and with their environment. The living component together with the non-living part of the environment make up an .... A field is an example of a ...... , and will have a ...... of several species of herbivorous insects. The insects may be preyed on by a ...... of one species of spider. As conditions in the field change, different plants and their herbivores may become established, and the herbivores may be preyed on by different carnivores. This gradual process of change is called ...... and the end point is known as a ...... community.

32. Use terms from the following list to complete the table listing the processes involved in the nitrogen cycle:

glucose; nitrate; nitrogen; carbon dioxide; amination; maltose; photosynthesis; respiration; nitrite; carbon dioxide

Process	Substrate	Product
Nitrification		Nitrate
Nitrogen fixation	Nitrogen	
Denitrification		
	Carbon dioxide	
	Glucose	
	Amino groups	Amino acids
Putrefaction	Starch	

33. Use words from the following list to complete the paragraph below. Each word may be used once, more than once, or not at all.

histamine; B-lymphocyte; T-lymphocyte; cytokines; T-helper; antibodies; cell-mediated; macrophage; antigen; humoral; T-cytotoxic

The body of a mammal is protected by an immune response. This has two parts – a ..... response which depends on the release of protein molecules called .... from plasma cells, and a ...... response controlled by a number of .... The plasma cells are one type of ...... , activated when another cell called a ..... 'presents' a piece of the invading organism to it. The plasma cells secrete ...... which may remove the invading ...... by several methods, including agglutination and precipitation. The cell-mediated response involves a number of interactions, controlled by chemicals called ...... These chemicals are released by ...... cells which may, for example, 'instruct' ...... cells to attack body cells infected with a virus or bacterium.

34. The diagram represents a single nephron from a mammalian kidney:

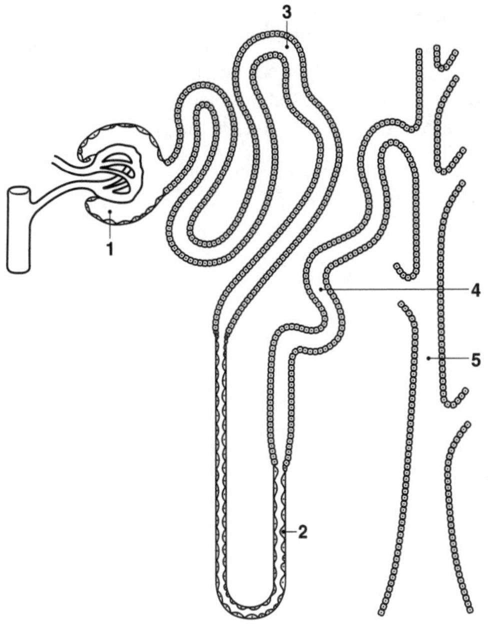

(a) Identify which of the numbered regions is
  (i) the site of ultrafiltration
  (ii) particularly sensitive to ADH
  (iii) the main site for the reabsorption of glucose and amino acids
  (iv) largely responsible for adjustment of blood pH.
(b) Which of the numbered regions would be particularly lengthy in a desert mammal? Explain your answer.

35. Use words or phrases from the following list to complete the paragraph below. Each word or phrase may be used once, more than once, or not at all.

ammonia; soluble; by diffusion; in solution; insoluble; toxic; water conservation; urea; uric acid; as a precipitate

The nitrogenous waste of most aquatic animals is ...... which is extremely ...... in water. It is also extremely ...... so much water is consumed in diluting it. Mammals excrete ...... which is quite soluble and so can be passed out ..... in the urine. Most insects have problems of ...... and so excrete ...... which is ...... and so consumes very little water.

36. The diagram illustrates structures involved in the iris reflex. Match labels from the following list with the letters on the diagram:

sensory neurone; retina; inhibitory neurone; excitatory neurone; circular muscle in iris; visual centre; radial muscle in iris

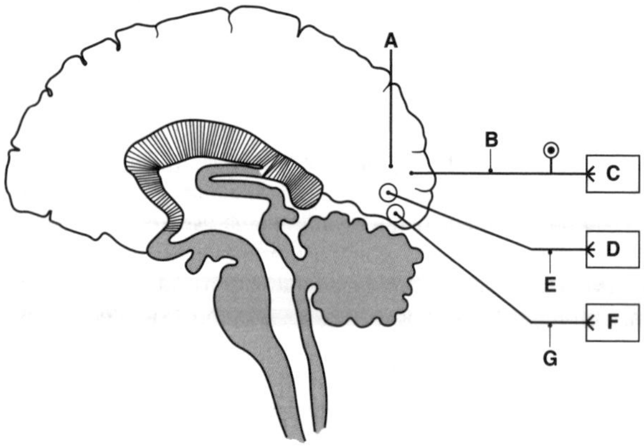

(a) What is the value of this reflex?
(b) Explain how a conditioned reflex differs from this example.

37. The following graphs were obtained during glucose tolerance tests on two hospital patients. Match the following labels with the letters on the graphs:

diabetic patient; glucose injection; increased insulin secretion; non-diabetic patient

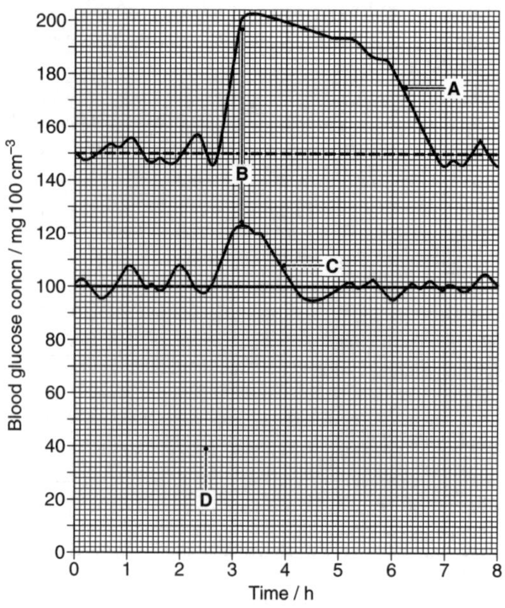

(a) Give *two* differences between the response shown by the diabetic and non-diabetic patients.
(b) Calculate the maximum percentage increase in blood glucose concentration following glucose injection in the diabetic patient. Show your working.
(c) Describe the differences in *cause* and in *method of treatment* for Type I and Type II diabetes.

38. The diagram shows light absorption by three cell types:

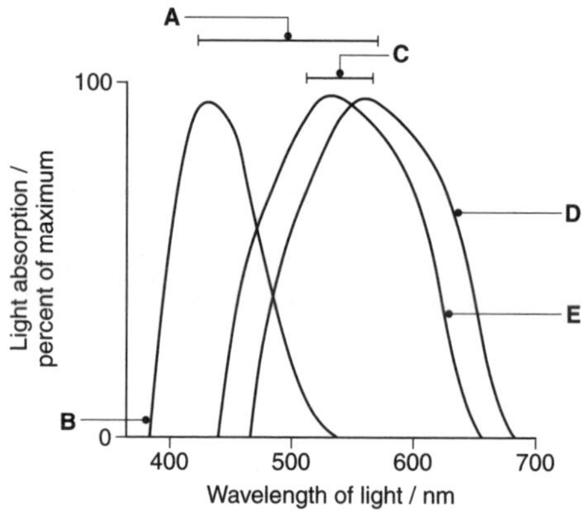

(a) Where, exactly, are these cell types found?
(b) Match labels from this list with the letters on the diagram:

orange colour perceived; red cone; blue cone; white colour perceived; green cone

(c) One defect in colour vision results from inheritance of a sex-linked mutant allele. What is meant by *sex-linkage*?

39. The diagram represents changes in the membrane potential of a nerve cell during an action potential.

Use the numbers on the diagram to identify

(a) the point at which sodium gates are opened
(b) the point at which potassium gates are opened
(c) the main period of sodium ion influx
(d) the period during which the Na–K pump begins to work
(e) the main period of potassium ion outflow.

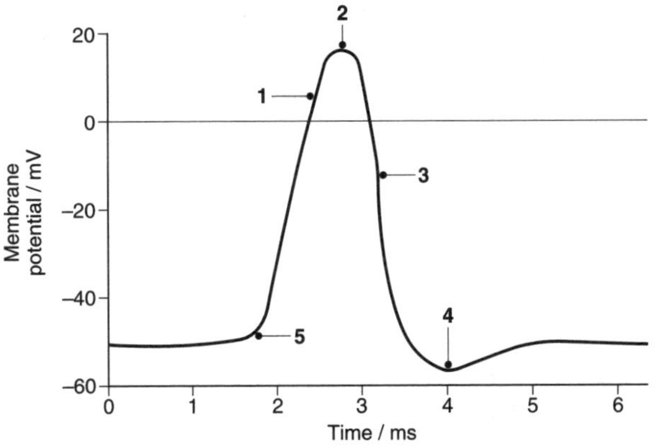

40. The table below describes some features of different skeletal systems. Complete the table using terms from the following list. Each term may be used once, more than once, or not at all.

ladybird; contains chitin; earthworm; fluid-filled muscular sac; annelid; cat; exoskeleton; endoskeleton; mollusc; chordate

Type of skeleton	Brief description	Phylum in which this type of skeleton typically occurs	Example organism
hydrostatic			
	internal framework of bones and/or cartilage		
		arthropod	

41. Use labels from this list to identify the regions on the structure shown in the diagram:

I band; H zone; M line; Z line; sarcomere; A band

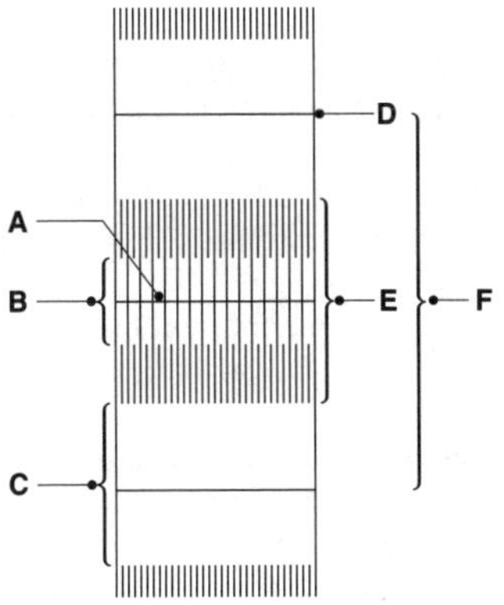

42. The following diagram shows some of the changes in blood hormone concentrations which occur during the menstrual cycle.

(a) Complete the diagram using labels from the following list:

oestrogen; ovulation; repair of endometrium; luteinising hormone; menstruation; luteal phase; progesterone; ovarian phase

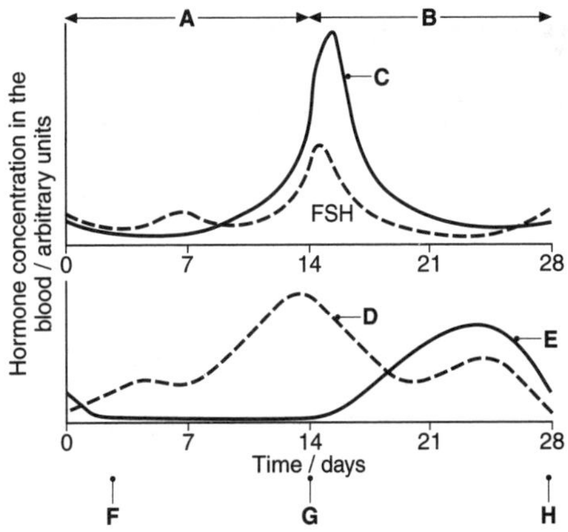

(b) Explain the hormonal basis of the contraceptive pill.

(c) State and explain *two* differences between male and female gametes.

43. The diagram shows a family tree of individuals with red-green colour blindness. This condition is X-linked – there are five possible genotypes:

X^RY, X^rY, X^rY^r, X^rY^R, X^RY^R R is dominant and normal-sighted, r is recessive and colour blind.

There are three possible phenotypes: carrier, normal, colour-blind.

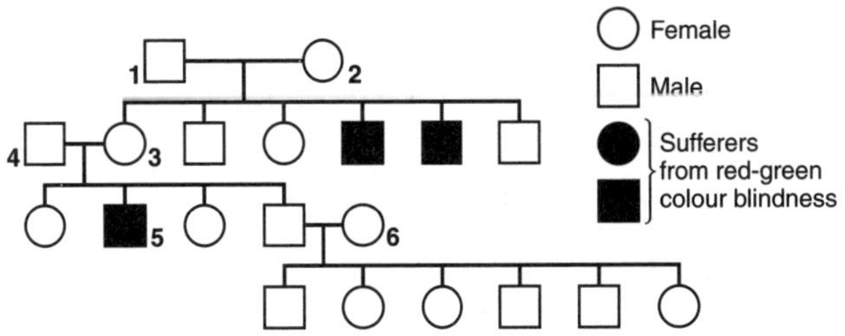

Use the family tree to work out the genotypes and phenotypes of the individuals labelled **1–6**.
Present your answers in the table below:

	Genotype	Phenotype
1		
2		
3		
4		
5		
6		

44. The diagram shows the inheritance of the condition cystic fibrosis.

(a) Is the condition sex-linked? Explain your answer.

(b) Is the condition dominant or recessive? Explain your answer.

(c) Why are antibiotics often prescribed for individuals with this condition?

(d) Suggest why individuals with cystic fibrosis often have difficulty with digestion of fats.

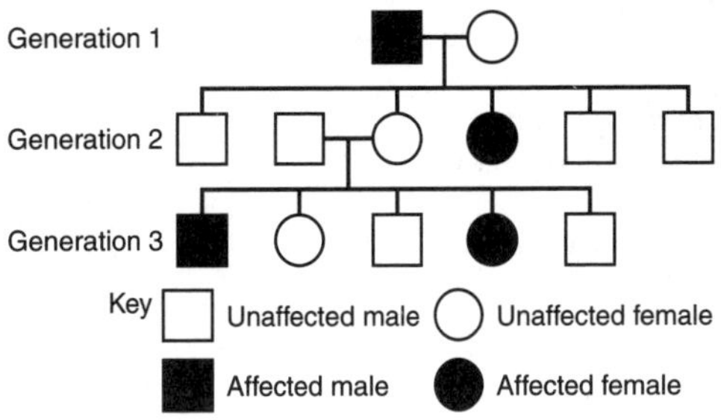

Answers to self assessment questions

1. A = Golgi body; B = nucleus; C = mitochondrion; D = ribosome; E = chloroplast

2. Prokaryotic: – + + + – + – Eukaryotic: + + + – + + +

3. **(a)** **A** = cellulose cell wall; **B** = mitochondrion; **C** = rough endoplasmic reticulum; **D** = cell surface membrane/plasmamembrane; **E** = chloroplast; **F** = nucleolus
   (b) 10 μm

4. **A** = cholesterol; **B** = glycocalyx; **C** = hydrophilic head of phospholipid; **D** = hydrophobic tail of phospholipid; **E** = transmembrane protein; **F** = intrinsic protein
   XY = 8 nm

5. **(a) C (b) A (c) G (d) D (e) F**

6. **(a)** **A** = triglyceride; **B** = α-glucose; **C** = β-glucose; **D** = amino acid; **E** = purine; **F** = steroid
   (b) (i) B (ii) D (iii) E (iv) C (v) D

7. **A** = hydrogen peroxide; **B** = E–S complex; **C** = catalase; **D/E** = oxygen/water

8. **(a)** A
   (b) **(i)** Succinate dehydrogenase/succinate/fumarate
   (ii) Cytochrome oxidase/hydrogen ions – oxygen/cyanide ion

9. **(a)** DNA
   (b) **A** = deoxyribose; **B** = phosphate; **C** = A–T base pair; **D** = hydrogen bond; **E** = C–G base pair; **F** = 5′ end; **G** = 3′ end

10. **A** = site of K^+ pump; **B** = cellulose cell wall; **C** = stomatal pore; **D** = epidermal cell; **E** = nucleus; **F** = chloroplast; **G** = guard cell
    (a) This reduces water loss as stomata are protected from direct sunlight/heat
    (b) Provides a local humid atmosphere and so reduces water potential gradient between internal surfaces of leaf and drying atmosphere
    (c) Increased carbon dioxide necessary as carbon dioxide concentration is a limiting factor in photosynthesis
    (d) Light becomes the limiting factor in photosynthesis so that there is no requirement for stomata to remain open for carbon dioxide uptake – stomata close to limit water loss

11. **(a)** **A** = sieve plate; **B** = sieve tube; **C** = companion cell; **D** = plasmodesma; **E** = mitochondrion; **F** = cytoplasmic strand
    (b) Phloem – transport of organic solutes/products of photosynthesis
    (c) Mitochondrion–epithelial cell–epithelium–ileum–digestive system

12. **(a)** Chromatography
    (b) Glucose and fructose
    (c) 0.5
    (d) Surface of gut epithelia cells
    (e) Hydrolysis

13. **(a)** (70 – 32)/20 = 1.9 g per day
    (b) Vitamins stored in body (liver) following transfer across placenta
    (c) Difference in mass = 80 – 50.5 = 29.5
    Percentage difference = 29.5/50.5 × 100 = 58%

14. **(a)** **A**; the countercurrent system provides the greater surface with a concentration gradient which favours oxygen transfer from water to blood
    (b) *Three from* large surface area/thin/moist/close to blood vascular system

15. **A** = trachea; **B** = relaxation of external intercostal muscles; **C** = relaxation of diaphragm; **D** = air exhaled down pressure gradient; **E** = expiration; **F** = contraction of external intercostal muscles; **G** = contraction of diaphragm; **H** = air inhaled down pressure gradient; **I** = inspiration

16. Plasma; red blood cells; white cells; platelets; stem cells; haemopoiesis; bone marrow; sodium; calcium; hydrogencarbonate; albumin; solute potential; serum; urea; glucose

17. **(a)** **A** = arteriole; **B** = venule; **C** = vein; **D** = capillary; **E** = artery
    (b) (i) E (ii) C (iii) E (iv) D

18. **(a)** **A** = aortic valve; **B** = superior vena cava; **C** = wall of right atrium; **D** = A–V node; **E** = tricuspid valve; **F** = wall of right ventricle; **G** = carotid artery; **H** = pulmonary artery; **I** = pulmonary vein; **J** = chordae tendinae; **K** = wall of left ventricle; **L** = interventricular septum
    (b) (i) 5 (ii) 2 (iii) 3 (iv) 1 (v) 4

19. **(a)** **A** = pressure change in left ventricle; **B** = aortic valve closes; **C** = pressure changes in aorta; **D** = opening of aortic valve; **E** = pressure change in left atrium; **F** = bicuspid valve opens; **G** = bicuspid valve closes
    (b) 118/80 mmHg
    (c) 1 complete beat takes (0.9 – 0.1) s = 0.8 s
    Number of beats per minute = 60/0.8 = 72
    (d) 118/5 = 23.6 mmHg; right ventricle needs to generate only sufficient pressure to propel blood to lungs, left ventricle must generate enough pressure to propel blood around body in the systemic circulation

20. Interphase; chromatids; centromere; nucleolus; spindle; metaphase; spindle; centromere; pole; cytokinesis

21. **A** = chromosome; **B** = cell wall; **C** = spindle fibre; **D** = centromere; **E** = chromatid; **F** = spindle organiser

22. **A** = amino acid attachment site; **B** = unpaired folds; **C** = hydrogen bonds; **D** = anticodon

23. **(a)** **A** = haploid; **B** = DNA replication; **C** = separation of homologous chromosomes; **D** = separation of chromatids; **E** = cytokinesis
    (b) Restores diploid number following fertilisation; variation during gamete formation

24. Mutation; gene; aneuploidy; Down's; X; polyploidy; assortment; crossing over; fusion

25. **A** = reverse transcriptase; **B** = DNA polymerase; **C** = plasmid; **D** = restriction endonuclease; **E** = plasmid with 'sticky ends'; **F** = DNA ligase; **G** = recombinant DNA

26. (a) **A** = TCA cycle; **B** = electron transport; **C** = pyruvate transport; **D** = glycolysis
(b) **A** = ATP; **B** = ADP; **C** = NAD; **D** = $NADH_2$; **E** = $NADH_2$; **F** = NAD
(c) The oxygen debt

27. (a) (i) AC (ii) BD (iii) DE (iv) EF
(b) Compensation point

28. (a) **A** = $NADPH_2$; **B** = ATP synthesis; **C** = light energy; **D** = excited electrons; **E** = protons; **F** = light energy; **G** = water; **H** = photosystem II; **I** = photosystem I
(b) **(i)** Chloroplast **(ii)** 5 **(iii)** 5 μm **(iv)** carbonic anhydrase – combination of carbon dioxide and ribulose bisphosphate

29. Auxins: + + – + – – gibberellins: + – – – + – abscisic acid: – – – – – + ethene: – – + – – –

30. A – 4; B – 3; C – 6; D – 5; E – 1; F – 2; G – 8; H – 7

31. Ecology; ecosystem; habitat; community; population; succession; climax

32. nitrite; nitrate; nitrate; nitrogen; photosynthesis; glucose; respiration; carbon dioxide; amination; glucose

33. humoral; antibodies; cell-mediated; T-lymphocytes; B-lymphocyte; macrophage; antibodies; antigen; cytokines; T-helper; T-cytotoxic

34. (a) (i) 1 (ii) 5 (iii) 3 (iv) 4
(b) 2 (loop of Henle) – very lengthy to create high solute concentration in medulla so that water potential gradient favours reabsorption of water from collecting duct. This is vital in an animal living in an area of water shortage

35. Ammonia; soluble; toxic; urea; in solution; water conservation; uric acid; insoluble

36. **A** = visual centre; **B** = sensory neurone; **C** = retina; **D** = circular muscle; **E** = excitatory neurone; **F** = radial muscle; **G** = inhibitory neurone (**D/E** and **F/G** can be exchanged)
(a) Prevents bleaching of retina and consequent loss of vision
(b) Conditioned reflex replaces normal stimulus with another one not directly related to survival

37. **A** = diabetic patient: **B** = increased insulin secretion; **C** = non-diabetic patient; **D** = glucose injection
(a) Difference in height and duration of glucose peak
(b) (202 – 146) = 56, 56/146 × 100 = 38%
(c)

	Cause	Treatment
Type I	Deficient insulin secretion	Insulin injection
Type II	Liver insensitivity to insulin secretion	Control dietary intake of glucose/fat

38. (a) In the retina
(b) **A** = white colour perceived; **B** = blue cone; **C** = orange colour perceived; **D** = green cone; **E** = red cone
(c) A condition caused by a gene carried on one of the sex chromosomes (usually the X-chromosome)

39. (a) 1 (b) 2 (c) 1 (d) 4 (e) 3

40. Fluid-filled muscular sac; annelid; earthworm; endoskeleton; chordate; cat; exoskeleton; contains chitin; ladybird

41. **A** = M line; **B** = H zone; **C** = I band; **D** = Z line; **E** = A band

42. (a) **A** = ovarian phase; **B** = luteal phase; **C** = luteinising hormone; **D** = oestrogen; **E** = progesterone; **F** = repair of endometrium; **G** = ovulation; **H** = menstruation
(b) Progesterone acts as feedback inhibitor of FSH and so prevents ovulation – no ovulation means no conception
(c)

Difference	Reason
Male gamete much smaller	More cytoplasm in female gamete provides more food reserves for early embryonic development
Male gamete has flagellum	Male gamete must swim to female gamete

43.

	Genotype	Phenotype
1	X^RY	Normal (male)
2	X^RX^r	Carrier (female)
3	X^RX^r	Carrier (female)
4	X^RY	Normal (male)
5	X^RY	Colour blind (male)
6	X^RX^r or X^RX^R	Normal/carrier (female)

44. (a) Unlikely – there are similar numbers of males and females with the condition
(b) Recessive – two 'normal' parents in generation 2 produce a CF child in generation 3
(c) Thick mucus in lungs becomes infected with bacteria – antibiotics control bacterial multiplication
(d) Thick mucus blocks pancreatic duct and so limits release of lipase in pancreatic juice

Disease may be caused by infectious pathogens or may reflect the effects of a poor lifestyle

Pathogens cause disease: many organisms may colonise the body of a human - the body is WARM, MOIST and a GOOD FOOD SOURCE. Pathogens damage human cells by:

- competing with the human cells for nutrients;
- releasing poisonous compounds (TOXINS);
- producing by-products of their metabolism e.g. lactic acid.

There are are three major groups of pathogens:

Type of pathogen	Disease	Symptoms
Virus 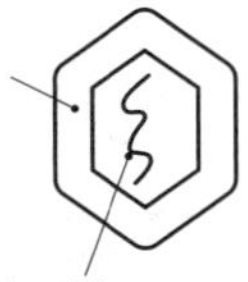	Influenza	Fever (raised body temperature). Aching joints. Breathing problems. CONTROL BY PAIN RELIEF/REST/ DRINKING FLUIDS
Bacterium 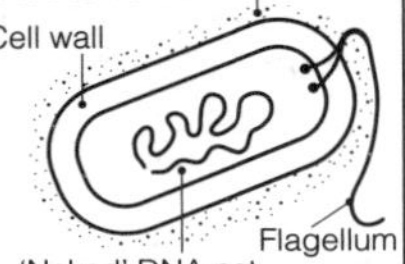	Cholera	Diarrhoea (very watery faeces). Loss of water and salts. Dehydration and weakness. CONTROL BY ORAL REHYDRATION.
Fungus 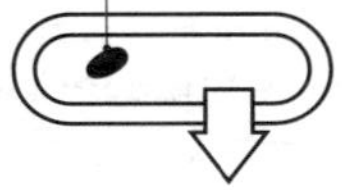	Athlete's foot	Irritation to moist areas of skin (e.g. between toes): cracked skin may become infected. CONTROL WITH FUNGICIDE/ DRYING POWDERS

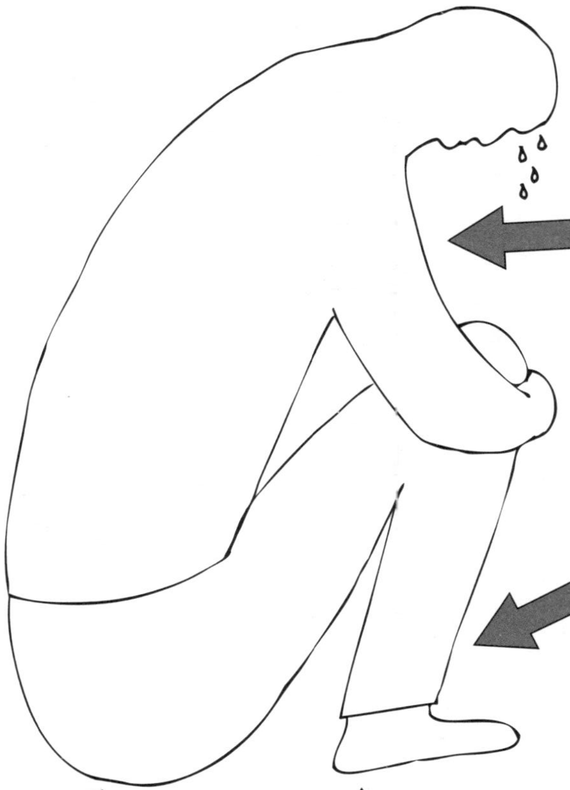

Some diseases are not infectious i.e. they are NOT caused by pathogens. They are often the result of a POOR LIFESTYLE, e.g.

- lung cancer, bronchitis and emphysema caused by smoking of cigarettes;
- coronary heart disease is made worse by a diet high in saturated fats and salt, smoking of cigarettes and by lack of exercise;
- cirrhosis of the liver is caused by alcohol abuse;
- skin cancer is caused by over exposure to ultraviolet light during sunbathing.

There are also inherited, metabolic diseases caused by some failure of the body's normal set of chemical reactions:

sickle cell anaemia (abnormal haemoglobin);
cystic fibrosis (overproduction of mucus);
diabetes (failure to produce enough insulin).

These conditions are due to alterations in the genes.

Infectious diseases are caused by pathogens

- **these must cross an** interface with the environment (e.g. gut lining or lungs);
- **these may spread from** one human to another in a number of ways ....

via **droplets in the air** e.g. Influenza virus (which infects cells of the gas exchange system)

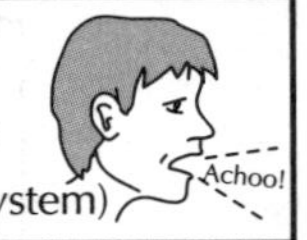

in **infected water** e.g. Cholera bacterium (which infects cells of the digestive system)

by **direct contact (contagious)** e.g. Athlete's foot fungus (which infects cells of the skin)

During disease the body's activities deviate from their normal levels by an amount which is more than can be counteracted by the usual homeostatic mechanisms.

Typical symptoms of disease include:

- **sweating/fever** due to resetting of body's thermostat;
- **vomiting/diarrhoea** due to attempt to 'clear' gut of irritants;
- **pain** due to release of toxins by pathogens.

Parasites and pathogens

Are all parasites pathogens?

It depends **how** dependent they are on their host … the most dependent (such as viruses) **do** cause their host some harm or discomfort. The best adapted parasites cause little harm - in this way they keep their food source for a long time!

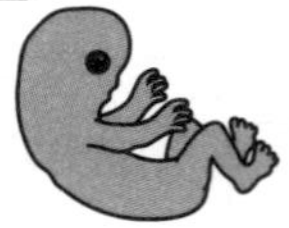

Is a foetus a parasite? … it is certainly 100% dependent on its host!

Parasites obtain their nutrients from another **living** organism, called the **host**.

Viruses: all viruses are parasitic – they cannot multiply outside the cells of a living host. They show some adaptations to their way of life:

- once inside the cells of the host they may be 'hidden' from the immune system;
- they may mutate so that their surface antigens change so frequently that antibody production cannot keep up and the virus escapes destruction.

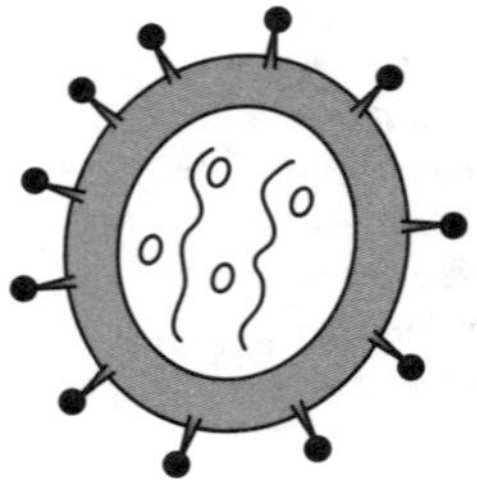

HIV shows both adaptations – it 'hides' inside cells of the immune system!

Some important viral diseases are Influenza, Rabies and Foot-and-mouth.

Bacteria: some bacteria are parasitic but most (e.g. bacteria of the carbon and nitrogen cycles, and symbiotic intestinal bacteria) are not. Some important human pathogens are parasitic bacteria.

Species	Site of infection	Disease
Mycobacterium tuberculosis	Lungs	TB
Chlamydia trachomatis	Urethra	Urethritis
Neisseria meningitidis	Spinal cord	Meningitis

Multicellular organisms

Endoparasites ('endo' means inside) feed inside the body e.g. tapeworms in the gut. They show many adaptations …

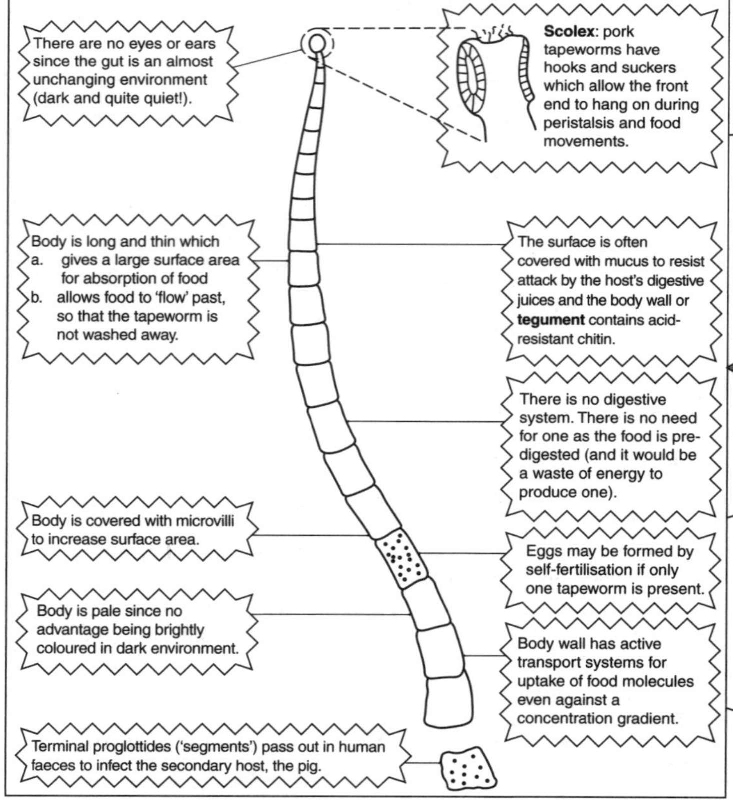
There are no eyes or ears since the gut is an almost unchanging environment (dark and quite quiet!).
Scolex: pork tapeworms have hooks and suckers which allow the front end to hang on during peristalsis and food movements.
Body is long and thin which
a. gives a large surface area for absorption of food
b. allows food to 'flow' past, so that the tapeworm is not washed away.
The surface is often covered with mucus to resist attack by the host's digestive juices and the body wall or **tegument** contains acid-resistant chitin.
There is no digestive system. There is no need for one as the food is pre-digested (and it would be a waste of energy to produce one).
Body is covered with microvilli to increase surface area.
Eggs may be formed by self-fertilisation if only one tapeworm is present.
Body is pale since no advantage being brightly coloured in dark environment.
Body wall has active transport systems for uptake of food molecules even against a concentration gradient.
Terminal proglottides ('segments') pass out in human faeces to infect the secondary host, the pig.

Many plants have ectoparasites (e.g. aphids): they pierce the phloem and feed on the solution of sucrose and amino acids transported in this tissue.

Ectoparasites ('ecto' = outside) feed on the body surface:

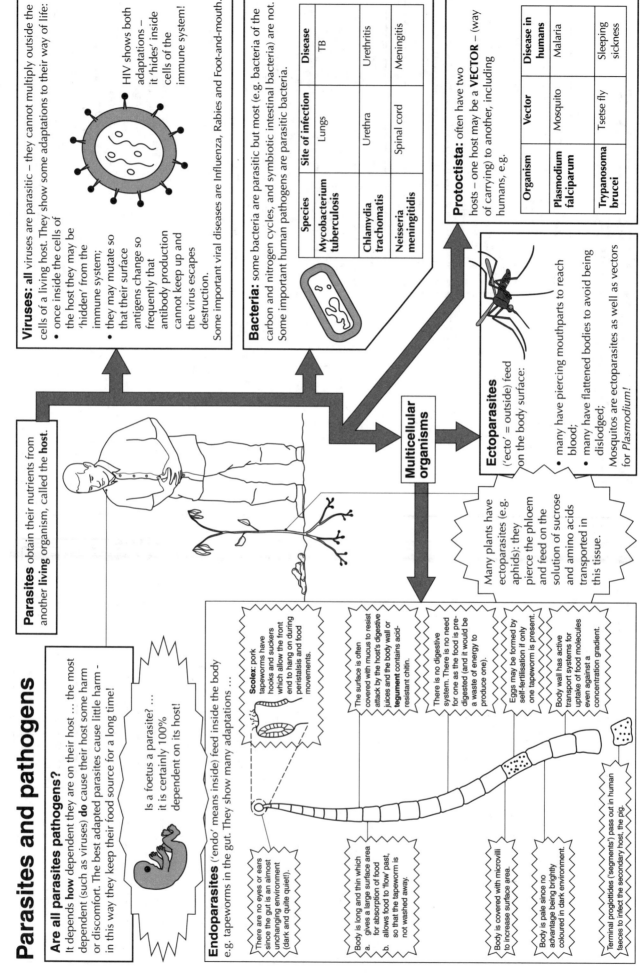

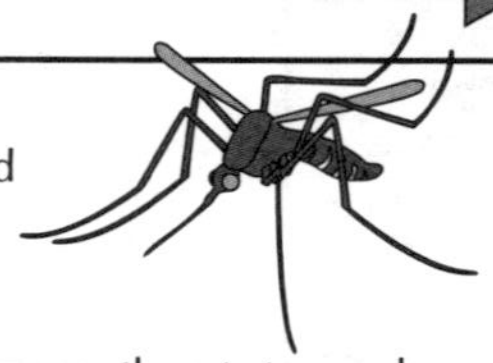

- many have piercing mouthparts to reach blood;
- many have flattened bodies to avoid being dislodged;

Mosquitos are ectoparasites as well as vectors for *Plasmodium!*

Protoctista: often have two hosts – one host may be a **VECTOR** – (way of carrying) to another, including humans, e.g.

Organism	Vector	Disease in humans
Plasmodium falciparum	Mosquito	Malaria
Trypanosoma brucei	Tsetse fly	Sleeping sickness

Section B Molecules and cells

Biological molecules

are largely based upon the ability of carbon to form **covalent bonds** with other atoms, usually of carbon, oxygen, hydrogen or nitrogen. The molecules may adopt long chain forms, or a variety of ring structures.

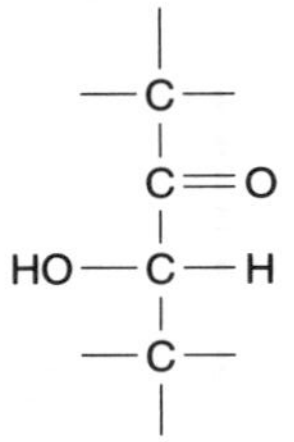

Nucleic Acids and Nucleotides

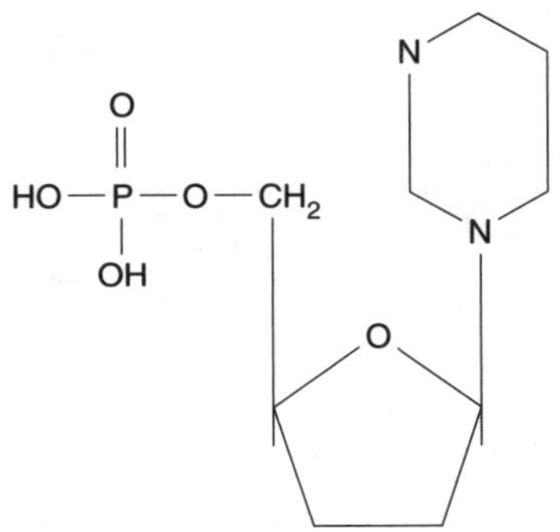

- Contain C,H,O,N and P.
- The phosphate group gives an ACIDIC REACTION.
- After hydrolysis the pentose sugar gives a positive BENEDICT'S TEST.

Principle: Copper (II) sulphate, which forms a **blue solution** under alkaline conditions is reduced to copper (I) oxide, which forms a **brick-red precipitate** under the same conditions.

$$\underset{\text{BLUE}}{Cu^{2+}} + e^- \rightarrow \underset{\text{RED}}{Cu^+}$$

The test is *semi-quantitative* – the greater the concentration of reducing sugar, the darker and heavier the precipitate.

Concentration of reducing sugar	Colour of solution and precipitation
0	Blue
Increasing ↓	Green
	Yellow
	Pale brown
	Orange red

Carbohydrates

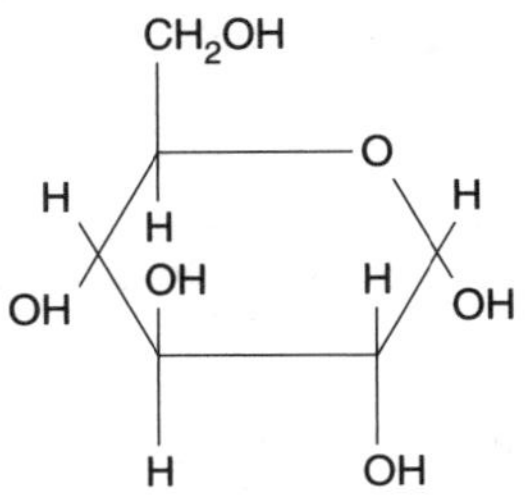

- Usually contain C,H and O only.
- Empirical formula is $C_nH_{2n}O_n$.
- The usual chemical test for the simpler carbohydrates (*reducing sugars*) is heating with **Benedict's reagent.**
- Many carbohydrates are soluble in water.

Amino Acids and Proteins

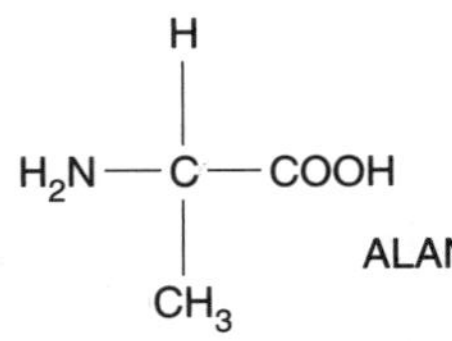

ALANINE

- Contain C,H,O,N and, sometimes, S.
- Are usually soluble in water.
- Proteins give a positive **biuret test.**

Principle: the alkali (sodium hydroxide) hydrolyses the protein into fragments containing the peptide linkage -CO–NH- which then forms coloured complexes with copper (II) from the copper sulphate.

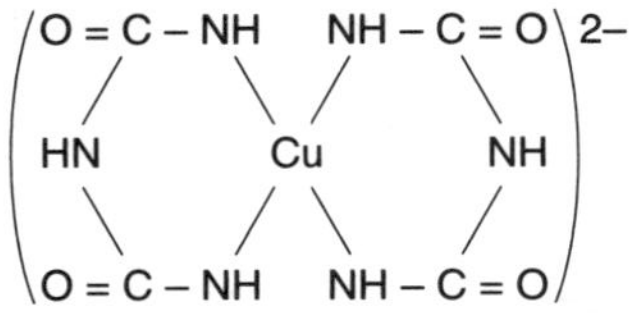

Lipids

- are usually insoluble in water, but soluble in organic solvents;

$CH_3\,(CH_2)_n\,COOH$

Fatty acid

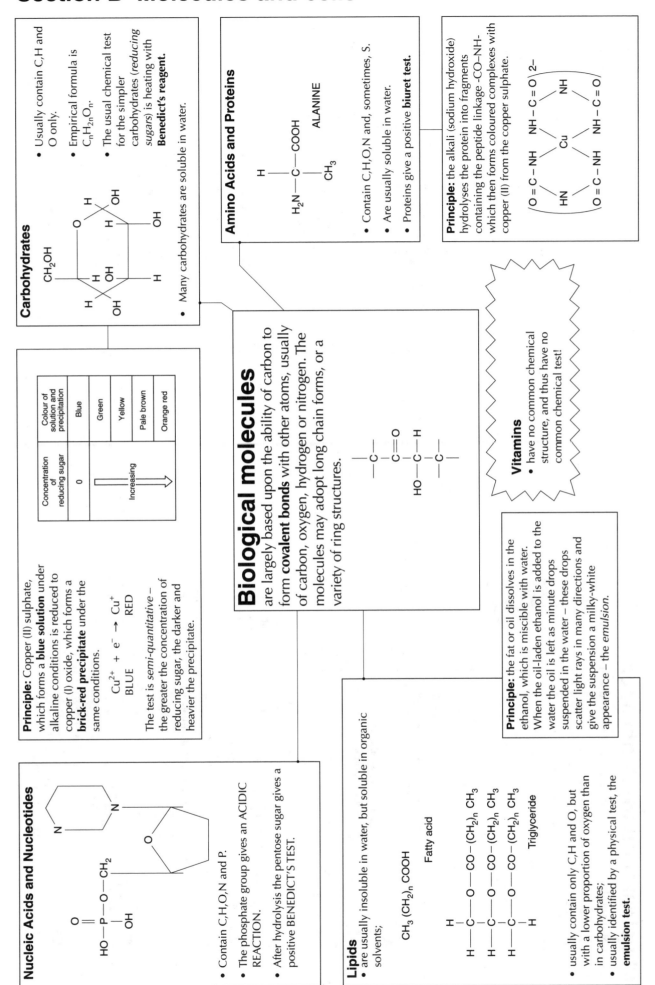

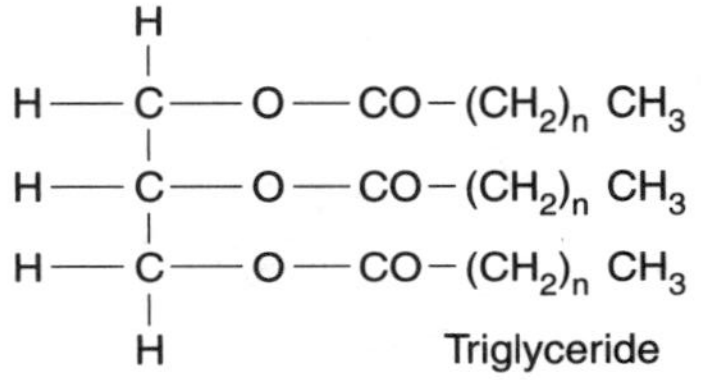

Triglyceride

- usually contain only C,H and O, but with a lower proportion of oxygen than in carbohydrates;
- usually identified by a physical test, the **emulsion test.**

Principle: the fat or oil dissolves in the ethanol, which is miscible with water. When the oil-laden ethanol is added to the water the oil is left as minute drops suspended in the water – these drops scatter light rays in many directions and give the suspension a milky-white appearance – the *emulsion*.

Vitamins

- have no common chemical structure, and thus have no common chemical test!

Physical properties of water

are explained by hydrogen bonding between the individual molecules.

Solvent properties: the polarity of water makes it an excellent solvent for other polar molecules ...

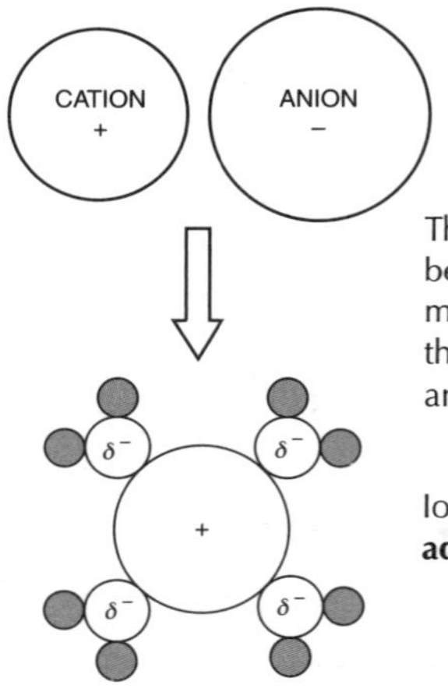

The electrostatic attractions between polar water molecules and ions are greater than those between the anion and cation.

Ions become **hydrated** in **aqueous solution.**

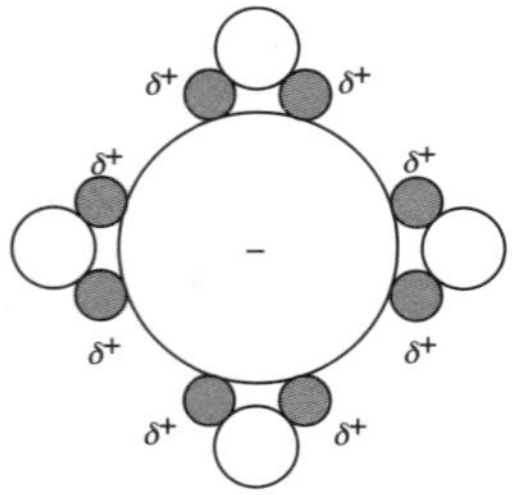

Such polar substances, which dissolve in water, are said to be **hydrophilic** ('water-loving').

... but means that non-polar (**hydrophobic** or 'water-hating') substances do not readily dissolve in water.

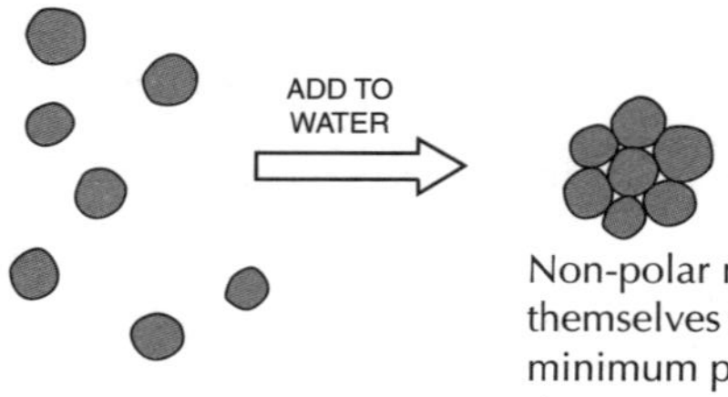

Non-polar molecules arrange themselves to expose the minimum possible surface to the water molecules.

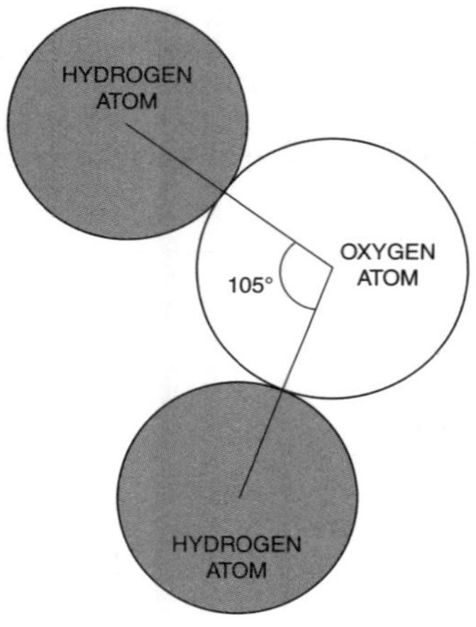

Because hydrogen and oxygen atoms are different in **size** and **electronegativity** the water molecule (H_2O) is **non-linear** and **polar.**

Hydrogen bond: one water molecule may form hydrogen bonds with up to **four** other water molecules.

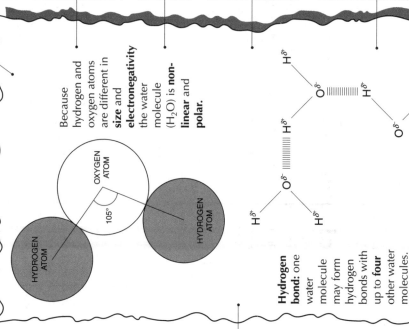

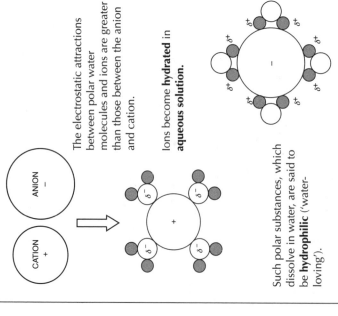

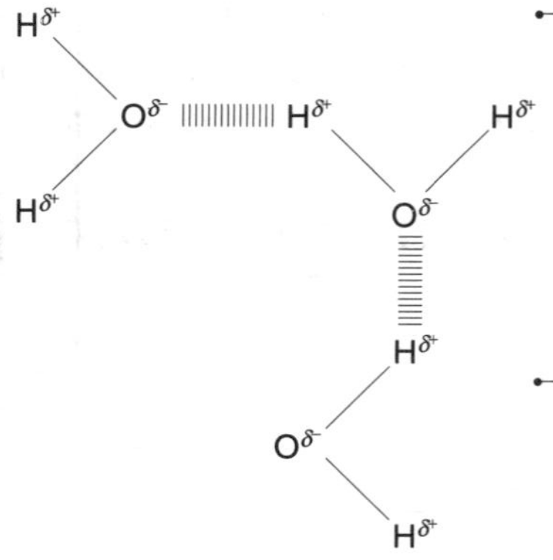

This polarity means that individual water molecules can form **hydrogen bonds** with other water molecules. Although these individual hydrogen bonds are weak, collectively **they make water a much more stable substance than would otherwise be the case.**

High specific heat capacity: the specific heat capacity of water (the amount of heat, measured in joules, required to raise 1 kg of water through 1°C) is very high: much of the heat absorbed is used to break the hydrogen bonds which hold the water molecules together.

High latent heat of vaporisation: hydrogen bonds attract molecules of liquid water to one another and make it difficult for the molecules to escape as vapour: thus a relatively high energy input is necessary to vaporise water and water has a much higher boiling point than other molecules of the same size.

Molecular mobility: the weakness of individual hydrogen bonds means that individual water molecules continually jostle one another when in the liquid phase.

Cohesion and surface tension: hydrogen bonding causes water molecules to 'stick together', and also to stick to other molecules - the phenomenon of **cohesion**. At the surface of a liquid the inwardly-acting cohesive forces produce a 'surface tension' as the molecules are particularly attracted to one another.

Density and freezing properties: as water cools towards its freezing point the individual molecules slow down sufficiently for each one to form its maximum number of hydrogen bonds. To do this the water molecules in liquid water must move further apart to give enough space for all four hydrogen bonds to fit into. As a result water expands as it freezes, so that ice is less dense than liquid water and therefore floats upon its surface.

Colloid formation: some molecules have strong intramolecular forces which prevent their solution in water, but have charged surfaces which attract a covering of water molecules. This covering ensures that the molecules remain dispersed throughout the water, rather than forming large aggregates which could settle out. The dispersed particles and the liquid around them collectively form a **colloid**.

The biological importance of water depends on its physical properties.

Volatility/stability: is balanced at Earth's temperatures so that a water cycle of evaporation, transpiration and precipitation is maintained.

Solvent properties: allow water to act as a transport medium for polar solutes.
For example:
- movements of minerals to lakes and seas;
- transport via blood and lymph in multicellular animals;
- removal of metabolic wastes such as urea and ammonia in urine.

Transpiration stream: the continuous column of water is able to move up the xylem because of cohesion between water molecules and adhesion between water and the walls of the xylem vessels.

Molecular mobility: the rather weak nature of individual hydrogen bonds means that water molecules can move easily relative to one another - this allows *osmosis* (vital for uptake and movement of water) to take place.

Expansion on freezing: since ice floats it forms at the surface of ponds and lakes - it therefore insulates organisms in the water below it, and allows the ice to thaw rapidly when temperatures rise. Changes in density also maintain circulation in large bodies of water, thus helping nutrient cycling. Floating ice also means that penguins and polar bears have somewhere to stand!

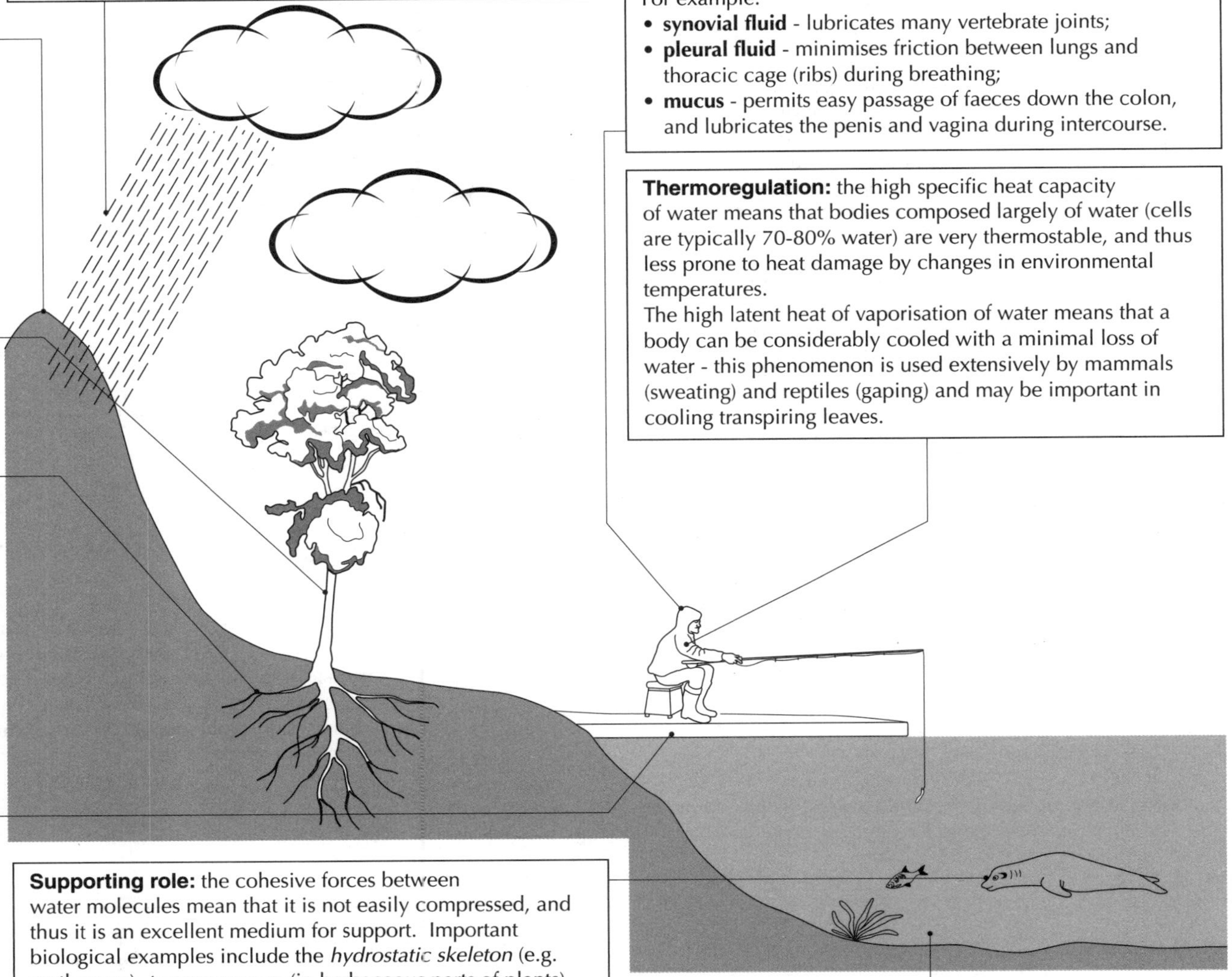

Lubricant properties: water's cohesive and adhesive properties mean that it is viscous, making it a useful lubricant in biological systems.
For example:
- **synovial fluid** - lubricates many vertebrate joints;
- **pleural fluid** - minimises friction between lungs and thoracic cage (ribs) during breathing;
- **mucus** - permits easy passage of faeces down the colon, and lubricates the penis and vagina during intercourse.

Thermoregulation: the high specific heat capacity of water means that bodies composed largely of water (cells are typically 70-80% water) are very thermostable, and thus less prone to heat damage by changes in environmental temperatures.
The high latent heat of vaporisation of water means that a body can be considerably cooled with a minimal loss of water - this phenomenon is used extensively by mammals (sweating) and reptiles (gaping) and may be important in cooling transpiring leaves.

Metabolic functions
Water is used directly …
- as a reagent (source of reducing power) in photosynthesis;
- to hydrolyse macromolecules to their subunits, in digestion for example;

… and is also the medium in which all biochemical reactions take place.

Supporting role: the cohesive forces between water molecules mean that it is not easily compressed, and thus it is an excellent medium for support. Important biological examples include the *hydrostatic skeleton* (e.g. earthworm), *turgor pressure* (in herbaceous parts of plants), *amniotic fluid* (which supports and protects the mammalian foetus) and as a *general supporting medium* (particularly for large aquatic mammals such as whales).

Transparency: water permits the passage of visible light. This means that photosynthesis (and associated food chains) is possible in relatively shallow aquatic environments.

Functions of soluble carbohydrates include transport, protection, recognition and energy release.

Sugar derivatives include *sugar alcohols,* e.g. glycerol, *sugar acids,* e.g. ascorbic acid, and *mucopolysaccharides,* which are important components of connective tissues, synovial fluid, cartilage and bone. Heparin (anticoagulant in blood) is derived from mucopolysaccharides and has a protective function.

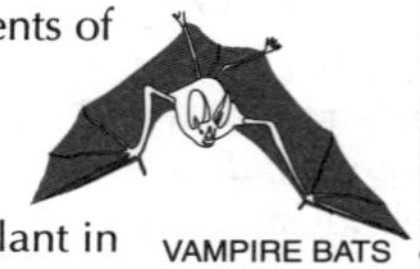

Oligosaccharides are short (often 6-12 units) condensation products which combine with protein (*glycoprotein*) or lipid (*glycolipid*) and form the outer coat (*glycocalyx*) of animal cells. They are important in *cell-cell recognition* and the *immune response.*

Glucose is the most common substrate for respiration (energy release). **Fructose** is a constituent of nectar and sweetens fruits to attract animals and aid seed dispersal.

Glucose and fructose are both **monosaccharides** (single sugar units) with the typical formula $C_nH_{2n}O_n$. They each have **six carbon atoms** and are thus called **hexoses** (**pentoses** have 5 carbon atoms and **trioses** have 3). **Glucose** and **fructose** are isomers of $C_6H_{12}O_6$.

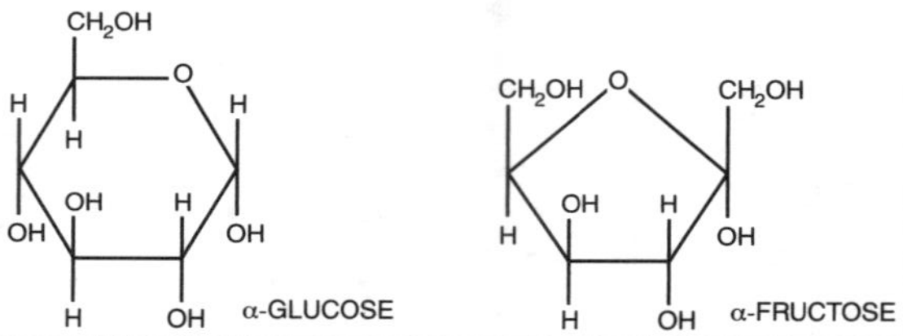

In naturally occurring **disaccharides** monosaccharide rings are joined together by *glycosidic bonds.*

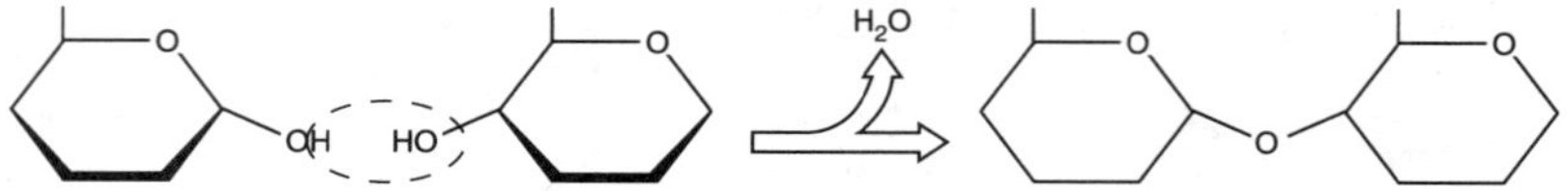

This most usually occurs between *aldehyde* or *keto group* (i.e. the reducing group) of one monosaccharide and an *hydroxyl group* of another monosaccharide, e.g. *lactose.*

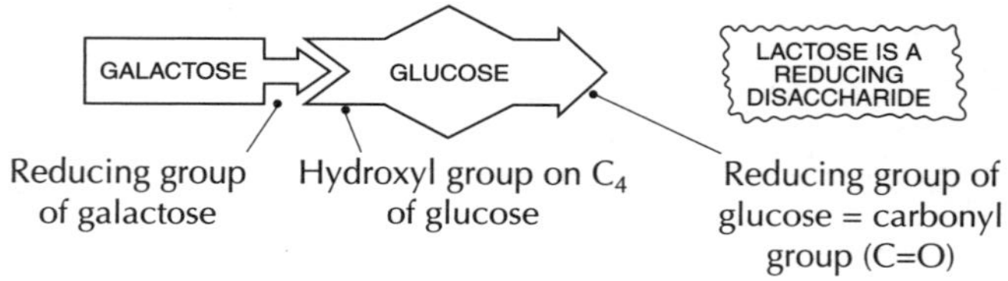

(*Maltose* is a reducing disaccharide formed from two molecules of α-glucose.)

More rarely it can happen between *reducing groups of adjacent monosaccharides,* e.g. *sucrose.*

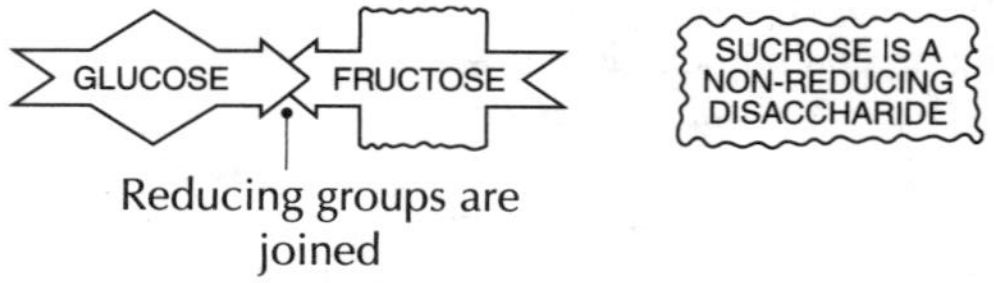

α-glucose and β-glucose are isomers. These two molecules only differ in the arrangement of –H and –OH at the first C atom in the 'ring'.

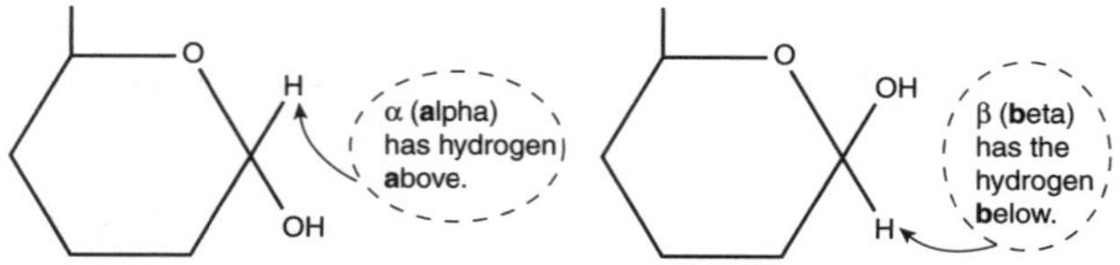

Lactose intolerance occurs in many adults. This results from a deficiency in *lactase* so that dietary lactose accumulates in the lumen of the small intestine. This lowers the water potential of the gut contents causing an influx of fluid into the small intestine - this results in abdomenal distension, nausea, pain and diarrhoea. The condition is much more common in adult populations for whom milk is an unusual or uncommon food.

Sucrose (*glucose-fructose*) is the main transport compound in plants. Commonly extracted from sugar cane and sugar beet and used as a sweetener.

Lactose (*glucose-galactose*) is the carbohydrate source for suckling mammals - milk is about 5% lactose.

Protein function depends on structure

Carrier proteins play an important part in *transport across membranes,* e.g. Na/K pumps are globular proteins with binding sites which recognise and transport ions across nerve cell membranes in preparation for transmission of an action potential.

Fibrinogen and **prothrombin** are *protective proteins* essential for the clotting of blood by forming the fibres making the 'network' of a scab.

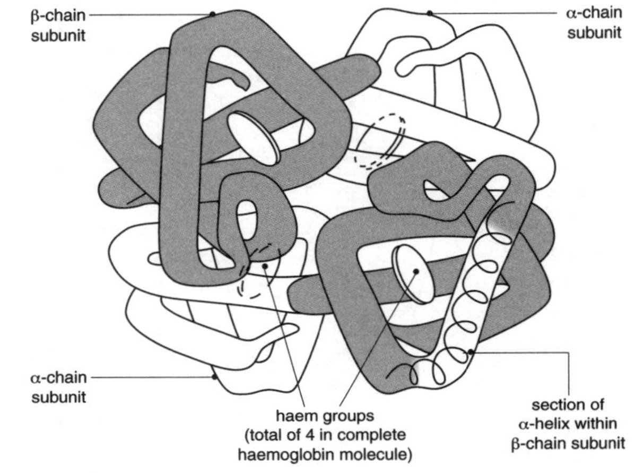

Haemoglobin transports oxygen inside red blood cells:
- four chains can move to cause co-operative binding of oxygen;
- inside region of each subunit provides ideal environment for oxygen to bind to haemoglobin;
- outside of each subunit has charged R groups to help solubility of haemoglobin in cytoplasm.

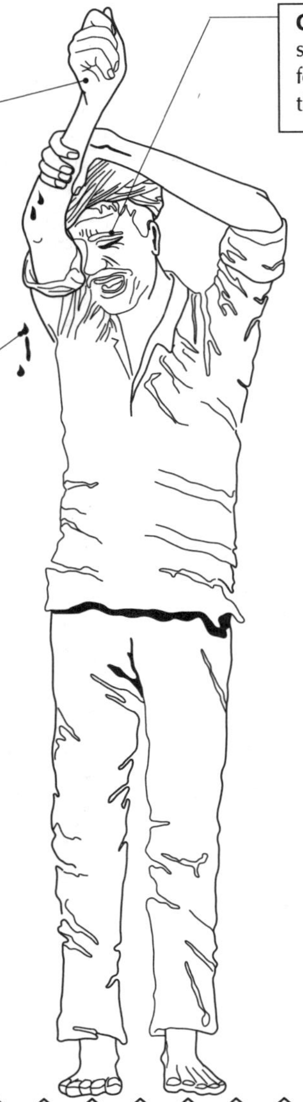

Opsin is a part of the light-sensitive pigment *rhodopsin* found in rod cells of the retina.

Collagen is the most abundant of all animal proteins. It is found in the connective tissue of skin, tendons and ligaments.

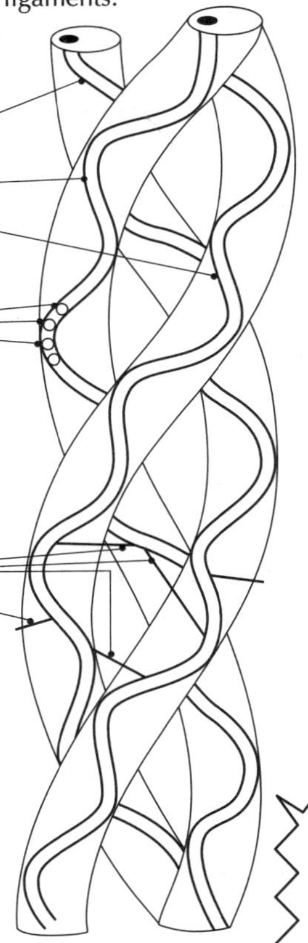

- the long fibres provide a 'framework' for tissues (like iron in reinforced concrete);
- presence of glycine allows three chains to pack together, giving strength to the molecule;
- side chains of other amino acids are hydrophobic, so the molecule is insoluble in water.

Any mutation which causes glycine to be replaced with a different amino acid means that the triple helix cannot pack properly and becomes weakened. This happens in **osteogenesis imperfecta** - collagen in bones is unstable and bones become brittle.

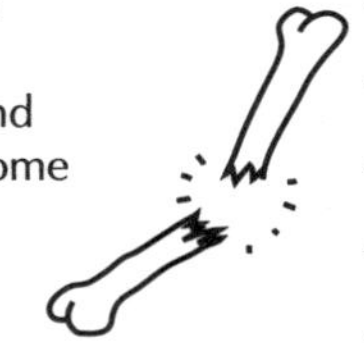

In **Scurvy**, lack of vitamin C means that the hydrogen bonds do not form properly. The collagen weakens so skin lesions (tears) occur and the teeth fall out from the gums.

In **Sickle cell disease**, a mutation causes a change in haemoglobin shape. The molecules stick together under acidic conditions and the red blood cells take on a 'sickle' shape.

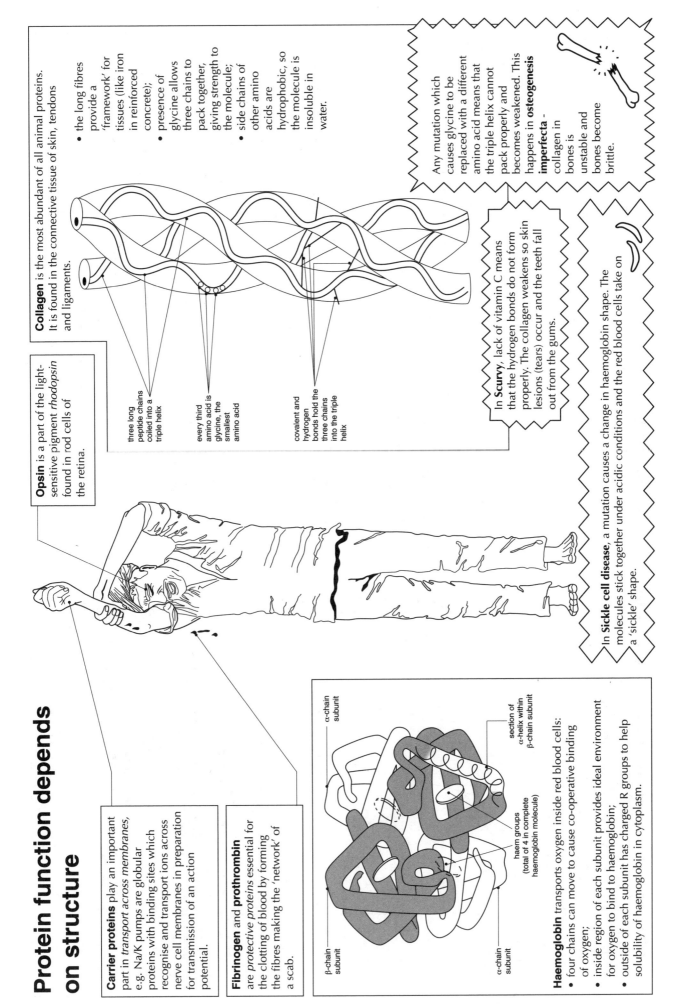

Catalysis by enzymes

An important step in enzyme catalysis is substrate binding to the active sites.

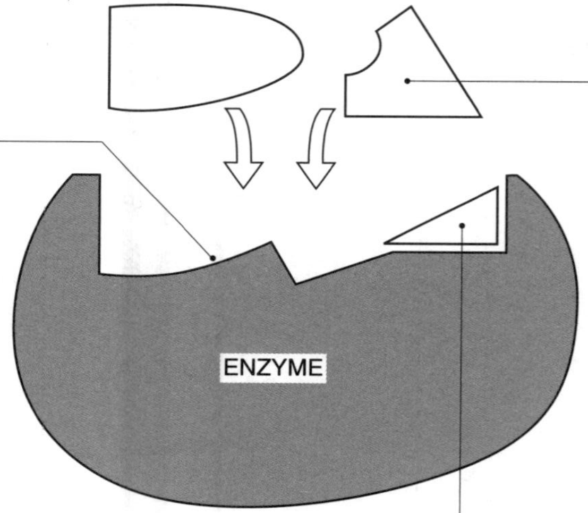

Enzymes form **enzyme-substrate complexes** which reduce the activation energy for reactions which they catalyse.

Consider the reaction: SUBSTRATE (S) $\rightleftharpoons$ PRODUCT (P)

which can be illustrated by a **reaction profile.**

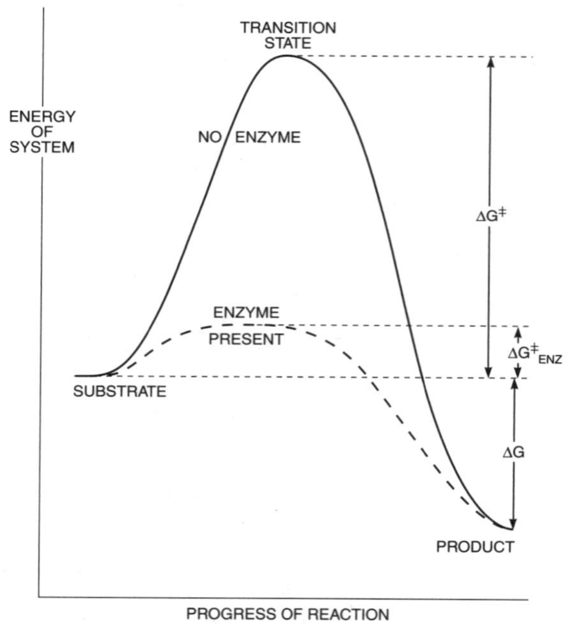

Effect of enzyme on activation energy

- Rate of forward reaction, S→P, depends on activation energy and temperature.
- Enzymes act as catalysts by lowering the activation energy ($\Delta G^{\ddagger}$). They do this by providing alternative reaction pathways.
- Enzymes **do not** reduce the overall free energy change (ΔG) for the reaction.
- Large temperature changes cannot be used by cells to change rates of reaction because of possible denaturation.

Summary for overall reaction:

$$E + S \rightleftharpoons [E - S] \rightleftharpoons E + P$$

Enzyme-substrate complex

Stereospecificity: relationship of substrate(s) to active site

Emil Fischer's **lock and key hypothesis** suggested that the active site and the substrate were **exactly complementary,** i.e. the substrate fits exactly into the active site.

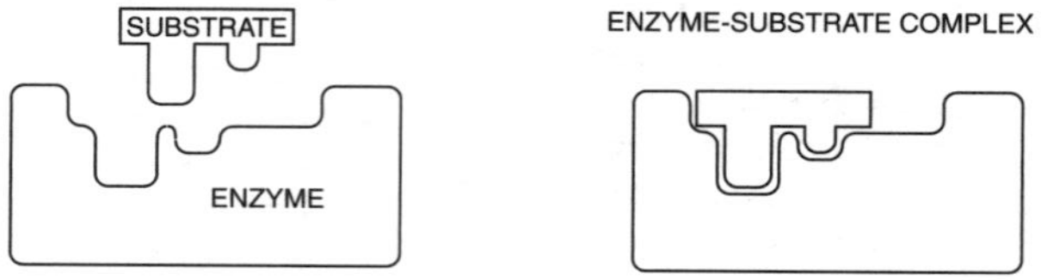

More recent work allowed **Koshland** to propose the **induced fit hypothesis** which suggests that active site and substrate are only fully complementary **after the substrate is bound.**

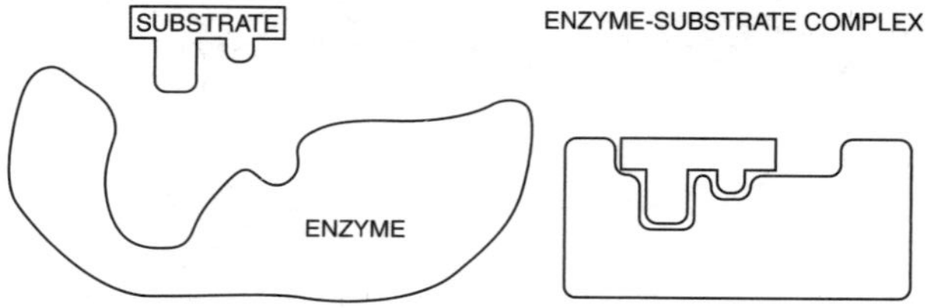

This latter process of **dynamic recognition** is now the more widely accepted hypothesis. It is supported by modern techniques of imaging which show the enzyme changing shape as the substrate binds it.

Cofactors are essential for enzyme activity

Some, such as Zn^{2+} or Mg^{2+}, or porphyrin groups such as the **haem** in catalase, may form part of the active site and cannot easily be separated from the enzyme protein: these are commonly called **prosthetic groups.**

Some, such as NAD (nicotinamide adenine dinucleotide), bind temporarily to the active site and actually take part in the reaction.

e.g. lactate + NAD $\xrightleftharpoons[\text{DEHYDROGENASE}]{\text{LACTATE}}$ pyruvate + $NADH_2$

Such **coenzymes** shuttle between one enzyme system and another - most are formed from dietary components called **vitamins** (e.g. NAD is formed from niacin, one of the B vitamin complex).

Enzymes are released once the reaction is complete, and are ready for use again. Eventually the enzyme-proteins do lose their shape and must be replaced by the cells.

Some enzymes (e.g. those involved in protein synthesis) are INTRACELLULAR (they are used **inside** the cell). Others (e.g. digestive enzymes) are EXTRACELLULAR (they are secreted and used **outside** the cell.

Factors affecting enzyme activity

Factors affecting enzyme activity exert their effects by altering the ease with which an enzyme-substrate complex is formed.

Any factor which alters the conformation (dependent on tertiary structure) of the enzyme will alter the shape of the active site, affect the frequency of enzyme-substrate complex formation and thus influence the rate of the enzyme-catalysed reaction.

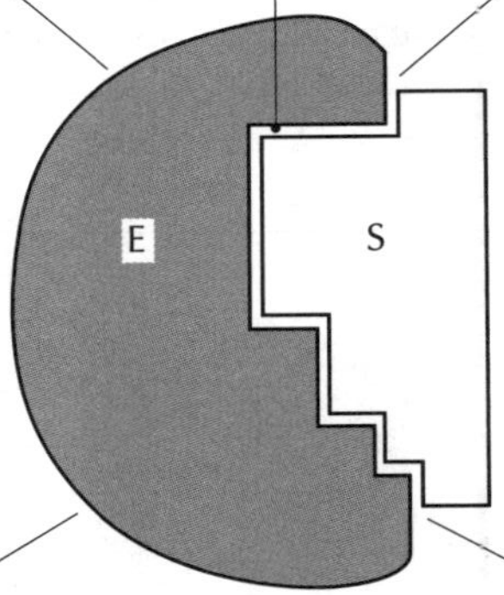

Effect of temperature

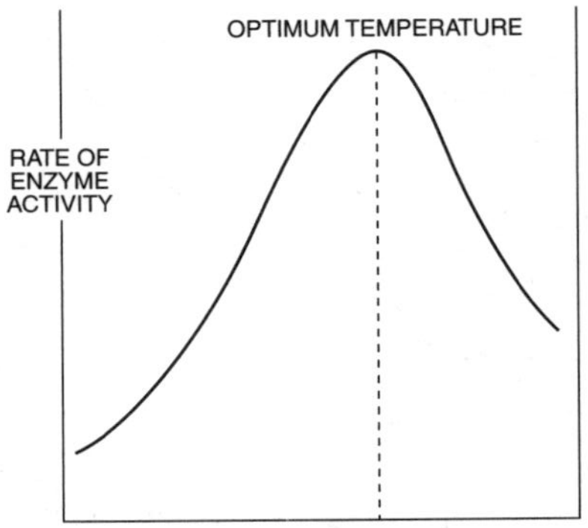
RATE OF ENZYME ACTIVITY
OPTIMUM TEMPERATURE
TEMPERATURE /°C

Enzymes have an **optimum temperature** which represents a compromise between **activation** due to increased rate of collision between E and S and **loss of activity** due to denaturation of E molecules and consequent distortion of the active site.

Effect of enzyme concentration

Enzymes are not used up during catalysis, and enzyme molecules can be used over and over again; enzymes therefore work very well at low concentrations. Increasing the concentration of enzyme provides more active sites so the rate of enzyme activity increases **so as an excess of substrate molecules is available.**

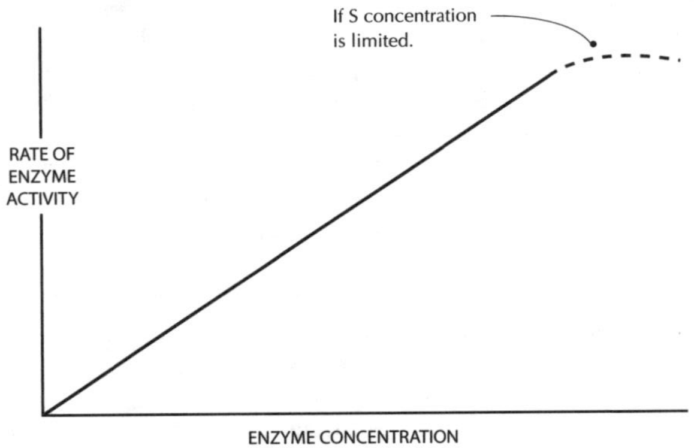
If S concentration is limited.
RATE OF ENZYME ACTIVITY
ENZYME CONCENTRATION

Effect of substrate concentration

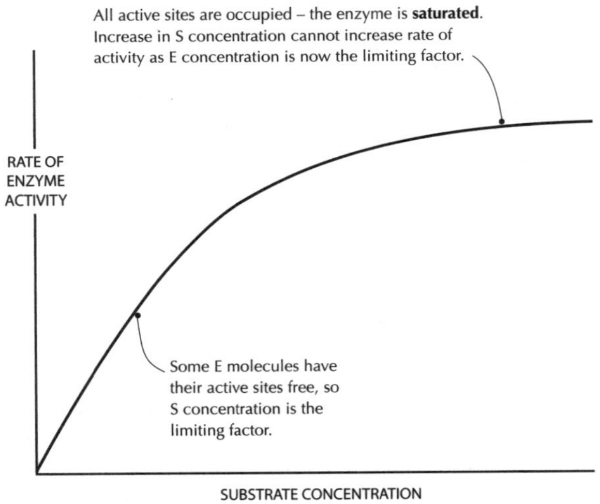
All active sites are occupied – the enzyme is saturated. Increase in S concentration cannot increase rate of activity as E concentration is now the limiting factor.
RATE OF ENZYME ACTIVITY
Some E molecules have their active sites free, so S concentration is the limiting factor.
SUBSTRATE CONCENTRATION

Effect of pH

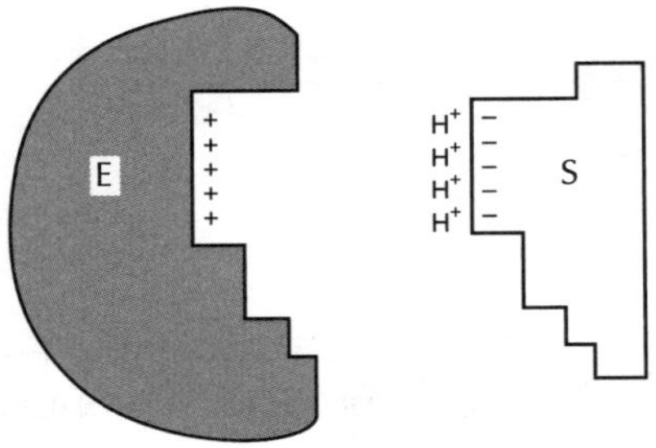

An enzyme has an **optimum pH** which results from the effects of hydrogen ion concentration on the 3-dimensional shape of the enzyme in the active site region.

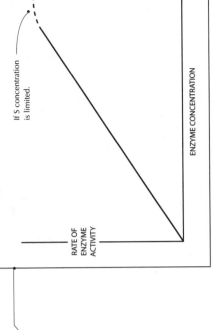

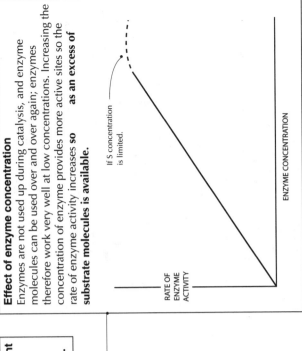

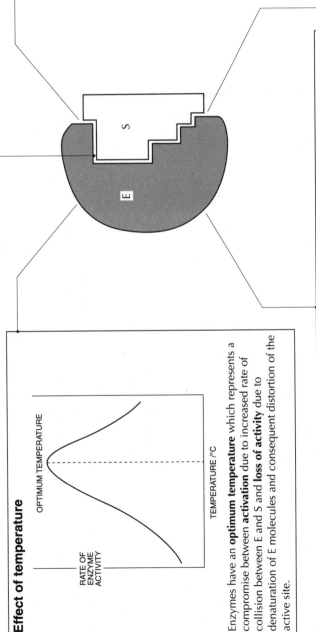

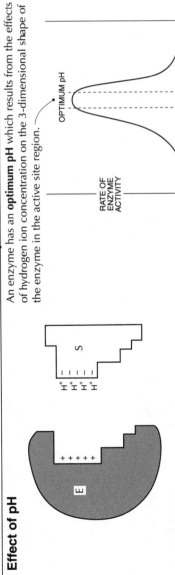

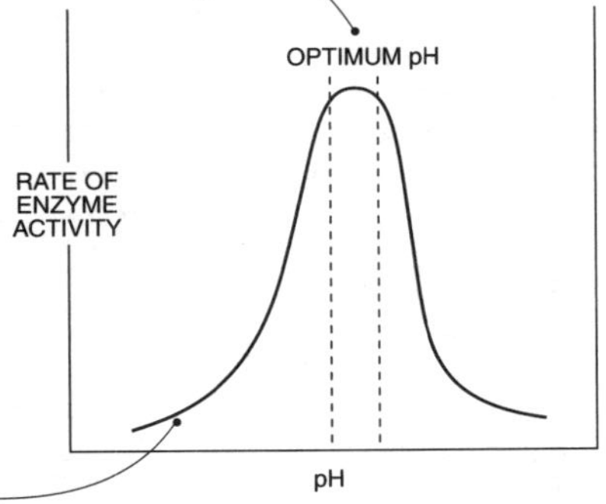
OPTIMUM pH
RATE OF ENZYME ACTIVITY
pH

Greater changes in pH can distort the whole enzyme - DENATURATION - and cause huge loss of enzyme activity.

Inhibitors affect enzyme activity

by an alteration in the binding of enzyme to substrate molecules.

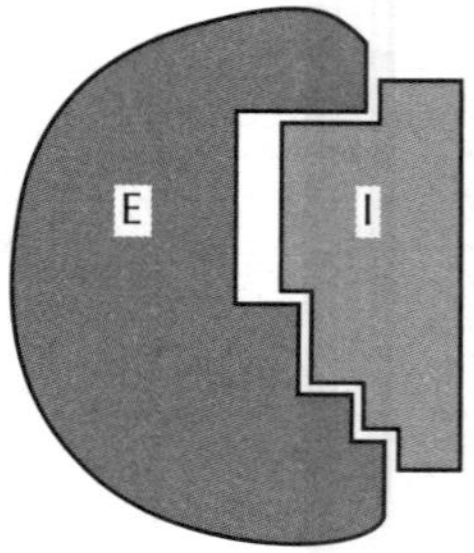

Competitive inhibitors compete for the active site with the normal substrate. These inhibitors therefore must have a similar structure to the natural substrate.
The success of the binding of I to the active site depends on the relative concentrations of I and S, and such inhibition is therefore **reversible by an increase in substrate concentration,** e.g. malonate competes with succinate for the active site on the enzyme **succinate dehydrogenase.**

ACE INHIBITORS
ACE (ANGIOTENSIN CONVERTING ENZYME) inhibitors are powerful drugs for relieving high blood pressure because they inhibit the conversion of angiotensin I → angiotensin II.

Absence of drug: angiotensin I → angiotensin II - vasoconstriction which causes resistance to blood flow and impaired circulatory function.

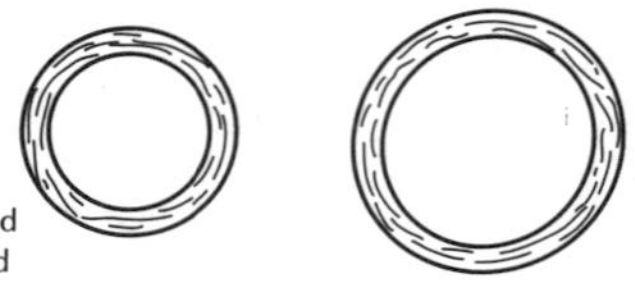

ACE inhibitor present: no conversion to angiotensin II - vasodilation which improves blood flow and circulatory function.

Irreversible inhibition occurs if the enzyme-inhibitor binding is covalent and the distortion of the active site may be permanent, e.g. cyanide (CN^-) acts as a poison because it binds irreversibly to the active site of the enzyme **cytochrome oxidase.**

Competitive or non-competitive

The effect of an inhibitor can be identified as competitive or non-competitive by any change to the substrate concentration curve.

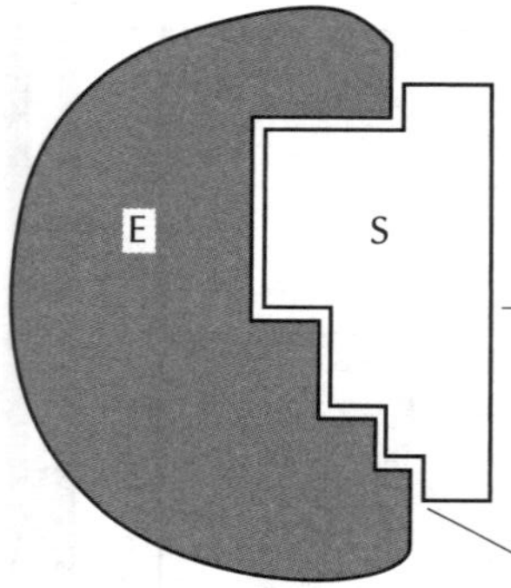

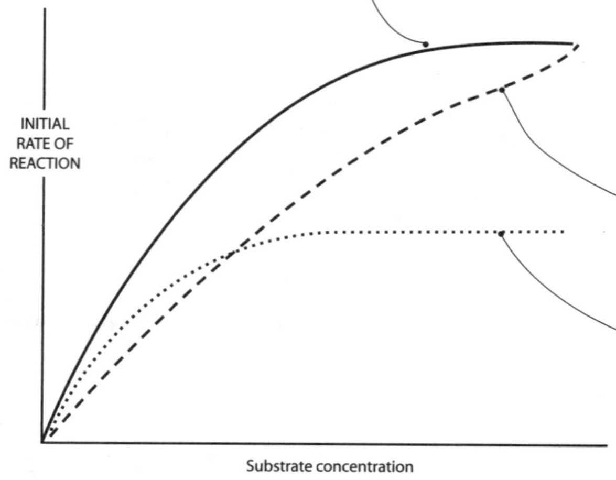

Competitive inhibitor: S and I compete for active sites – maximum rate can be achieved as high [S] 'outcompetes' the I molecules.

Non-competitive inhibitor: even high [S] will not overcome inhibition, as S and I are not competing for the active sites.

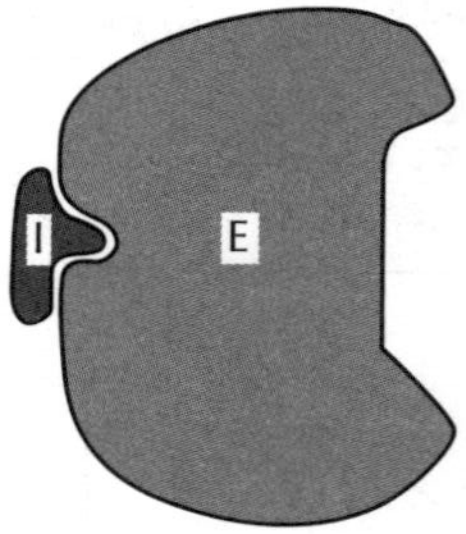

Non-competitive inhibitors reduce enzyme activity by distortion of enzyme conformation caused by binding to some site **other than the active site.** If the binding is non-covalent the inhibition may be **reversible if the inhibitor concentration is diminished.** Many such inhibitors are natural **allosteric regulators** of metabolism, e.g. ATP controls the rate of respiration by inhibition of the enzyme **phosphofructokinase.**

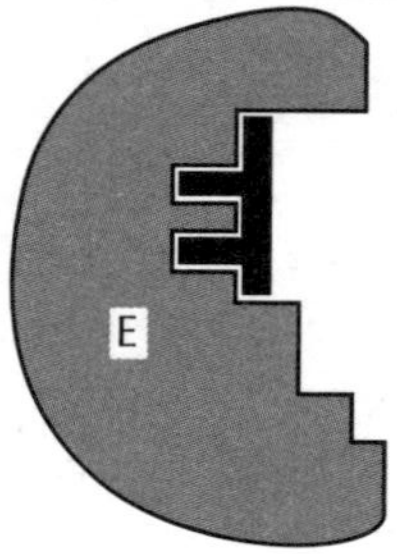

Activators may be necessary to complete the structural relationship between active site and substrate, e.g. chloride ions (Cl^-) are required for activity of the enzyme **salivary amylase.**

There are also **allosteric activators** which enhance enzyme-substrate binding by alteration of enzyme conformation when binding to another (**'allosteric'**) site on the enzyme.

Lipid structure and function

True lipids are esters of fatty acids and alcohols, formed by condensation reactions. Many of their properties result from their insolubility in water.

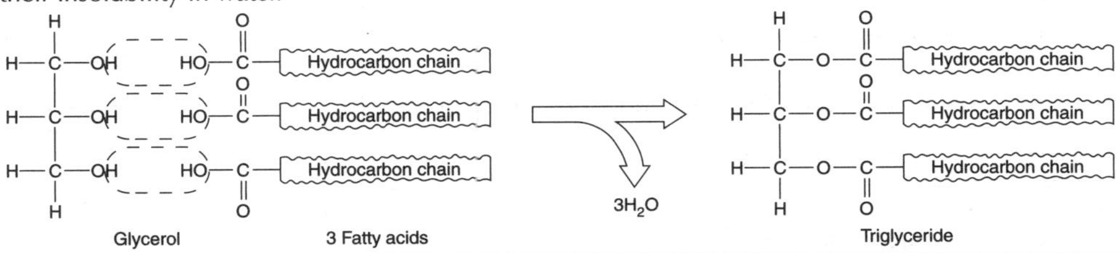

Since the hydrocarbon chains are long (19 C in arachidonic acid) most of the weight of the triglyceride is fatty acid.

Water-repellent properties: oily secretions of the sebaceous glands help to waterproof the fur and skin. The preen gland of birds produces a secretion which performs a similar function on the feathers.

Cell membranes: phospholipids (phosphatides) are found in all cell membranes. These molecules have a polar 'phosphate-base' group substituted for one of the fatty acids in a triglyceride.

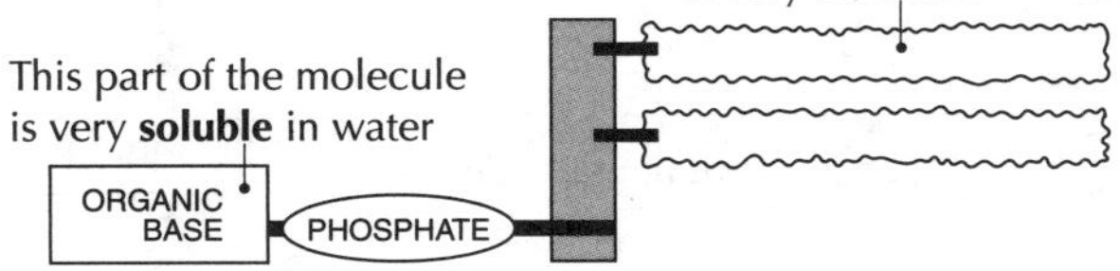

Electrical insulation: myelin is secreted by Schwann cells and insulates some neurones in such a way that impulse transmission is made much more rapid.

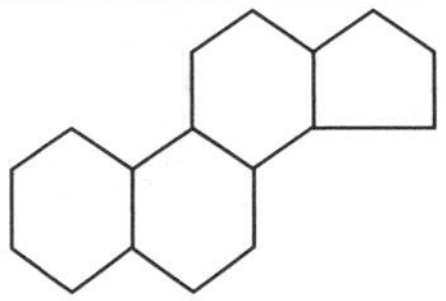

Basic steroid nucleus

Hormones: an important group of hormones, including cortisone, testosterone and oestrogen, are **steroids**. Steroids are not true esters but have the same solubility properties as them.

Physical protection: the shock-absorbing ability of subcutaneous fat stores protects delicate organs such as the kidneys from mechanical damage.

Thermal insulation: fats conduct heat very poorly – subcutaneous fat stores help heat retention in endothermic animals. Incompressible blubber is an important insulator in diving mammals.

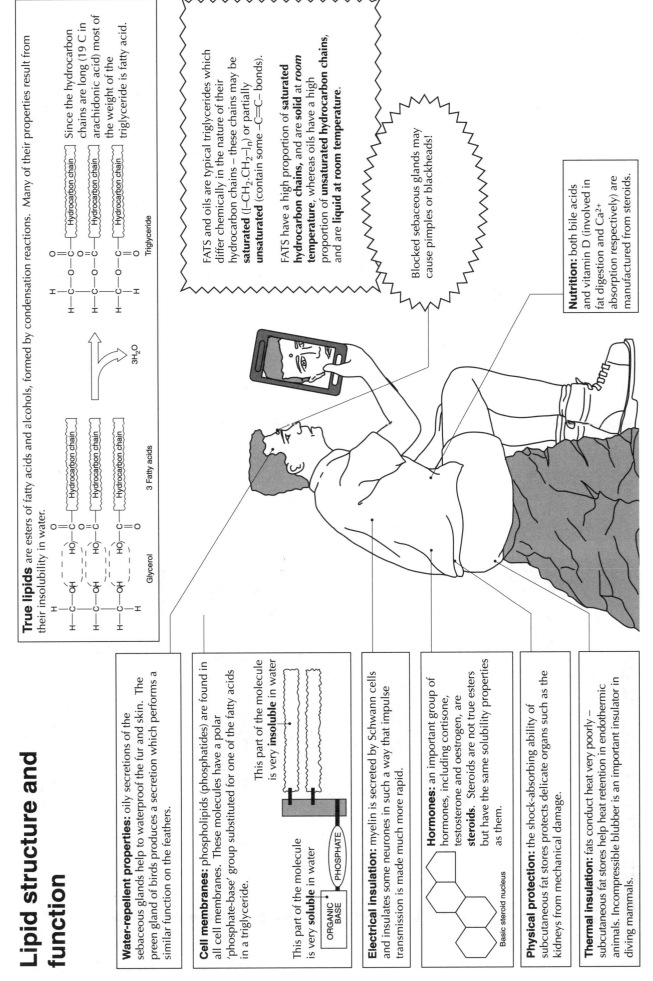

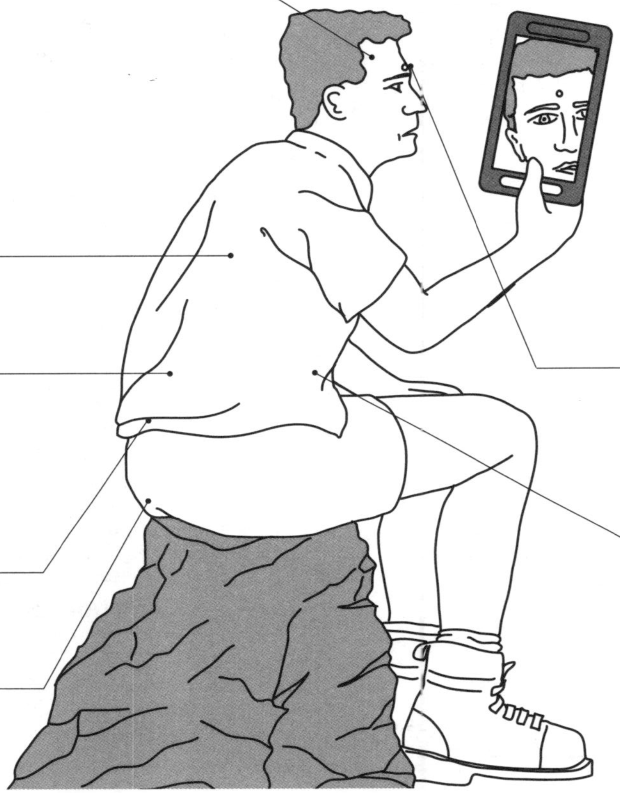

FATS and oils are typical triglycerides which differ chemically in the nature of their hydrocarbon chains – these chains may be **saturated** ($[-CH_2.CH_2-]_n$) or partially **unsaturated** (contain some –C=C– bonds).

FATS have a high proportion of **saturated hydrocarbon chains,** and are **solid** at ***room*** **temperature**, whereas oils have a high proportion of **unsaturated hydrocarbon chains**, and are **liquid at room temperature**.

Blocked sebaceous glands may cause pimples or blackheads!

Nutrition: both bile acids and vitamin D (involved in fat digestion and Ca^{2+} absorption respectively) are manufactured from steroids.

Use of the light microscope

Preparation for light microscopy

- **Fixation** – preserves material in a life-like condition with minimum distortion
- **Dehydration** – removes traces of water from the fixed material
- **Clearing** – removes dehydrating alcohol so that material is made transparent
- **Embedding** – supports the material so that it is firm enough for sectioning
- **Sectioning** – prepares slices of material which are thin enough to allow light to pass through
- **Staining** – improves contrast between different structures (most biological material is transparent)
- **Mounting** – embeds and protects material so that it is suitable for viewing over a long period

Magnification and resolution

- increases image size **but no more detail**
- dependent on 'power' of lenses
- degree of magnification $= \frac{\text{image size}}{\text{object size}}$

- the ability to separate objects that are close together i.e. **more detail**
- depends on **lens quality** and **wavelength of illumination**.

Resolution of a light microscope is limited by wavelength: better resolution requires the shorter wavelength of an electron beam.

Methylene blue
- nuclei and nucleolus stain blue

Iodine solution
- starch grains stain a deep blue-purple

Acetic orcein
- chromosomes stain deep red

Gram stain
- some bacteria stain deep blue-black

Eyepiece: produces a **'real image'**;

magnifies but does not **resolve** the image produced by the objective lens;

the eyepiece may be dismantled so that an **eyepiece graticule** may be inserted if the microscope is to be **used for measurement.**

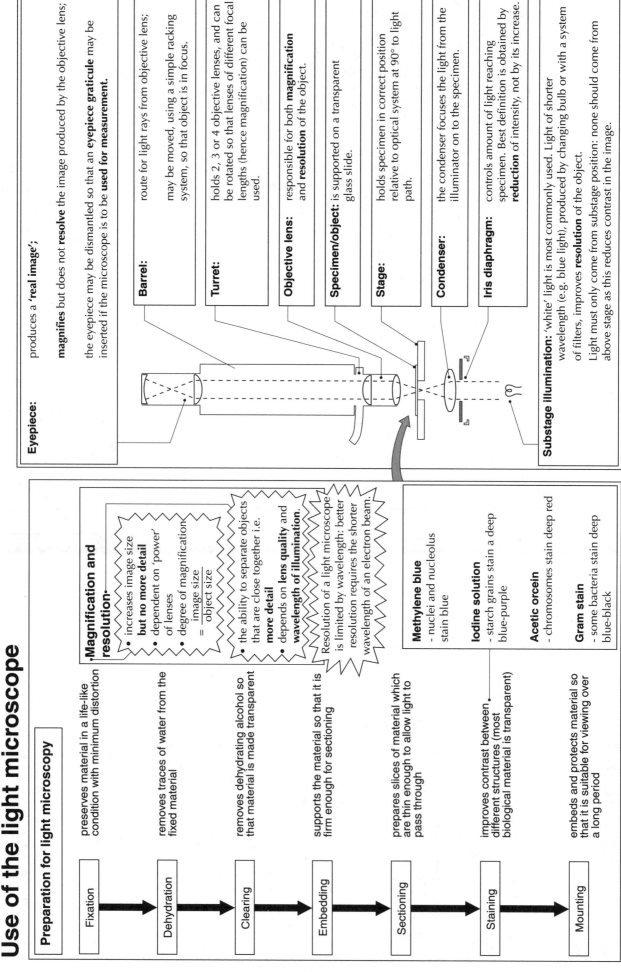

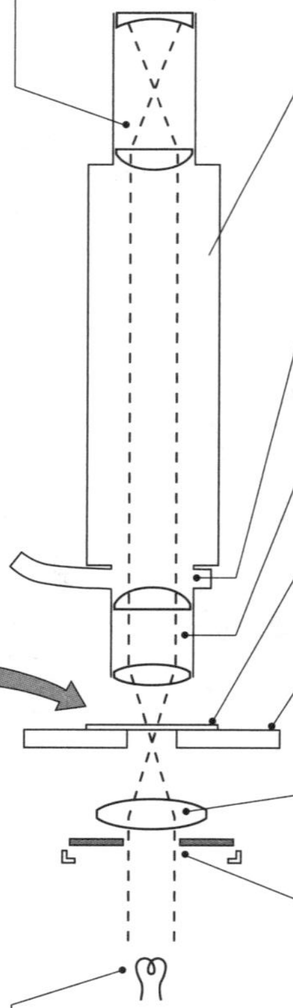

Barrel: route for light rays from objective lens;

may be moved, using a simple racking system, so that object is in focus.

Turret: holds 2, 3 or 4 objective lenses, and can be rotated so that lenses of different focal lengths (hence magnification) can be used.

Objective lens: responsible for both **magnification** and **resolution** of the object.

Specimen/object: is supported on a transparent glass slide.

Stage: holds specimen in correct position relative to optical system at 90° to light path.

Condenser: the condenser focuses the light from the illuminator on to the specimen.

Iris diaphragm: controls amount of light reaching specimen. Best definition is obtained by **reduction** of intensity, not by its increase.

Substage illumination: 'white' light is most commonly used. Light of shorter wavelength (e.g. blue light), produced by changing bulb or with a system of filters, improves **resolution** of the object.

Light must only come from substage position: none should come from above stage as this reduces contrast in the image.

Measuring length with a microscope: micrometry

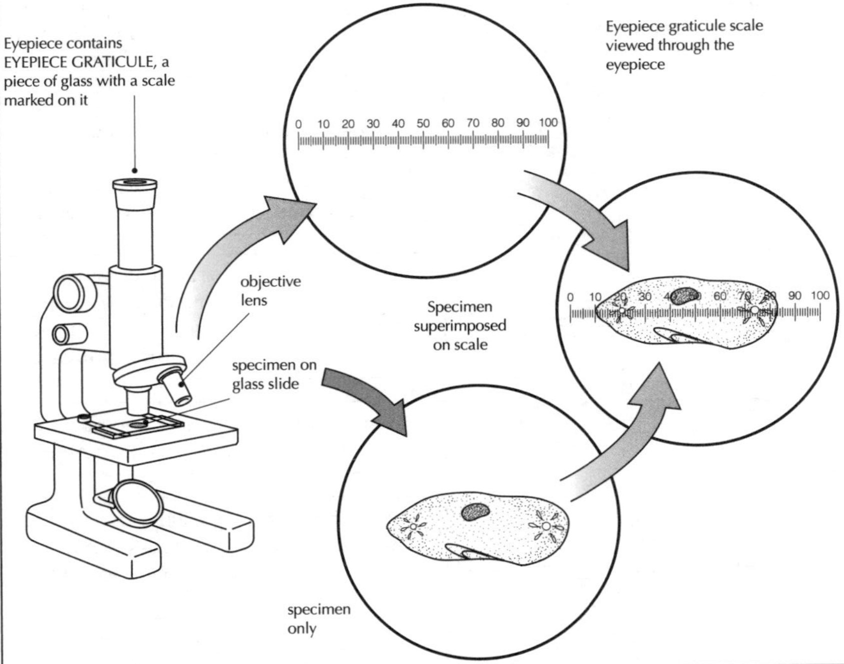

Specimen could be compared with others, but to measure its actual length the graticule must be **calibrated** (we must know the actual length represented by each division on the scale).

Using a stage micrometer: this is a microscope slide with a scale, of known length, marked on it, i.e. we know exactly how much each division on the micrometer measures (usually each division = 0.1 mm).

The stage micrometer is then used to calculate how much each division on the **eyepiece** graticule measures.

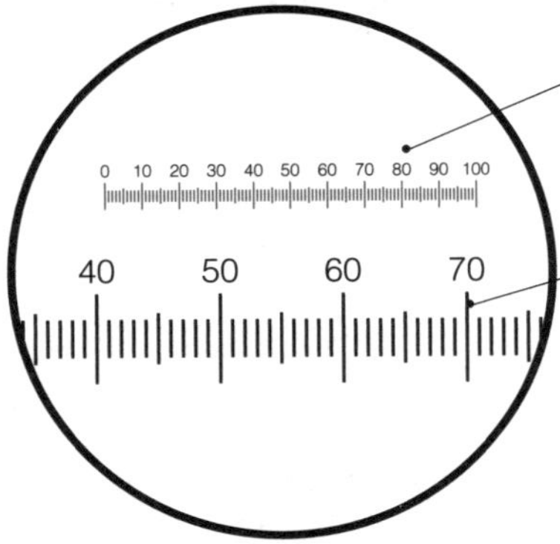

Objective lens: × 40

The higher the magnification, the fewer micrometer divisions will 'line up' with the graticule.
In this example: at total magnification × 400
(× 10 eyepiece, × 40 objective)

100 graticule units	=	25 micrometer divisions
	=	25 × 0.1 mm
	=	2.5 mm = 2 500 μm
So 1 graticule unit	=	25 μm

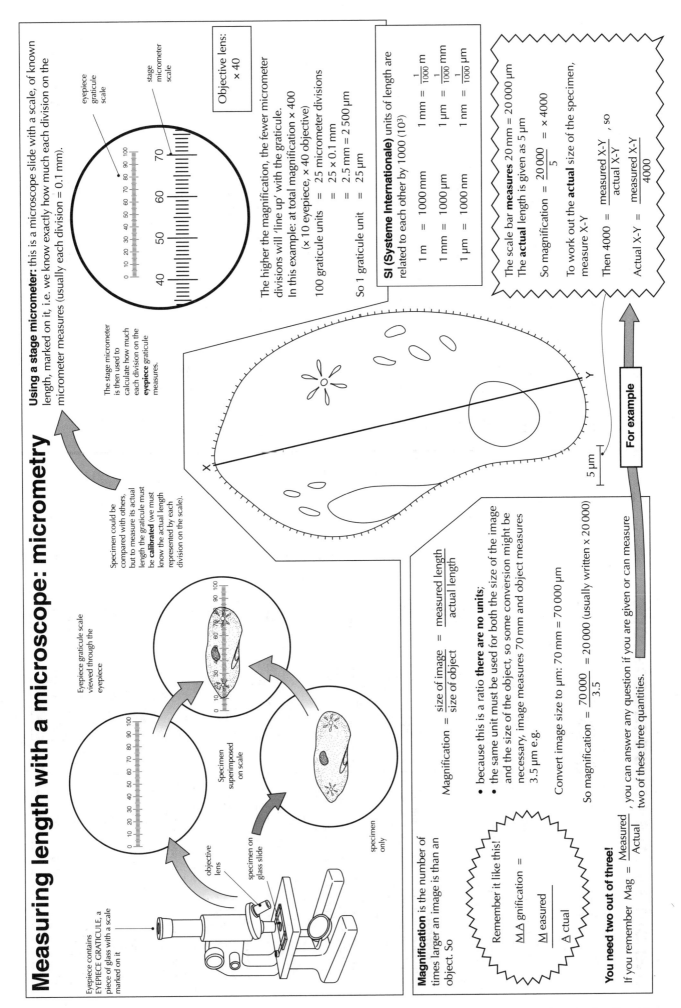

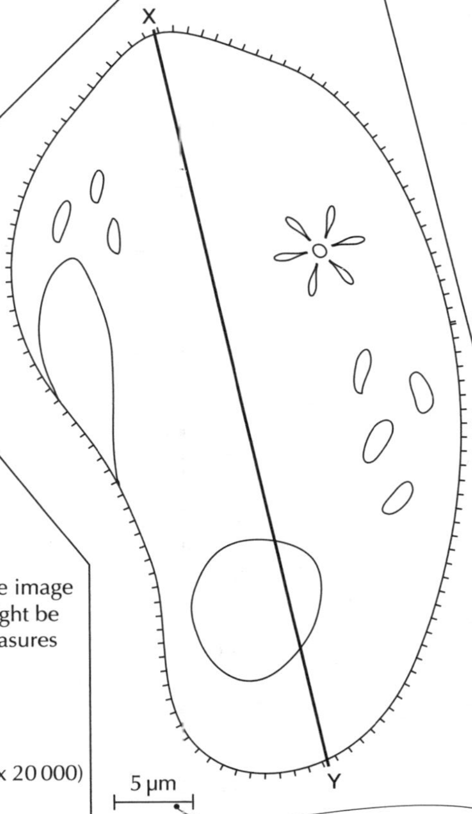

SI (Systeme Internationale) units of length are related to each other by 1000 (10^3)

1 m = 1000 mm		1 mm = $\frac{1}{1000}$ m	
1 mm = 1000 μm		1 μm = $\frac{1}{1000}$ mm	
1 μm = 1000 nm		1 nm = $\frac{1}{1000}$ μm	

Magnification is the number of times larger an image is than an object. So

$$\text{Magnification} = \frac{\text{size of image}}{\text{size of object}} = \frac{\text{measured length}}{\text{actual length}}$$

- because this is a ratio **there are no units**;
- the same unit must be used for both the size of the image and the size of the object, so some conversion might be necessary, image measures 70 mm and object measures 3.5 μm e.g.

Convert image size to μm: 70 mm = 70 000 μm

$$\text{So magnification} = \frac{70\,000}{3.5} = 20\,000 \text{ (usually written x 20 000)}$$

Remember it like this!

<u>M</u> <u>A</u> gnification = $\frac{\underline{M}\text{easured}}{\underline{A}\text{ctual}}$

You need two out of three!

If you remember $\text{Mag} = \frac{\text{Measured}}{\text{Actual}}$, you can answer any question if you are given or can measure two of these three quantities.

For example

The scale bar **measures** 20 mm = 20 000 μm
The **actual** length is given as 5 μm

$$\text{So magnification} = \frac{20\,000}{5} = \times 4000$$

To work out the **actual** size of the specimen, measure X-Y

$$\text{Then } 4000 = \frac{\text{measured X-Y}}{\text{actual X-Y}}, \text{ so}$$

$$\text{Actual X-Y} = \frac{\text{measured X-Y}}{4000}$$

Transmission electron microscope

Cathode: metal electrode (commonly platinum) which emits high velocity electron beam. Electrons are negatively charged particles (e^-).

Anode: positively charged electrode at potential of 50 kV with respect to cathode - accelerates the electron beam.

Condenser: electromagnetic lens which focuses the electron beam on to the specimen.

Air lock/specimen port: allows the introduction of the specimen into the microscope without the loss of vacuum.

Objective: electromagnetic lens which focuses and magnifies (depending on applied voltage) the first image.

Projector: further magnification by selection of region of image to be viewed.

To vacuum pump: creation of vacuum to minimise electron scattering and any heating due to electron/air molecule collision.

Fluorescent, swing out, screen: coated with electron sensitive compounds - necessary since deflected electron beam (the image) cannot be viewed directly.

Photographic plate: allows a black and white permanent record of the image to be made. Printing may offer further magnification.

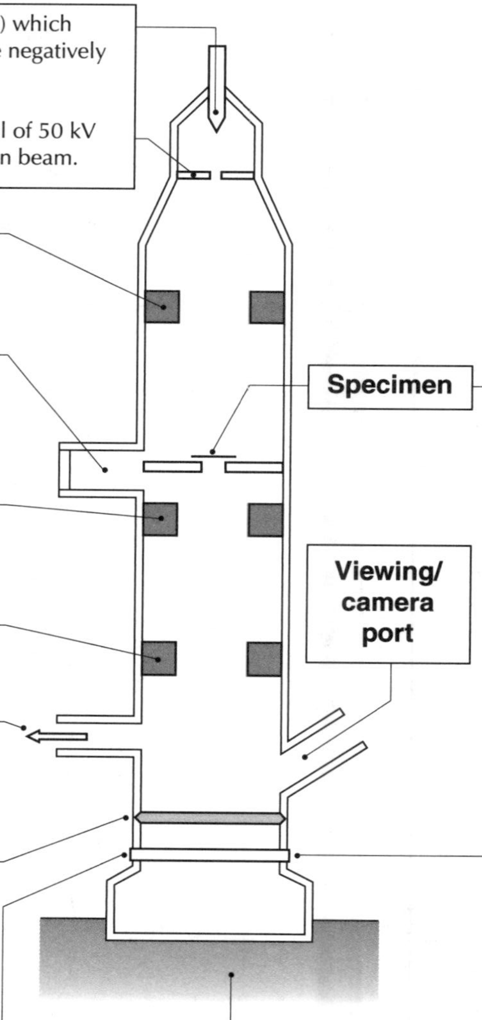

Concrete base: stable support which minimises vibration and thus eliminates unwanted deflection of electron beam.

Advantages of electron microscopy

- the short wavelength of the electron beam means that very high **resolution** can be achieved;
- the high resolution, together with the high voltages applied at the objective, means that high **useful magnification** (often x 100,000 or more) can be obtained.

Disadvantages of electron microscopy

- electrons are easily deflected, so a vacuum must be set up inside the microscope. As a result **no living specimens can be viewed**;
- the human eye cannot 'see' electrons, so a fluorescent screen or photographic plate is needed to form an image;
- preparing the specimen involves harsh procedures, often with corrosive chemicals. This may produce misleading 'artefacts' - structures not present in the living cell.

Sectioning
The material must be cut into **ultrathin sections** (20-100 nm thick) since the electron beam has very low penetrating power.

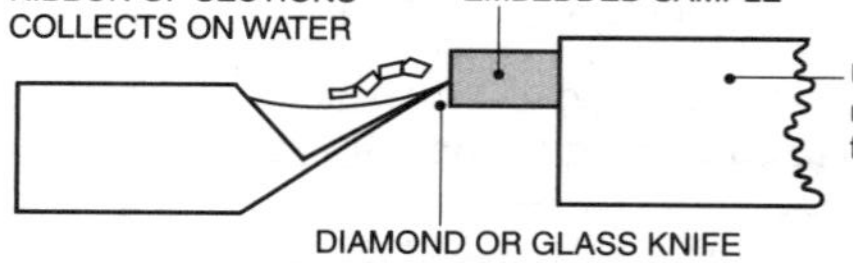

Staining: biological structures are transparent, or nearly so, to electrons. To increase electron beam deflection (i.e. contrast between different structures) sections are treated with **solutions of heavy metal salts** such as **uranyl** or **lead acetate.**

Mounting:

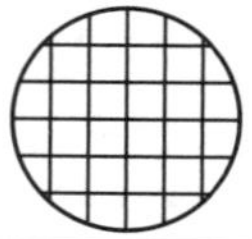

sections are supported on a small copper grid (~3 mm diameter). The electron beam may pass through the gaps in the grid (a glass slide would not permit transmission of electron beam).

Image interpretation

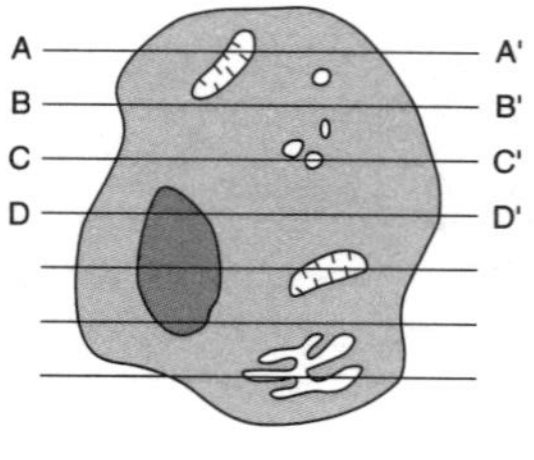

A number of ultrathin sections, e.g. A-A', B-B', must be examined to provide a true three-dimensional representation of the sample.

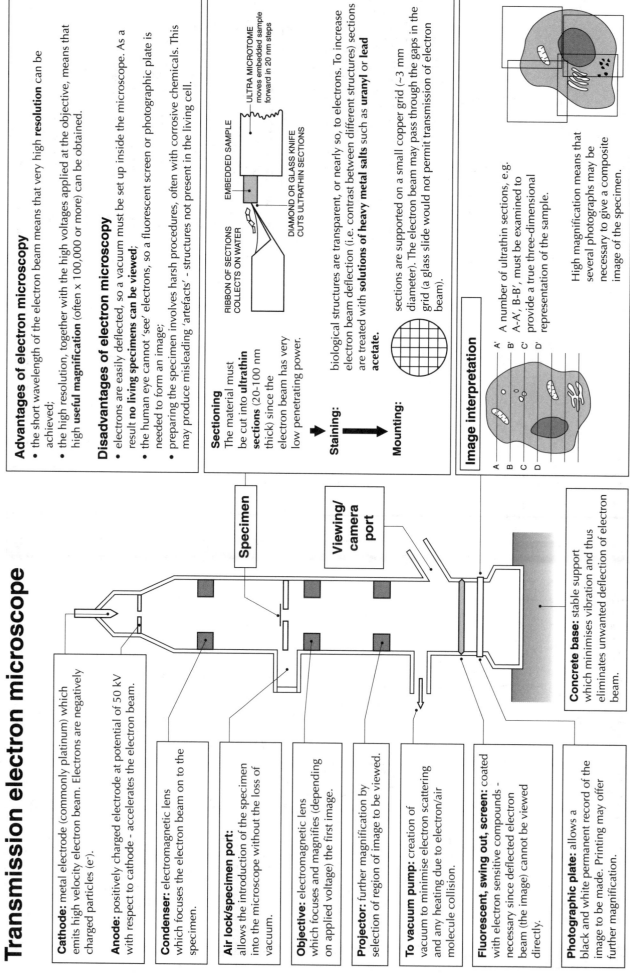

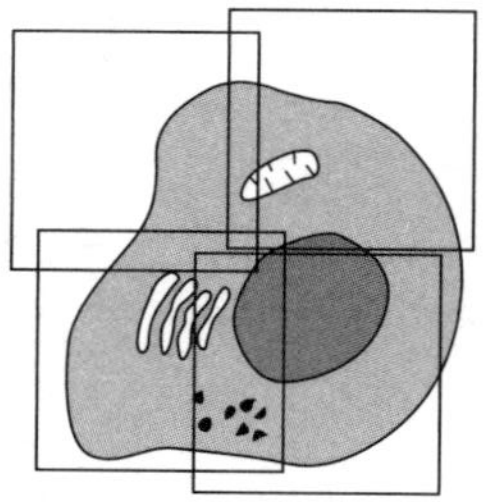

High magnification means that several photographs may be necessary to give a composite image of the specimen.

Scanning electron microscope (s.e.m)

Cathode: metal electrode (commonly platinum) which emits high velocity electron beam. Electrons are negatively charged particles (e^-)

Anode: positively charged electrode at potential of 50 kV with respect to cathode – accelerates the electron beam.

Condenser: electromagnetic lens which focuses the electron beam onto the specimen.

To vacuum pump: creation of vacuum to minimise electron scattering and any heating due to electron/air molecule collision.

Second condenser: further focusing of beam onto specimen.

Objective: electromagnetic lens which focuses and magnifies (depending on applied voltage) the first image.

Air lock/specimen port: allows the introduction of the specimen into the microscope without the loss of vacuum.

Concrete base: stable support which minimises vibration and thus eliminates unwanted deflection of electron beam.

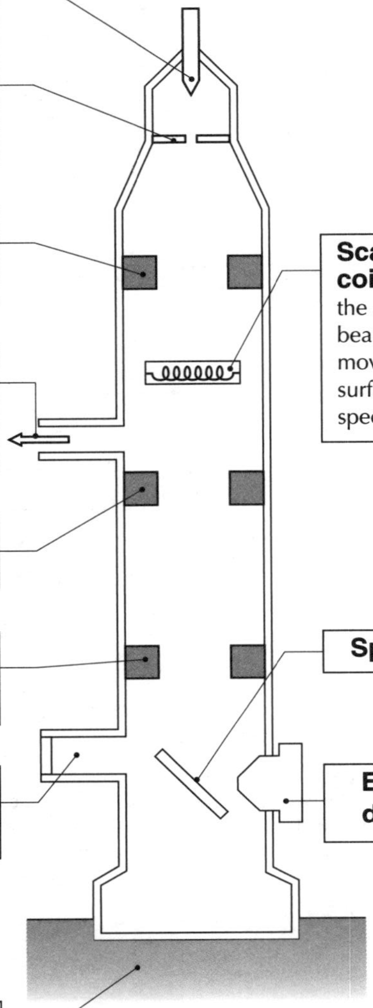

Scanning coil: Deflects the electron beam so that it moves over the surface of the specimen.

Specimen

Electron detector

Amplifier: boosts signal from electron detector.

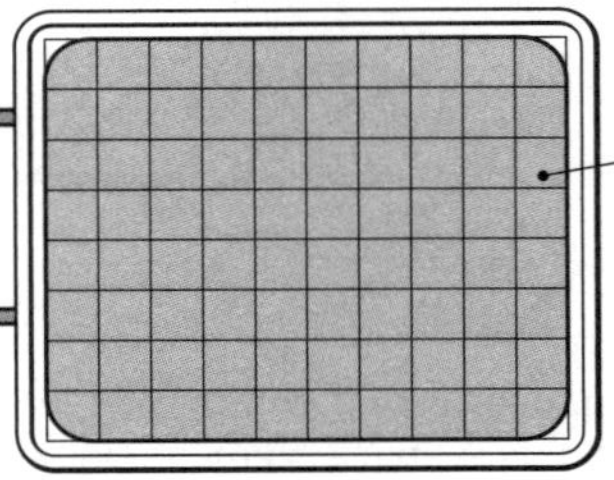

Cathode ray tube: presents visual image of scanned surface of specimen by combining input from scanning coil and electron detector.

Sample is Fixed → **Dehydrated** → **Cleared** → **Embedded**

The s.e.m is especially valuable for providing 3-D images such as:

- whole viruses or virus capsids;
- membrane surfaces, including membrane proteins.

Larger specimens can be viewed then with the t.e.m, but the resolution is not always great.

These techniques are potentially damaging to the specimen and may cause the formation of *artefacts* (misleading products of preparation). Thus a specimen for *scanning electron microscopy* is usually prepared by *freeze fracture* which exposes surfaces on or within the specimen.

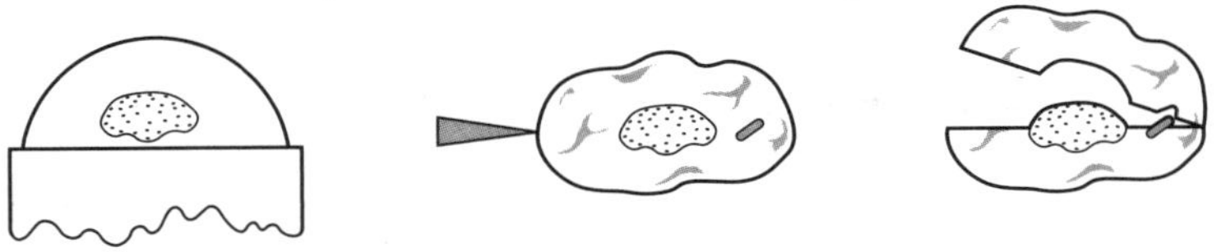

Specimen frozen in ice — Blade used to fracture frozen specimen — Fracture surface is exposed

→ **Staining**

Biological structures are transparent, or nearly so, to electrons. To increase electron beam deflection (i.e. contrast between different structures) the surface is shadowed with a spray of a metal (commonly gold). This improves the three-dimensional resolution of the surface.

→ **Mounting**

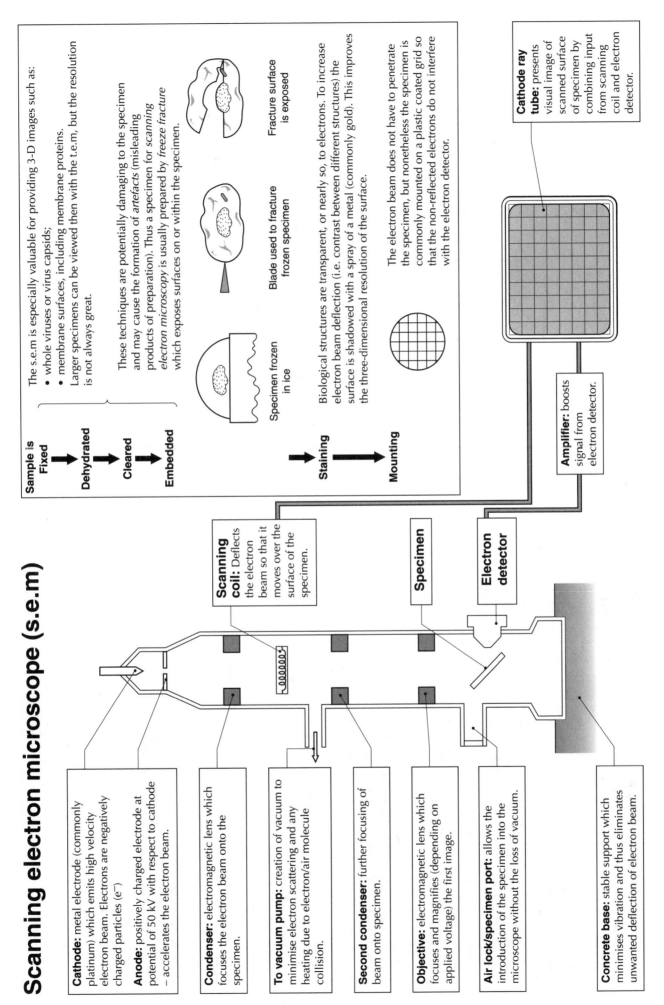

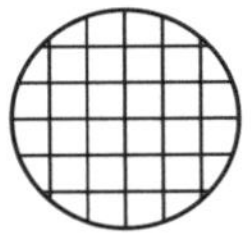

The electron beam does not have to penetrate the specimen, but nonetheless the specimen is commonly mounted on a plastic coated grid so that the non-reflected electrons do not interfere with the electron detector.

The size of cells is limited

Cell surface membrane is fragile

- All cell membranes, including the plasmamembrane, are made up of phospholipids, cholesterol and various proteins.
- The membrane must be flexible ('fluid') to allow exocytosis and endocytosis to occur, and to allow structures such as microvilli to develop.
- Even a slight increase in temperature can reduce the hydrophobic interactions between hydrocarbon tails of phospholipids, and can lead to loss of protein function.
- A slight decrease in the water potential of the cytoplasm can cause endosmosis: 'excess' water in the cytoplasm could cause an animal cell to burst.

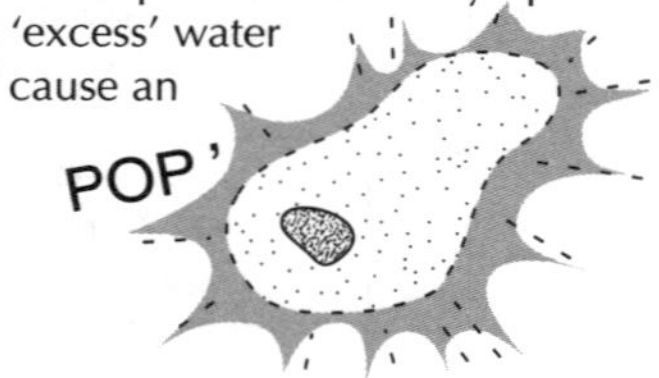

So
the maximum size of a cell is limited because of the danger that the cell membranes would be damaged.

Because of these problems, large organisms are made up of **more** cells not **bigger** cells. An elephant has about 10^6 times as many cells as a mouse. But a mouse liver cell is the same size as an elephant liver cell!

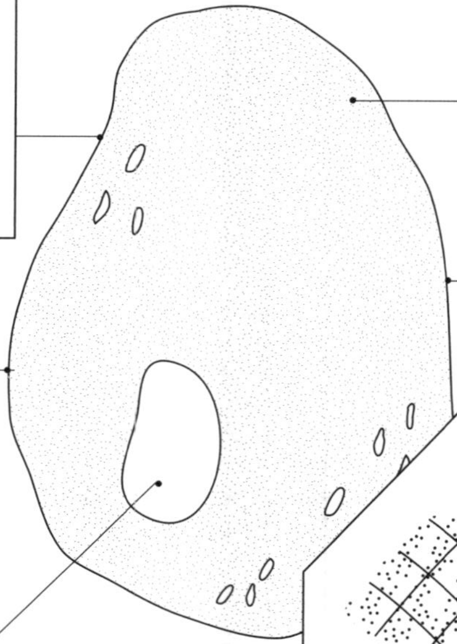

Surface area to volume ratio limits the size of cells, as more cytoplasm creates more **demands** for the **supply** possible through the surface membrane.

Nucleo-cytoplasmic ratio

The nucleus controls the activities of the rest of the cell because it carries the information for protein synthesis. One nucleus can only 'control' a certain volume of cytoplasm. Some cells overcome this problem by having more than one nucleus: they are **coenocytic**. Examples include muscle cells in animals, and the hyphae of fungi.

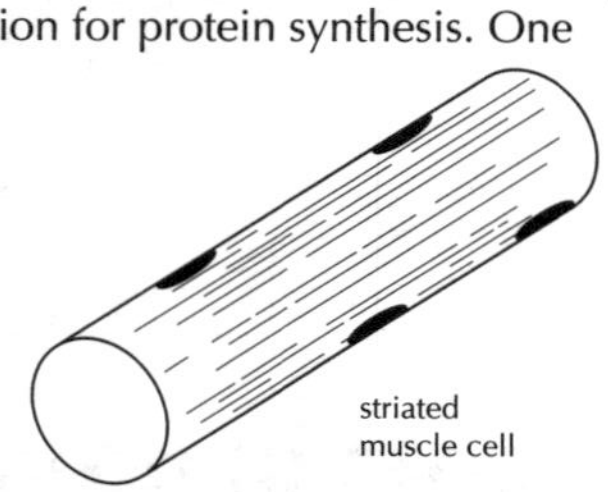

Support from within: the cytoskeleton

The cell surface membrane, the organelles and the cytoplasm are linked to one another by a series of protein structures called the **cytoskeleton**.

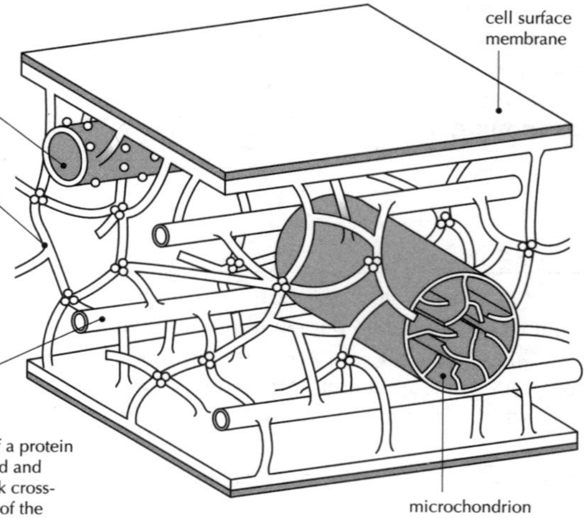

Endoplasmic reticulum: allows closely-controlled movement of molecules throughout the cell.

Microfilaments: solid fibres of **actin** and **myosin**. responsible for movement of organelles inside cell (e.g. chloroplasts can move in response to changes in light intensity), as well as for movements of whole cells - **cell motility** (e.g. leucocytes moving to a site of infection).

Microtubules: tiny tubes made of a protein called **tubulin**. Can be lengthened and shortened, and can build or break cross-bridges. They allow the structure of the cytoplasm to be continuously modified.

Support from outside: cell walls and the glycocalyx

Most cells have some sort of **extracellular** ('outside the cell') strengthening.

Plant cell wall structure:

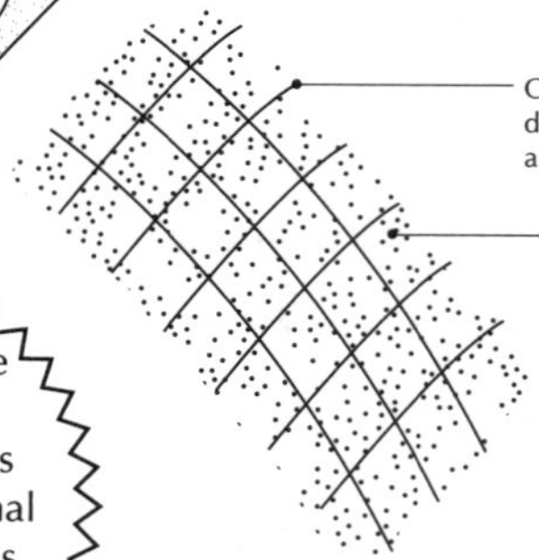

Cellulose microfibrils: laid down at right angles to form a mesh.

Matrix of polysaccharides: contains **pectins** (a very sticky glue) and **hemicelluloses** (more watery, allowing some flexibility).

= **Composite material** (like reinforced concrete)

↓

much greater structural strength

The glycocalyx, made up of short-chain polysaccharides, gives some support to animal cells. This is **much** less than the support given by a true cell wall.

The presence of a cell wall means that expansion of the cytoplasm due to endosmosis does not damage the cell membrane. This extra support is one reason that plant cells are usually larger than animal cells.

Differential centrifugation may be used to isolate cell components.

Isotonic solution prevents osmotic damage to cells and organelles.

Motor-driven homogeniser forces cells between wall of tube and rotating pestle. Shearing forces developed are just sufficient to rupture the cells but not damage the organelles.

Buffer solution prevents pH changes which might otherwise denature proteins, especially enzymes and membrane proteins.

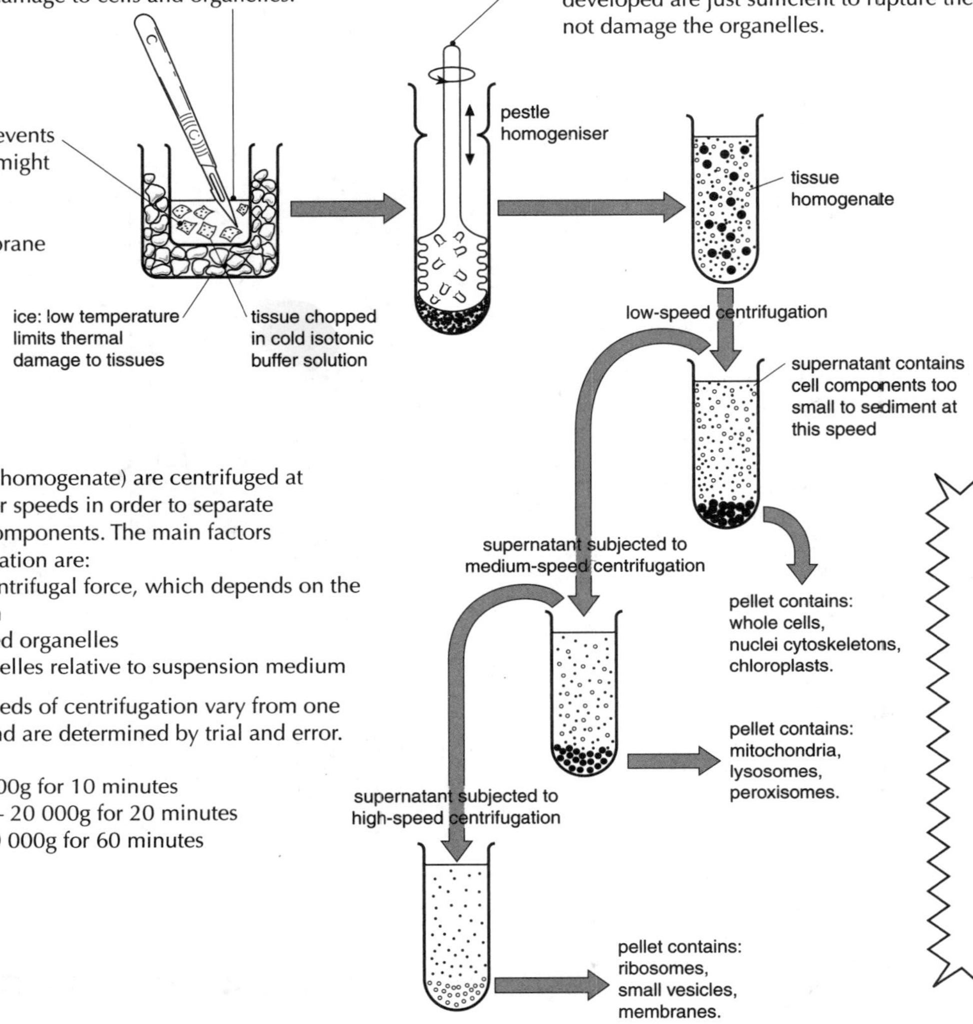

Cell extracts (tissue homogenate) are centrifuged at progressively greater speeds in order to separate (fractionate) their components. The main factors governing sedimentation are:

1 Magnitude of centrifugal force, which depends on the speed of rotation
2 Size of suspended organelles
3 Density of organelles relative to suspension medium

Exact times and speeds of centrifugation vary from one tissue to another, and are determined by trial and error. Typical values:

Low Speed – 1000g for 10 minutes
Medium Speed – 20 000g for 20 minutes
High Speed – 80 000g for 60 minutes

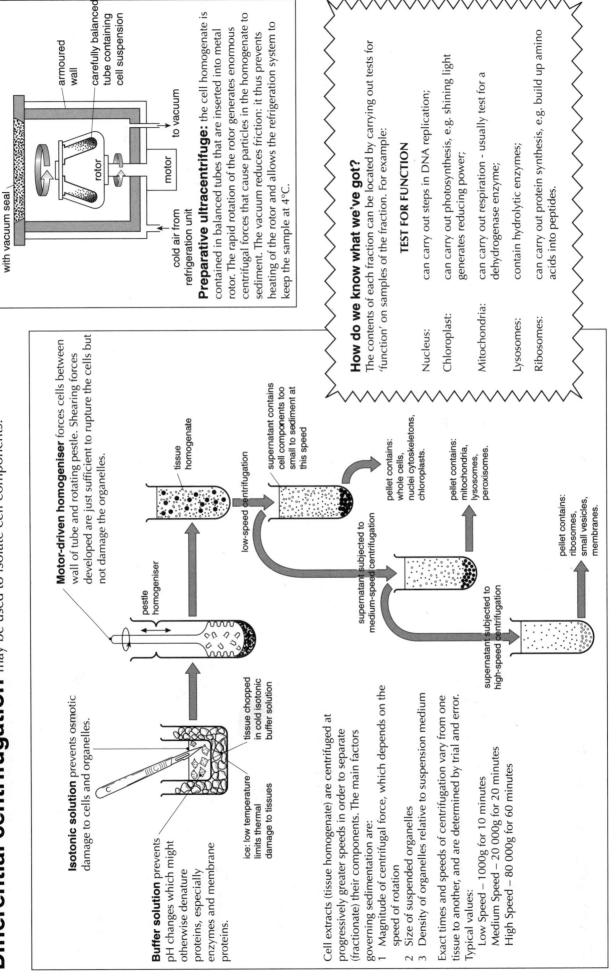

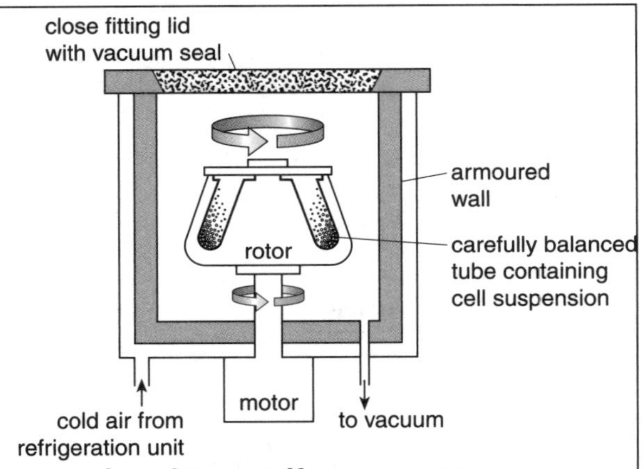

Preparative ultracentrifuge: the cell homogenate is contained in balanced tubes that are inserted into metal rotor. The rapid rotation of the rotor generates enormous centrifugal forces that cause particles in the homogenate to sediment. The vacuum reduces friction: it thus prevents heating of the rotor and allows the refrigeration system to keep the sample at 4°C.

How do we know what we've got?

The contents of each fraction can be located by carrying out tests for 'function' on samples of the fraction. For example:

	TEST FOR FUNCTION
Nucleus:	can carry out steps in DNA replication;
Chloroplast:	can carry out photosynthesis, e.g. shining light generates reducing power;
Mitochondria:	can carry out respiration - usually test for a dehydrogenase enzyme;
Lysosomes:	contain hydrolytic enzymes;
Ribosomes:	can carry out protein synthesis, e.g. build up amino acids into peptides.

Plant and animal cells

Animal cell features often relate to heterotrophic nutrition and high rates of metabolic activity.

Microvilli ,which provide an increased surface area for absorption or secretion, are found only on animal cells.

Secretory vesicles containing cell products such as hormones or enzymes are much more common in animal cells.

Cytoplasm of animal cells is often denser, with many more organelles and dissolved substances.

Vacuoles are small and temporary. They can be involved with digestion (e.g. in phagocytes) or with excretion (contractile vacuoles may remove excess water).

Glycogen is the storage form of carbohydrates.

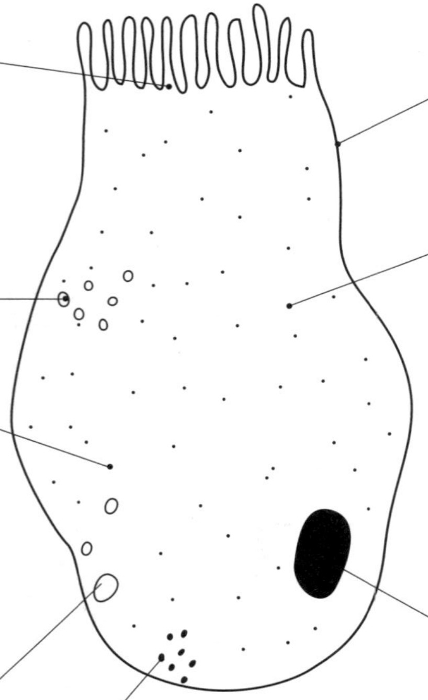

Plant and animal cells have common features which relate to maintaining the characteristics of life.

Cell membrane which surrounds the cytoplasm. It controls the **entry and exit** of dissolved substances and is therefore responsible for separating the cells contents from its surroundings.

Cytoplasm contains water, dissolved substances such as amino acids and sugars, and supports the various organelles (for example, mitochondria, ribosomes). It is within the cytoplasm and organelles that the various metabolic reactions needed to sustain life take place (for example, respiration).

Nucleus contains the genetic material (**DNA** which makes up **genes** or the **chromosomes**) which carries the coded instructions controlling the activities and characteristics of the cell. The chromosomes only become visible during cell division.

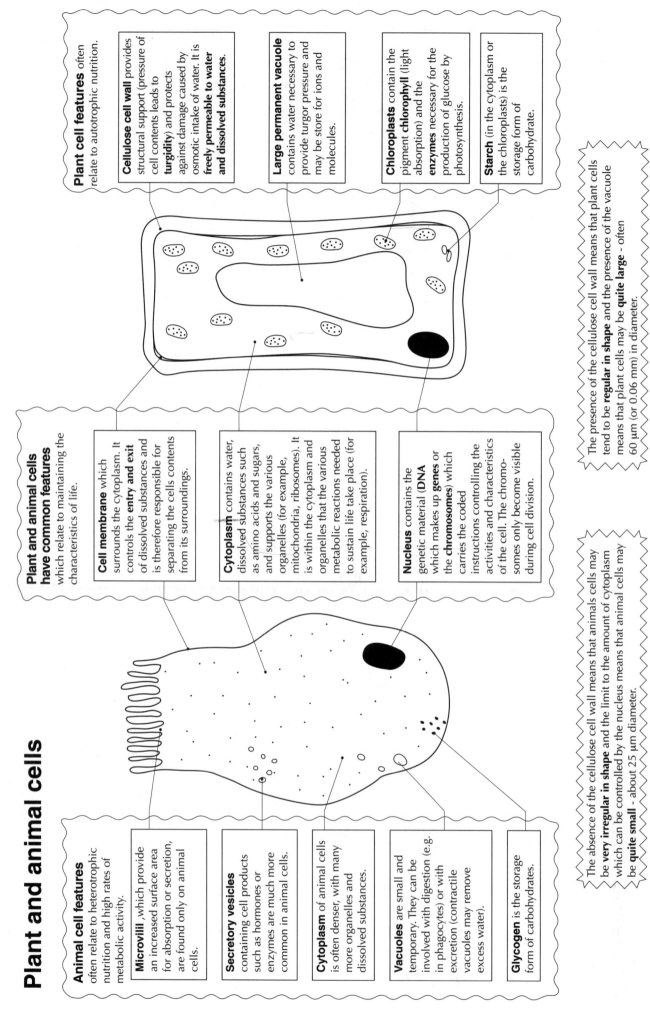

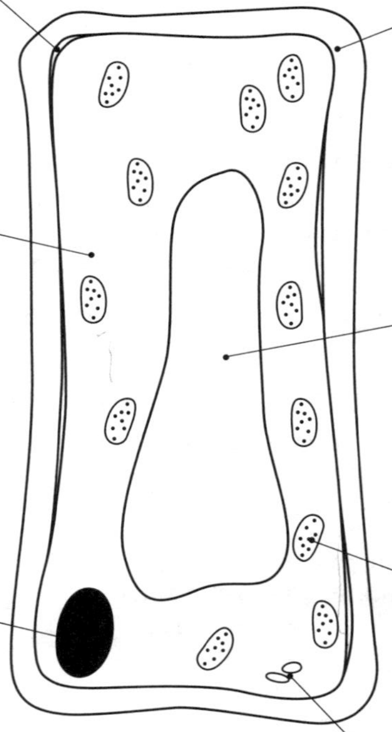

Plant cell features often relate to autotrophic nutrition.

Cellulose cell wall provides structural support (pressure of cell contents leads to **turgidity**) and protects against damage caused by osmotic intake of water. It is **freely permeable to water and dissolved substances**.

Large permanent vacuole contains water necessary to provide turgor pressure and may be store for ions and molecules.

Chloroplasts contain the pigment **chlorophyll** (light absorption) and the **enzymes** necessary for the production of glucose by photosynthesis.

Starch (in the cytoplasm or the chloroplasts) is the storage form of carbohydrate.

The absence of the cellulose cell wall means that animals cells may be **very irregular in shape** and the limit to the amount of cytoplasm which can be controlled by the nucleus means that animal cells may be **quite small** - about 25 μm diameter.

The presence of the cellulose cell wall means that plant cells tend to be **regular in shape** and the presence of the vacuole means that plant cells may be **quite large** - often 60 μm (or 0.06 mm) in diameter.

Animal cell ultrastructure

Microvilli are extensions of the plasmamembrane which increase the cell surface area. They are commonly abundant in cells with a high absorptive capacity, such as epithelial cells of the small intestine or cells of the *first coiled tubule of the nephron.* Collectively the microvilli make up a *brush border* to the cell.

Free ribosomes are the sites of protein synthesis, principally for proteins destined for intracellular use. There may be 50 000 or more in a typical eukaryote cell.

Endocytic vesicle may contain molecules or structures too large to cross the membrane by active transport or diffusion.

Nucleus is the centre of the regulation of cell activities since it contains the hereditary material, DNA, carrying the information for protein synthesis. The DNA is bound up with histone protein to form chromatin. The nucleus contains one or more nucleoli in which ribosome subunits, ribosomal RNA, and transfer RNA are manufactured. The nucleus is surrounded by a double nuclear membrane, crossed by a number of nuclear pores. The nucleus is continuous with the endoplasmic reticulum. There is usually only one nucleus per cell, although there may be many in very large cells such as those of striated (skeletal) muscle. Such multinucleate cells are called coenocytes.

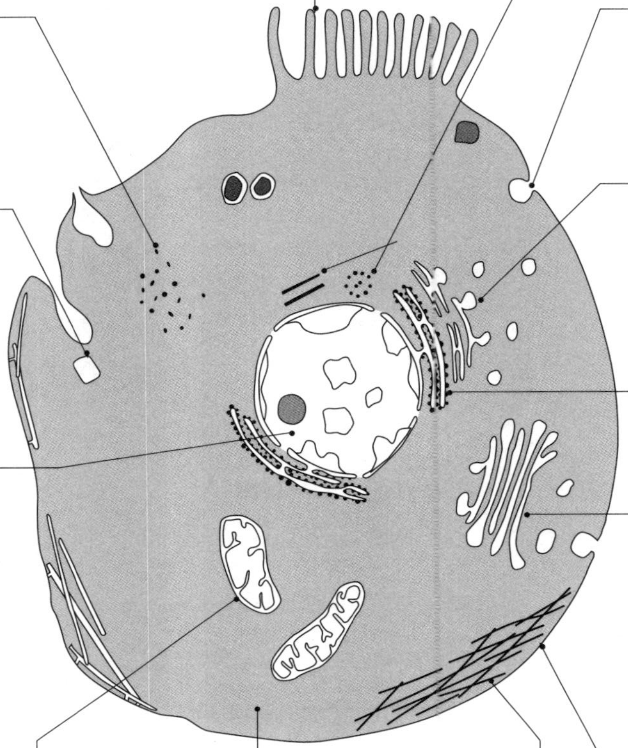

Centrioles are a pair of structures, held at right angles to one another, which act as organisers of the nuclear spindle in preparation for the separation of chromosomes or chromatids during nuclear division.

Secretory vesicle undergoing exocytosis. May be carrying a synthetic product of the cell (such as a protein packaged at the Golgi body) or the products of degradation by lysosomes. Secretory vesicles are abundant in cells with a high synthetic activity, such as the cells of the *Islets of Langerhans.*

Smooth endoplasmic reticulum is a series of flattened sacs and sheets that are the sites of synthesis of steroids and lipids.

Rough endoplasmic reticulum is so-called because of the many ribosomes attached to its surface. This intracellular membrane system aids cell compartmentalisation and transports proteins synthesised at the ribosomes towards the Golgi bodies for secretory packaging.

Golgi apparatus consists of a stack of sacs called *cisternae.* It modifies a number of cell products delivered to it, often enclosing them in vesicles to be secreted. Such products include trypsinogen (from *pancreatic acinar cells*), insulin (from *beta-cells of the Islets of Langerhans*) and mucin (from *goblet cells in the trachea*). The Golgi is also involved in lipid modification in cells of the ileum, and plays a part in the formation of lysosomes.

Mitochondrion (pl. mitochondria) is the site of aerobic respiration. Mitochondria have a highly folded inner membrane which supports the proteins of the electron transport chain responsible for the synthesis of ATP by oxidative phosphorylation. The mitochondrial matrix contains the enzymes of the TCA cycle, an important metabolic 'hub'. These organelles are abundant in cells which are physically *(skeletal muscle)* and metabolically *(hepatocytes)* active.

Cytoplasm is mainly water, with many solutes including glucose, proteins and ions. It is supported by the **cytoskeleton**, made up of microtubules and microfilaments.

Microfilaments are threads of the protein *actin.* They are usually situated in bundles just beneath the cell surface and play a role in endo- and exocytosis, and possibly in cell movement.

Plasmalemma (plasmamembrane) is the surface of the cell and represents its contact with its environment. It is differentially permeable and regulates the movement of solutes between the cell and its environment. There are many specialisations of the membrane, often concerning its protein content.

Typical plant cell contains chloroplasts and a permanent vacuole, and is surrounded by a cellulose cell wall.

Cell wall is composed of long cellulose molecules grouped in bundles called **microfibrils** which, in turn, are twisted into rope-like **macrofibrils.** There may be a secondary cell wall containing **lignin** (gives strength to xylem) or **suberin** (makes a waterproof layer in the endodermis).
The function of the cell wall is a mechanical one - pressure from the cell protoplast maintains cell turgidity. The wall is freely permeable to water and most solutes so that the cell wall represents an important transport route - the **apoplast system** - throughout the plant body.

Chloroplast is the site of photosynthesis. It is one of a number of plastids, all of which develop from **proplastids** which are small, pale green or colourless organelles.

Vacuole may occupy 90% of the volume of a mature plant cell. It is filled with cell sap (a solution of salts, sugars and organic acids) and helps to maintain turgor pressure inside the cell. The vacuole also contains anthocyanins, pigments responsible for many of the red, blue and purple colours of flowers. Vacuoles also contains enzymes involved in recycling of cell components such as chloroplasts. The vacuolar membrane is called the **tonoplast.**

Microtubules are hollow structures (about 25 nm in diameter) composed of the protein tubulin. They occur just below the plasmamembrane where they may aid the addition of cellulose to the cell wall. They are also involved in the cytoplasmic streaming of organelles such as Golgi bodies and chloroplasts, and they form the spindles and cell plates of dividing cells.

Plasmamembrane (plasmalemma, cell surface membrane) is the differentially-permeable cell surface, responsible for the control of solute movements between the cell and its environment. It is flexible enough to move close to or away from the cell wall as the water content of the cytoplasm changes. The membrane is also responsible for the synthesis and assembly of cell wall components.

Golgi body (dictyosome) synthesises polysaccharides and packages them in vesicles which migrate to the plasmamembrane for eventual incorporation in the cell wall.

Mitochondrion contains the enzyme systems for ATP synthesis by oxidative phosphorylation. May be abundant in sieve tube companion cells, root epidermal cells and dividing meristematic cells.

Plasmodesmata are minute strands of cytoplasm which pass through pores in the cell wall and connect the protoplasts of adjacent cells. This represents the **symplast** pathway for the movement of water and solutes throughout the plant body. These cell-cell cytoplasm connections are important in cell survival during periods of drought. The E.R. of adjacent cells is also in contact through these strands.

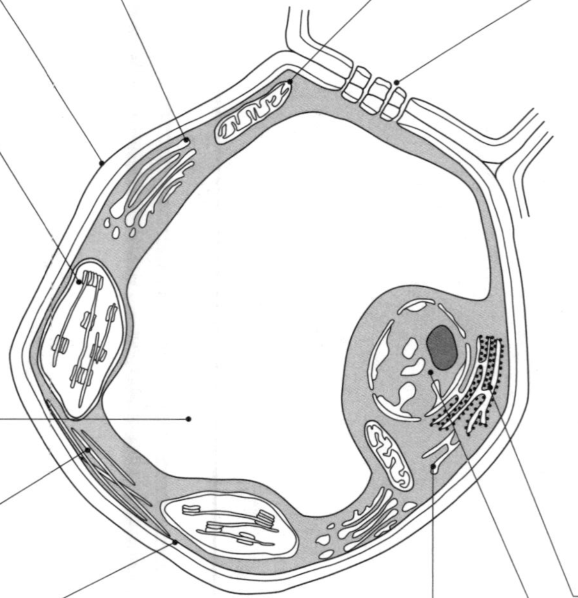

Plant, animal and bacterial cells

Feature	Plant	Animal	Bacterium
Cell wall	✓ (cellulose)	✗	✓ (murein)
Nucleus	✓	✓	✗
Plasmids	✗	✗	✓
Mitochondria	✓	✓	✗
Ribosomes	✓	✓	✓ (but small)
Chloroplasts	✓	✗	✗
Permanent vacuole	✓	✗	✗

Rough endoplasmic reticulum is the site of protein synthesis (on the attached ribosomes), storage and preparation for secretion. The endoplasmic reticulum (E.R.) also plays a part in the compartmentalisation of the cell.

Smooth endoplasmic reticulum is the site of lipid synthesis and secretion.

Nucleus is surrounded by the nuclear envelope and contains the genetic material, DNA, associated with histone protein to form chromatin. The nucleus thus controls the activity of the cell through its regulation of protein synthesis. The nucleolus is the site of synthesis of transfer RNA, ribosomal RNA, and ribosomal subunits.

The Cholera bacterium (*Vibrio cholerae*)

is a prokaryotic cell and so has no true organelles.

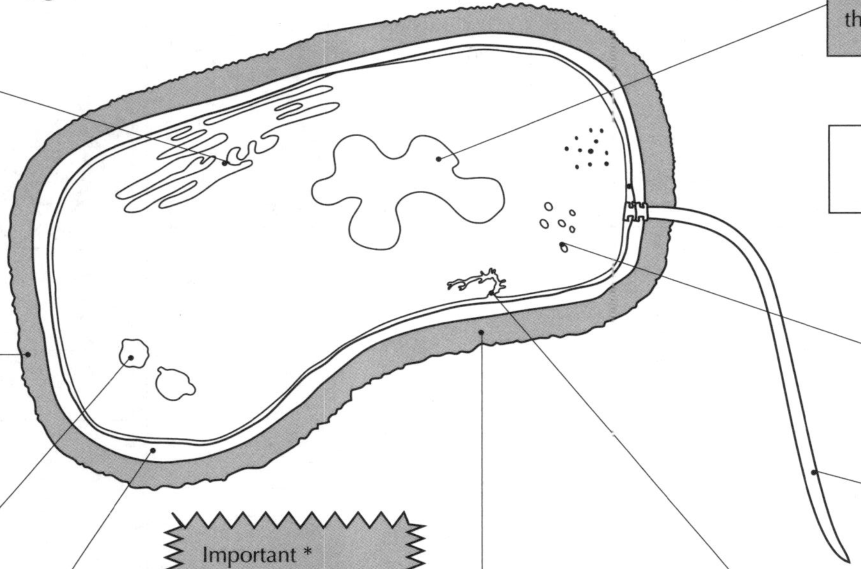

SCALE

0.1 μm

***Photosynthetic membranes** are surfaces for light-absorbing pigments, but there are no chloroplasts N.B. Bacterial photosynthesis does not evolve oxygen.

Capsule is a gummy layer of mucilage which may join bacteria into colonies (e.g. *Bacillus anthracis*) or give protection (e.g. rough strain of *D. pneumoniae*).

***Plasmids** are short pieces of circular DNA which replicate independently of the cell genome. They have been widely exploited in recombinant DNA technology.

Cell wall has a rigid framework of **murein**, a polysaccharide cross-linked by peptide chains.

***Genetic material** is composed of a circle of double-stranded DNA **which is not enclosed within a nuclear membrane.** There are typically about 2000 genes, much less than the number found in a eukaryotic cell.

***Ribosomes** smaller than those in eukaryotes. They are scattered throughout the cytoplasm, not supported on an endoplasmic reticulum.

Plasmamembrane is a typical phospholipid bilayer.

Food stores are typically lipid globules or glycogen granules.

***Flagellum** is responsible for motility of many bacteria. It is much simpler than the flagellum of a eukaryotic cell, being composed of a single cylinder of protein subunits (flagellin). The flagellum does not 'beat' but instead rotates about a 'bearing' anchored in the cell wall to produce a corkscrew motion which drives the cell along.

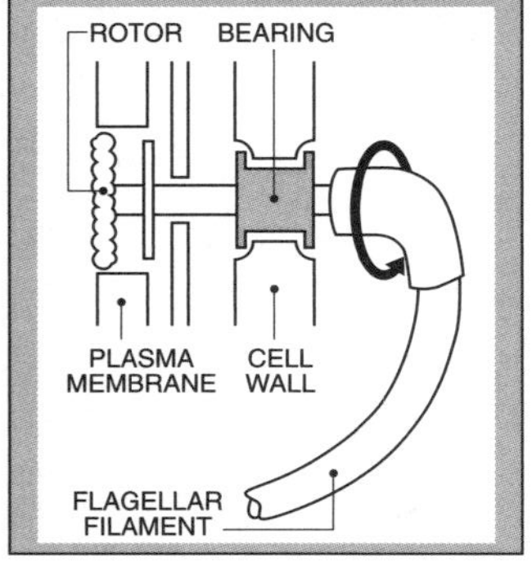

Important * comparisons with eukarytic cells.

Oral rehydration solutions (ORS)

These are solutions of ions and glucose. If sufficient is available they can reduce deaths by over 90%. Rice or other cereals in the solution are also very effective at reducing the problems of **dehydration**.

Cholera bacterium causes severe diarrhoea

- The bacterium releases a toxin.
- Toxin increases loss of Cl^- and water.
- Toxin blocks reuptake of Cl^- and water.

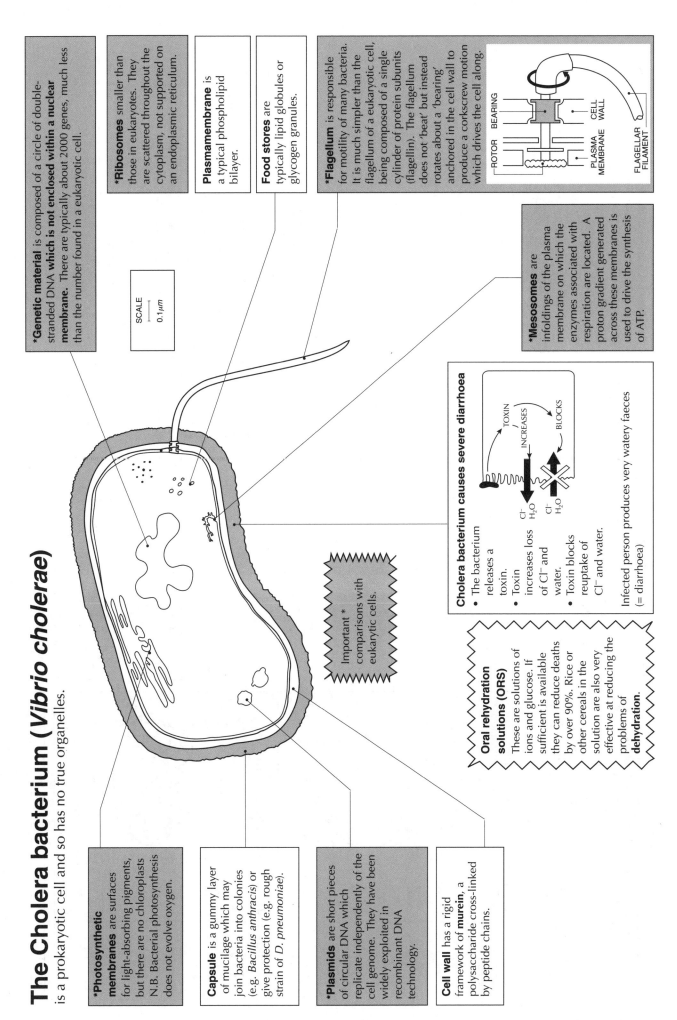

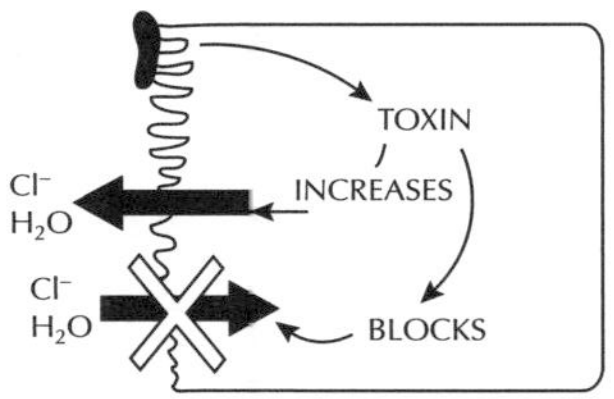

Infected person produces very watery faeces (= diarrhoea)

***Mesosomes** are infoldings of the plasma membrane on which the enzymes associated with respiration are located. A proton gradient generated across these membranes is used to drive the synthesis of ATP.

Polysaccharides are polymers formed by glycosidic bonding of monosaccharide subunits. The structure of these molecules affects their functions in living organisms.

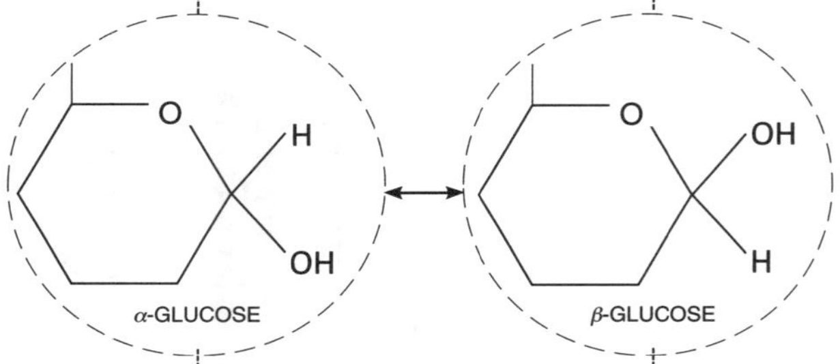

Starch is a mixture of two polymers of α-glucose. The most common is **amylose** which usually contains about 300 glucose units joined by α 1,4 **glycosidic bonds**

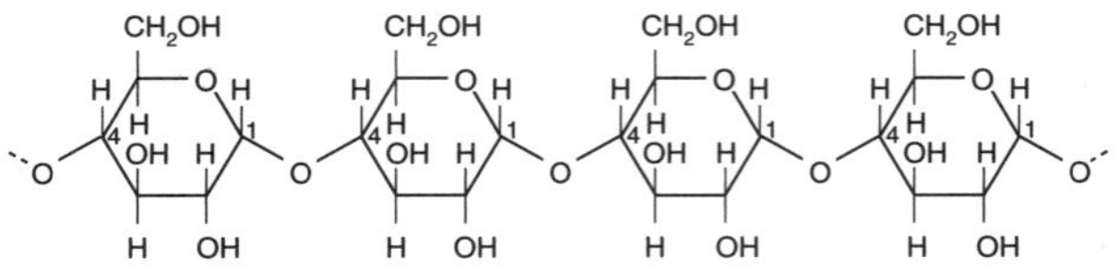

The bulky $-CH_2OH$ side chains cause the molecule to take up a helical shape (excellent for packing many subunits into a limited space).

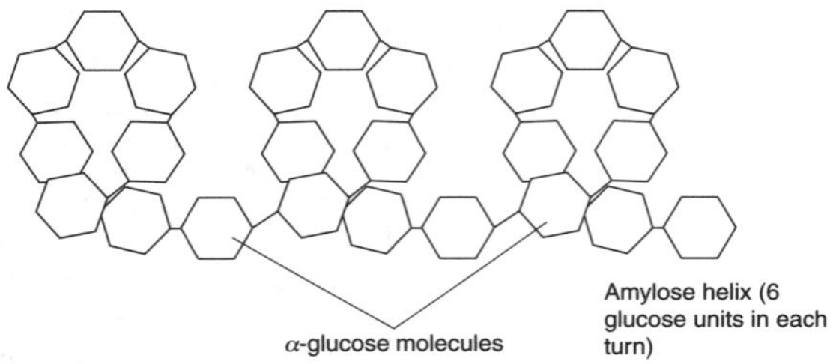

Because there are so few 'ends' within the starch molecule there are few points to begin hydrolysis by the enzyme **amylase**. Starch is therefore an excellent long-term **storage compound**.

All those coils let you test for starch!

Add a drop of iodine solution (iodine in potassium iodide solution)

positive result (blue-black colouration) = ***starch present***

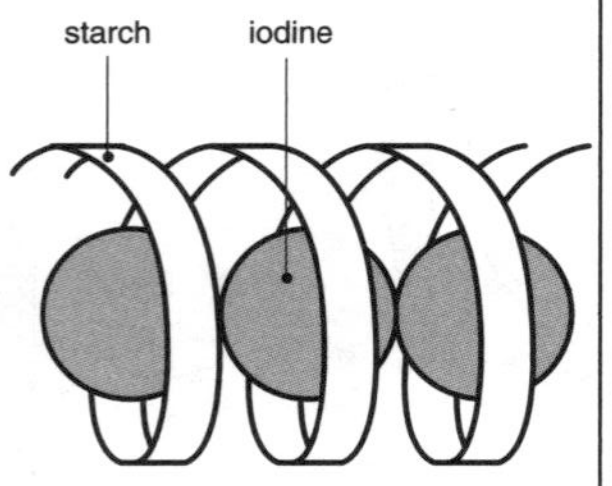

Principle: iodine binds to the centre of the starch helix, forming a starch-iodine complex which is intense blue-black in colour.

Cellulose is a polymer of glucose linked by β 1,4 **glycosidic bonds.** The β-conformation inverts successive monosaccharide units so that a straight chain polymer is formed.

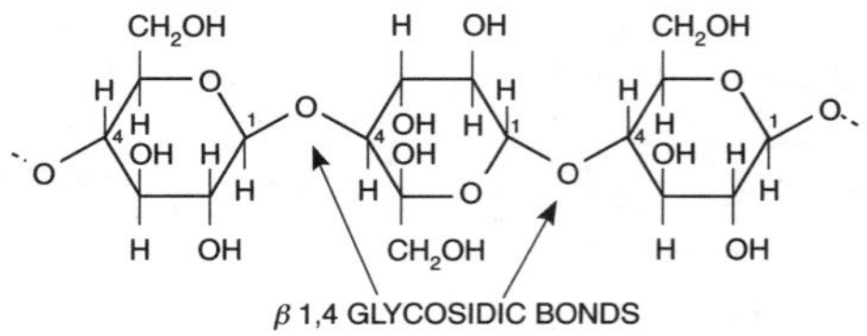

The parallel polysaccharide chains are then cross-linked by **hydrogen bonds.**

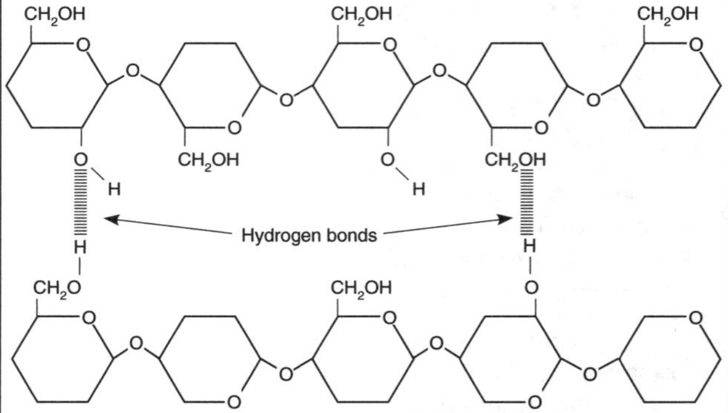

This cross-linking prevents access by water, so that cellulose is very resistant to hydrolysis and is therefore an excellent **structural molecule** (cellulose cell walls): ideal in plants which can readily synthesise excess carbohydrate.

Glycogen is an α-glucose polymer, but with many cross-links and shorter α 1,4 chains.

This gives more 'ends' to the molecule which is ideal for animal cells which may need to hydrolyse food reserves more rapidly than plant cells would do.

Subunits are joined by **condensation** (**removal** of the elements of water) and separated by **hydrolysis** (bond breakage by **adding** the elements of water)

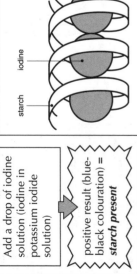

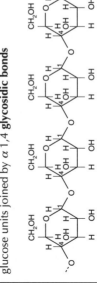

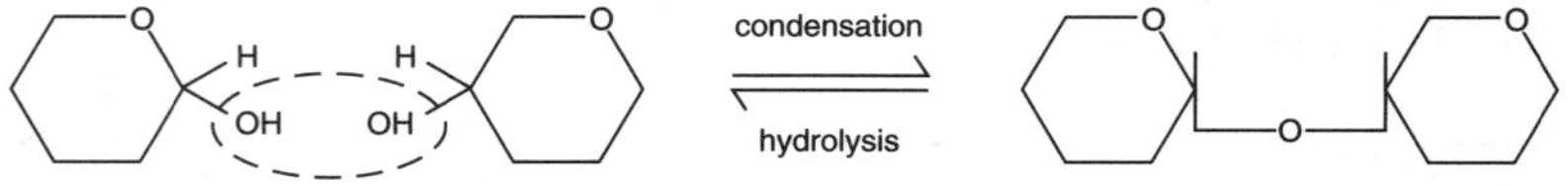

The four levels of protein structure

Individual amino acids are joined together by **peptide bonds** by **condensation reactions** catalysed by enzymes. Hydrolysis occurs here during digestion.

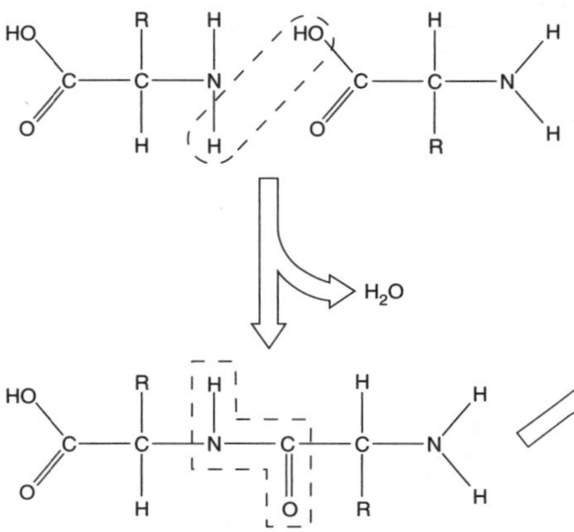

CONH is a very important linkage referred to as **peptide linkage.**

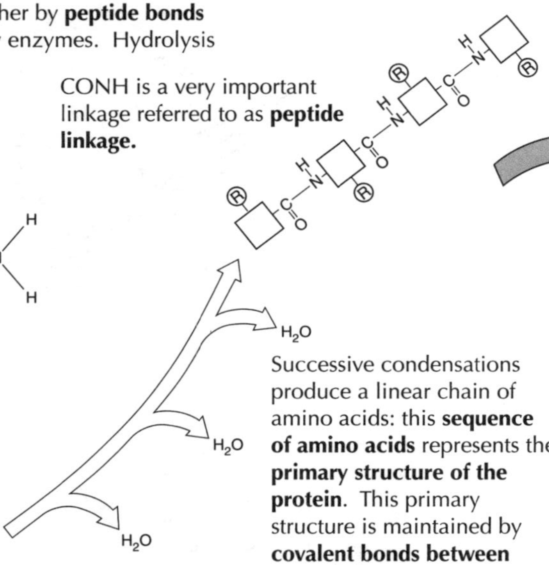

Successive condensations produce a linear chain of amino acids: this **sequence of amino acids** represents the **primary structure of the protein**. This primary structure is maintained by **covalent bonds between adjacent amino acids.**

The polypeptide chains may take on regular arrangements called the **secondary structure** of the protein e.g. the **α-helix**. This secondary structure is maintained by **hydrogen bonds between the >C=O and >N-H groups of every fourth peptide link.**

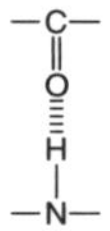

An alternative secondary structure - the **β-pleated sheet** - has hydrogen bonds between peptide links of adjacent polypeptide chains.
Proteins with a well-developed secondary structure are **fibrous proteins**. Examples are **keratin** in hair, **collagen** in skin and **fibrin** in blood clots.

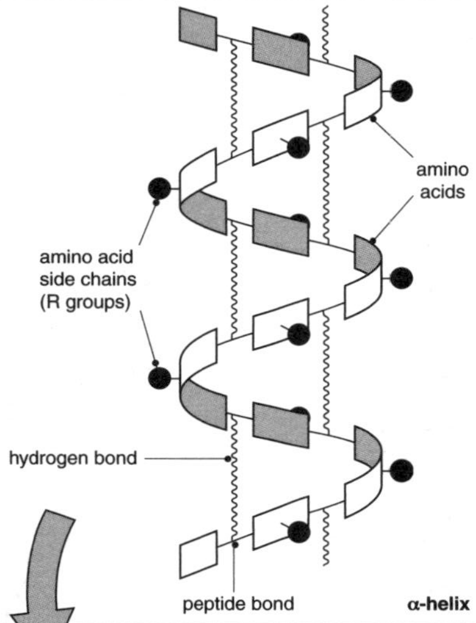

Sections of α-helix may be folded on themselves: this **supercoiling of the α-helix** represents the **tertiary structure of the protein**. This three-dimensional shape or **conformation** of the protein is maintained by a **series of interactions between -R groups on the polypeptide chain.**

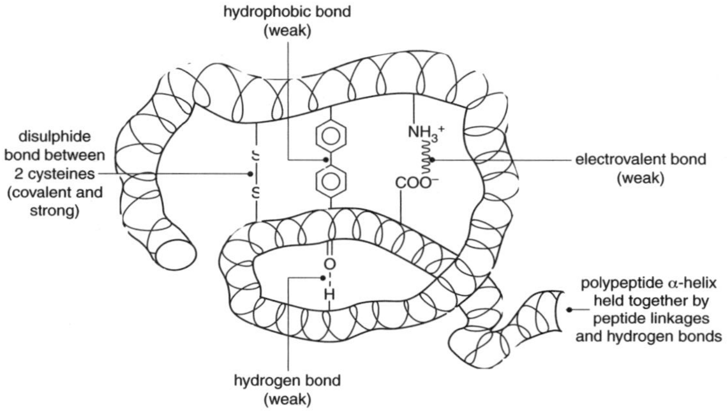

These interactions are very weak so that the conformation of such globular proteins can be easily altered by local physical changes - these alterations are reversible and are essential for the biological function of these molecules. Proteins with a well-developed tertiary structure are **globular proteins**. Examples are **enzymes**, **regular molecules** in membranes and **serum albumin**.

Several **polypeptide chains (tertiary structures) may be fitted together** to produce the **quaternary structure of the protein.** The stability of the quaternary structure is maintained by **weak interactions between -R groups of adjacent polypeptide chains** and by **Van der Waal's forces between subunits.**

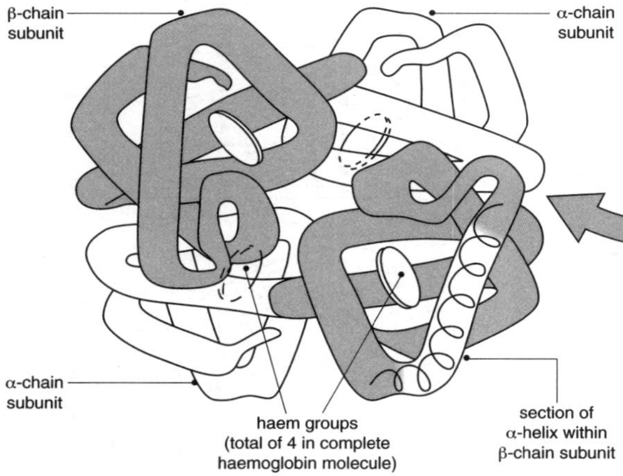

The relative movement of the polypeptide chains may be critical to the function of the protein. The oxygen transporter **haemoglobin** is an example of a protein with a quaternary structure.

Cells, tissues, and organs

The component parts of the human body are arranged in increasing levels of complexity.

The **human organism** contains ten different body systems.

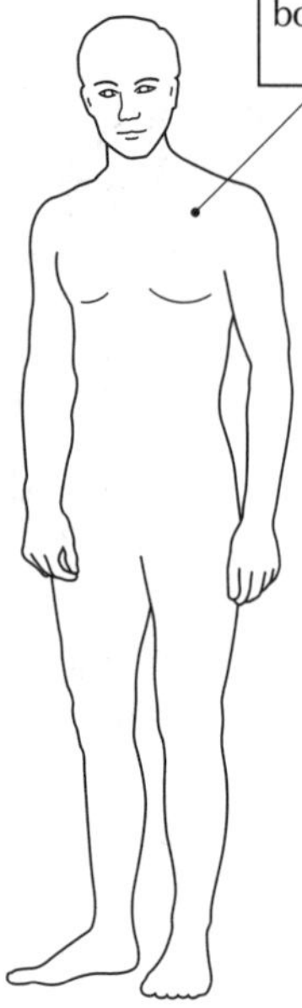

An **organ system** is a group of organs and tissues working in a co-ordinated way to perform a gross function. The **digestive system** is made up of mouth, oesophagus, stomach, small and large intestines together with associated glands, nerves, and blood vessels all contributing to the ingestion and processing of food.

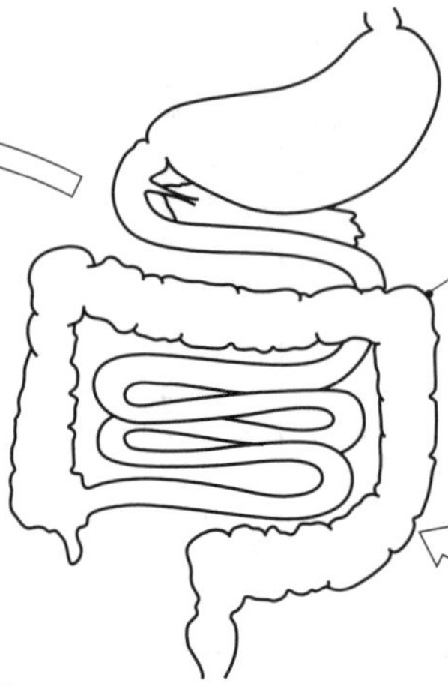

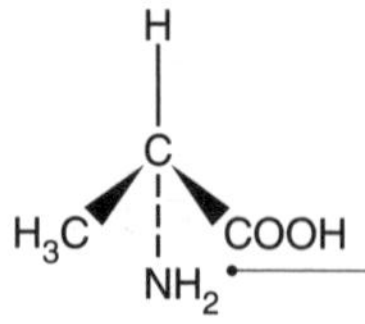

Alanine

Molecules are the building blocks of all organisms, including humans. The molecules must be obtained in the diet or be manufactured from dietary components. For example the amino acid **alanine** is a sub-unit incorporated into proteins such as **actin**, **myosin**, and the enzyme **ATPase**.

Organs are groups of physically associated tissues which together perform a major function with great efficiency. For example, the **ileum** is composed of smooth muscle, nervous tissue, blood vessels, and epithelium all bound together by connective tissue and performing both digestive and absorptive functions.

Ileum

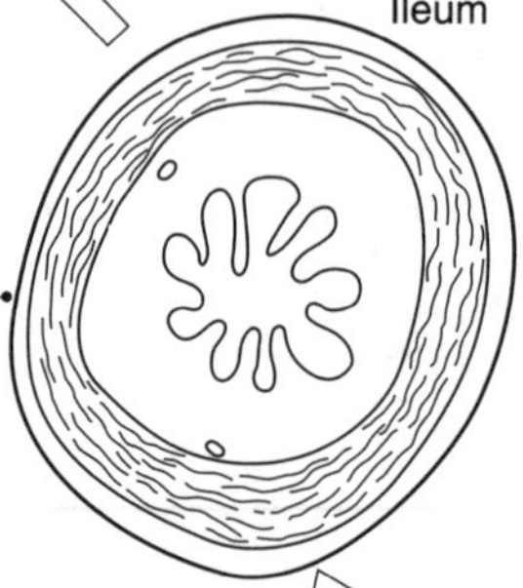

Organelles are subcellular compartments specialised for specific cellular functions. For example the **mitochondrion** has a particular arrangement of membranes, membrane-bound proteins, and matrix enzymes which together carry out the oxidation of food molecules and the consequent synthesis of ATP. Mitochondria are thus particularly abundant in cells with a high ATP/energy demand such as muscle.

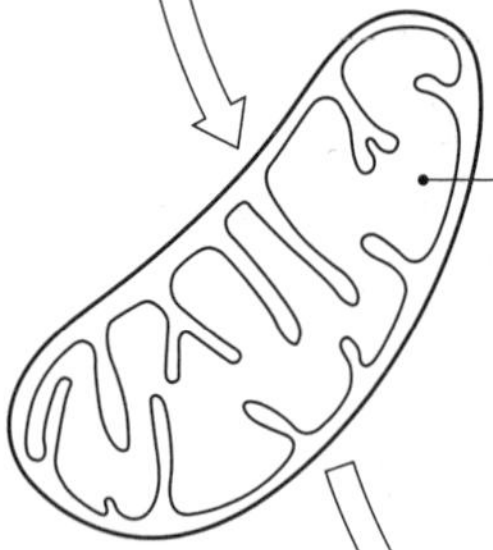

Mitochondrion

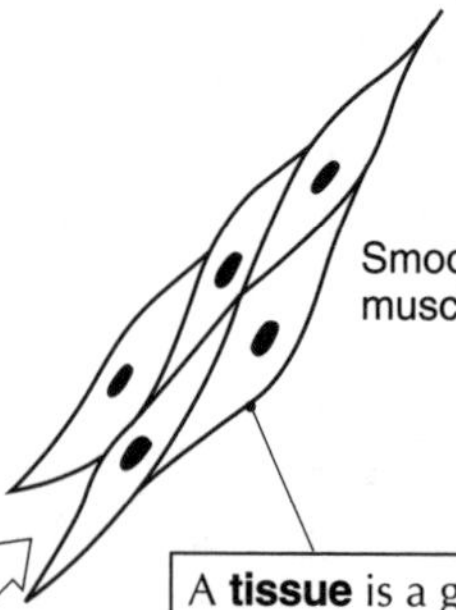

Smooth muscle

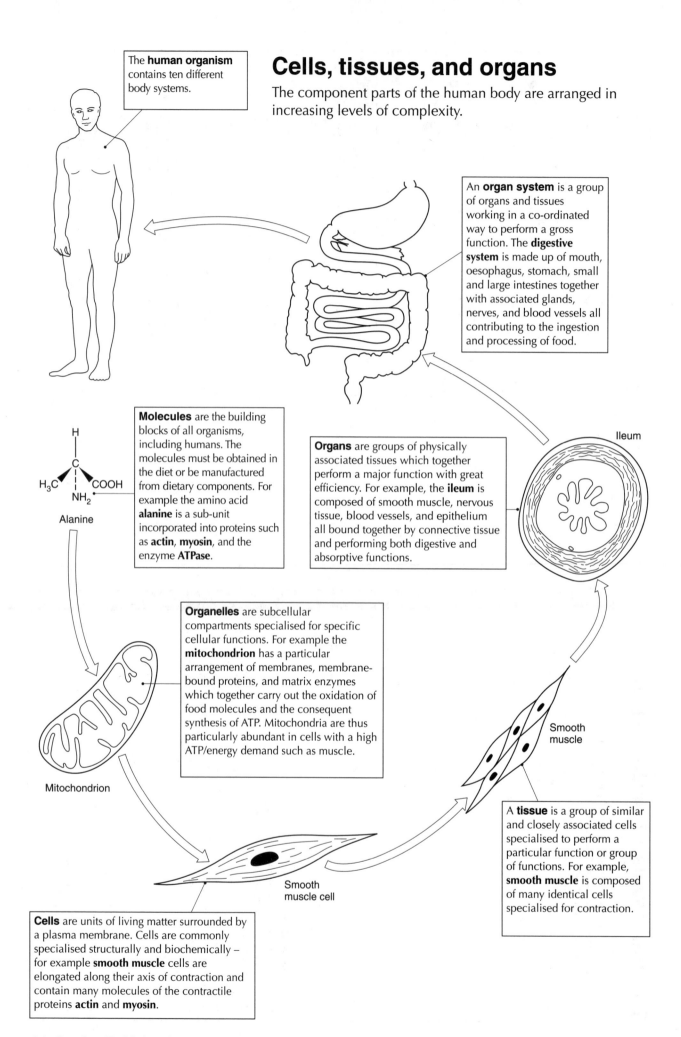

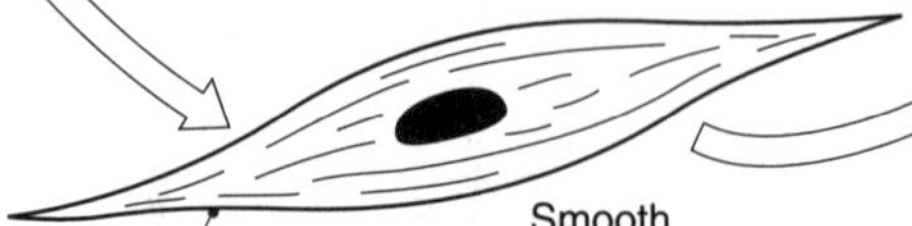

Smooth muscle cell

Cells are units of living matter surrounded by a plasma membrane. Cells are commonly specialised structurally and biochemically – for example **smooth muscle** cells are elongated along their axis of contraction and contain many molecules of the contractile proteins **actin** and **myosin**.

A **tissue** is a group of similar and closely associated cells specialised to perform a particular function or group of functions. For example, **smooth muscle** is composed of many identical cells specialised for contraction.

Specialisation of cells: animals

Fertilisation: fusion of male and female gametes forms a **zygote**.

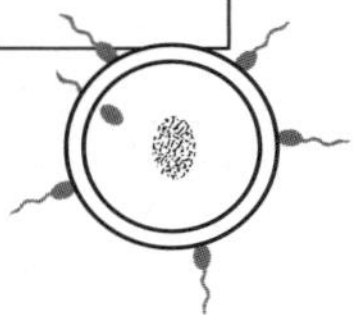

Mitotic cell division produces many smaller, identical cells.

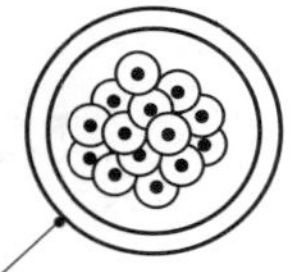

Zona pellucida

Morula: a ball of cells formed 3-4 days after fusion.

Further mitotic cell division and first **cell migrations**.

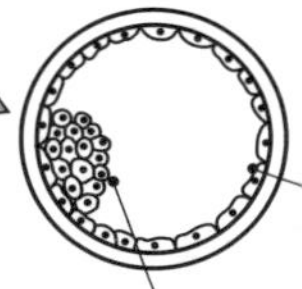

Blastocyst Day 5

Trophoblast

Inner cell mass

Ball of cells undergoes mitosis and more cell migration to produce three **germ layers**. Each layer contains basic **stem cells:** these cells are **totipotent** (i.e. have the ability to develop into any type of cell in the animal concerned). They undergo differentiation (develop specialised structures and biochemical processes to perform specific functions).

A mature human body contains over 200 types of specialised cells.

Ectoderm

For example:

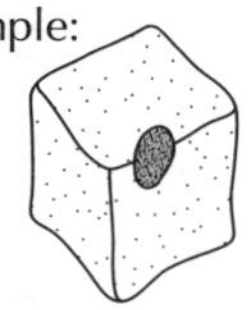

Epithelial cell: shape can be adapted for covering/absorption

Neurone

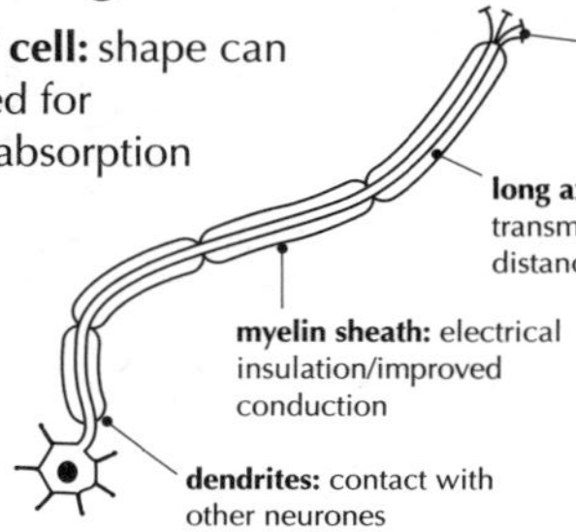

- **nerve endings**: can secrete neurotransmitters
- **long axon:** impulse transmission over great distance
- **myelin sheath:** electrical insulation/improved conduction
- **dendrites:** contact with other neurones

Mesoderm

For example:

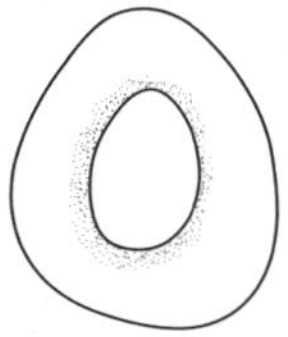

Erythrocyte (red blood cell)
- no nucleus - very flexible;
- packed with haemoglobin.

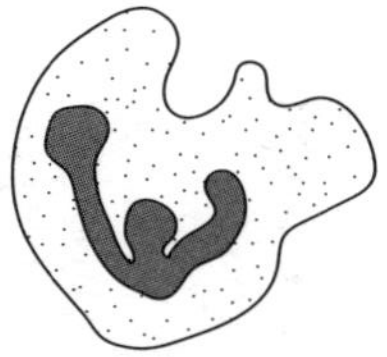

Neutrophil
- lobed nucleus - cell can change shape during movement;
- many lysosomes contain digestive (hydrolytic) enzymes.

Sperm cell

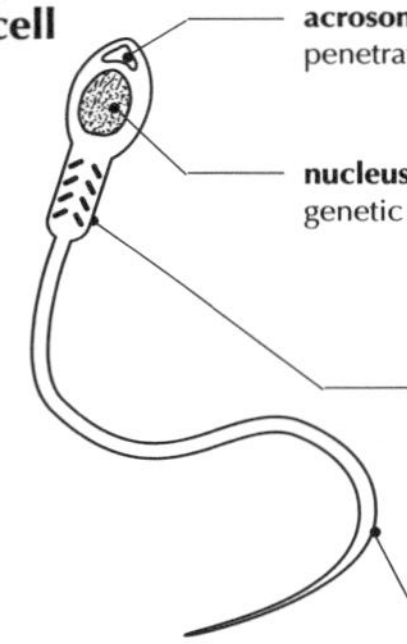

- **acrosome:** enzymes to penetrate egg
- **nucleus:** contains male genetic material
- **mid-piece:** many mitochondria to power movement
- **tail**: beats to move sperm through fluids in vagina and uterus

Endoderm

For example:

Goblet cell (in trachea or gut)

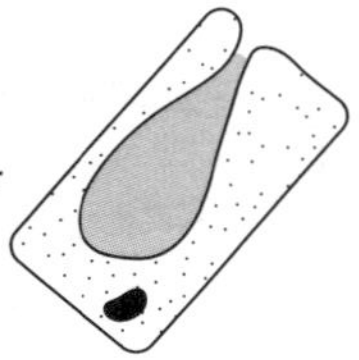

mucus for lubrication

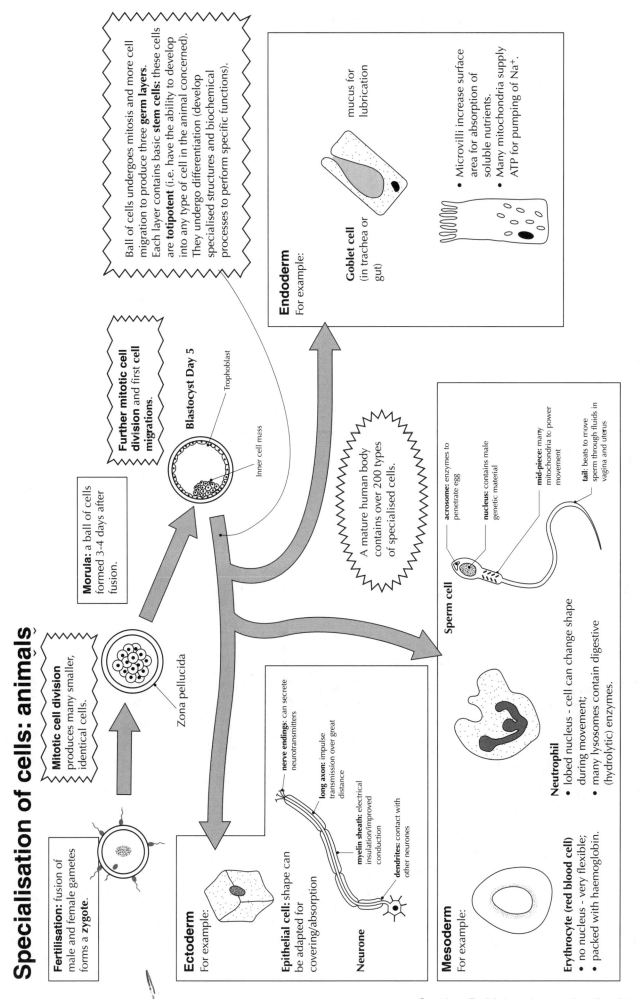

- Microvilli increase surface area for absorption of soluble nutrients.
- Many mitochondria supply ATP for pumping of Na^+.

Specialisation of cells: plants

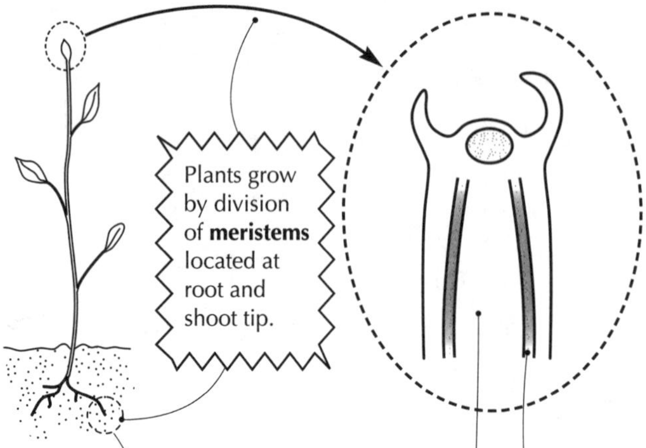

Plants grow by division of **meristems** located at root and shoot tip.

Meristem: cells here are totipotent (have the ability to become any type of cell in the species). They divide by mitosis.

Epidermis: the outer covering of the plant. Includes **guard cells**.

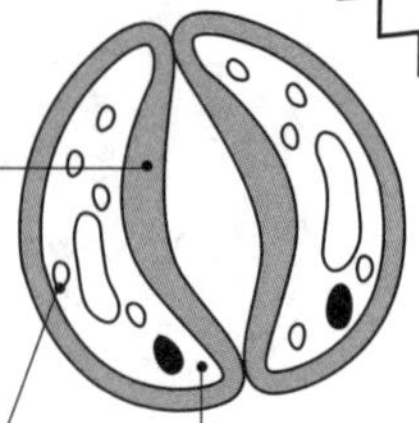

Epidermis

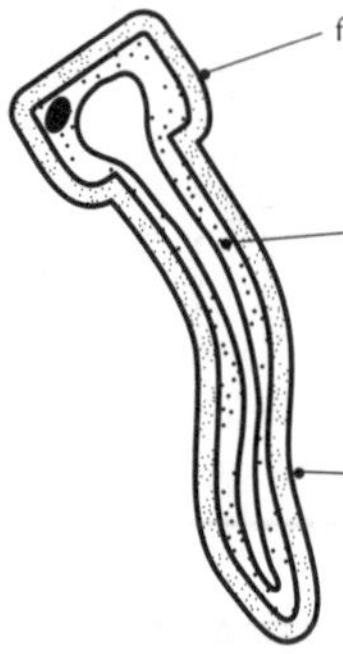

Vascular strand

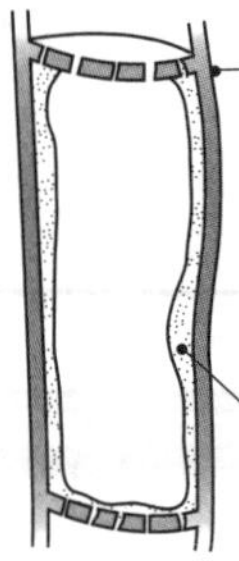

phloem sieve tube

xylem vessel

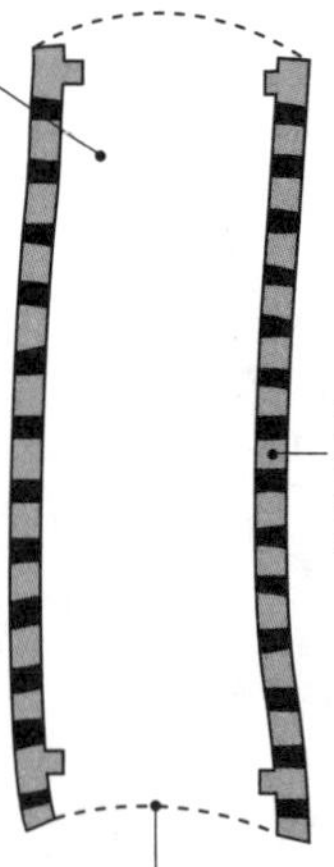

Cortex/pith/parenchyma:

includes **palisade cells.**

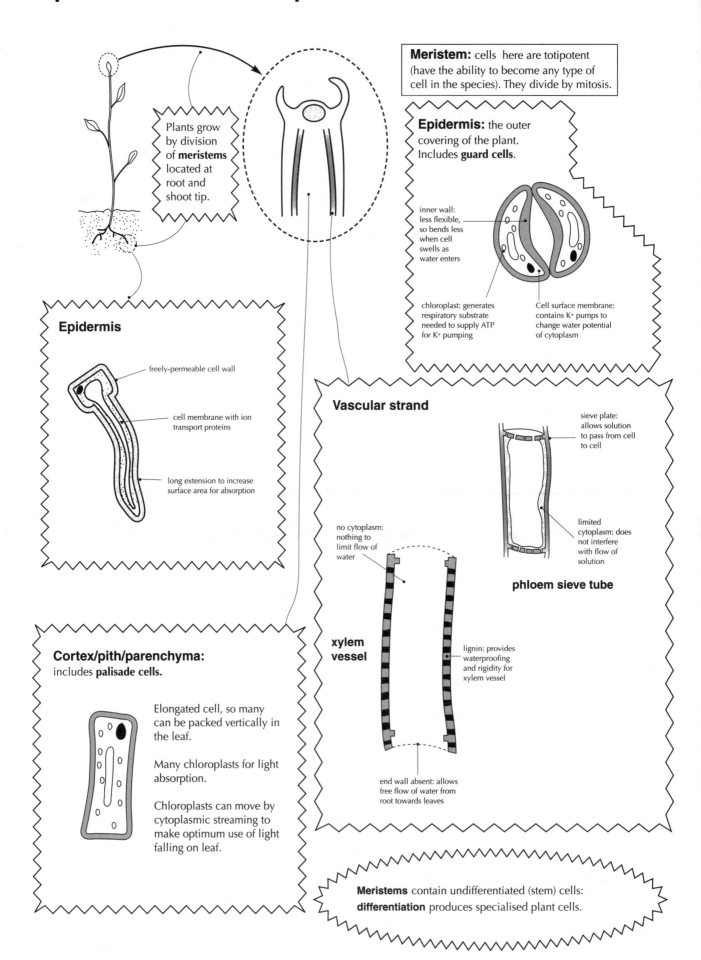

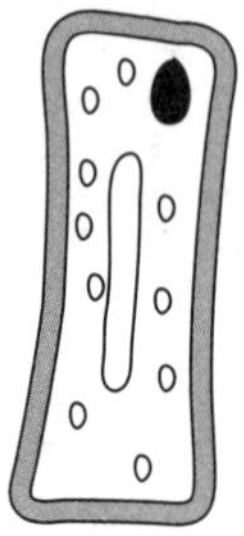

Elongated cell, so many can be packed vertically in the leaf.

Many chloroplasts for light absorption.

Chloroplasts can move by cytoplasmic streaming to make optimum use of light falling on leaf.

Meristems contain undifferentiated (stem) cells: **differentiation** produces specialised plant cells.

Commercial applications of enzymes

Analysis

Glucose oxidase
The reaction catalysed by glucose oxidase is

$$\beta\text{-D-glucose} + O_2 \longrightarrow \text{gluconic acid} + H_2O_2$$

The quick and accurate measurement of glucose is of great importance both medically (in sufferers from diabetes, for example) and industrially (in fermentation reactions, for example). A simple quantitative procedure can be devised by coupling the production of hydrogen peroxide to the activity of the enzyme **peroxidase.**

$$\underset{\substack{\text{chromagen}\\ \text{a hydrogen donor}\\ \text{(colourless)}}}{DH_2} + H_2O_2 \xrightarrow{\text{peroxidase}} 2H_2O + \underset{\substack{\text{coloured}\\ \text{compound}\\ \text{(colour)}}}{D}$$

Peroxidase can oxidise an organic chromagen (DH_2) to a coloured compound (D) utilising the hydrogen peroxide – the amount of the coloured compound D produced is a direct measure of the amount of glucose which has reacted. It can be measured quantitatively using a colorimeter or, more subjectively, by comparison with a colour reference card.

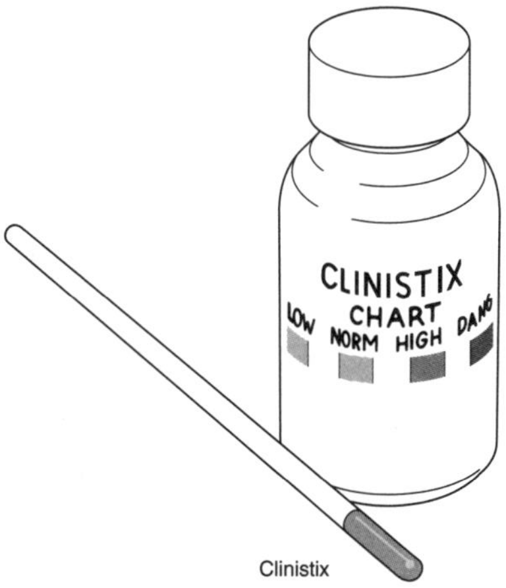

Clinistix

This method of glucose analysis is **highly specific** and has the enormous advantage over chemical methods in that this specificity allows glucose to be assayed **in the presence of other sugars**, e.g. in a biological fluid such as blood or urine, without the need for an initial separation.

Both of the enzymes glucose oxidase and peroxidase, and the chromagen DH_2, can be immobilised on a cellulose fibre pad. This forms the basis of the glucose dipsticks ('Clinistix') which were developed to enable diabetics to monitor their own blood or urine glucose levels.

Textiles **Subtilisin** is a bacterial protease (protein → amino acids) which is used in bioactive detergents to remove protein stains from clothes.

Food production **Amylase** and **glucose isomerase** convert starch → high fructose syrups. these syrups have enhanced sweetening power and lowered energy content.

There are many applications of enzyme technology to industry. Enzyme technology has several advantages over 'whole-organism' technology.

1. **No loss of substrate due to increased biomass.** For example, when whole yeast is used to ferment sugar to alcohol it always 'wastes' some of the sugar by converting it into cell wall material and protoplasm for its own growth.
2. **Elimination of wasteful side reactions.** Whole organisms may convert some of the substrate into irrelevant compounds or even contain enzymes for degrading the desired product into something else.
3. **Optimum conditions for a particular enzyme may be used.** These conditions may not be optimal for the whole organism – in some organisms particular enzymes might be working at less than maximum efficiency.
4. **Purification of the product is easier.** This is especially true using immobilised enzymes.

Medicine

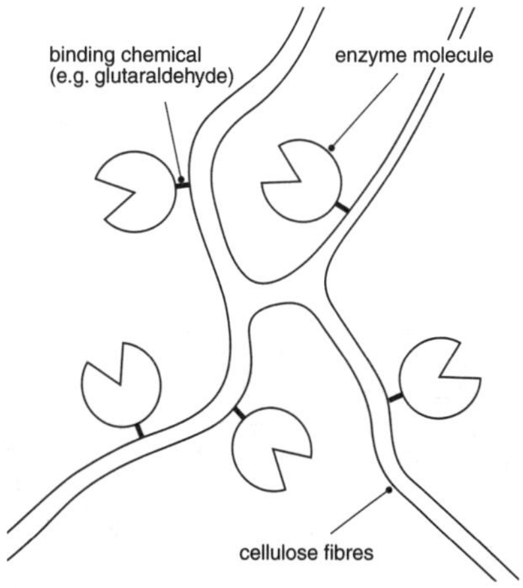

Enzyme immobilisation
Immobilisation means physically or chemically trapping enzymes or cells onto surfaces or inside fibres. The benefits can be considerable:

- the same enzyme molecules can be used again and again, since they are not lost;
- the enzyme does not contaminate the end product;
- the enzymes may be considerably more stable in immobilised form – for example, glucose isomerase is stable at 65 °C when immobilised.

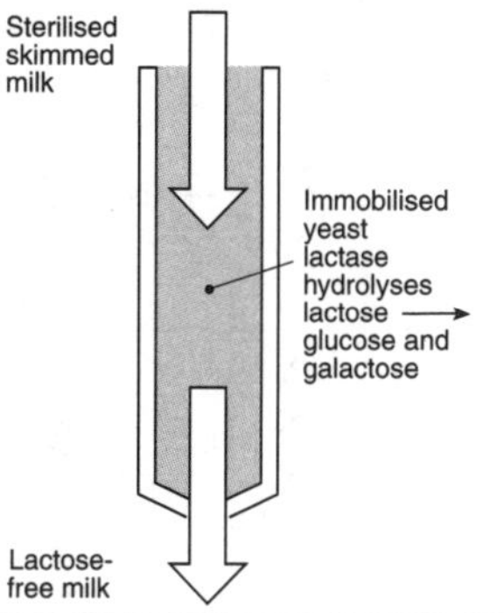

An important medical application of an immobilised enzyme

Some adults are **lactose-intolerant** since they lack an intestinal lactase, and undigested lactose in the gut is metabolised by bacteria causing severe abdominal pain and diarrhoea.

Milk is an important dietary component and can be made **lactose-free** by passage down a column packed with **yeast lactase** immobilised on fibres of cellulose acetate.

Immobilisation of enzymes

Cross linking can damage some enzymes but enzymes that are not damaged remain very active.

Cross linkage

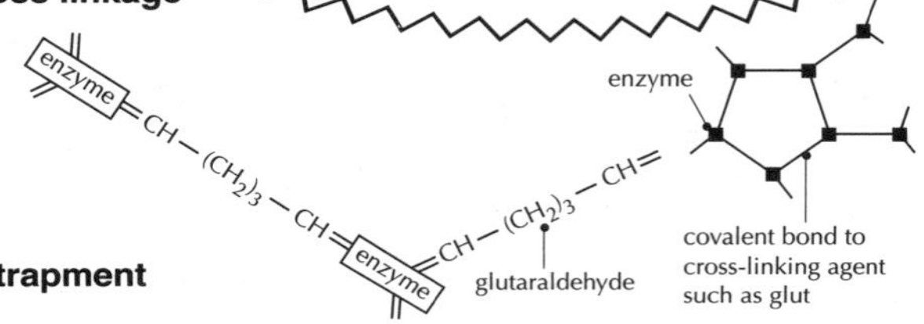

Entrapment

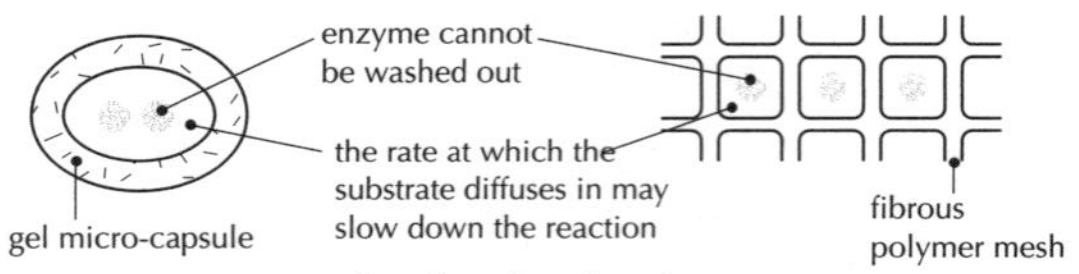

Entrapment is the most gentle method of immobilisation and does not damage enzymes.

Adsorption

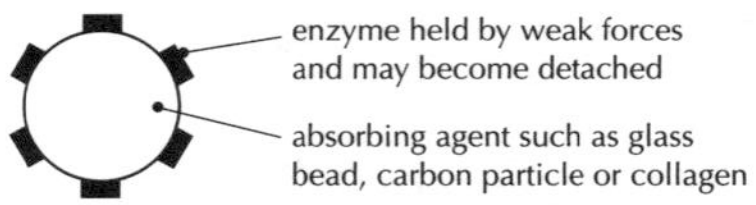

Adsorption makes it easy for the enzyme to come into contact with its substrate but the process is expensive.

Good news or bad?

- Enzymes are not mixed with product, so recovery costs are low.
- Enzymes are more stable because conditions can be controlled close to the enzyme.
- Process is continuous – enzyme is re-used and reactor only requires occasional cleaning.

- Equipment is expensive, and control systems are very complex.
- Some immobilised enzymes are less active.
- Contamination is very expensive because the complete reactor must be closed down.
- Substrate and product may block the column.

Commercial production of enzymes

Strain selection e.g. choose bacteria able to secrete a PROTEASE able to function at pH 10/50°C

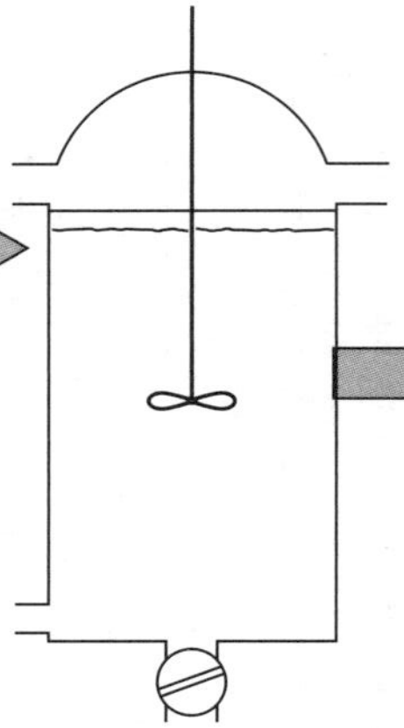

Bioreactor: cells in stationary phase produce much protease to make maximum use of limited nutrient protein.

Disrupt by:
- ultrasound;
- using lysozyme;
- grinding.

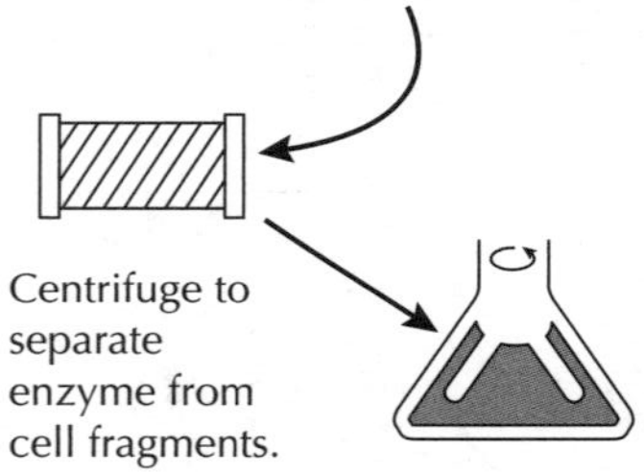

Centrifuge to separate enzyme from cell fragments.

Concentration: remove water by reverse osmosis.

Drying/powder production Heating to evaporate water: also add stabilisers.

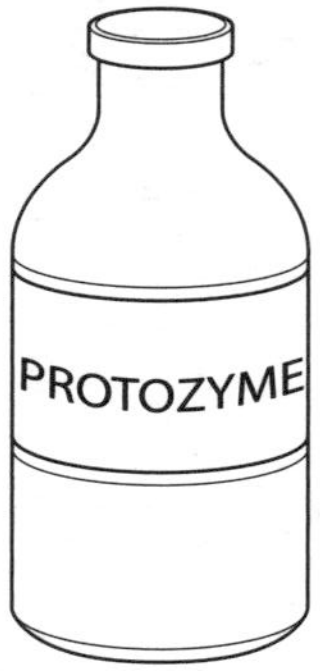

Now ready for market:
- food and industrial;
- diagnostic and analytical;
- treatment of disease.

Intracellular or extracellular?

- Kept within cell so costly to extract
- Enzyme may be very delicate
- Fewer molecules may be synthesised

- Secreted from cell so no extraction needed
- Enzyme robust, since exposed to 'outside' environment
- More molecules since not taking up space in cell
- Few types, so less purification needed

Section C Membranes and transport

Cell membranes and organelles

are involved in the production and secretion of proteins.

Large molecules such as proteins cross membranes by a process called CYTOSIS. Energy (as ATP) is needed, and the process is made possible by the flexibility of the membrane.

Exocytosis: a vesicle containing the molecule fuses with the inside of the plasmamembrane and the molecule is expelled.

Endocytosis: the membrane recognises and binds to a molecule in its environment. The fluid membrane then forms a vesicle (sac) around the molecule, and the sac enters the cell. Phagocytosis (uptake of solids) and pinocytosis (uptake of fluids) are examples of endocytosis.

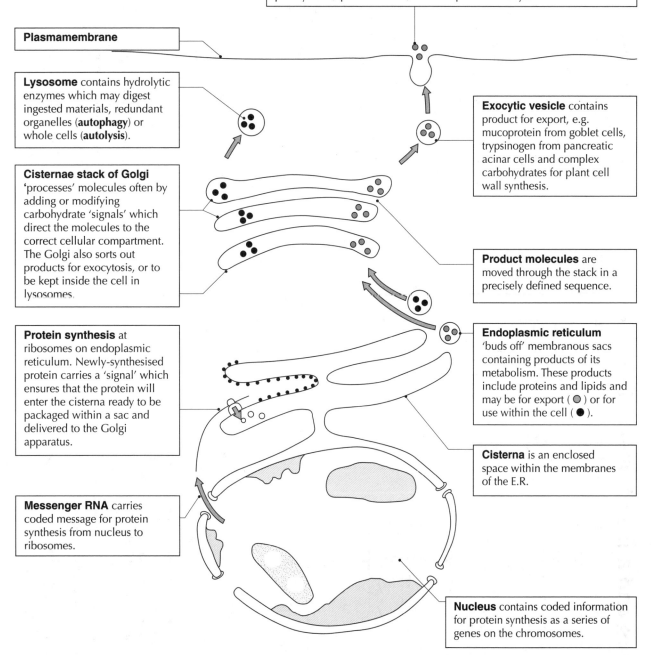

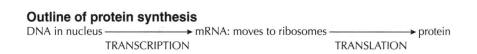

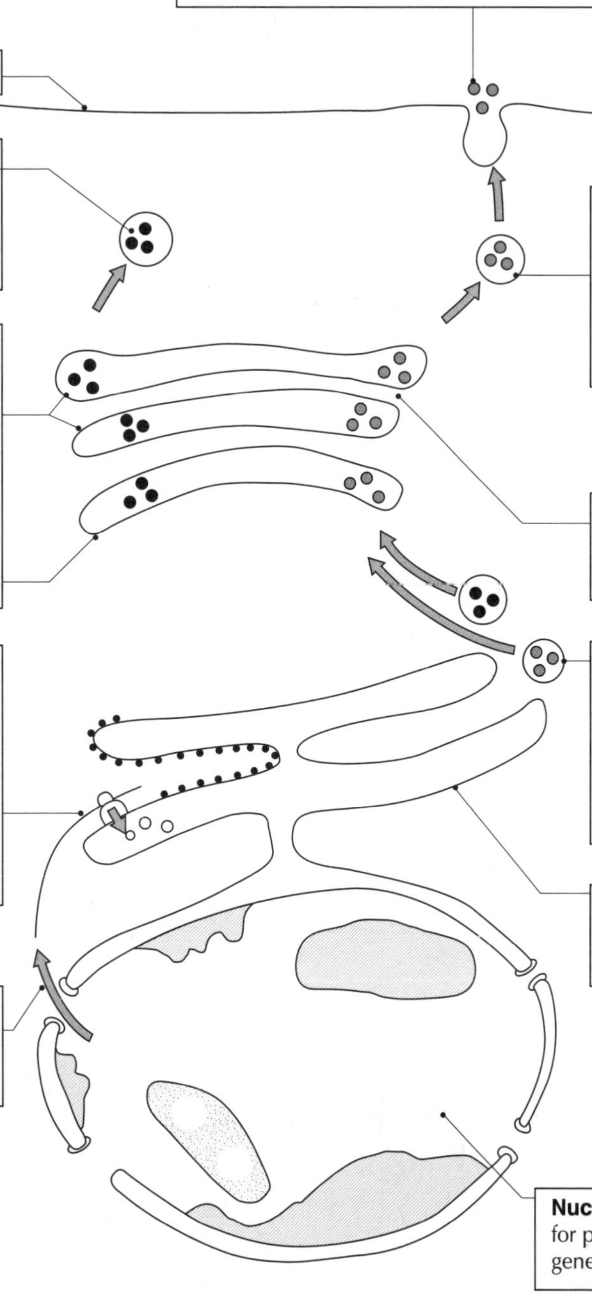

Plasmamembrane

Lysosome contains hydrolytic enzymes which may digest ingested materials, redundant organelles (**autophagy**) or whole cells (**autolysis**).

Cisternae stack of Golgi 'processes' molecules often by adding or modifying carbohydrate 'signals' which direct the molecules to the correct cellular compartment. The Golgi also sorts out products for exocytosis, or to be kept inside the cell in lysosomes.

Protein synthesis at ribosomes on endoplasmic reticulum. Newly-synthesised protein carries a 'signal' which ensures that the protein will enter the cisterna ready to be packaged within a sac and delivered to the Golgi apparatus.

Messenger RNA carries coded message for protein synthesis from nucleus to ribosomes.

Exocytic vesicle contains product for export, e.g. mucoprotein from goblet cells, trypsinogen from pancreatic acinar cells and complex carbohydrates for plant cell wall synthesis.

Product molecules are moved through the stack in a precisely defined sequence.

Endoplasmic reticulum 'buds off' membranous sacs containing products of its metabolism. These products include proteins and lipids and may be for export (● grey) or for use within the cell (● black).

Cisterna is an enclosed space within the membranes of the E.R.

Nucleus contains coded information for protein synthesis as a series of genes on the chromosomes.

Outline of protein synthesis

DNA in nucleus —TRANSCRIPTION→ mRNA: moves to ribosomes —TRANSLATION→ protein

Diffusion, osmosis and active transport are processes by which molecules are moved. Diffusion and osmosis are passive, but active transport requires energy.

Diffusion:
the movement of ions or molecules down a concentration gradient i.e. from a region of higher concentration to one of lower concentration.

This is a physical process which depends on the energy possessed by the molecules, thus:
- small molecules diffuse faster than large molecules
- diffusion speeds up as temperature increases.

Fick's law states that

Rate of diffusion is proportional to

$$\frac{\text{Surface area} \times \text{difference in concentration}}{\text{thickness of membrane}}$$

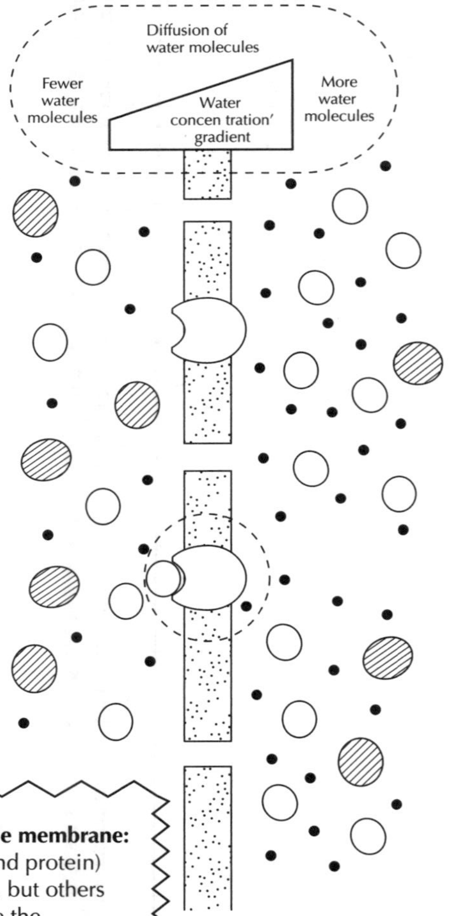

For living cells the principle (the movement of molecules down a concentration gradient) is the same, but there is one problem → the cell is surrounded by a **cell membrane** which can restrict the free movement of the molecules →

This is **a selectively permeable membrane:** the composition of the membrane (lipid and protein) allows some molecules to cross with ease, but others with difficulty or not at all. In this example the membrane is permeable to water ● but not to the larger sucrose molecule - the simplest sort of selection is based on the **size** of the molecules.

Important examples are:
- oxygen from air sacs in the lung to blood, and from blood to cells;
- soluble foods from gut to blood;
- carbon dioxide from air to spaces inside leaf.

Osmosis is the diffusion of water

Water crosses membranes very freely and always tends to move, by diffusion, down the water 'concentration' gradient. the term 'concentration' can be confusing when used to describe water molecules, and is better replaced by the term 'potential'.

Thus osmosis is:
- the movement of water
- across a selectively permeable membrane
- down a water potential gradient.

Osmosis is responsible for water movement:
- from tissue fluid to cells
- from soil water to root hairs
- from xylem to leaf mesophyll cells.

Active transport may move molecules against a concentration gradient

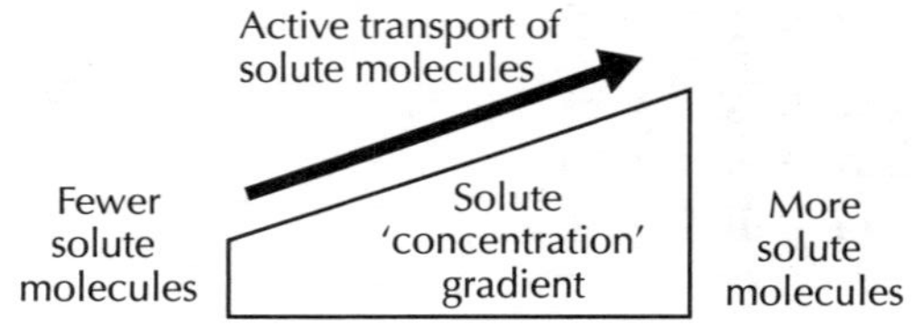

In this example there are more amino acid molecules on the right side of the membrane than on the left – to move any more from left to right will be 'uphill', **against** the amino acid gradient. This active transport:
- requires energy to 'drive' the molecules 'uphill' – this energy is supplied as ATP from respiration;
- is affected by any factor which affects respiration, e.g. temperature and oxygen concentration;
- is carried out by 'carrier proteins' in the membrane, which bind to the solute molecule, change shape, and carry the molecule across the membrane.

Important examples are:
- uptake of mineral ions from soil by root hair cells;
- movement of sodium ions to set up nerve impulses.

The human digestive system prepares ingested food for absorption

Tongue: manoeuvres food for chewing and rolls food into a bolus for swallowing. Mixes food with saliva.

Ingested food must be made ready for digestion

Salivary glands (3 pairs of them!)
These produce SALIVA, which is 99% water plus:
- mucin – lubricates food
- salivary amylase – begins starch digestion
- lysozyme – destroys bacterial cells
- hydrogen carbonate – provides optimum pH for amylase
- chloride ions – activate salivary amylase.

Teeth (4 different types in humans!)
Cut and grind food so that solids are reduced to smaller particles for swallowing, and food has a larger surface area for the action of enzymes.

Pancreas
Produces digestive secretions, including pancreatic amylase, as well as hormones controlling blood glucose concentration.

Duodenum
- The first 30 cm of the small intestine.
- Pancreatic and other enzymes complete carbohydrate, protein and lipid digestion.
- Releases an alkaline mucus which neutralises chyme to provide pH optimum for enzyme action, and protects wall of duodenum from digestive enzymes.

Ileum
Up to 6 m in length, and the main part of the small intestine. Provides the great surface area for the absorption of the soluble products of digestion.

Colon
Large intestine:
- absorbs water from faeces to produce more solid 'pooh'!
- some B vitamins and vitamin K are made here by bacteria
- glands produce mucus to lubricate faeces as water is removed.

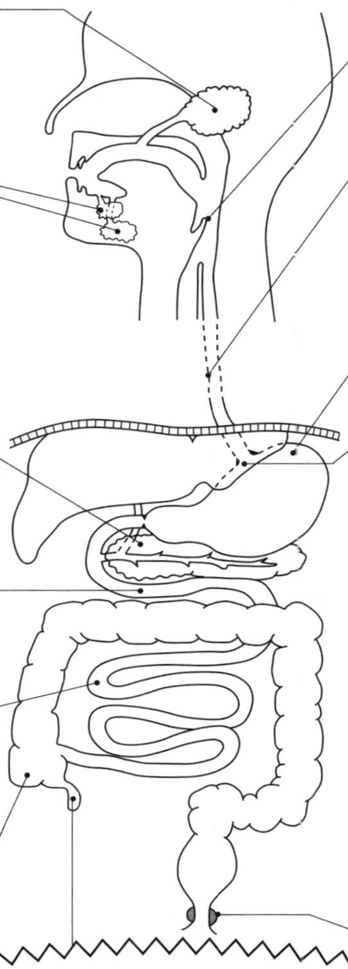

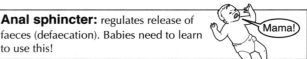

Appendix: has no function in humans, but may become infected if bacteria feed on trapped food.

Going down the right tube!
The epiglottis: a flap of muscle which is moved by a reflex during swallowing. This prevents food entering the respiratory system.

Oesophagus: muscular tube which lies behind the trachea and connects the buccal cavity (mouth) to the stomach. Muscular to generate peristaltic waves, which drive bolus of food downwards, and glandular to lubricate bolus with mucus. Semi-solid food passes to stomach in 4–8 seconds, very soft foods and liquids take only 1 second.

Stomach: A muscular bag which can stretch to allow storage of large quantities of food. Lining has gastric pits which release hydrochloric acid and protein-digesting enzymes. There are three muscle layers which contract and relax to churn the stomach contents to make sure they mix thoroughly to form **chyme**. The muscles eventually push the chyme through to the duodenum.

Cardiac sphincter: allows entry of food to stomach. Helps to retain food in stomach.

Pyloric sphincter: opens to permit passage of chyme into duodenum and closes to prevent backflow of food from duodenum to stomach.

Sphincter muscles: allow the contents of the gut into the correct regions for the most efficient processing of food.

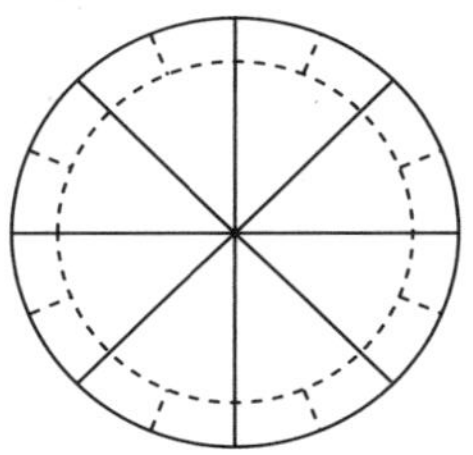

There are only **radial** muscle fibres in sphincters (including the bladder sphincter).
When the fibres are **relaxed** the sphincter is **closed.**
When the fibres are **contracted** the sphincter is **open.**
Since the sphincter is closed for most of the time, the muscle fibres are relaxed and there is no fatigue.

Anal sphincter: regulates release of faeces (defaecation). Babies need to learn to use this!

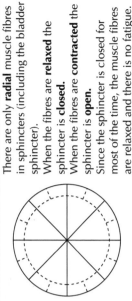

Digestive enzymes hydrolyse insoluble starch to soluble sugars ready for absorption

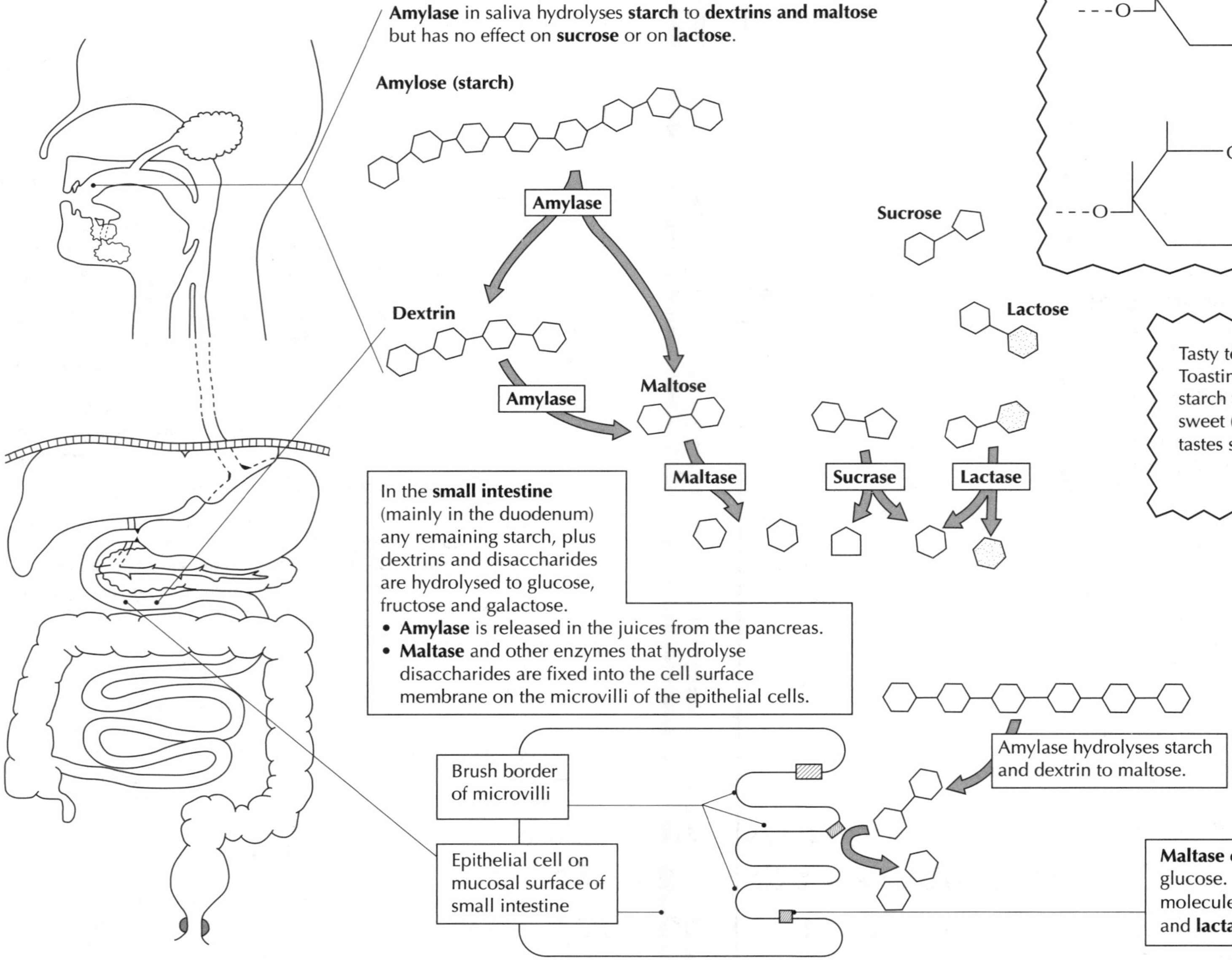
Amylase in saliva hydrolyses starch to dextrins and maltose but has no effect on sucrose or on lactose.
Amylose (starch)
Amylase
Dextrin
Amylase
Maltose
Sucrose
Lactose
Maltase
Sucrase
Lactase
In the small intestine (mainly in the duodenum) any remaining starch, plus dextrins and disaccharides are hydrolysed to glucose, fructose and galactose.
• Amylase is released in the juices from the pancreas.
• Maltase and other enzymes that hydrolyse disaccharides are fixed into the cell surface membrane on the microvilli of the epithelial cells.
Brush border of microvilli
Epithelial cell on mucosal surface of small intestine
Amylase hydrolyses starch and dextrin to maltose.
Maltase on cell surface - hydrolyses maltose to glucose. NB there are also membrane-bound molecules of sucrase [sucrose → glucose + fructose] and lactase [lactose → glucose + galactose].

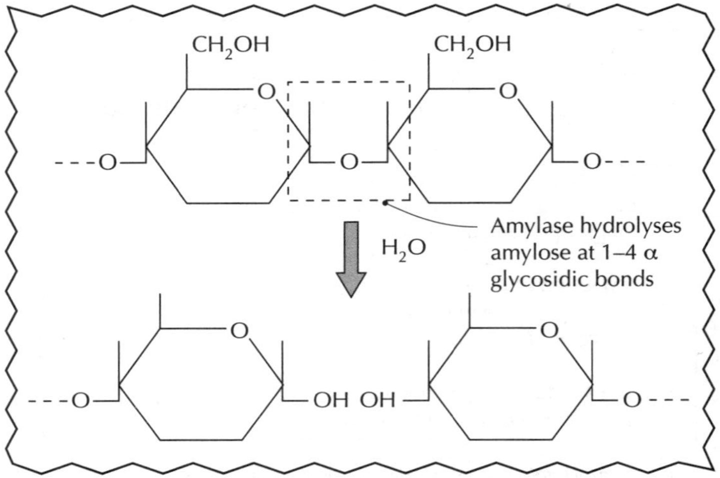
CH₂OH
CH₂OH
Amylase hydrolyses amylose at 1–4 α glycosidic bonds
H₂O
OH OH

Tasty toast!
Toasting bread converts some starch to dextrins: these are sweet (starch is not) so toast tastes sweet!

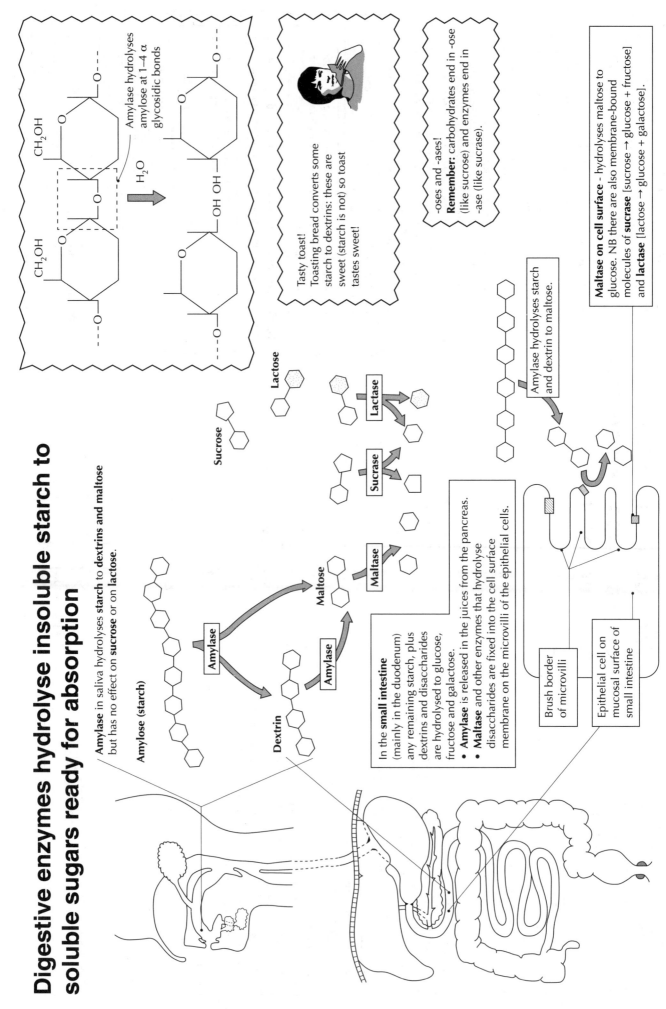

-oses and -ases!
Remember: carbohydrates end in -ose (like sucrose) and enzymes end in -ase (like sucrase).

Molecules in the plasmamembrane

Allow:

- recognition;
- signalling;
- adhesion;
- fluidity.

Because of the different solubility properties of the two ends of **phospholipid** molecules …

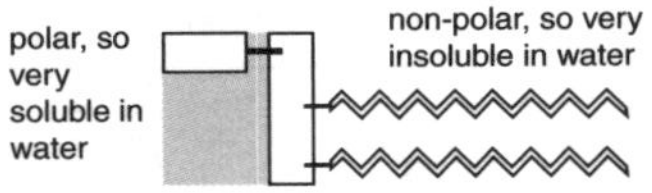

… these molecules form a layer at a water surface.

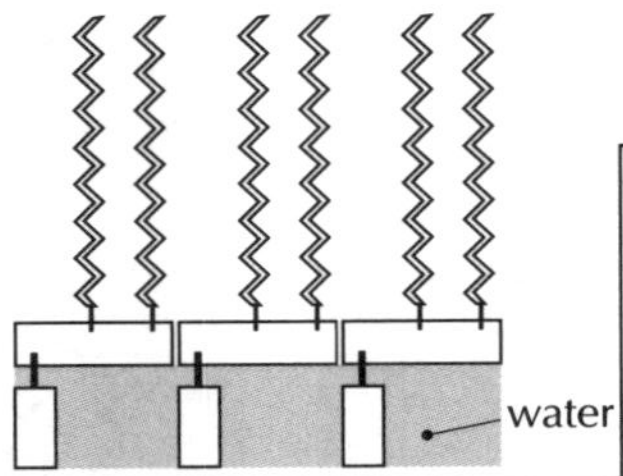

and a **phospholipid bilayer** can act as a barrier between two aqueous (watery) environments.

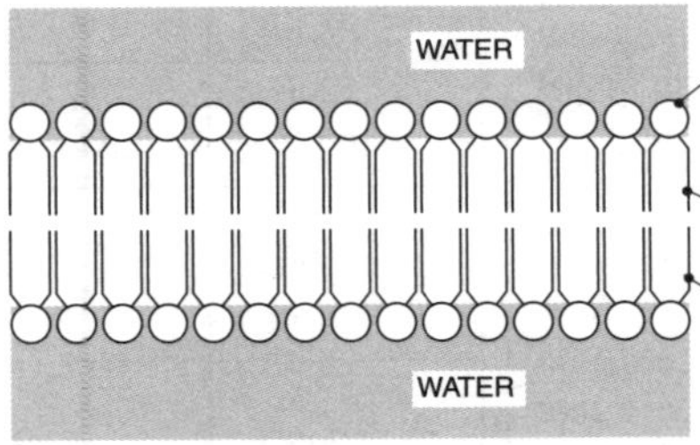

Hydrophilic heads point outwards: form hydrogen bonds with water

Hydrophobic tails point towards one another: this maximises hydrophobic attractions and excludes water

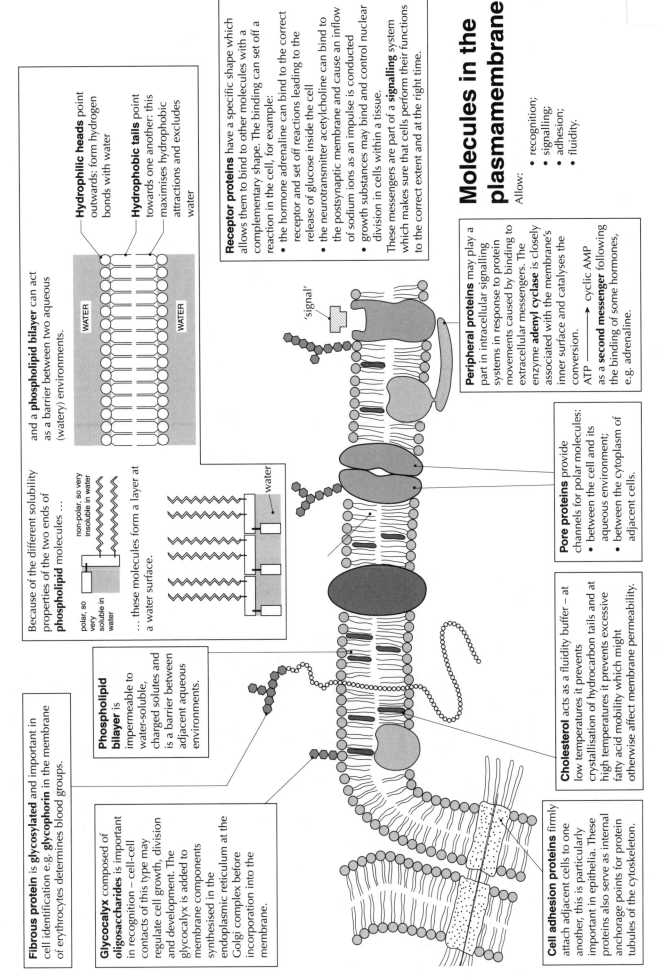

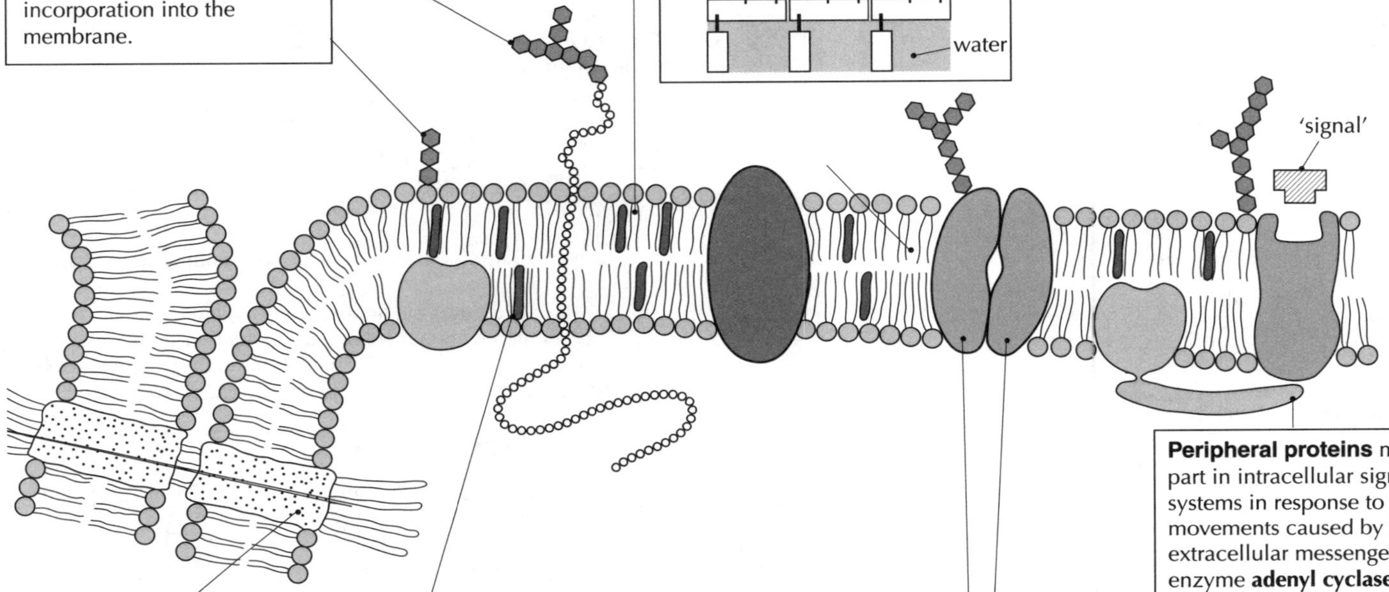

Fibrous protein is **glycosylated** and important in cell identification e.g. **glycophorin** in the membrane of erythrocytes determines blood groups.

Glycocalyx composed of **oligosaccharides** is important in recognition – cell-cell contacts of this type may regulate cell growth, division and development. The glycocalyx is added to membrane components synthesised in the endoplasmic reticulum at the Golgi complex before incorporation into the membrane.

Phospholipid bilayer is impermeable to water-soluble, charged solutes and is a barrier between adjacent aqueous environments.

Receptor proteins have a specific shape which allows them to bind to other molecules with a complementary shape. The binding can set off a reaction in the cell, for example:

- the hormone adrenaline can bind to the correct receptor and set off reactions leading to the release of glucose inside the cell
- the neurotransmitter acetylcholine can bind to the postsynaptic membrane and cause an inflow of sodium ions as an impulse is conducted
- growth substances may bind and control nuclear division in cells within a tissue.

These messengers are part of a **signalling** system which makes sure that cells perform their functions to the correct extent and at the right time.

Cell adhesion proteins firmly attach adjacent cells to one another, this is particularly important in epithelia. These proteins also serve as internal anchorage points for protein tubules of the cytoskeleton.

Cholesterol acts as a fluidity buffer – at low temperatures it prevents crystallisation of hydrocarbon tails and at high temperatures it prevents excessive fatty acid mobility which might otherwise affect membrane permeability.

Pore proteins provide channels for polar molecules:

- between the cell and its aqueous environment;
- between the cytoplasm of adjacent cells.

Peripheral proteins may play a part in intracellular signalling systems in response to protein movements caused by binding to extracellular messengers. The enzyme **adenyl cyclase** is closely associated with the membrane's inner surface and catalyses the conversion.

ATP ⟶ cyclic AMP

as a **second messenger** following the binding of some hormones, e.g. adrenaline.

Transport across membranes

Active transport uses a **carrier protein** to transport a solute across a membrane but **energy is required** since transport may be **against a concentration gradient.** ATP is hydrolysed and the binding of the phosphate group to the carrier changes the protein's conformation in such a way that the solute molecule is moved across the membrane.

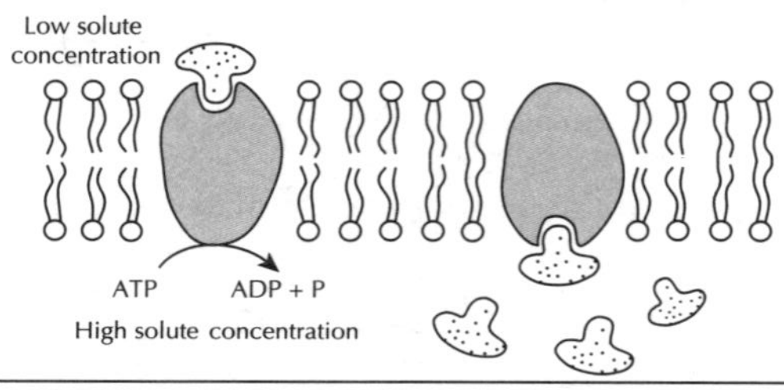

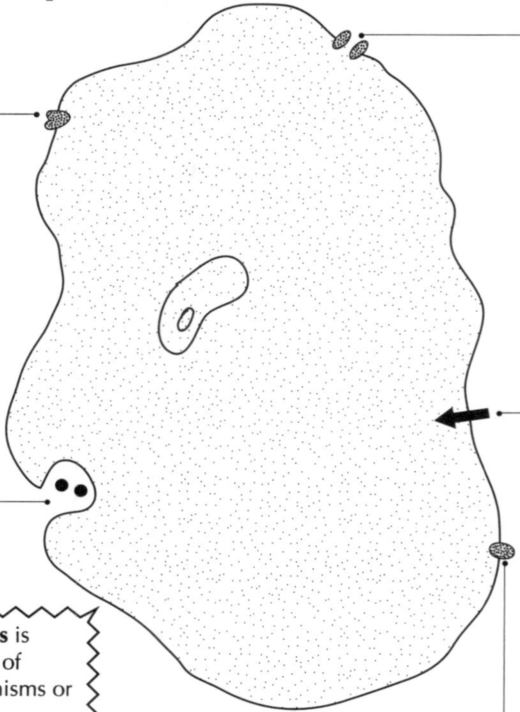

Diffusion through aqueous channels in pore proteins: transmembrane proteins may have aqueous channels through which charged molecules may pass and thus avoid the hydrophobic tails of the phospholipid molecules.

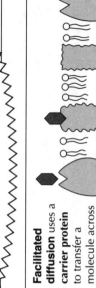

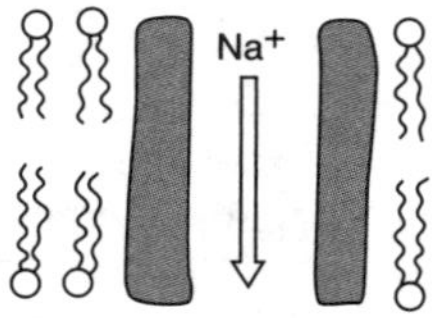

Some channels are open all of the time, but others are **gated** (they open and close only in response to a stimulus, such as a change in the membrane's electrical potential). Such **gated channels** are vital to the operation of nerve and muscle, where movements of Na^+, K^+ and Ca^{2+} set off information transfer.

Diffusion across the lipid bilayer is responsible for the movement of **small, uncharged molecules.**

Thus O_2, H_2O, CO_2, urea and ethanol cross rapidly (they 'squeeze between') the polar phospholipid heads then dissolve in the lipid on one side of the membrane and emerge on the other.

Large or **charged molecules** cannot cross the lipid bilayer.

Thus Na^+, K^+, Cl^-, HCO_3^- and glucose do not cross in this way.

OSMOSIS is the diffusion of water.

Cytosis allows substances to cross membranes through 'gaps' in the bilayer: the fluidity of the membrane allows it to fuse with sacs (vesicles) containing the substance to be transported.

Endocytosis is the movement of substances **into** the cell:

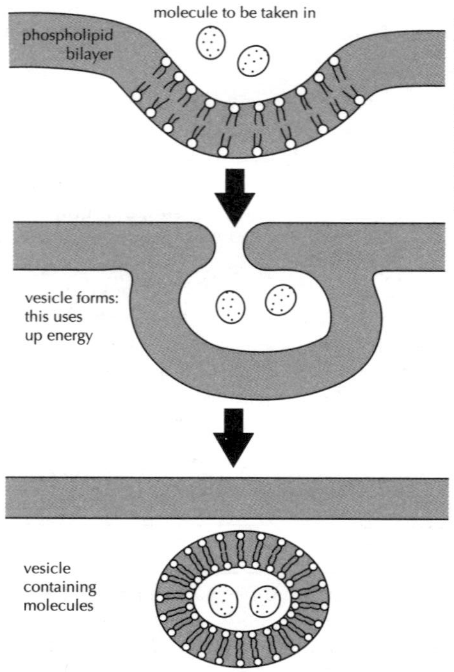

Phagocytosis is endocytosis of whole organisms or very large molecules: carried out by **phagocytes.**

Process	Proteins	ATP/energy	Concentration gradient	For example
Diffusion	May be pore proteins	No	Down	Water, Na^+
Facilitated diffusion	Yes	No	Down	Glucose, amino acids in gut
Active transport	Yes	Yes	Up or down (usually up)	Na^+ out of neurones
Endocytosis	Sometimes, to recognise the molecule	Yes	Up or down (usually down)	Milk proteins across baby's gut wall

Facilitated diffusion uses a **carrier protein** to transfer a molecule across a membrane **along** its electrochemical gradient. The binding of the solute alters the conformation of the carrier so that its position in the membrane changes and the solute molecule is discharged on the other side of the membrane.

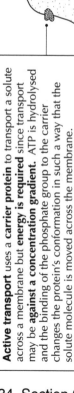

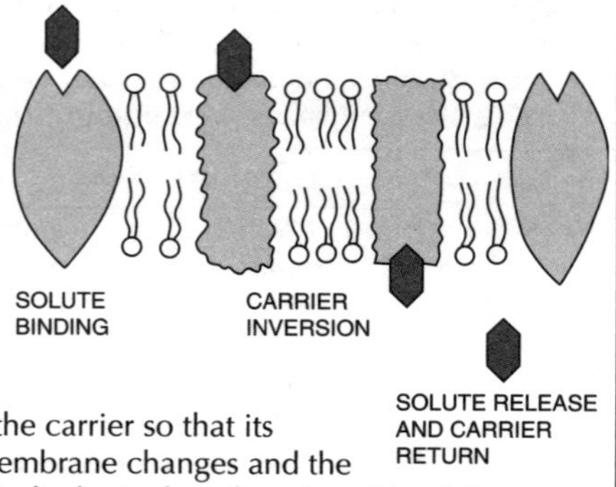

Glucose uptake by erythrocytes occurs in this way.

N.B. There is **no requirement for ATP**, as there is **no energy consumption.**

Water potential explains the direction of osmosis

Water potential is a measure of the free kinetic energy of water in a system, or the tendency of water to leave a system. It is measured in units of pressure (kPa) and is given the symbol ψ ('psi').

For pure water the water potential is arbitrarily given the value 0: this is a reference point, rather like the redox potential system used in chemistry.

i.e. for pure water $\psi = 0$

In a solution the presence of molecules of solute prevents water molecules leaving. Thus

ψ solution < 0

(In the solution the solute molecules 'hinder' the movement of the water molecules, thus the kinetic energy of the water molecule is reduced and ψ becomes negative.)

Water moves down a gradient of water potential, i.e. from a less negative (e.g. -500 kPa) to a more negative (e.g. -1000 kPa) water potential.

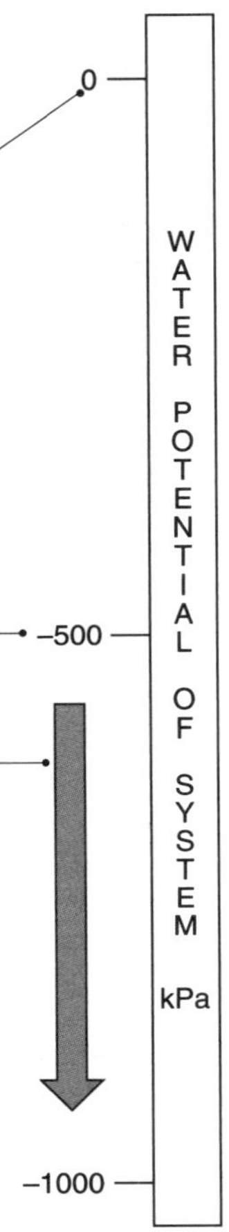

The advantages of the water potential nomenclature:

- the movement of water is considered from the 'system's' point of view, rather than from that of the environment;
- comparison between different systems can be made, e.g. between the atmosphere, the air in the spaces of a leaf and the leaf mesophyll cells.

We should remember that:

Osmosis is the movement of water, through a partially permeable membrane, along a water potential gradient.

The presence of solutes in the cell sap makes ψ_s lower (more negative). The ψ for pure water = 0, so a plant cell placed in pure water will experience an inward movement of water, by osmosis, along a water potential gradient.

The cellulose cell wall is freely permeable to water and to solutes.

The plasma membrane is freely permeable to water but of limited permeability to solutes.

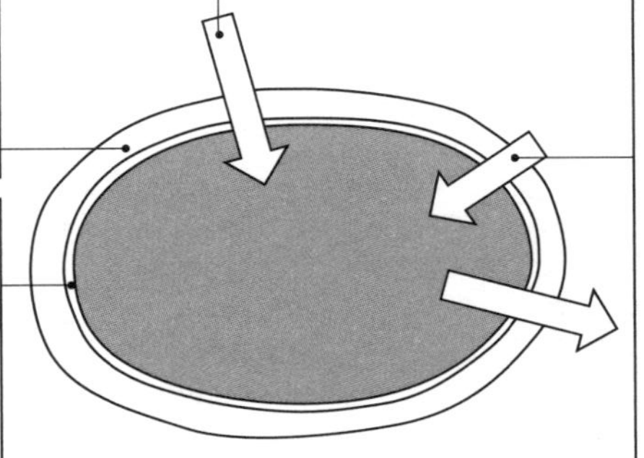

The entry of water will cause swelling of the protoplast (the cell contents inside the plasma membrane) causing a pressure (**turgor pressure**) to be exerted on the cell wall. Further expansion is resisted by a force **equal, but opposite in sign,** to the turgor pressure. This is called the **wall pressure** or **pressure potential** (ψ_p).

The pressure potential tends to resist the entry of water or to force water out of the cell.

$$\psi_{Cell} = \psi_s + \psi_p$$

ψ_{Cell}: Water potential of cell, i.e. the tendency of water to leave the cell: sometimes written as ψ_w, ψ_o or simply as ψ.

ψ_s: Effect of solute concentration on water potential – this is always negative in value.

ψ_p: Effect of wall pressure and turgor pressure on water potential – it represents the tendency for water to be forced out of the cell, and is either zero or positive in value.

Animal cells and osmosis

Animal cell cytoplasm is a solution of salts and other molecules in water.

The solution surrounding an animal cell can be:

- **Hypertonic** – has a higher solute concentration (so a lower water potential) than the cell cytoplasm;
- **Isotonic** – has the same solute concentration (so the same water potential) as the cell cytoplasm;
- **Hypotonic** – has a lower ('hypo' sounds like 'low') solute concentration (higher ψ) than the cell cytoplasm).

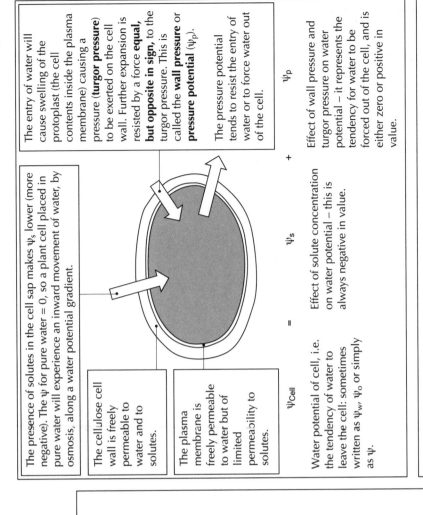

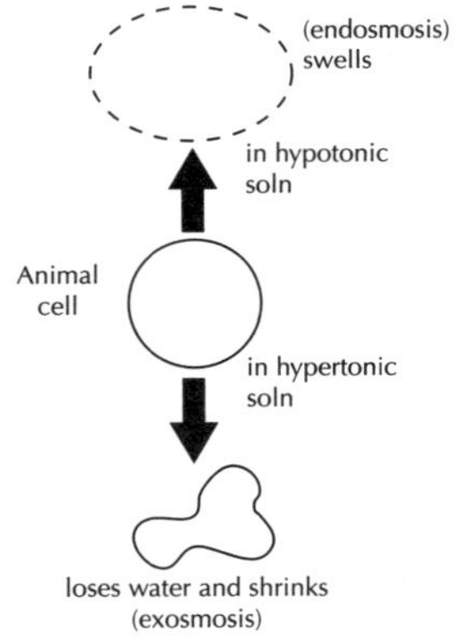

Water relationships of plant cells

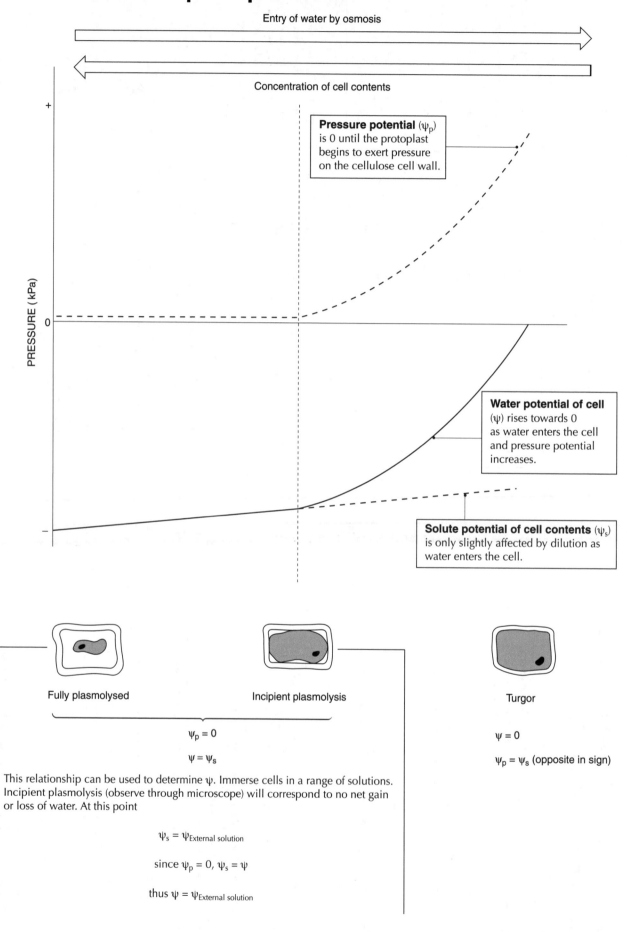

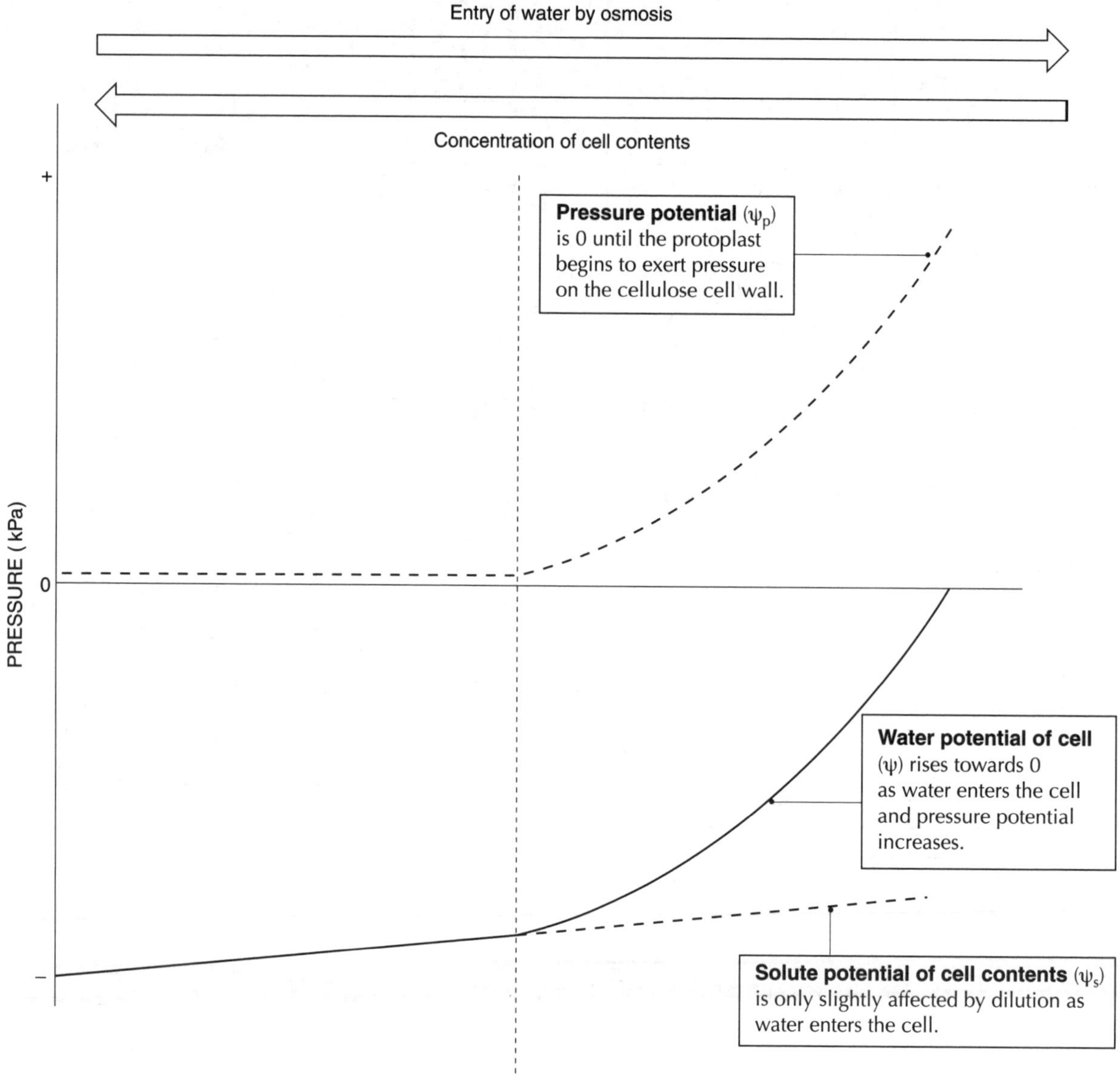

Entry of water by osmosis
Concentration of cell contents
PRESSURE (kPa)
+
0
–
Pressure potential (ψ_p) is 0 until the protoplast begins to exert pressure on the cellulose cell wall.
Water potential of cell (ψ) rises towards 0 as water enters the cell and pressure potential increases.
Solute potential of cell contents (ψ_s) is only slightly affected by dilution as water enters the cell.

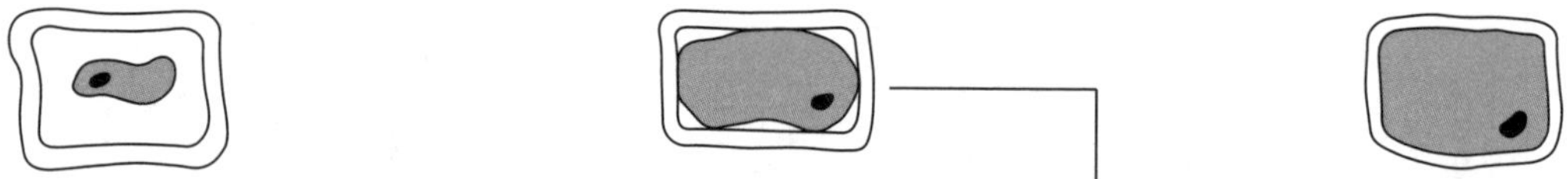

Fully plasmolysed | Incipient plasmolysis | Turgor

Fully plasmolysed and incipient plasmolysis:

$\psi_p = 0$

$\psi = \psi_s$

This relationship can be used to determine ψ. Immerse cells in a range of solutions. Incipient plasmolysis (observe through microscope) will correspond to no net gain or loss of water. At this point

$$\psi_s = \psi_{\text{External solution}}$$

$$\text{since } \psi_p = 0,\ \psi_s = \psi$$

$$\text{thus } \psi = \psi_{\text{External solution}}$$

Turgor:

$\psi = 0$

$\psi_p = \psi_s$ (opposite in sign)

Absorption of glucose

Sports drinks contain Na^+ as well as sugars, to improve uptake of glucose.

Absorption of glucose is helped by

- specific transport systems;
- a large surface area;
- a well-developed transport network.

Mitochondria carry out aerobic respiration to provide ATP for the sodium pump.

In **coeliac disease** an immune response to the protein **gluten** in wheat causes a reduction in the number of villi. Absorption is poor, and the sufferer becomes tired and loses weight. The situation quickly improves on a **gluten-free diet**. About 1 in 2000 people in the UK have this, but it is very rare in black African people.

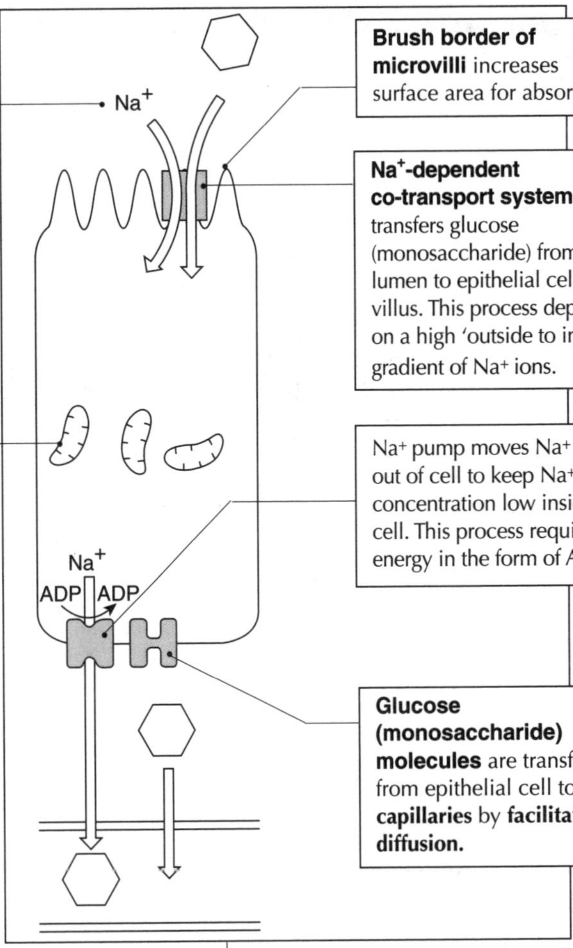

Brush border of microvilli increases surface area for absorption.

Na^+-dependent co-transport system transfers glucose (monosaccharide) from gut lumen to epithelial cell of villus. This process depends on a high 'outside to inside' gradient of Na^+ ions.

Na^+ pump moves Na^+ ions out of cell to keep Na^+ concentration low inside the cell. This process requires energy in the form of ATP.

Glucose (monosaccharide) molecules are transferred from epithelial cell to **capillaries** by **facilitated diffusion.**

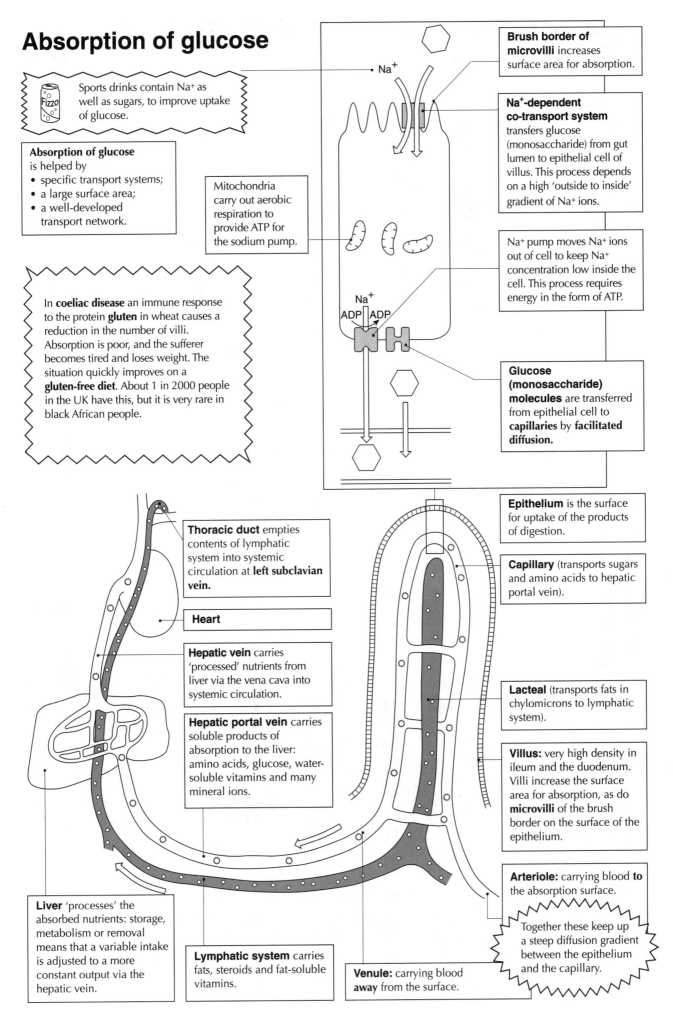

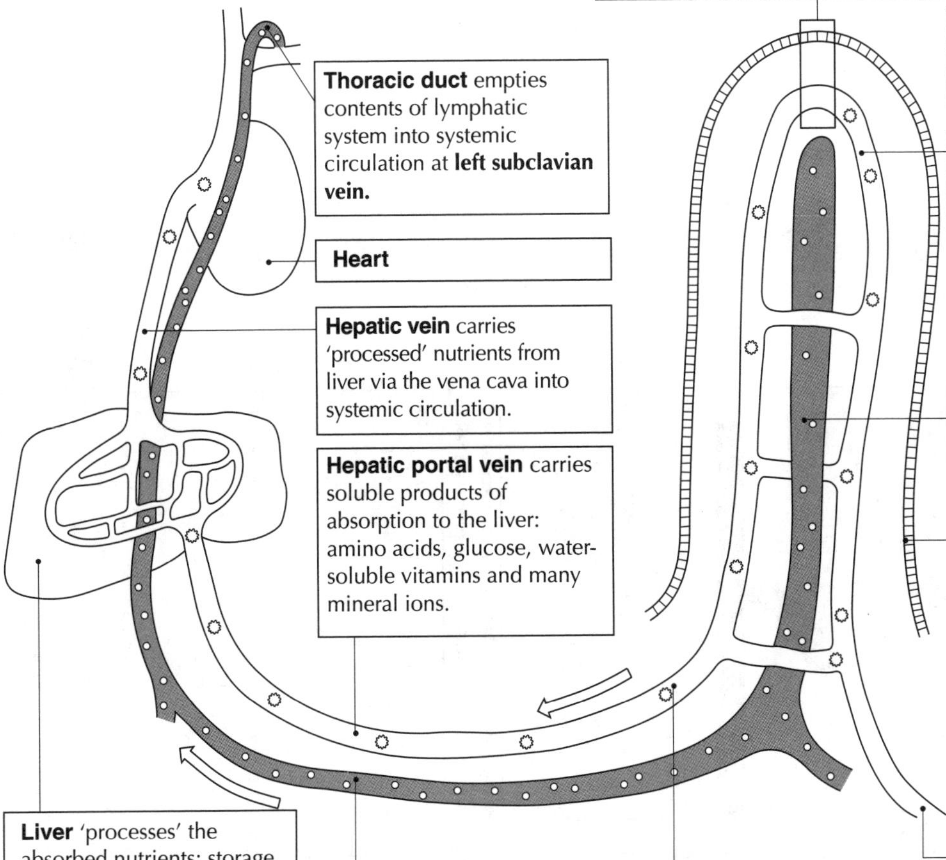

Thoracic duct empties contents of lymphatic system into systemic circulation at **left subclavian vein.**

Heart

Hepatic vein carries 'processed' nutrients from liver via the vena cava into systemic circulation.

Hepatic portal vein carries soluble products of absorption to the liver: amino acids, glucose, water-soluble vitamins and many mineral ions.

Epithelium is the surface for uptake of the products of digestion.

Capillary (transports sugars and amino acids to hepatic portal vein).

Lacteal (transports fats in chylomicrons to lymphatic system).

Villus: very high density in ileum and the duodenum. Villi increase the surface area for absorption, as do **microvilli** of the brush border on the surface of the epithelium.

Arteriole: carrying blood **to** the absorption surface.

Together these keep up a steep diffusion gradient between the epithelium and the capillary.

Liver 'processes' the absorbed nutrients: storage, metabolism or removal means that a variable intake is adjusted to a more constant output via the hepatic vein.

Lymphatic system carries fats, steroids and fat-soluble vitamins.

Venule: carrying blood **away** from the surface.

Surface area to volume ratio

Surface area to volume ratio

Surface area:
- the amount of surface an organism has;
- is shown as (units of length)^2;
- represents the opportunities which the organism for exchange of substances with its environment: it it offers the chance to **supply** the requirements of the organism.

Volume:
- the total space taken up by an organism;
- is shown as (units of length)^3
- represents the amount of each substance an organism will need or will produce: it shows the **demands** the organism puts on its environment.

THE PROBLEM IS MATCHING SUPPLY AND DEMAND

Think of a cell as a cube: the cube has six sides.

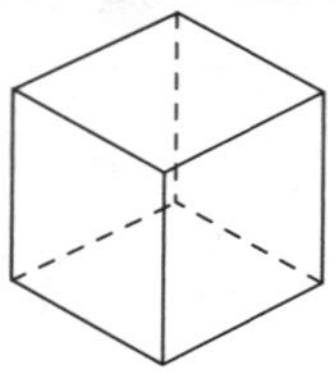

Volume = length × breadth × height

Surface area = length × height × 6

Cube side length (mm)	Surface area (mm^2)	Volume (mm^3)	Surface area to volume ratio
1	6	1	6 : 1
2	24	8	3 : 1
5	150	125	1.2 : 1
7	294	343	0.8 : 1
50	15 000	125 000	0.12 : 1

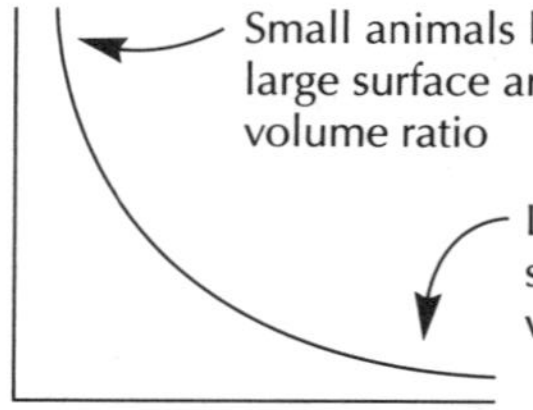

SO → As size increases, surface area to volume ratio falls

AND → As **demands** on the environment increase the opportunity to **supply** requirements falls

This affects the **size** and **shape** of living organisms:

- oxygen must be supplied for respiration
- heat generated by metabolism must be lost to the environment
- wastes such as urea and carbon dioxide must be removed from cells
- cells must be supplied with nutrients such as glucose, amino acids and mineral ions.

WHAT HAPPENS IS

- large organisms are **multicellular** - many small cells have a much greater combined surface area than one large one of the same total volume
- large organisms develop **specialised exchange surfaces** to improve the S.A. to Vol. ratio
- large organisms develop **mass flow transport systems** so that the exchange surfaces can work most efficiently.

And as for fat-hydrolysing enzymes!

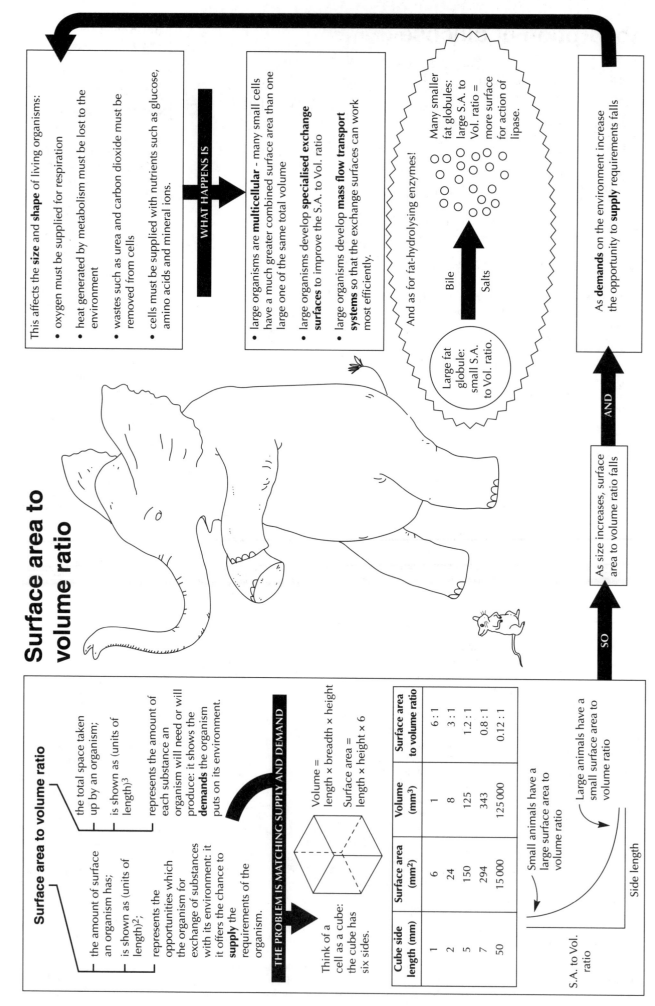

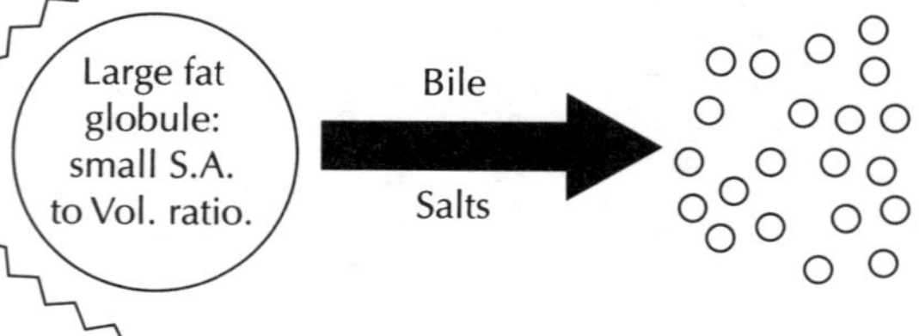

Section D Gas exchange

Lung structure and function may be affected by disease

Larynx containing vocal cords which snap shut during hiccoughs.

Trachea with C-shaped rings of cartilage to support trachea in open position when thoracic pressure falls.

Lining of ciliated epithelium has **goblet cells**, which secrete mucus, to trap and remove pathogens and particles of dust and smoke.

Bronchitis – inflammation of the linings of the bronchi

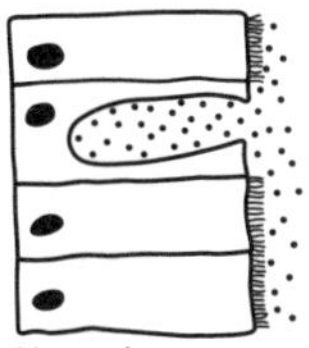

Normal

- increased growth and activity of mucus-secreting cells
- loss or inactivation of ciliated epithelium.

coughing and **production of sputum**

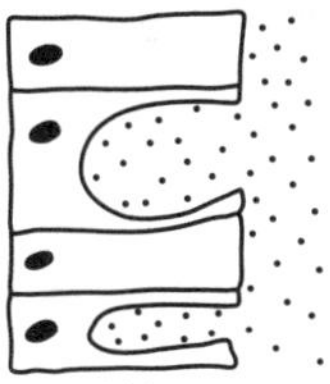

Bronchitic

Mechanical damage to lung tissue

+

Provision of nutrient supply for bacteria

greater risk of lung infections

- Mortality from chronic bronchitis is about 40 × higher in 25-per-day smokers than in non-smokers.

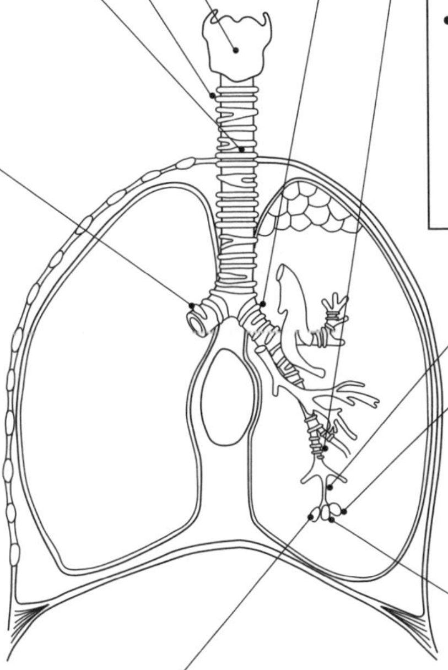

Bronchus with cartilaginous rings to prevent collapse during inspiration.

Asthma (Bronchospasm) reduces the flow of air in the bronchioles:

- caused by the contraction of smooth muscle in bronchiole walls;
- may be triggered by tobacco smoke, pet fur and by physical stress;
- relieved by bronchodilators (drugs which relax the smooth muscle). These often mimic the hormone **adrenaline**.

normal, relaxed bronchiole

asthmatic, constricted bronchiole

Terminal bronchiole with no cartilage support.

Alveolus, which is the actual site of gas exchange between air and blood.

Lung fibrosis occurs when the alveoli become inflamed and scar tissue develops as the lung attempts to repair itself:

- hardening of lung tissue prevents free flow of air, causing shortness of breath
- caused by inhaled particles (e.g. of coal dust or asbestos) which phagocytes ingest but cannot digest
- complications include right-sided heart failure as oxygenation of blood is reduced.

Cystic fibrosis is an inherited condition in which an overproduction of sticky mucus reduces airflow through bronchi and bronchioles, and may cause lung infections.

Emphysema – destruction of lung tissue, thus reducing surface area available for oxygen uptake / CO_2 excretion:

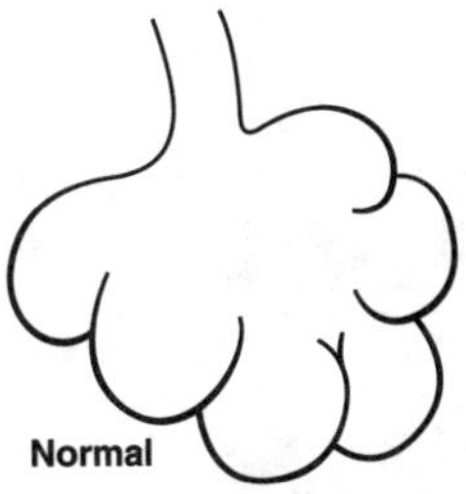

Normal

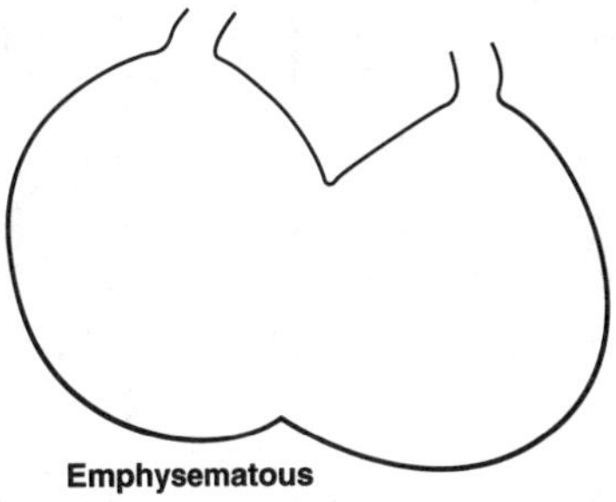

Emphysematous

- some component of tobacco smoke encourages neutrophils to accumulate in lung tissue
- neutrophils move through lung tissue by secreting enzymes (proteases and elastases) – 'repair' mechanisms, including α_1-antitrypsin, are inhibited by tobaco smoke
- infection of damaged tissue stimulates further invasion by neutrophils, making the situation worse
- emphysema is 20 × more common in 25-per-day smokers than in non-smokers.

Epidemiology: patterns in the distribution of disease.

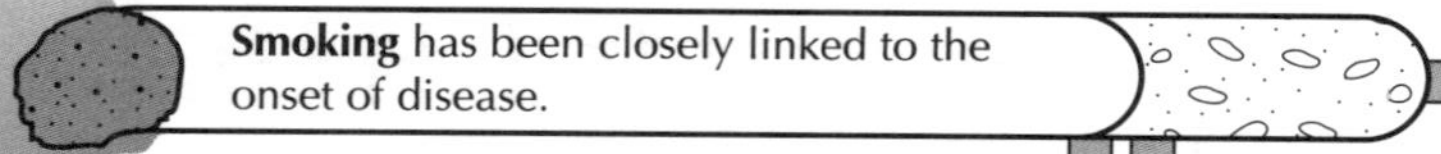

Smoking has been closely linked to the onset of disease.

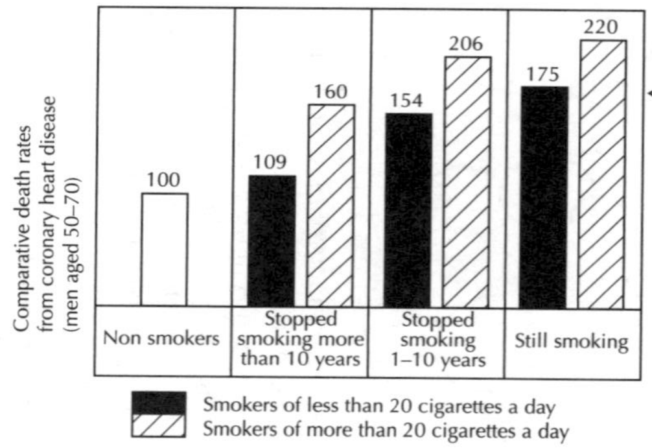

The relationship between smoking and coronary heart disease

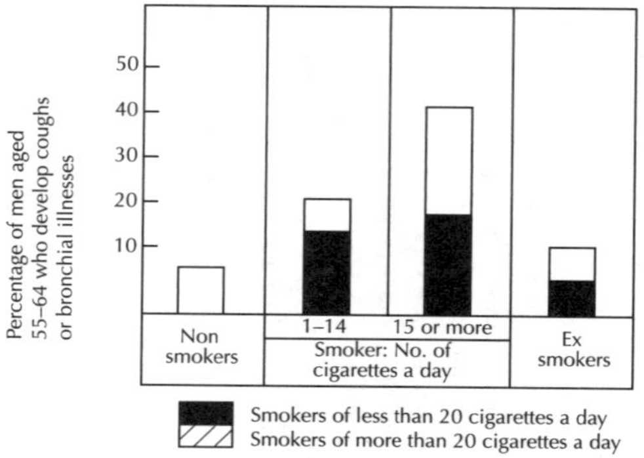

The effect of smoking on the chance of developing bronchitis or coughing with phlegm

Epidemiology is essential!
It would be unethical to force a human to smoke in order to record whether disease resulted!

Evidence linking smoking with lung cancer

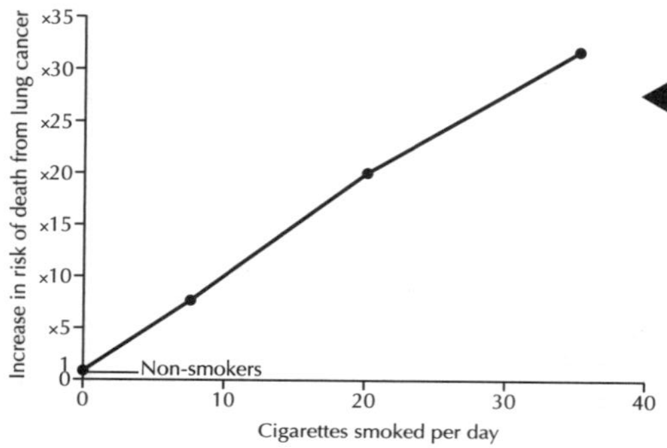

Results of a **prospective study** – looking forward and suggesting that if smoking was an important cause of lung disease and death, then early death would be more likely in smokers than in non-smokers.

A correlation between two variables does not necessarily prove that one of the variables causes the other – it might be that heavy smokers develop lung cancer because they are exposed to some other environmental factor, or that individuals with a genetic make-up that carries a risk of lung cancer are also 'genetically' more likely to smoke. The work of Richard Doll and others is so convincing because it has eliminated many other environmental factors. However, it is likely that the development of lung cancer is a complex process. It is probably multifactorial, involving both genetic and lifestyle factors.

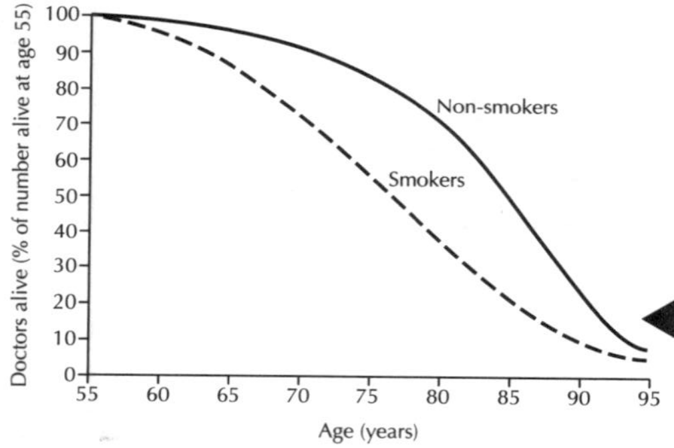

Results of a **retrospective study** – looking back at the lifestyle, occupation and environment of people who have died from lung cancer.

In 1962 the Royal College of Physicians published a report on smoking and health, which suggested a clear link between cancer and smoking. For example, among a sample of doctors living in similar-sized cities, those who smoked regularly were more likely to develop cancer of the lung.

Epidemiology has also linked smoking to low sperm counts in UK men!

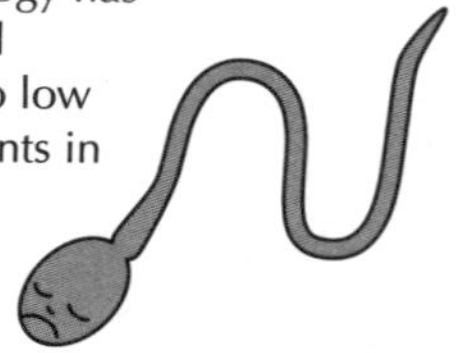

Some disturbing data

Fewer boys are taking up smoking in the UK than 20 years ago (good news) but more girls are now smoking (bad news). Can you suggest reasons for these changes?

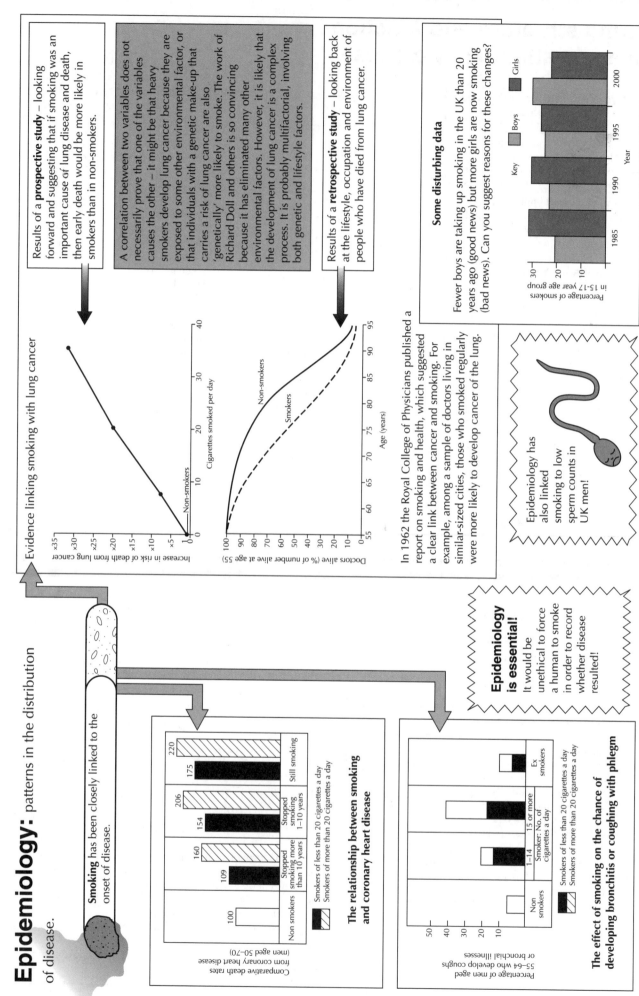

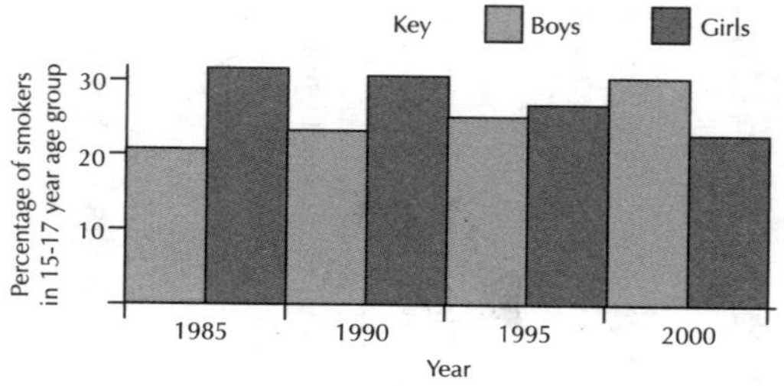

Principles of respiration:

a number of processes are involved in the provision/consumption of **oxygen** and the excretion/production of **carbon dioxide.**

- **Pulmonary ventilation** moves gases between atmosphere and respiratory surface.
- **External respiration** occurs when gases diffuse across the respiratory surface.
- **Internal respiration** occurs when gases diffuse between circulating blood and respiring cells.
- **Tissue/cell respiration** occurs when oxygen is consumed and carbon dioxide is produced during the oxidation of foods to release energy.

Respiratory medium supplies **oxygen** and accepts excreted **carbon dioxide.**

Respiratory surface across which **oxygen absorption** and **carbon dioxide release** occurs is **thin** (minimum diffusion distance only 0.5 µm in humans), **moist** (so that diffusion may occur in solution), and of **great surface area** (about 70 m² in humans).

Ventilation is the movement of the respiratory medium to and from the respiratory surface: this helps to maintain adequate concentration gradients of O_2 and CO_2 so that diffusion may take place across the respiratory surface.

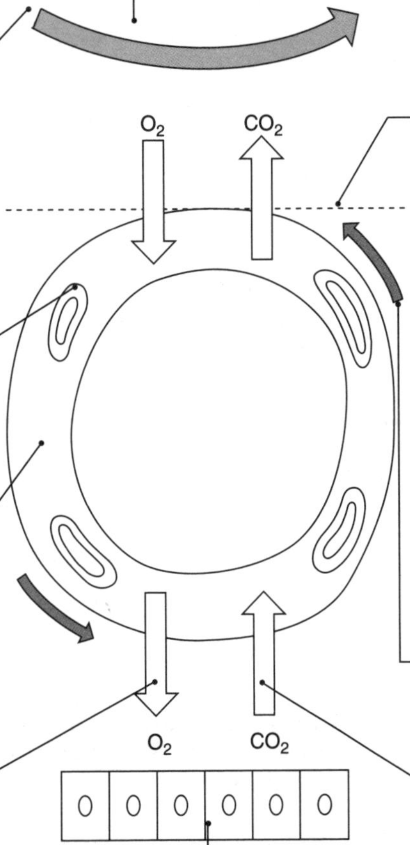

Erythrocytes (red blood cells) contain haemoglobin, which plays a part in both O_2 and CO_2 transport.

In fish, countercurrent flow of blood and water improves gas exchange across gills:

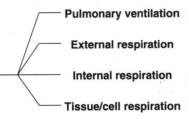

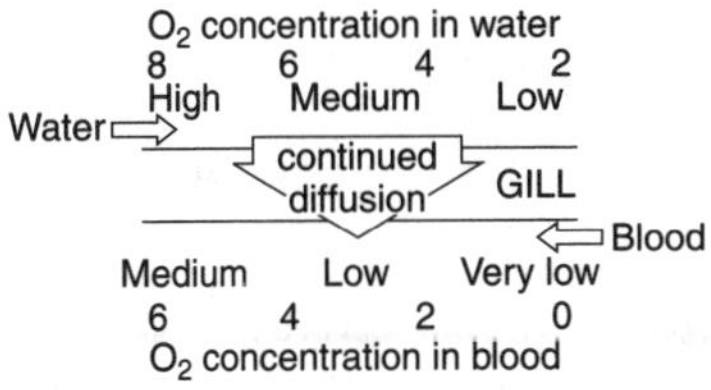

Because blood and water flow in opposite directions, there is an oxygen concentration gradient over the whole gill lamella.

Oxygen is transported in the circulating blood: approx. 98% as $Hb(O_2)_4$ **(oxyhaemoglobin)** in the erythrocytes; approx. 2% in solution in the plasma.

Circulating blood returns CO_2 to respiratory surface: approx. 85% as HCO_3^- **carried** in the plasma but **formed** in erythrocytes; approx. 10% as $Hb\text{-}CO_2$ **(carbamino-haemoglobin)** carried in erythrocytes; approx. 5% dissolved in the plasma.

Oxygen is released from oxyhaemoglobin under conditions of low oxygen tension and diffuses to respiring cells via the tissue fluid.

Respiration in cells: most efficient energy release demands oxygen and produces carbon dioxide.
e.g. **Glucose + oxygen → energy release + carbon dioxide + water**

Carbon dioxide is a product of cell respiration and diffuses from the respiring cell, via the tissue fluid, to the circulating blood.

Feature	Amoeba	Insect	Mammal	Fish	Compare with plants obtaining CO_2
Respiratory medium	Water	Air	Air	Water	Some air/ some water
Surface	Cell membrane	Spiracles/Tracheoles contact cells directly	Alveoli of lungs	Filaments/Lamellae of gills	Mesophyll in leaf
Transport system	None	None	Blood in **double circulation**	Blood in **single circulation**	Diffusion directly to cells
Ventilation	None	Little – some	Negative pressure	Muscular **abdominal**	None
system	movements drive	**movement**	initiates **tidal** low of air	**one-way** flow of water	

Fine structure of the lung: exchange of gases in the alveolus requires:

1. A tube for the movement of gases to and from the atmosphere (e.g. **bronchiole).**
2. A surface across which gases may be transported between air and blood (i.e. **alveolar membrane).**
3. A vessel which can take away oxygenated blood or deliver carboxylated blood (i.e. a **branch of the pulmonary circulation).**

Inspired air

Expired air

Terminal bronchiole has no rings of cartilage and collapses when external pressure is high – dangerous when diving as trapped air in alveolus may give up nitrogen to blood, where it forms damaging bubbles.

Alveolar duct (atrium)

Alveolus (air sac)

Elastic fibres in alveolus permit optimum extension during inspiration – properties are adversely affected by tobacco smoke → **emphysema.**

Branch of pulmonary artery delivers deoxygenated blood to the alveolar capillaries.

The lung is an excellent gas exchange surface:

- diffusion distance is short
- there is an enormous surface area, so more gas molecules can cross in a given amount of time
- an efficient ventilation system, and an extensive circulation, combine to keep up large diffusion gradients in the right direction.

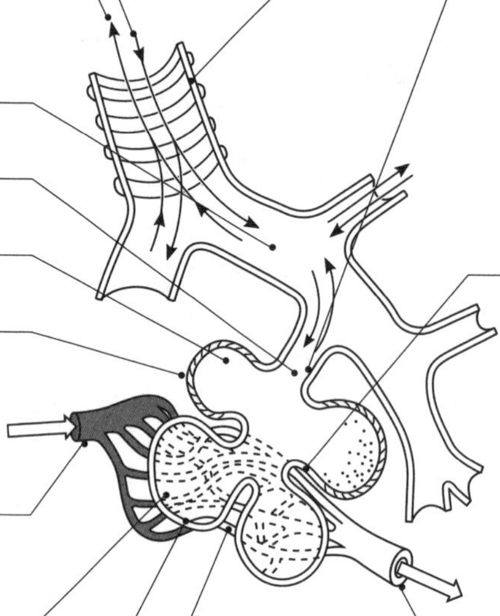

Bronchiole has supporting rings of cartilage to prevent collapse during low pressure phase of breathing cycle.

Smooth muscle in the bronchioles allows the diameter to be controlled. Adrenaline causes relaxation and allows better airflow during exercise.

Alveolar-capillary (respiratory) membrane consists of:

- **Alveolar wall:** squamous epithelium and alveolar macrophages;
- **Epithelial** and **capillary basement membranes;**
- **Endothelial cells of the capillary wall.**

Despite the number of layers this membrane averages only 0.5 μm in thickness.

Tributary of pulmonary vein returns oxygenated blood to the four pulmonary veins and thence to the left atrium of the heart.

Alveolar capillaries adjacent to the alveolus are the site of oxygen and carbon dioxide transfer between the air in the alveolus (air sac) and the circulating blood.

Stretch receptors provide sensory input, which initiates the **Hering-Breuer reflex** control of the breathing cycle.

Changes in composition of inspired and expired air

	Inspired	Alveolar	Expired	
O_2	20.95	13.80	16.40	Oxygen diffuses from alveoli into blood: expired air has an increased proportion of oxygen due to additional oxygen added from the anatomical dead space.
CO_2	0.04	5.50	4.00	Carbon dioxide concentration in alveoli is high because CO_2 diffuses from blood: the apparent fall in CO_2 concentration in expired air is due to dilution in the anatomical dead space.
N_2	79.01	80.70	79.60	The apparent increase in the concentration of nitrogen, a metabolically inert gas, is due to a **relative** decrease in the proportion of oxygen rather than an **absolute** increase in nitrogen.
$H_2O(g)$	**Vari–able**	**Saturated**		The moisture lining the alveoli evaporates into the alveolar air and is then expired unless the animal has anatomical adaptations to prevent this (e.g. the extensive nasal hairs in desert rats).
Temp.	**Atmos–pheric**	**Body**		Heat lost from the blood in the pulmonary circulation raises the temperature of the alveolar air.

Pulmonary ventilation is a result of changes in pressure inside the thorax (chest cavity).

	Inspiration (breathing in)	Expiration (breathing out)
External intercostal muscles	Contract – pulling ribs upwards and outwards	Relax – permitting rib cage to move downwards and inwards
Diaphragm	Contracts – moves downwards from domed position	Relaxes – elasticity returns to domed position
Liver piston	Moves downwards	Moves upwards

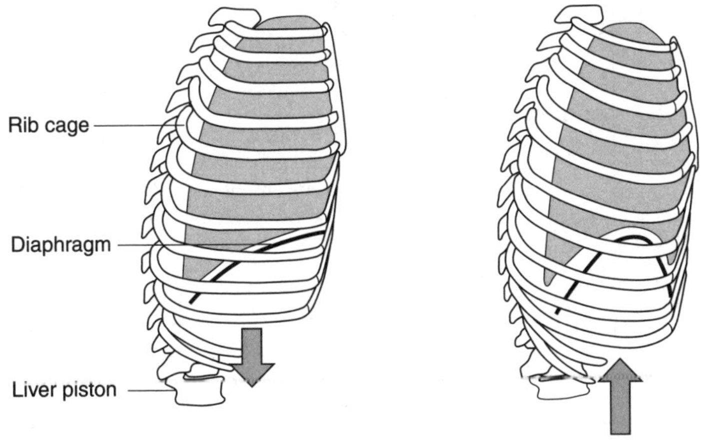

Air pressure in lungs	Decreases	Increases
Air movement along pressure gradient	Into lungs	Out of lungs
Lung volume	Increases	Decreases

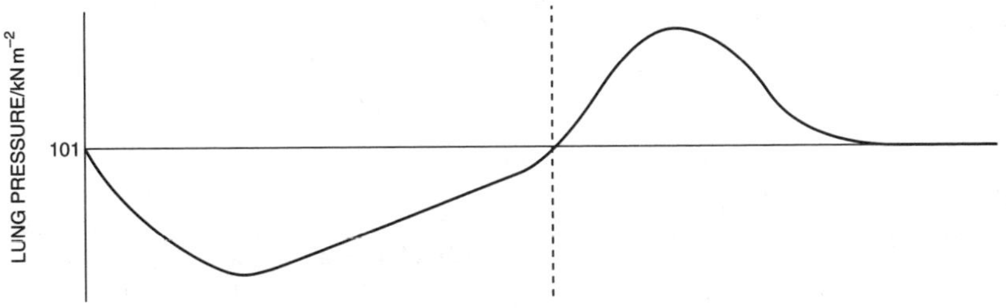

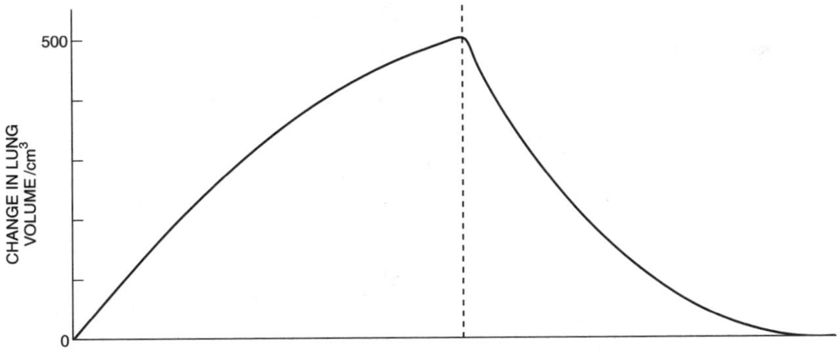

An **active** process which decreases lung pressure so that lungs fill by movement of air down a pressure gradient.

A **passive** process which increases lung pressure so that air moves out of lungs down a pressure gradient.

At rest, when air requirements are low, there is no need for violent breathing movements and the **liver piston** and **diaphragmatic** systems may be sufficient. Heavier demands, for example during exercise, will also require contraction of the external intercostal muscles.

Lungs are passive, elastic structures which are able to 'follow' the movements of the surrounding muscular structures because of the effective negative pressure in the **intrapleural space** between the **pleural membranes.**

Air has a very low density and can therefore respond very quickly to changes in pressure – this permits air-breathing animals to adopt a **tidal system of ventilation.**

The mammalian method involves a **negative pressure system,** i.e. air is **drawn into** the lungs by a fall in pressure within them rather than **forced in** by muscular movements of the mouth. This:
- requires less effort (compare breathing with blowing up a balloon);
- permits an animal to eat and breathe at the same time (gentle air stream does not pull food downwards).

The spirometer and the measurement of respiratory activity

Floating chamber which can be filled with air or oxygen. Chamber moves down when subject inhales and up when subject exhales.

Recording pen which draws a trace on the drum in response to movements of the floating chamber.

Counterbalance which can be adjusted so that floating chamber moves only in response to breathing movements and is unaffected by gravity.

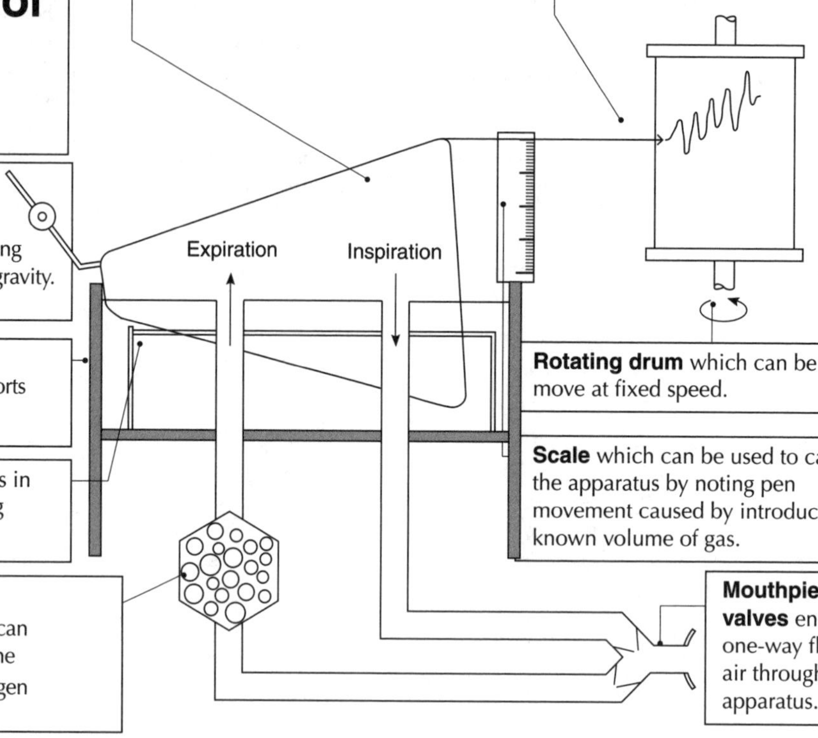

Lower chamber contains water which seals the system and supports the floating chamber.

Support to hold breathing tubes in fixed position within the floating chamber.

Carbosorb is a solid, non-toxic, carbon dioxide absorbant which can remove CO_2 from exhaled air if the apparatus is used to measure oxygen consumption.

Rotating drum which can be set to move at fixed speed.

Scale which can be used to calibrate the apparatus by noting pen movement caused by introduction of known volume of gas.

Mouthpiece with valves ensures one-way flow of air through the apparatus.

Vital capacity = maximum volume of air which can be exchanged from full inspiration to full expiration.

Tidal volume = volume of air exchanged during normal quiet breathing.

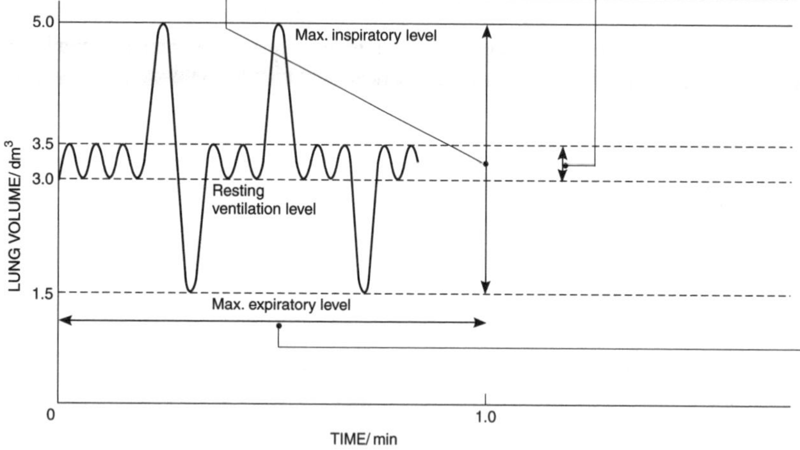

Pulmonary ventilation
(the volume of air available to the lungs per unit of time)

= tidal volume × breathing rate

P.V. can be increased by increasing either or both tidal volume and breathing rate.

Breathing rate = number of breaths in one minute. About 14–16 bpm in a resting adult human.

Measurement of lung volumes by spirometry

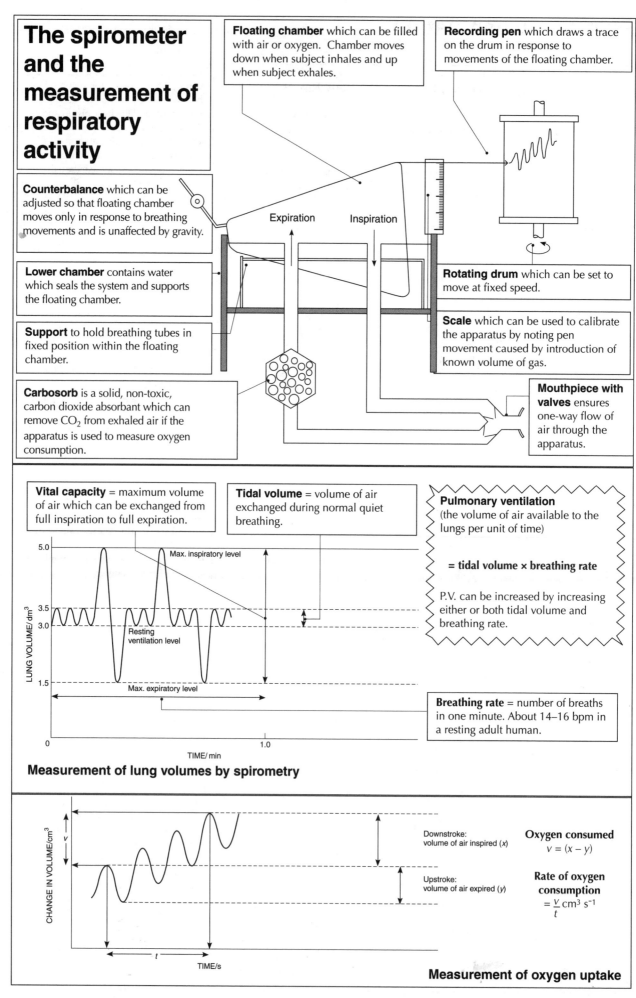

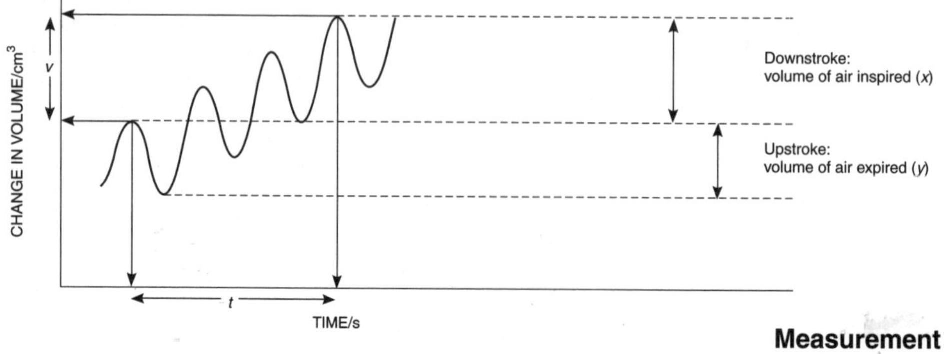

Oxygen consumed
$v = (x - y)$

Rate of oxygen consumption
$= \frac{v}{t}$ $cm^3\ s^{-1}$

Measurement of oxygen uptake

Tuberculosis (TB) is caused by a bacterium, and may be spread from the lungs to other organs.

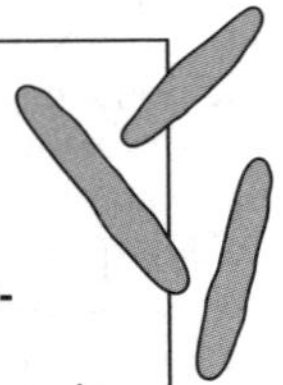

Organism

The most common cause of TB is *Mycobacterium tuberculosis (MTB)*:

- slow growing (divides every 16–20 h.) and aerobic
- has a single phospholipid cell membrane so is **gram-positive**
- cell wall contains complex lipids and glycolipids – one is called the **cord factor** and causes granulomas.

MTB is identified in sputum (phlegm coughed up) by the Ziehl-Nielsen stain. This stains MTB – an 'acid-fast bacillus' – bright red against a blue background.

Infection This occurs when droplets from an actively-infected person are inhaled:

- many (up to 80% of some Asian populations) people carry MTB, but it is suppressed by their immune system.

These people have a **latent infection**, and do not infect other people:

- infection is more likely in crowded conditions, and is very much reduced in areas with efficient ventilation.

Adults with TB have usually had a latent infection which has now overcome the immune system. This is particularly common in AIDS sufferers, whose immune system is weakened.

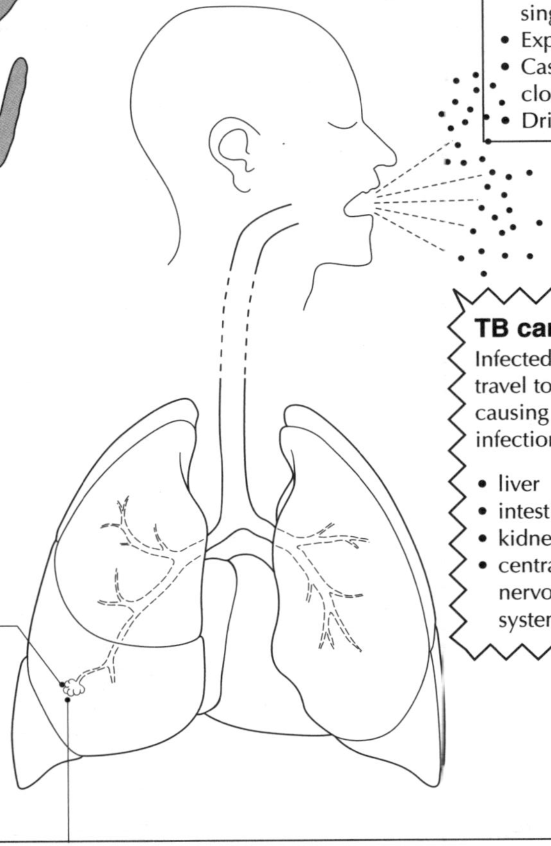

Transmission

- On aerosol droplets produced as lung cells disintegrate when infection is 'active'.
- Droplets are about 1–5μm in diameter, and 40 000 can be released in a single sneeze.
- Expelled when actively-infected people cough, sneeze, spit, speak or kiss!
- Casual contact is unlikely to spread the infection – usually contracted from close family or co-workers.
- Dried 'spit' can release infective droplets e.g. from pavements!

TB can spread!

Infected white blood cells can travel to other parts of the body, causing potentially fatal TB infections outside the lungs:

- liver
- intestines
- kidneys
- central nervous system
- skeleton, causing arthritis
- skin
- eyes
- lymph nodes

Prevention

- BCG (Bacille Calmette-Guérin) vaccine is effective in children, but not in adults.
- Identification of actively-infected people and their contacts. Isolation and treatment removes the source of infection.

Treatment

- Antibiotics (usually rifampicin and/or isoniazid).
- Course lasts from 6–12 months.
- Single antibiotic for latent TB, dual antibiotics for active TB.
- Some risk of liver damage, especially with dual antibiotic courses.
- Drug-resistant TB strains are evolving!

Symptoms and diagnosis

- most cases of active TB involve the lungs so are called **pulmonary TB**;
- give prolonged cough, chest pain and coughing up blood;
- cause weight loss, loss of appetite, pale skin and fatigue.

- X-ray may show 'shadow' on lungs;
- tuberculin skin test picks up antibodies present in infected people;
- 'contact history' with infected persons is investigated.

Invasion involves attachment to the mucous membranes (the surface of the epithelial cells lining the lungs):

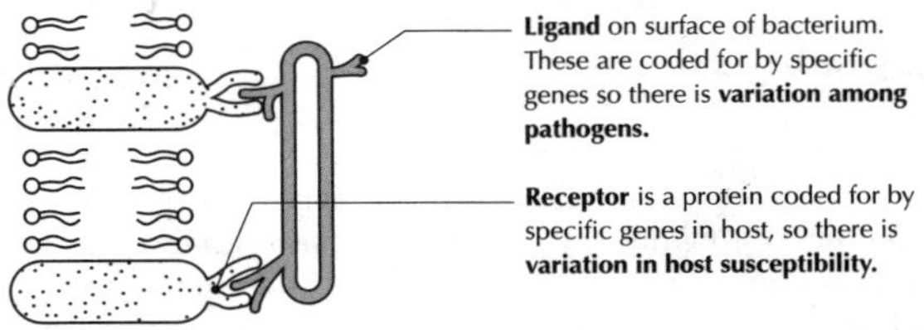
Ligand on surface of bacterium. These are coded for by specific genes so there is **variation among pathogens.**
Receptor is a protein coded for by specific genes in host, so there is **variation in host susceptibility.**

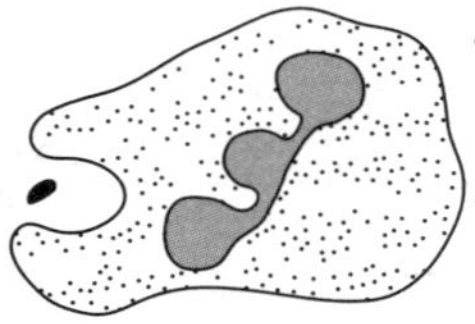

- Bacteria are picked up by white blood cells (**neutrophils** and **macrophages**) in the alveoli.
- Bacteria are resistant to digestive enzymes produced by white blood cells, and multiply inside these cells.
- Infected cells are surrounded by other cells of the immune system to form a **granuloma**, with a white 'cheesy' centre and a harder exterior.

Swollen lymph nodes may cause bronchi to close, and parts of the lung to collapse.
Granuloma cells produce antibodies to the pathogen.

Carbon dioxide transport

The red blood cell and haemoglobin both play a significant part in the transport of CO_2 from tissue to lung, as well as in the transport of oxygen.

At the respiring tissue: carbon dioxide produced in the mitochondria diffuses out of the cells, through the plasma, and into the erythrocytes (red blood cells), where it combines with water to produce carbonic acid, H_2CO_3, under the influence of the enzyme carbonic anhydrase. This can dissociate into H^+ and HCO_3^-. ①

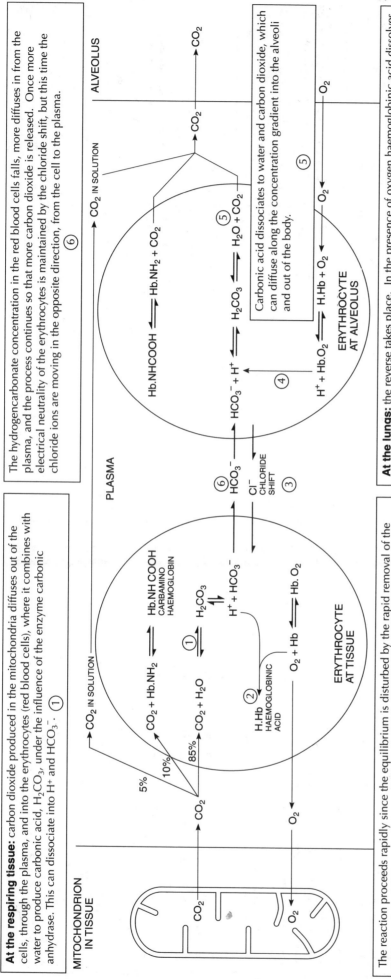

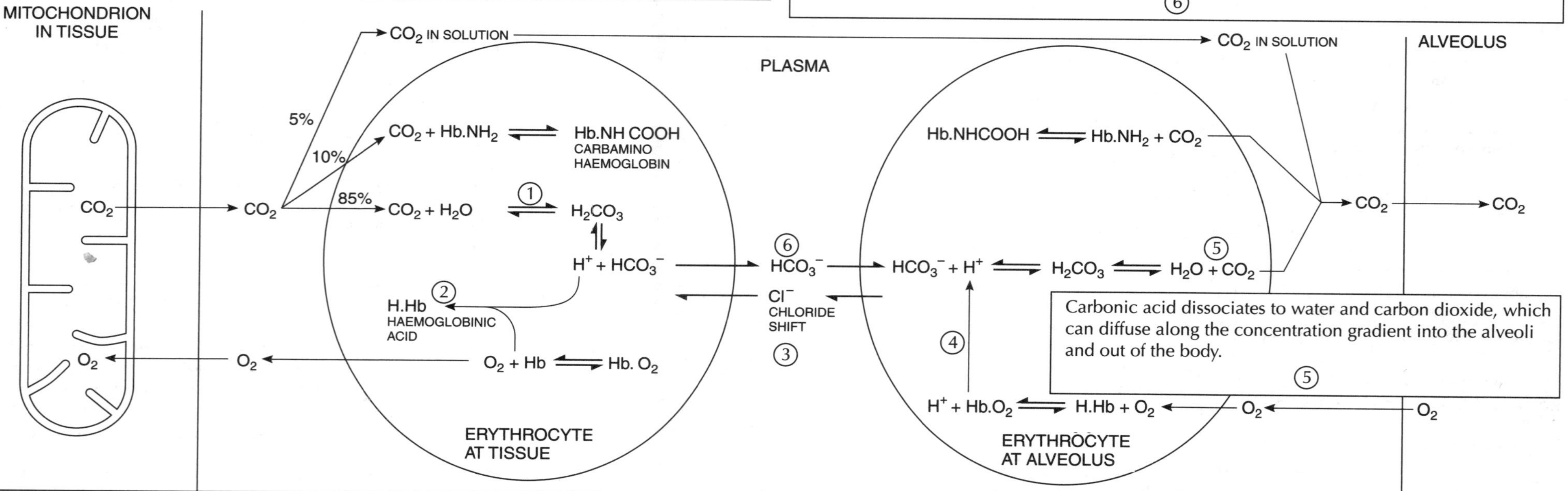

The reaction proceeds rapidly since the equilibrium is disturbed by the rapid removal of the hydrogen ions (H^+) by association with haemoglobin to form haemoglobinic acid (H.Hb). By accepting hydrogen ions in this way haemoglobin is acting as a buffer, permitting the transport of large quantities of carbon dioxide without any significant change in blood pH. ②

As a result of these changes the hydrogencarbonate concentration in the erythrocyte rises and these ions begin to diffuse down a concentration gradient into the plasma. However, this movement of negative ions is not balanced by an equivalent outward flow of positive ions since the membrane of the erythrocyte is relatively impermeable to sodium and potassium ions, which are therefore retained within the cell. This could potentially be disastrous, since positively charged erythrocytes would repel one another, a situation which would not enhance their function as oxygen carriers in the confines of a closed circulatory system! The situation is avoided, and electrical neutrality maintained, by an inward diffusion of chloride ions from the plasma sufficient to balance the HCO_3^- moving out. This movement of chloride ions to maintain erythrocyte neutrality is called the **chloride shift.** ③

At the lungs: the reverse takes place. In the presence of oxygen haemoglobinic acid dissolves so that oxyhaemoglobin ($Hb.O_2$) may be formed. This releases hydrogen ions which combine with hydrogencarbonate from the plasma to produce carbonic acid. ④

Carbonic acid dissociates to water and carbon dioxide, which can diffuse along the concentration gradient into the alveoli and out of the body. ⑤

The hydrogencarbonate concentration in the red blood cells falls, more diffuses in from the plasma, and the process continues so that more carbon dioxide is released. Once more electrical neutrality of the erythrocytes is maintained by the chloride shift, but this time the chloride ions are moving in the opposite direction, from the cell to the plasma. ⑥

$Hb.O_2$ is a simplified abbreviation for oxy-haemoglobin. In fact each haemoglobin molecule combines with $4O_2$: oxyhaemoglobin is really $Hb(O_2)_4$ or $Hb.O_8$.

Disorders of the heart and circulation

Aneurysms are thin, weakened sections of the wall of an artery or a vein that bulges outward forming a balloon-like sac. Aneurysms may be congenital or may be caused by atherosclerosis. The vessel may become so thin that it bursts causing massive haemorrhage, pain and severe tissue damage. Unruptured aneurysms may exert pressure on adjacent tissues e.g. aortic aneurysm may compress the oesophagus, causing difficulty in swallowing.

Thrombosis is obstruction of a blood vessel by an abnormal mass of material (a thrombus) such as a fragment of blood clot, atheromatous plaque, bone fragment or air bubble. A thrombus may lodge in the coronary vessels, the lungs or the liver causing restricted circulation and, perhaps, death.

Myocardial Infarction (commonly called 'heart attack'): Infarction means the death of an area of tissue following an interrupted blood supply. The tissue beyond the obstruction dies and is replaced by non-contractile tissue – the heart loses some of its strength. In **chronic ischaemic heart disease** many small infarcts form and gradually lead to myocardial weakness, but in **acute ischaemic heart disease** the infarct is large since a large artery is closed – death may occur suddenly for a number of reasons:

1 Acute cardiac failure due to shock
2 Severe arrhythmia due to disruption of the conducting system
3 Rupture of the ventricle wall, usually within two weeks of the infarction.

Recent studies suggest that much tissue damage occurs when blood flow is **restored** not when it is **interrupted.** When blood supply is interrupted the affected tissue releases an enzyme called **xanthoxidase** – when blood supply is restored the enzyme reacts with oxygen in the blood to produce **superoxide free radicals** which cause enormous tissue damage by protein oxidation. Drugs such as **Superoxide dismutase** and **Allopurinol**, are used to prevent the formation of the free radicals and limit tissue damage.

Heart disease affects 1 in 4 people between the ages of 30 and 60. high risk factors include:
- Gender (males more at risk) • Diet (high cholesterol, saturated fat and salt) • Cigarette smoking • Obesity
- Lack of regular exercise • High blood pressure.

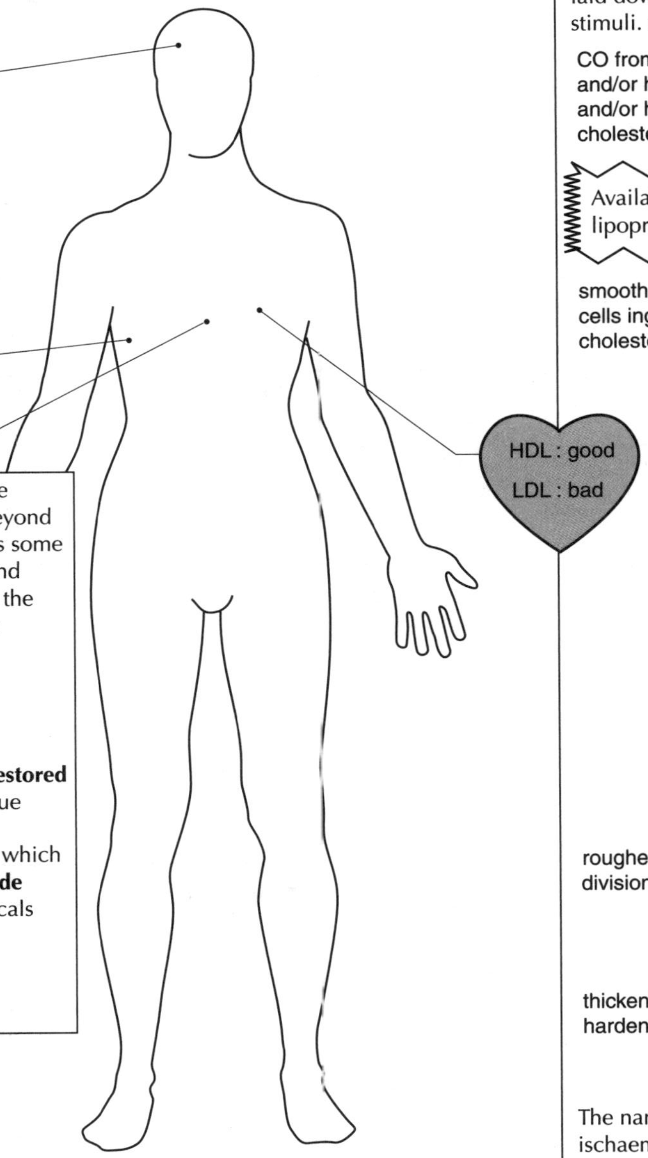

Atherosclerosis is a process in which fatty substances are laid down in the walls of arteries in response to certain stimuli. One possible sequence is:

CO from smoking and/or hypertension and/or high available cholesterol → damage to arterial endothelium → invasion of arterial wall by monocytes

Available cholesterol is reduced by high density lipoprotein (HDL): it is removed from circulation to liver.

smooth muscle cells ingest cholesterol + macrophages which take up cholesterol from low density lipoproteins (LDL) → Atheroma or atherosclerotic plaque

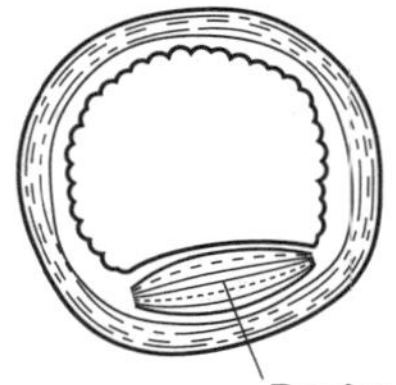

roughened surface causes division of smooth muscle cells → thickening of plaque (often hardened by Ca^{2+} deposition)

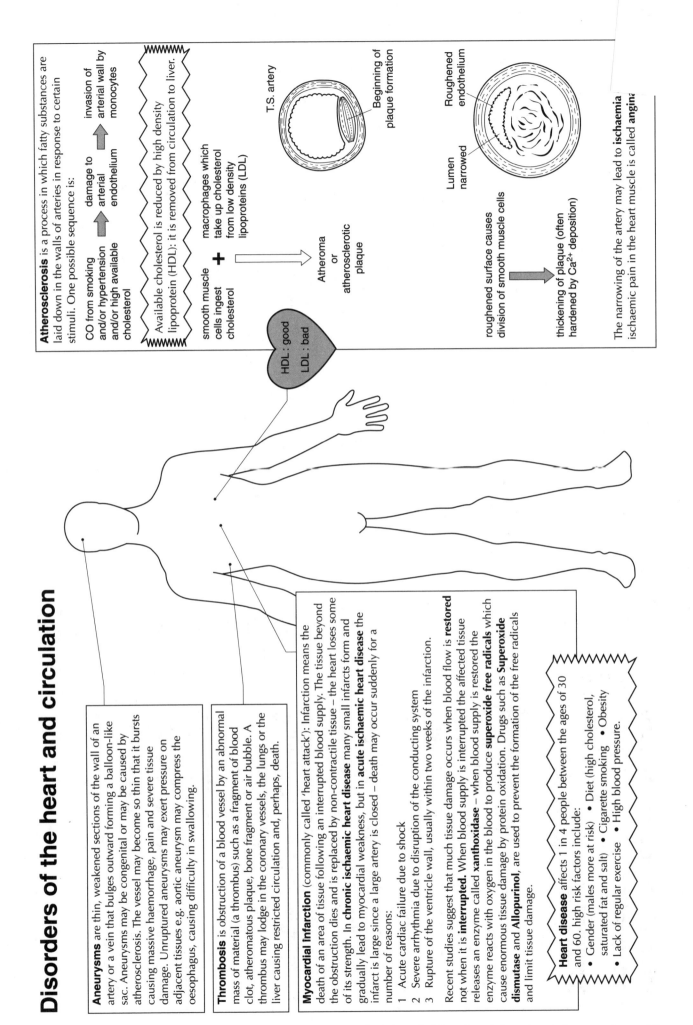

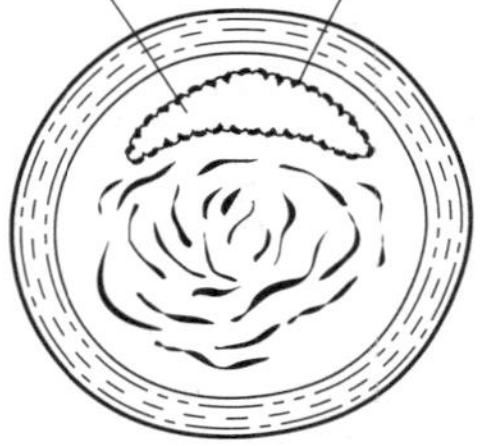

The narrowing of the artery may lead to **ischaemia** ischaemic pain in the heart muscle is called **angin**

Mammalian heart structure and function

Pulmonary (semilunar) valve: has three cusps or watchpocket flaps which are forced together, then the pressure in the pulmonary artery exceeds that in the right ventricle, preventing backflow of blood into the relaxing chambers of the heart.

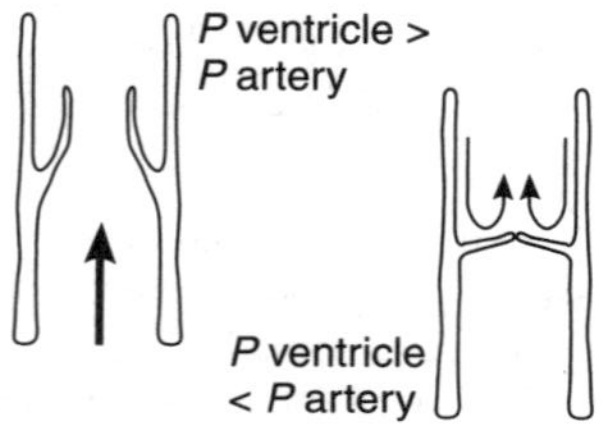

Tricuspid (right atrioventricular) valve: has three fibrous flaps with pointed ends which point into the ventricle. The flaps are pushed together when the ventricular pressure exceeds the atrial pressure so that blood is propelled past the inner edge of the valve through the pulmonary artery instead of through the valve and back into the atrium.

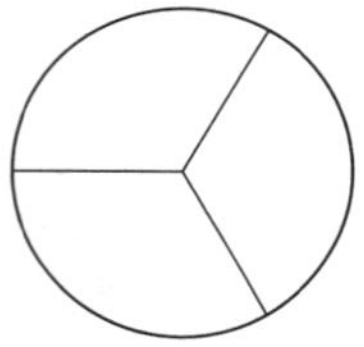

Superior (anterior) vena cava: carries deoxygenated blood back to the right atrium of the heart. As with other veins the wall is thin, with little elastic tissue or smooth muscle. In contrast to veins returning blood from below the heart there are no venous valves, since blood will return under the influence of gravity.

Volume: the same volume of blood passes through each side of the heart. Both ventricles pump the same volume of blood.

This is important to remember! If this wasn't true all the blood would end up in either the body (systemic circuit) or the lungs (pulmonary circuit).

Aorta: carries oxygenated blood from the left ventricle to the systemic circulation. It is a typical elastic (conducting) artery with a wall that is relatively thick in comparison to the lumen, and with more elastic fibres than smooth muscle. This allows the wall of the aorta to accommodate the surges of blood associated with the alternative contraction and relaxation of the heart - as the ventricles contract the artery expands and as the ventricle relaxes the elastic recoil of the artery forces the blood onwards.

Pulmonary arteries

Left atrium

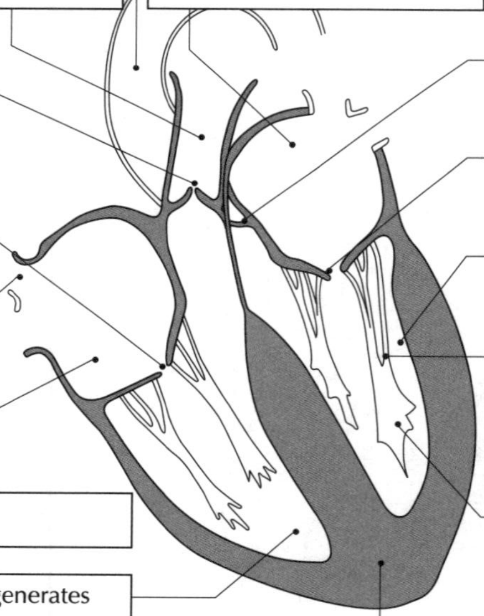

Right atrium

Right ventricle: generates pressure to pump deoxygenated blood to pulmonary circulation.

Myocardium is composed of cardiac muscle: **intercalated discs** separate muscle fibres, strengthen the muscle tissue and aid impulse conduction; **cross-bridges** promote rapid conduction throughout entire myocardium; **numerous mitochondria** permit rapid aerobic respiration. Cardiac muscle is **myogenic** (can generate its own excitatory impulse) and has a **long refractory period** (interval between two consecutive effective excitatory impulses), which eliminates danger of cardiac fatigue.

The pressure generated by the left ventricle must be greater than that generated by the right ventricle as the systemic circuit is more extensive than the pulmonary circuit.

The pressure generated by the atria can be less than that generated by the ventricles since the distance from atria to ventricles is less than that from ventricles to circulatory system.

Aortic (semilunar) valve: prevents backflow from aorta to left ventricle.

Bicuspid (mitral, left atrioventricular) valve: ensures blood flow from left ventricle into aortic arch.

Left ventricle: generates pressure to force blood into the systemic circulation.

Chordae tendinae: short, inextensible fibres – mainly composed of collagen – which connect to free edges of atrioventricular valves to prevent 'blow-back' of valves when ventricular pressure rises during contraction of myocardium.

Papillary muscles: contract as wave of excitation spreads through ventricular myocardium and tighten the chordae tendineae just before the ventricles contract.

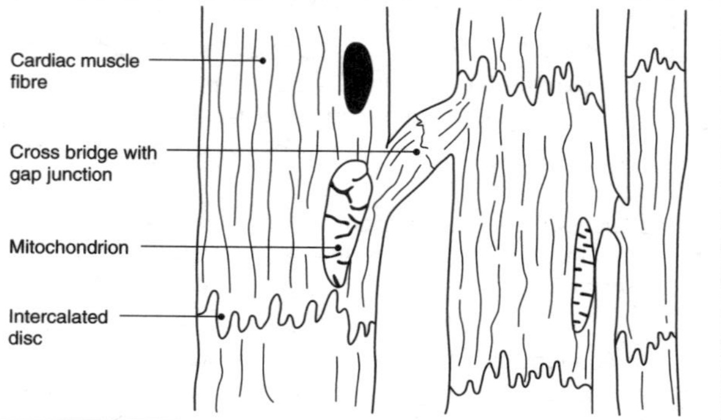

Control of the heartbeat

Autonomic neurones may alter the time it takes to complete a single cycle (heartbeat). A branch of the **vagus** (**parasympathetic**) nerve *decreases* the rate of discharge from the S-A node, whilst the **cardiac/accelerator (sympathetic) nerve** *increases* the rate of the discharge from both S-A and A-V nodes. A typical resting heart rate of 70 b.p.m. indicates that vagal activity dominates – the heart shows **vagal tone**.

Hormones such as thyroxine and adrenaline may increase the rate of discharge from the S-A node.
Drugs such as **beta-blockers** prevent the binding of adrenaline to **ß-receptors** and so may be prescribed to control heartbeat during times of stress, e.g. examinations.

Sino-atrial node: like all nodal cells the cells of this area of the myocardium of the right atrium are **self-excitable (myogenic)** – they may spontaneously and rhythmically generate action potentials *within themselves*. The membranes of cells of the S-A node are especially permeable to Na^+ ions, and are more easily depolarised – this depolarisation is triggered by extension of these cells as blood returns to the heart from the systemic circulation through the venae cavae. The cells of the S-A node have an inherent rate of depolarisation and repolarisation of about 78 cycles per minute which is faster than any other cells in the myocardium. The S-A node thus generates impulses which spread to other areas of the conduction system and myocardium so frequently that they are not able to generate impulses at their own inherent rate. In this way the sino-atrial node acts as the **'pacemaker'** for the rhythmical contraction of the heart.

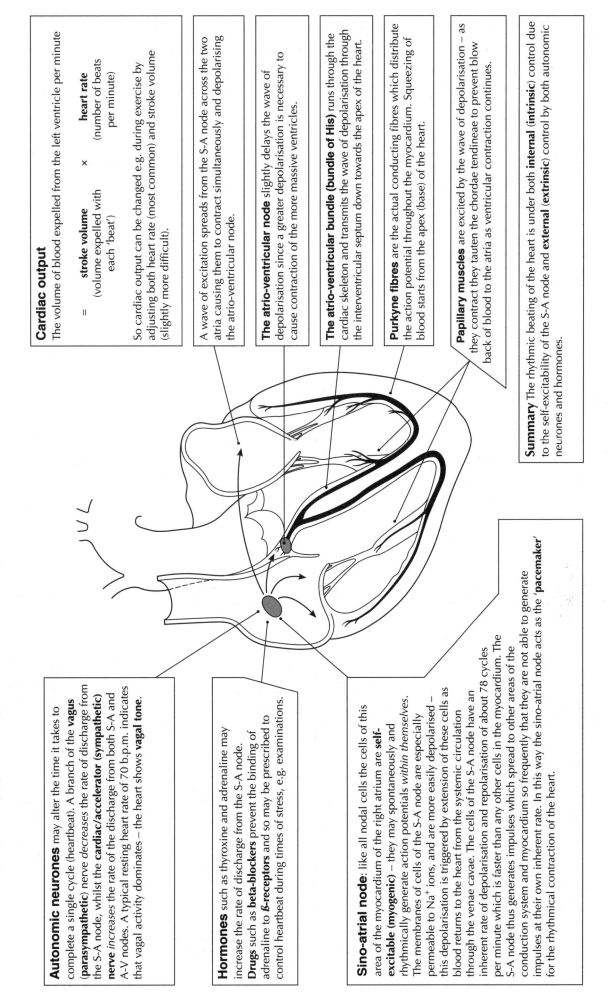

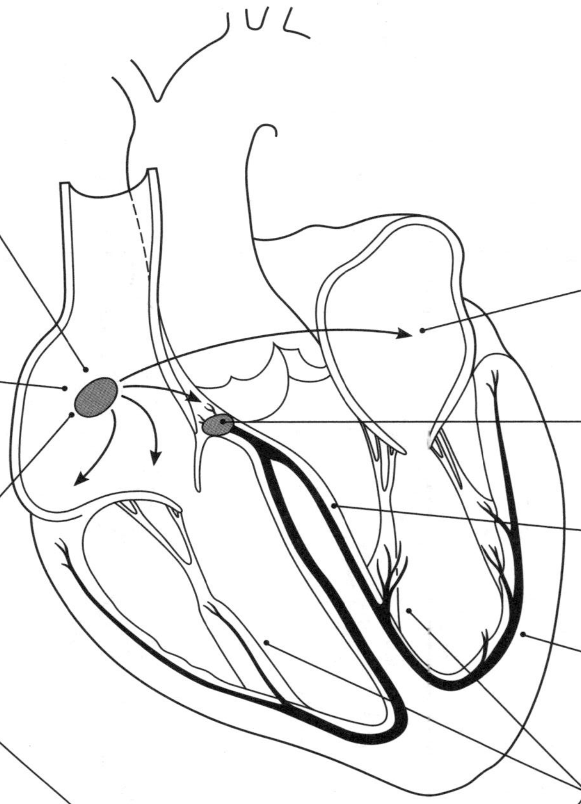

Cardiac output
The volume of blood expelled from the left ventricle per minute

= **stroke volume** (volume expelled with each 'beat') × **heart rate** (number of beats per minute)

So cardiac output can be changed e.g. during exercise by adjusting both heart rate (most common) and stroke volume (slightly more difficult).

A wave of excitation spreads from the S-A node across the two atria causing them to contract simultaneously and depolarising the atrio-ventricular node.

The atrio-ventricular node slightly delays the wave of depolarisation since a greater depolarisation is necessary to cause contraction of the more massive ventricles.

The atrio-ventricular bundle (bundle of His) runs through the cardiac skeleton and transmits the wave of depolarisation through the interventricular septum down towards the apex of the heart.

Purkyne fibres are the actual conducting fibres which distribute the action potential throughout the myocardium. Squeezing of blood starts from the apex (base) of the heart.

Papillary muscles are excited by the wave of depolarisation – as they contract they tauten the chordae tendineae to prevent blow back of blood to the atria as ventricular contraction continues.

Summary The rhythmic beating of the heart is under both **internal** (**intrinsic**) control due to the self-excitability of the S-A node and **external** (**extrinsic**) control by both autonomic neurones and hormones.

Blood flow through the heart is controlled by two phenomena: the opening and closing of the valves and the contraction and relaxation of the myocardium. Both activities occur without direct stimulation from the nervous system; the valves are controlled by the pressure changes in each heart chamber, and the contraction of the cardiac muscle is stimulated by its conduction system.

Inpulse transmission through the heart's conduction system generates electrical currents which can be recorded using an electrocardiograph and presented as an **electrocardiogram** (ECG).

The QRS wave (complex) represents **ventricular depolarisation** and is strong enough to mask atrial repolarisation.

The P wave indicates **atrial depolarisation** – the spread of an impulse from the SA node through the two atria. It is less powerful than the QRS complex since the atria are less massive than the ventricles.

As ventricular pressure exceeds atrial pressure the **semilunar values** open to allow blood from ventricles to arteries.

Contraction of the ventricles without any emptying (note no reduction in volume) causes a sharp rise in ventricular pressure.

Sharp rise in ventricular pressure above atrial pressure closes the **cuspid valves** to prevent backflow of blood from ventricles to atria

Ventricular volume rises as blood moves (passively and by atrial contraction) from atria.

Ventricular volume falls as ventricles contract and semilunar valves open allowing blood flow through to arteries.

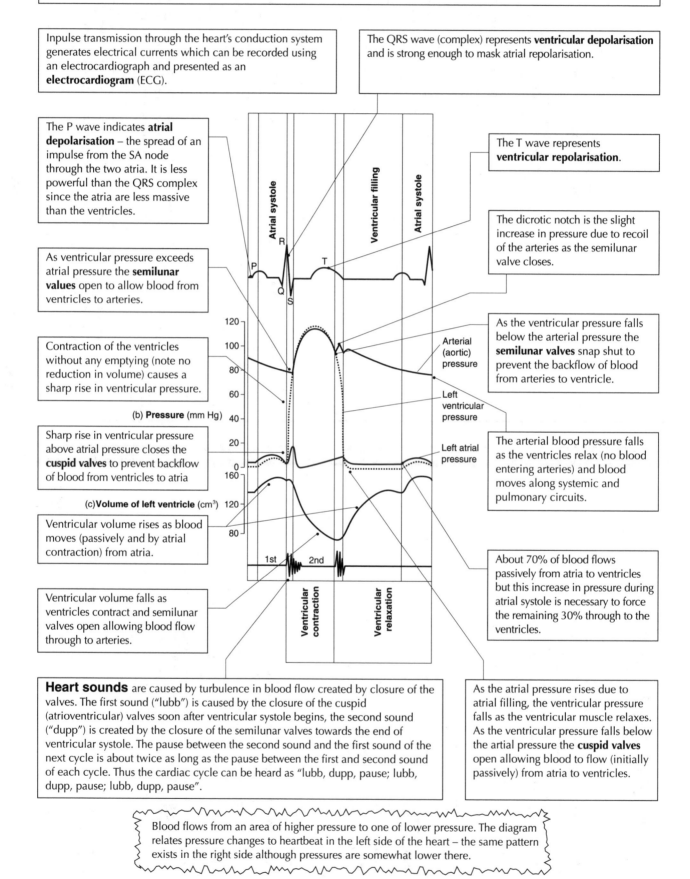

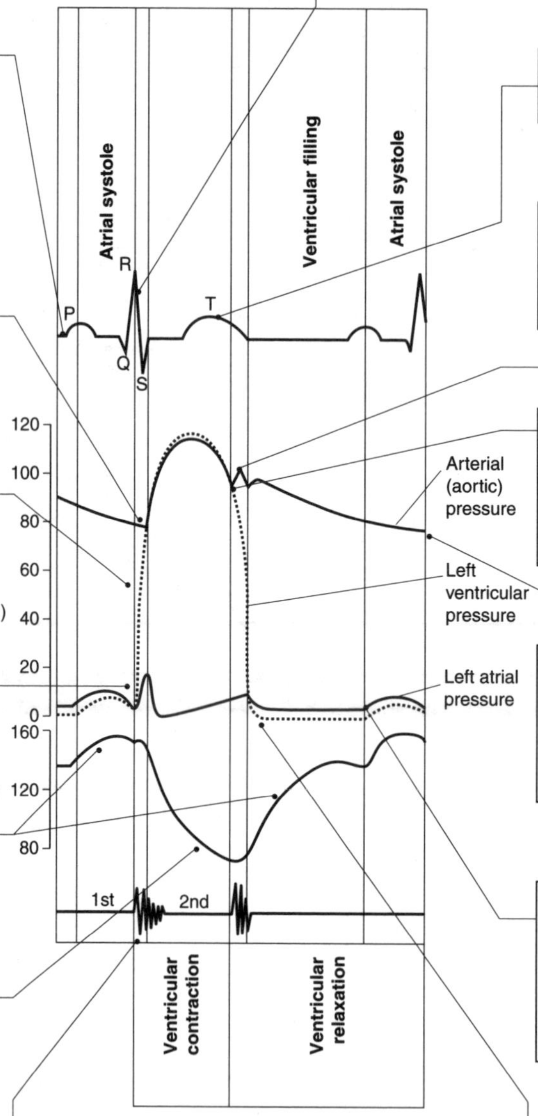

The T wave represents **ventricular repolarisation**.

The dicrotic notch is the slight increase in pressure due to recoil of the arteries as the semilunar valve closes.

As the ventricular pressure falls below the arterial pressure the **semilunar valves** snap shut to prevent the backflow of blood from arteries to ventricle.

The arterial blood pressure falls as the ventricles relax (no blood entering arteries) and blood moves along systemic and pulmonary circuits.

About 70% of blood flows passively from atria to ventricles but this increase in pressure during atrial systole is necessary to force the remaining 30% through to the ventricles.

Heart sounds are caused by turbulence in blood flow created by closure of the valves. The first sound ("lubb") is caused by the closure of the cuspid (atrioventricular) valves soon after ventricular systole begins, the second sound ("dupp") is created by the closure of the semilunar valves towards the end of ventricular systole. The pause between the second sound and the first sound of the next cycle is about twice as long as the pause between the first and second sound of each cycle. Thus the cardiac cycle can be heard as "lubb, dupp, pause; lubb, dupp, pause; lubb, dupp, pause".

As the atrial pressure rises due to atrial filling, the ventricular pressure falls as the ventricular muscle relaxes. As the ventricular pressure falls below the artial pressure the **cuspid valves** open allowing blood to flow (initially passively) from atria to ventricles.

Blood flows from an area of higher pressure to one of lower pressure. The diagram relates pressure changes to heartbeat in the left side of the heart – the same pattern exists in the right side although pressures are somewhat lower there.

Blood cells differ in structure and function.

If blood is spun for a few minutes in a high speed centrifuge it separates into two layers.

Serum is the name given to plasma from which the soluble protein fibrinogen (a protein involved in blood clotting) has been removed.

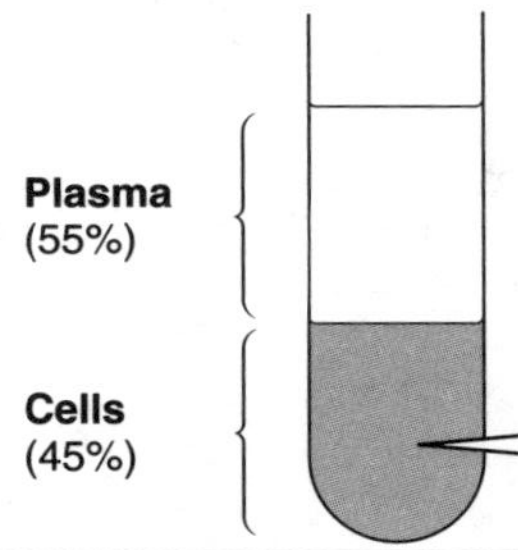

Blood cells originate from **stem cells** in the bone marrow by the process of **haemopoiesis.** This is an example of **differentiation:** cell structure is adapted to allow cells to perform specific functions.

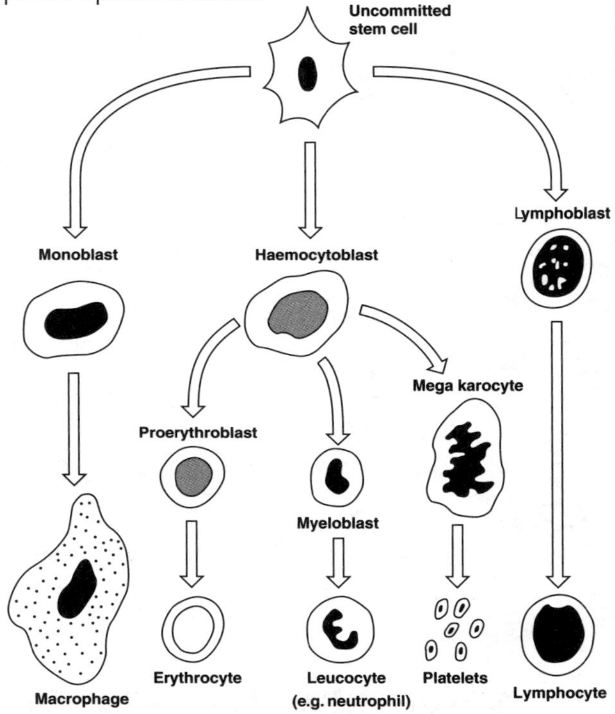

Erythrocytes (red blood cells) are the most numerous of blood cells – about 5 000 000 per mm^3 of adult blood. They function in the transport of O_2 and CO_2, and they contribute to the buffering capacity of the blood. The red colour is due to the presence of the pigment haemoglobin. There are several advantages in packing the haemoglobin into cells rather than leaving it free in the cytoplasm – it keeps the blood viscosity low, it allows the best arrangement of enzymes and solutes for functioning of haemoglobin and it prevents a dramatic reduction in blood water potential. The typical lifespan of a red cell is 90–120 days, before they are destroyed in the spleen.

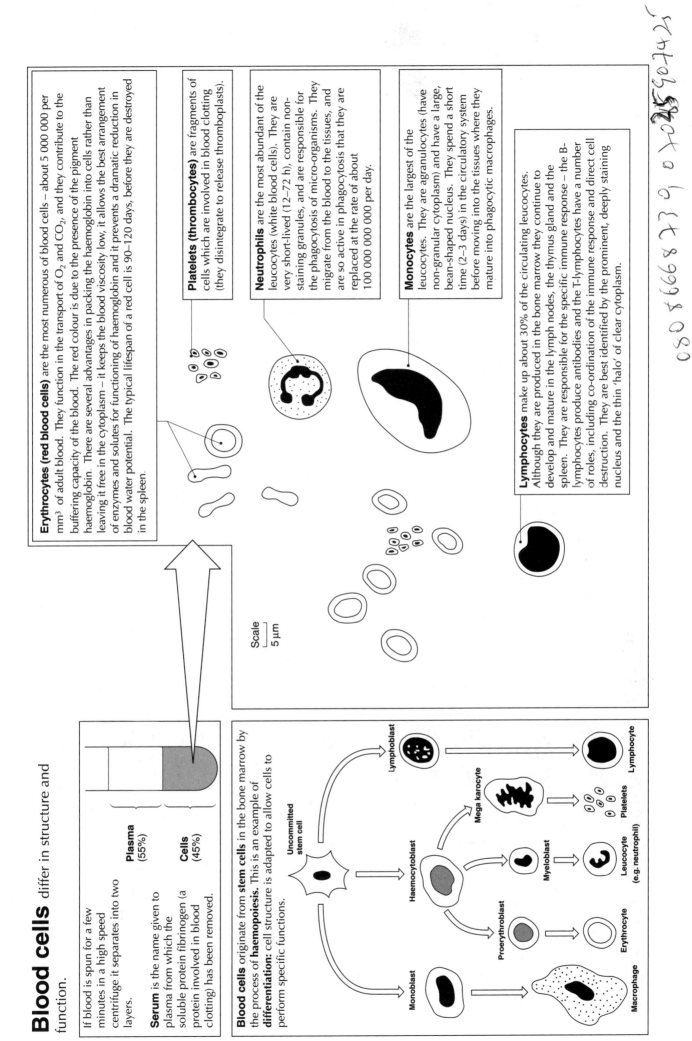

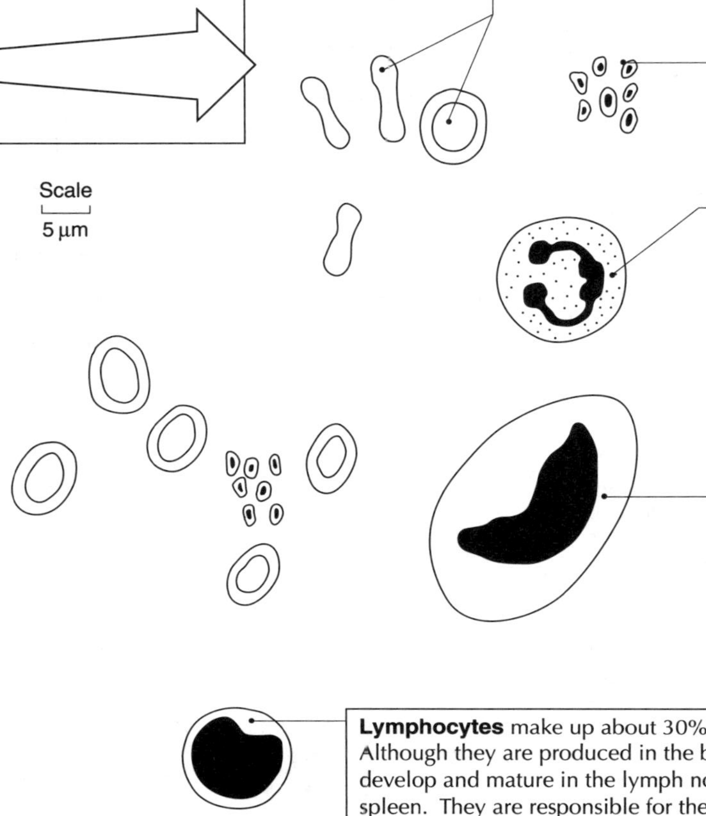

Platelets (thrombocytes) are fragments of cells which are involved in blood clotting (they disintegrate to release thromboplasts).

Neutrophils are the most abundant of the leucocytes (white blood cells). They are very short-lived (12–72 h), contain non-staining granules, and are responsible for the phagocytosis of micro-organisms. They migrate from the blood to the tissues, and are so active in phagocytosis that they are replaced at the rate of about 100 000 000 000 per day.

Monocytes are the largest of the leucocytes. They are agranulocytes (have non-granular cytoplasm) and have a large, bean-shaped nucleus. They spend a short time (2–3 days) in the circulatory system before moving into the tissues where they mature into phagocytic macrophages.

Lymphocytes make up about 30% of the circulating leucocytes. Although they are produced in the bone marrow they continue to develop and mature in the lymph nodes, the thymus gland and the spleen. They are responsible for the specific immune response – the B-lymphocytes produce antibodies and the T-lymphocytes have a number of roles, including co-ordination of the immune response and direct cell destruction. They are best identified by the prominent, deeply staining nucleus and the thin 'halo' of clear cytoplasm.

Monoclonal antibodies: production and applications.

Mouse injected with a mixture of antigens, including antigenO.

Culture of mutant myeloma cells. These cells do not have the machinery to synthesis nucleotides via a 'bypass' route.

B-lymphocytes isolated from spleen include some which can **produce antobody to antigen-O** (=X) B-lymphocytes die after a few days in culture.

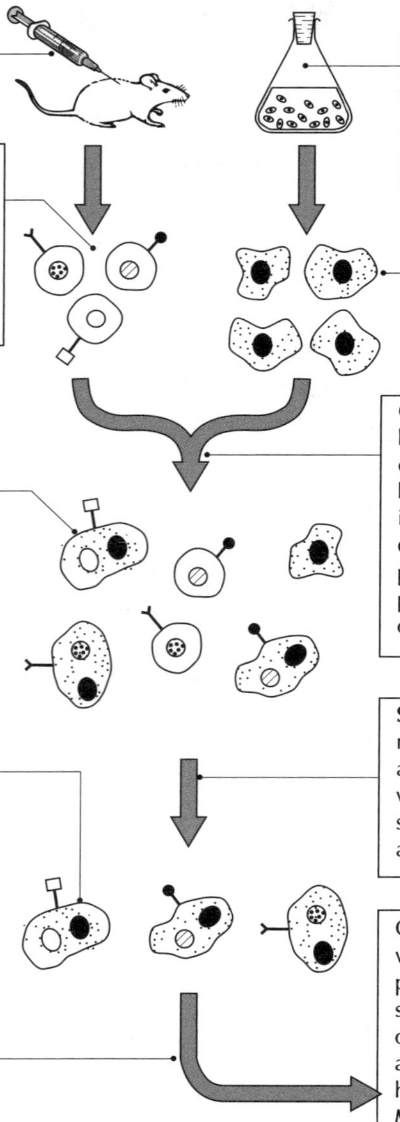

Mutant myeloma cells: **can grow indefinately in a culture medium**.

Mixture contains hybridomas, B-lymphocytes (which soon die) and myeloma cells (which are sensitive to nucleotide inhibitor).

Cell fusion to form **hybridomas** which combine the properties of both types of cell. Fusion is possible because of the dynamic nature of the plasma membrane, and is promoted by the presence of polyethylene glycol.

Only hybridomas are selected: these can **synthesise antibodies** and can **grow indefinately in culture medium.**

Selection by growth on medium which contains aminopterin, an inhibitor which blocks nucleotide synthesis, other than via a bypass pathway.

The hybridomas which can secrete the desired antibody are **selected** by diluting the mixture and **testing each cell line** for antibody production.

Clone of cells which produce and secrete only one type of antibody – hence **Monoclonal antibodies**.

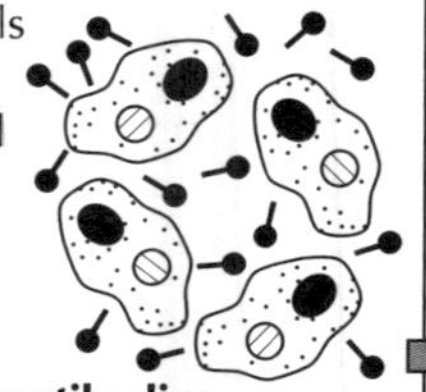

Monoclonal antibodies (MABs) have many applications.

Blood screening: undesirable proteins in blood to be used for blood transfusions may be removed by agglutination with MABs.

Drug targeting: diseased cells, particularly tumour cells, may express unique cell-surface proteins. These cells may be selectively destroyed by cytoxic drugs using an MAB-drug complex. This form of treatment can minimise the side effects of cancer chemotherapy.

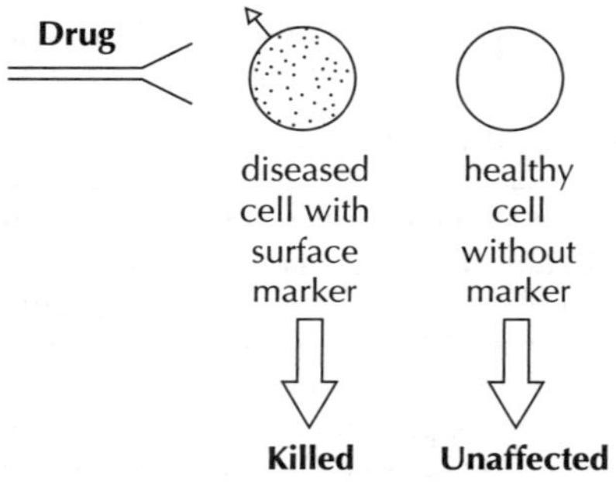

Diagnosis of infectious diseases: MABs can be used to detect antigens on pathogens in samples of body fluids. For example, the STD gonorrhoea is caused by **Neisseria gonorrhoea** which responds to antibiotic treatment, but has similar symptoms to the STD caused by the protozoan, **Chlamydia** which does not respond to **antibiotics**. MABs allow the correct diagnosis of these conditions within 20 minutes and thus allow the correct treatment to be given almost immediately.

Preganancy testing: is based on the interaction between an MAB and human chorionic gonadotrophin in the urine of the women suspected to be pregnant.

MAB here only gives + sign if HCG present in urine sample.

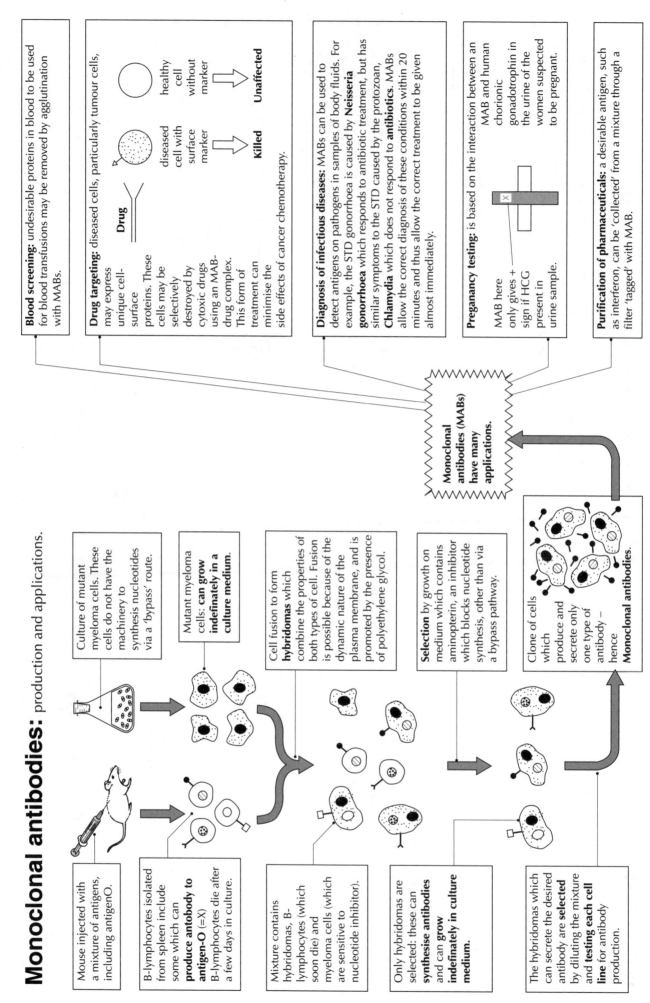

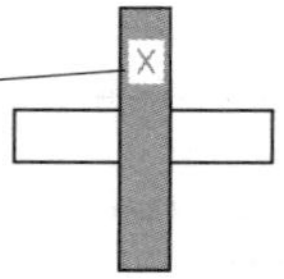

Purification of pharmaceuticals: a desirable antigen, such as interferon, can be 'collected' from a mixture through a filter 'tagged' with MAB.

Immune response I: cells

This involves a wide range of cells and their products in defence against diseases. Cells of the immune system produce **specific immune responses** to a stimulus from the environment.

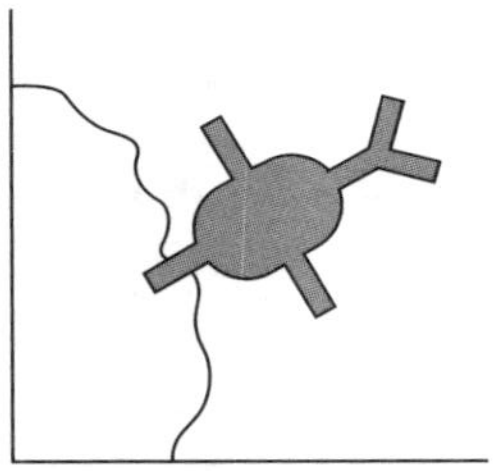

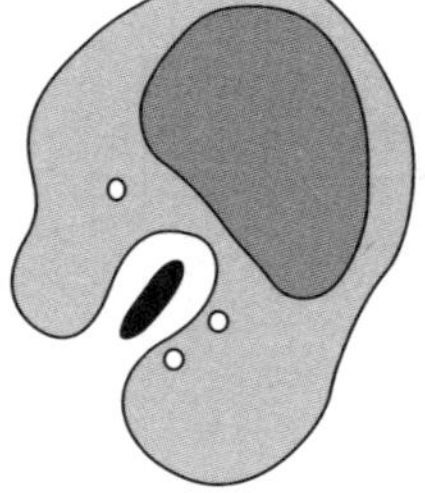

'Self' cells are protected by the presence of **MHC (major histocompatibility complex) protein** on their surface. Autoimmune diseases may result if the immune system fails to recognise self-cells properly.

Stimulus or challenge is an antigen

(molecule, usually protein or carbohydrate, recognised as 'non-self'). The antigen may be engulfed (phagocytosis) by a **macrophage**.

Macrophage may 'present' fragment of antigen (**epitope**) on surface.

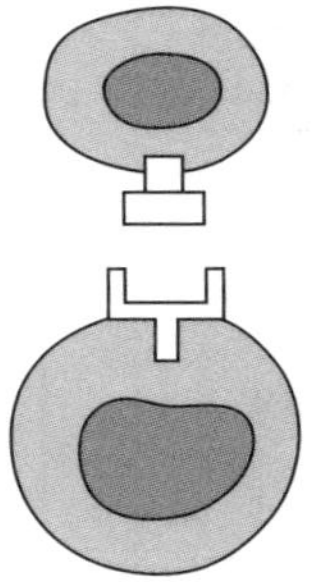

T-helper cell recognises antigen and initiates activity of T and B cells.

Lymphokines/Interleukins are chemical messengers.

Humoral (antibody-mediated) immune response, in which B-lymphocytes secrete **antibodies** to recognise and destroy bacteria, viruses, some fungi and protoctistans.

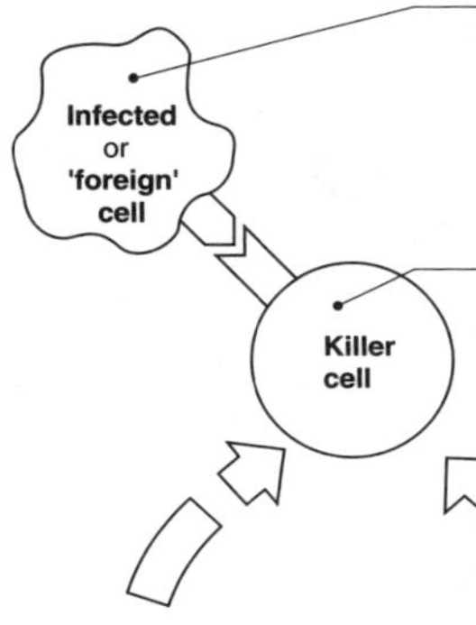

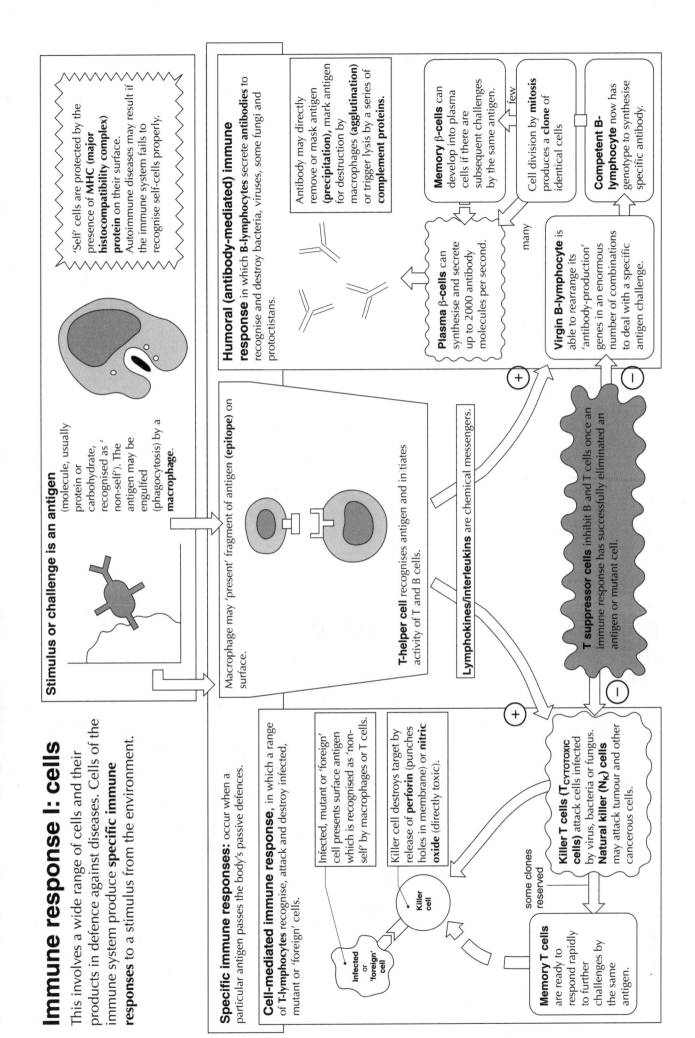

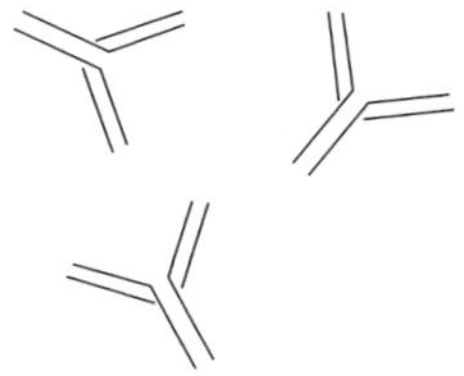

Antibody may directly remove or mask antigen (**precipitation**), mark antigen for destruction by macrophages (**agglutination**) or trigger lysis by a series of **complement proteins.**

Plasma β-cells can synthesise and secrete up to 2000 antibody molecules per second.

Memory β-cells can develop into plasma cells if there are subsequent challenges by the same antigen.

few

Cell division by **mitosis** produces a **clone** of identical cells.

Competent B-lymphocyte now has genotype to synthesise specific antibody.

many

Virgin B-lymphocyte is able to rearrange its 'antibody-production' genes in an enormous number of combinations to deal with a specific antigen challenge.

T suppressor cells inhibit B and T cells once an immune response has successfully eliminated an antigen or mutant cell.

Specific immune responses: occur when a particular antigen passes the body's passive defences.

Cell-mediated immune response, in which a range of T-lymphocytes recognise, attack and destroy infected, mutant or 'foreign' cells.

Infected, mutant or 'foreign' cell presents surface antigen which is recognised as 'non-self' by macrophages or T cells.

Killer cell destroys target by release of **perforin** (punches holes in membrane) or **nitric oxide** (directly toxic).

Infected or 'foreign' cell

Killer cell

some clones reserved

Killer T cells (T$_{CYTOTOXIC}$ cells) attack cells infected by virus, bacteria or fungus. **Natural killer (N$_K$) cells** may attack tumour and other cancerous cells.

Memory T cells are ready to respond rapidly to further challenges by the same antigen.

Inflammation and phagocytosis are part of the non-specific defence reactions of the body.

Inflammation: a local, non-specific response to tissue damage – the tissue damage may be due to microbial invasion, physical injury or exposure to irritant chemicals.

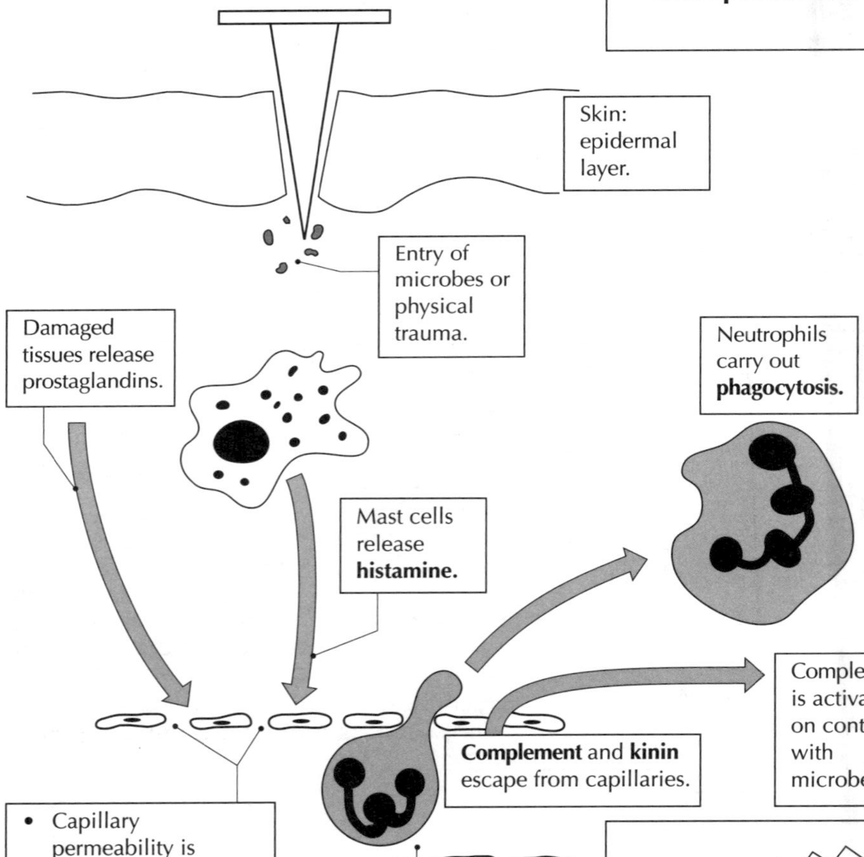

Skin: epidermal layer.

Entry of microbes or physical trauma.

Damaged tissues release prostaglandins.

Mast cells release **histamine.**

Neutrophils carry out **phagocytosis.**

Complement and **kinin** escape from capillaries.

Complement is activated on contact with microbe.

- Capillary permeability is increased.
- Arterioles vasodilate to increase blood flow to injured area

→ **Swelling, redness** and **heat**

Neutrophils are attracted to microbes (**chemotaxis**) by presence of **kinin**. They squeeze through capillary pores by **diapedesis**.

DANGER! Some substances cause **anaphylactic shock**: capillary walls become very leaky, and so much fluid leaves the circulation that blood pressure falls dangerously low. Nut fragments or insect stings can cause this potentially fatal reaction.

Phagocytosis: ingestion and destruction of microbes or particulate foreign matter by neutrophils and macrophages.

also called **polymorphonuclear** or **PMN** cells. Most abundant (10^{11} per kg of human tissue) and specialised of phagocytes, and the first to arrive at a site of infection.

Some are **wandering**, and travel through the body in the blood plasma as monocytes, until they leave by chemotaxis.

Some are **fixed**, and are confined by lymph nodes, spleen and other vulnerable organs e.g. lung and liver.

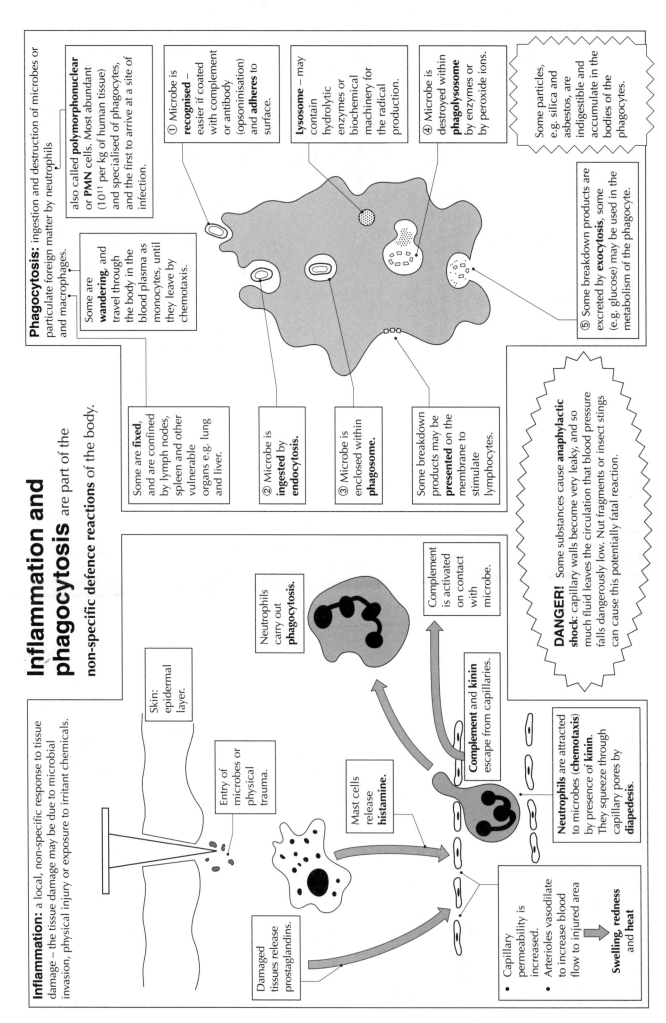

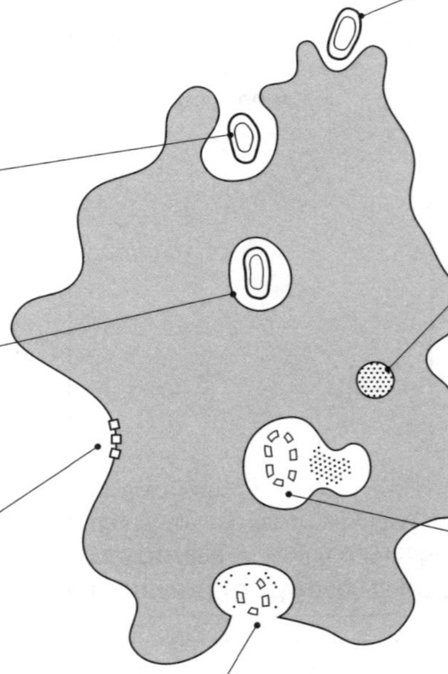

① Microbe is **recognised** – easier if coated with complement or antibody (opsoninisation) and **adheres** to surface.

② Microbe is **ingested** by **endocytosis.**

③ Microbe is enclosed within **phagosome.**

Lysosome – may contain hydrolytic enzymes or biochemical machinery for the radical production.

④ Microbe is destroyed within **phagolysosome** by enzymes or by peroxide ions.

Some breakdown products may be **presented** on the membrane to stimulate lymphocytes.

⑤ Some breakdown products are excreted by **exocytosis**, some (e.g. glucose) may be used in the metabolism of the phagocyte.

Some particles, e.g. silica and asbestos, are indigestible and accumulate in the bodies of the phagocytes.

Immune response II: the humoral response (antibodies and immunity)

An antibody is a protein molecule synthesised by an animal in response to a specific antigen.

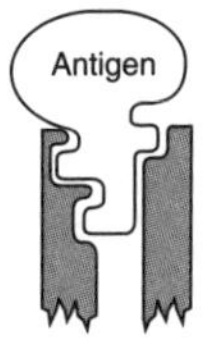

The **antigen-binding sites ('sticky ends')** have highly variable amino acid sequences which produce a huge number of possible recognition sites to bind to an enormous range of possible antigens.

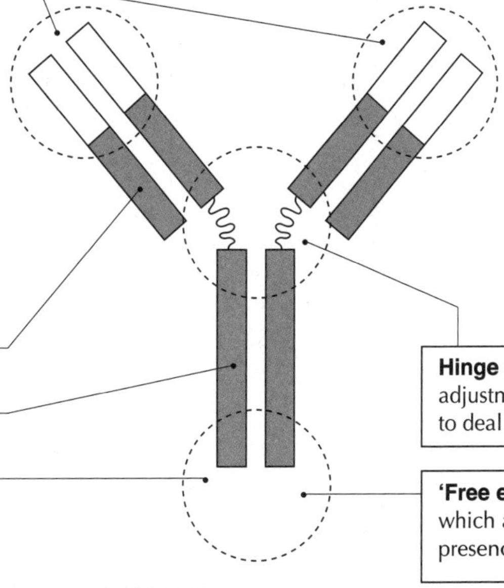

The basic structure of an antibody has the shape of the letter Y. Each molecule is composed of four polypeptide chains, two heavy and two light, all linked by disulphide bridges.

Constant (C) region of light chain

Constant (C) region of heavy chain

The constant regions determine the **general class** of the antibody:

IgG and IgM participate in the **precipitation, agglutination** and **complement** reactions.
IgA in tears, mucous secretions and saliva specifically binds to **surface antigens on bacteria.**
IgD helps to **activate lymphocytes.**
IgE is bound to mast cells and provokes **allergies.**

Hinge region permits some adjustment of position of sticky ends to deal with antigen.

'Free end' acts as a marker signal which alerts phagocytes to the presence of bound antigen.

Memory cells speed up immune response

Secondary response: curve is steeper, peak is higher (commonly 10^3 x the primary response), lag period is negligible **due to the presence of B-memory cells.** Dominant antibody is IgG which is more stable and has a greater affinity for the antigen.

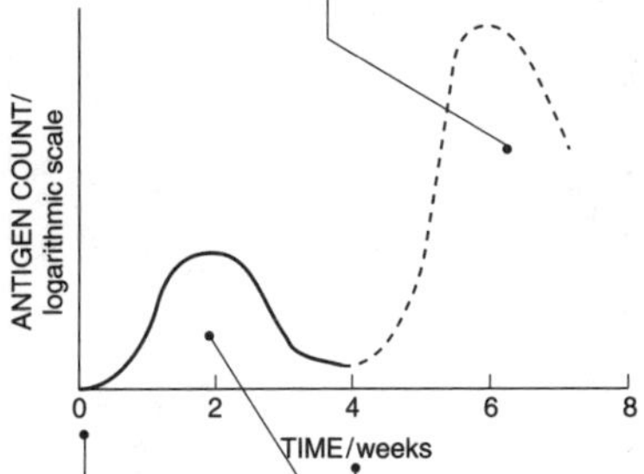

1st challenge with antigen

2nd challenge – with same antigen

Primary response: typical lag period is 3 days with peak at 11–14 days. Dominant antibody molecule is IgM.

Immunity may be enhanced

In **active immunity** an individual is provoked to **manufacture his or her own antibodies.**

a. **natural**

b. **artificial**

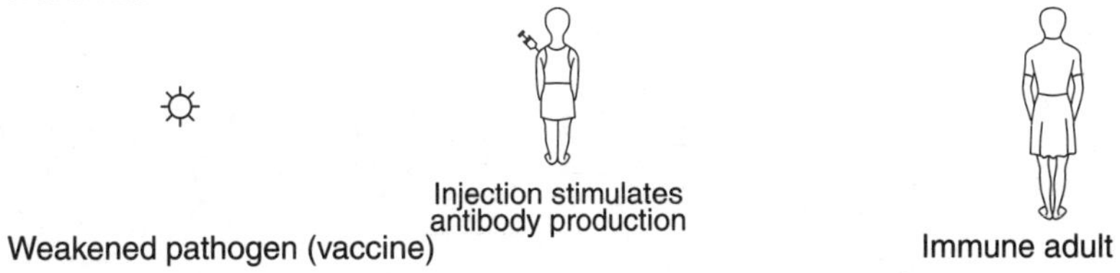

In **passive immunity** an individual is protected by a **supply of pre-formed antibodies.**

a. **natural**

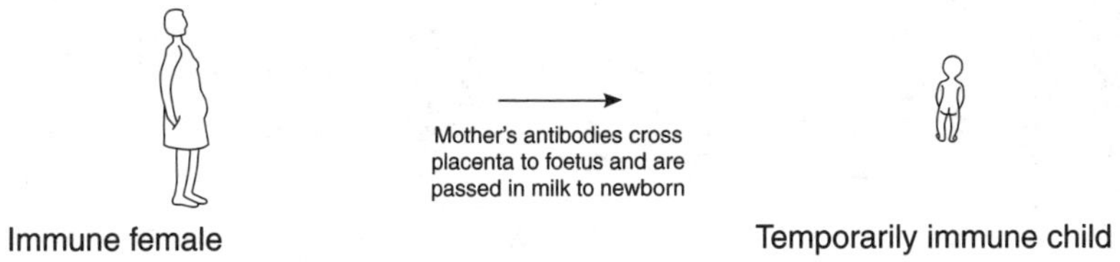

b. **artificial**

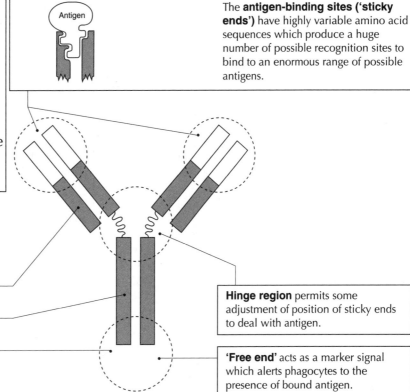

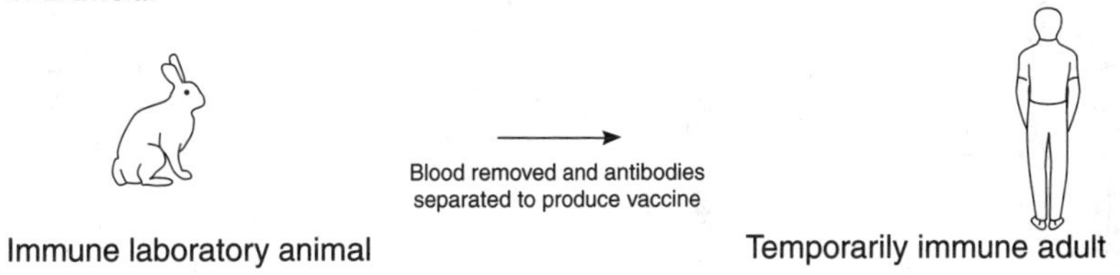

Haemoglobin–oxygen association curve

Shows the relationship between haemoglobin saturation with oxygen (i.e. the % of haemoglobin in the form of oxyhaemoglobin) and the partial pressure of oxygen in the environment (the pO_2 or oxygen tension).

Haemoglobin - the oxygen transporter

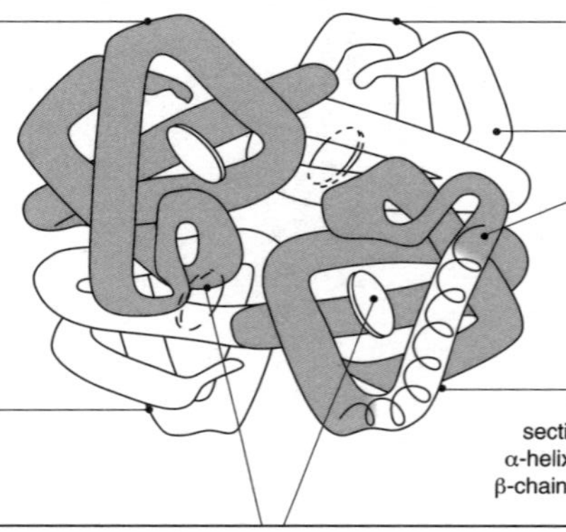

Globin (polypeptide) chains:
- provide correct environment for the binding of oxygen to the haem;
- keep haemoglobin soluble in the cytoplasm of the red blood cell.

Haemoglobin molecules are different from species to species:
- there may be different numbers of haem groups;
- polypeptide chains may have different amino acid sequences, so slightly different properties.

Haem groups:
- there are four in each molecule of human haemoglobin
- each has an iron^{II} ion at its centre.

The haem group iron^{II} can combine with oxygen – the four groups combine one at a time i.e.

$Hb + O_2 \leftrightharpoons HbO_2$

$HbO_2 + O_2 \leftrightharpoons HbO_4$

$HbO_4 + O_2 \leftrightharpoons HbO_6$

$HbO_6 + O_2 \leftrightharpoons HbO_8$

Oxyhaemoglobin
- this process is an **oxygenation**, not an oxidation – the iron in the haem group remains as iron^{II}.

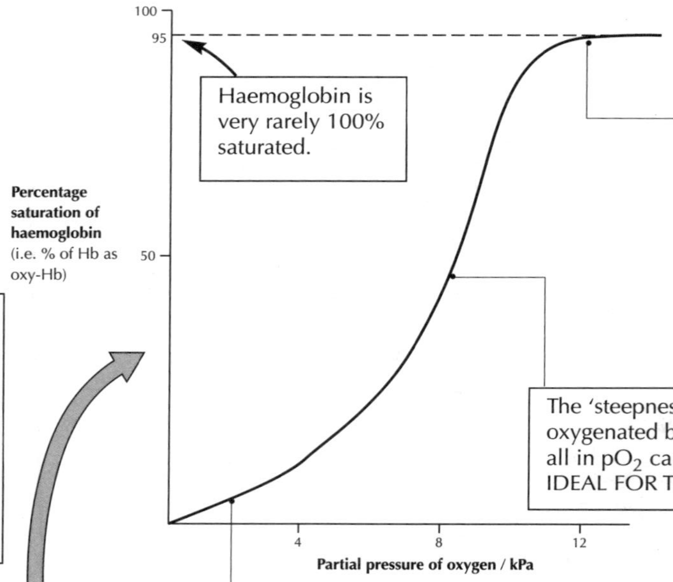

Haemoglobin is very rarely 100% saturated.

pO_2 corresponding to conditions in lungs, pulmonary veins and main arteries: haemoglobin is saturated with oxygen. The 'flatness' of the curve means that the Hb remains saturated (i.e. very little O_2 is released) despite a small reduction in pO_2 as blood enters the arteries: IDEAL FOR THE **TRANSPORT** OF OXYGEN.

The 'steepness' of the curve means that as oxygenated blood approaches the tissues a **small** all in pO_2 causes a **large** decrease in % saturation: IDEAL FOR THE **RELEASE** OF OXYGEN.

Low pO_2 corresponding to conditions in respiring tissues: almost complete dissociation of oxy-Hb.

The Hb-O_2 association curve is S-shaped

The reasons for this S-shapedness is that the haemoglobin and oxygen illustrate co-operative binding, that is, the binding of the first oxygen molecule to haemoglobin alters the shape of the haemoglobin molecule slightly so that the binding of a second molecule of oxygen is made easier, and so on until haemoglobin has its full complement of four molecules of oxygen. Conversely, when one molecule of oxygen dissociates from the oxyhaemoglobin, the haemoglobin shape is adjusted to make release of successive molecules of oxygen increasingly easy.

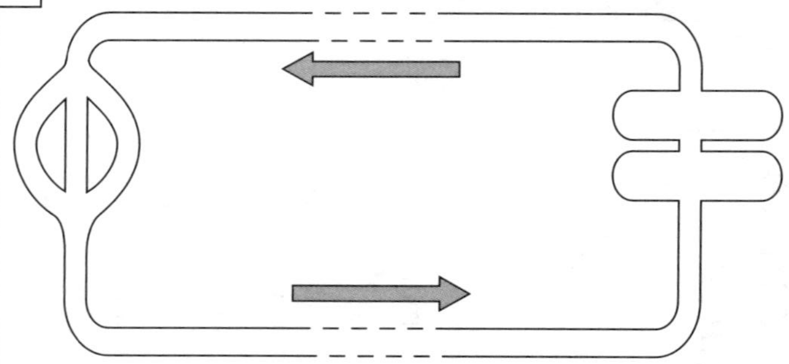

At the tissues:
Oxyhaemoglobin dissociates
$HbO_8 \leftrightharpoons Hb + 4O_2$.

In the lungs:
Oxygen and haemoglobin associate to form oxy-haemoglobin
$Hb + 4O_2 \leftrightharpoons HbO_8$.

BOHR SHIFT: The effect of carbon dioxide on the haemoglobin-oxygen association curve

This effect shows how the **supply** of oxygen by the dissociation of oxyhaemoglobin is well suited to the **demands** for oxygen by the respiring tissues.

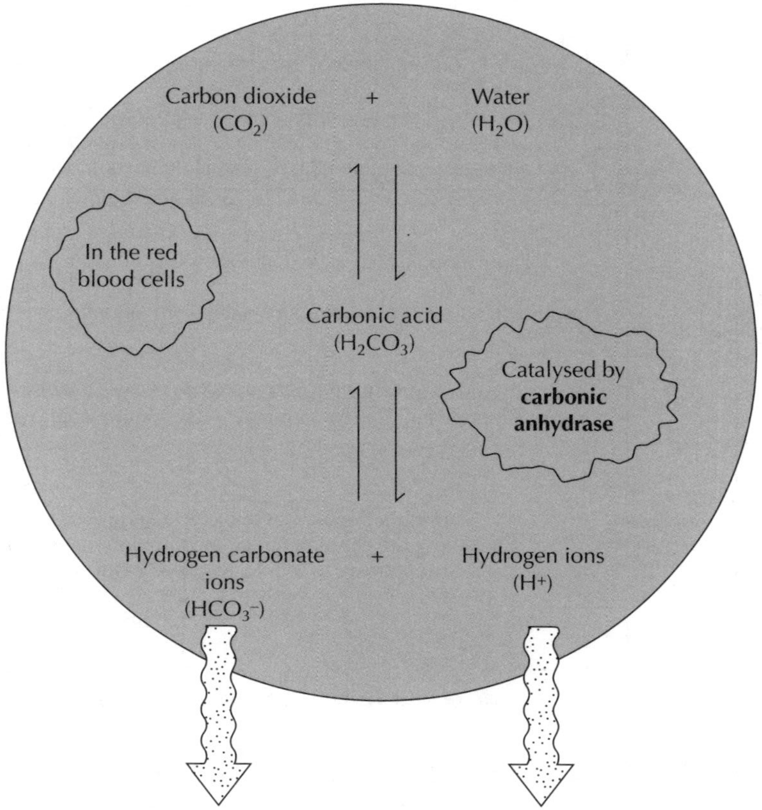

Diffuse into blood plasma and are transported to lungs

Taken up by haemoglobin:
- this removal of dangerous H^+ means haemoglobin is acting as a **buffer**
- the hydrogen ions change the shape of the haemoglobin molecule, so oxygen is released.

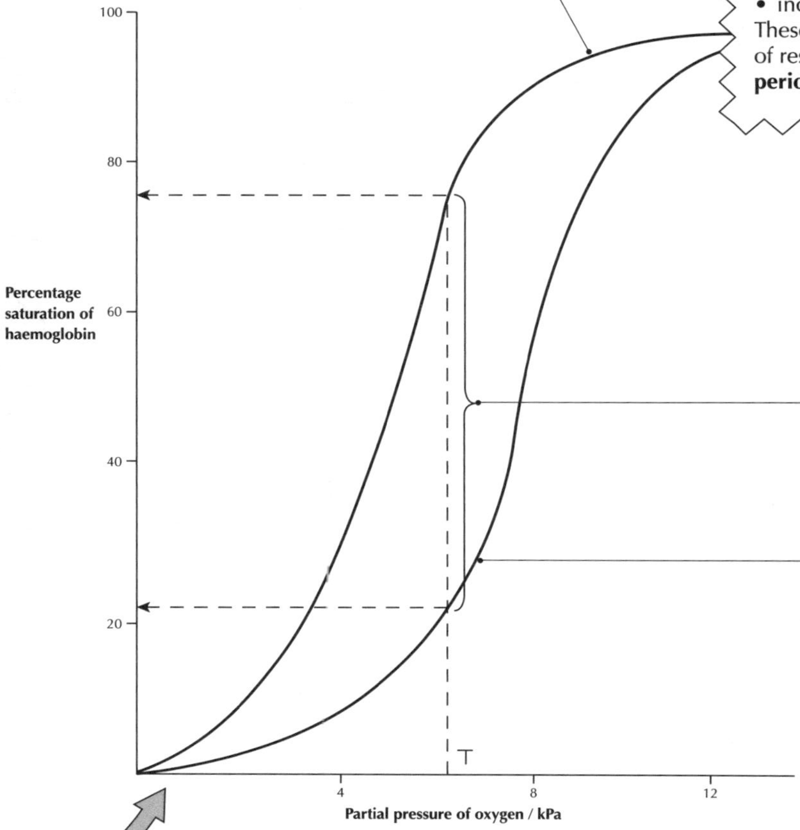

Low pCO_2 (at the lungs):
- curve is 'shifted' to the left
- Hb has increased affinity for oxygen
- more oxyhaemoglobin is formed.

A similar Bohr Shift of the haemoglobin-oxygen association curve **to the right** (i.e. towards more O_2 released as oxy-Hb dissociates) results from:
- falling pH;
- increased lactate concentration;
- increased temperature.

These conditions exist when the rate of respiration increases e.g. **during periods of heavy exercise**.

The 'difference' between the two curves represents the additional oxygen released from oxyhaemoglobin at any particular pO_2 (shown here as T kPa).

High pCO_2:
- curve is shifted to the right
- Hb has reduced affinity for oxygen
- oxyhaemoglobin dissociates.

Modified haemoglobins

help animals adapt to their environment.

Crocodile: is an air-breathing animal which spends long periods of time underwater where it cannot breathe

⇩

Requirement for haemoglobin to act as an oxygen **store**

⇩

Oxyhaemoglobin only dissociates at very low pO_2 thus conserving oxygen supplies.

The crocodile is not active for long periods, so overall oxygen demands are quite low.

This curve is similar to that of **myoglobin**, a different molecule which helps to 'pull' oxygen from circulating blood to working muscles.

Arenicola (lugworm – an annelid): lives in waterlogged burrows where the oxygen partial pressure is very low.

⇩

Haemoglobin must have very high affinity for oxygen, even at very low pO_2.
The lugworm is not very active, so oxygen demands for energy release in respiration are low.

Llama: this mammal lives at high altitudes, where the pO_2 is very low.

⇩

Problems with loading oxygen

⇩

Haemoglobin must have high affinity for oxygen so that oxyhaemoglobin is formed more readily.

This curve is very similar to that for **human fetal haemoglobin**. As a result the fetal form has higher affinity for oxygen and the gas is easily transferred across the placenta from the mother's blood to the fetus.

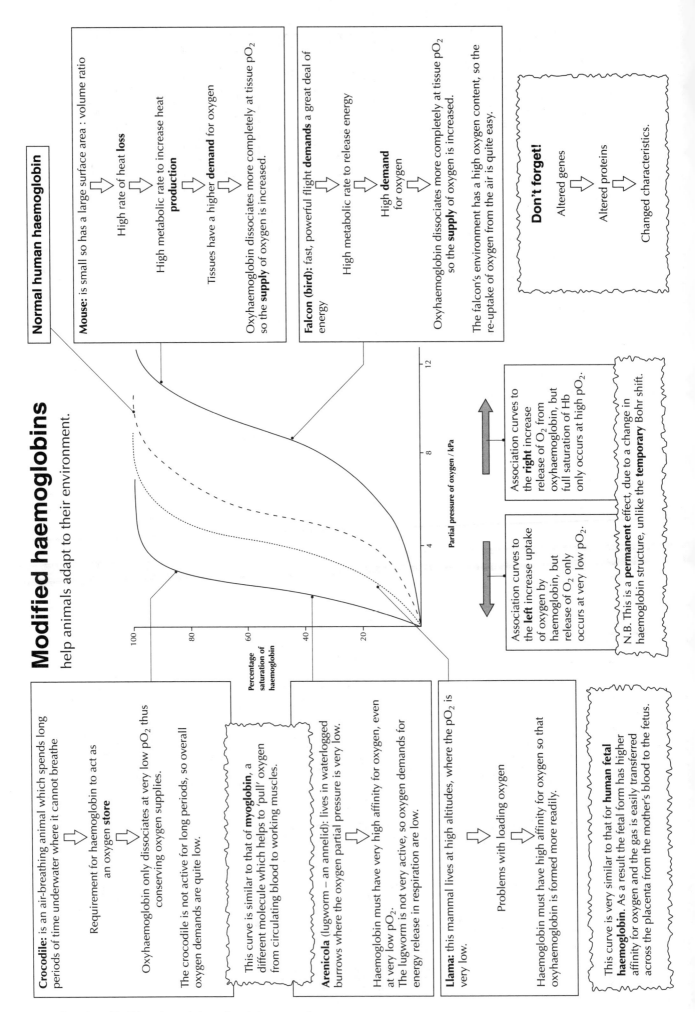

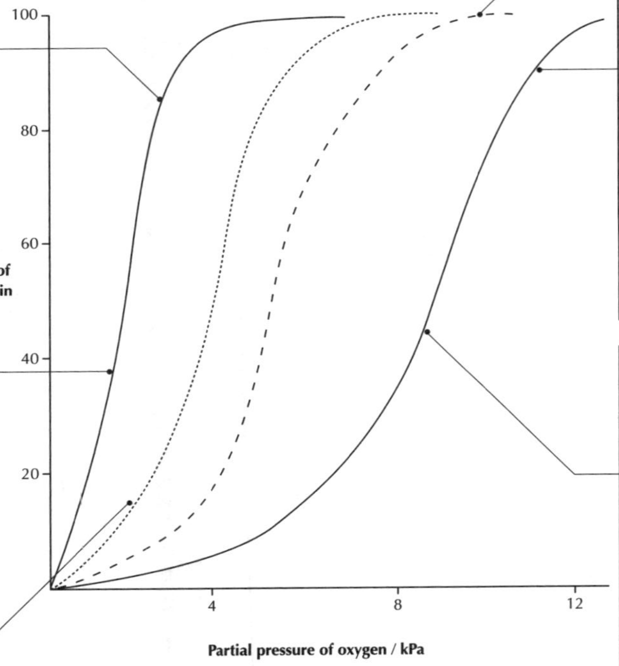

Association curves to the **left** increase uptake of oxygen by haemoglobin, but release of O_2 only occurs at very low pO_2.

Association curves to the **right** increase release of O_2 from oxyhaemoglobin, but full saturation of Hb only occurs at high pO_2.

N.B. This is a **permanent** effect, due to a change in haemoglobin structure, unlike the **temporary** Bohr shift.

Normal human haemoglobin

Mouse: is small so has a large surface area : volume ratio

⇩

High rate of heat **loss**

⇩

High metabolic rate to increase heat **production**

⇩

Tissues have a higher **demand** for oxygen

⇩

Oxyhaemoglobin dissociates more completely at tissue pO_2 so the **supply** of oxygen is increased.

Falcon (bird): fast, powerful flight **demands** a great deal of energy

⇩

High metabolic rate to release energy

⇩

High **demand** for oxygen

⇩

Oxyhaemoglobin dissociates more completely at tissue pO_2 so the **supply** of oxygen is increased.

The falcon's environment has a high oxygen content, so the re-uptake of oxygen from the air is quite easy.

Don't forget!

Altered genes

⇩

Altered proteins

⇩

Changed characteristics.

Tissue fluid

(interstitial or intercellular fluid) is the immediate environment of the cells, and represents the 'internal environment' described by Claude Bernard in his definition of homeostasis.

Plasma proteins do not move from plasma to tissue fluid (cannot cross capillary endothelium) - largely responsible for solute potential of plasma.

Movement from tissue fluid to plasma

- **Water**
- **Carbon dioxide**
- **Nitrogenous waste**
- **Hormones and other secretions**

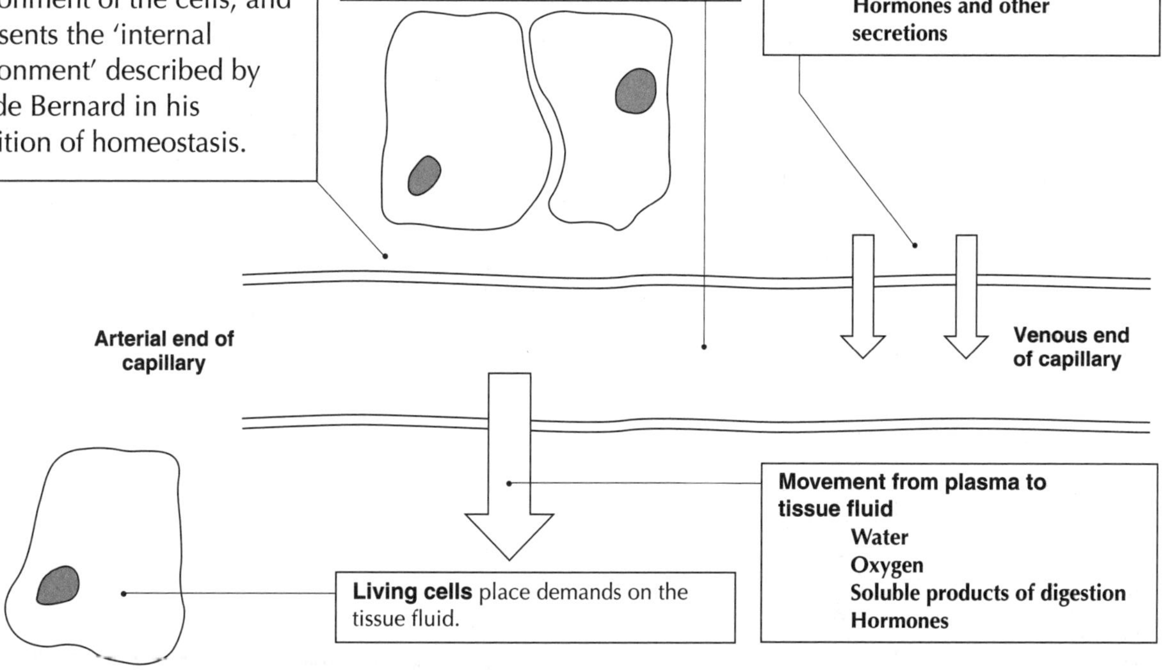

Movement from plasma to tissue fluid

- **Water**
- **Oxygen**
- **Soluble products of digestion**
- **Hormones**

Living cells place demands on the tissue fluid.

Forces which regulate the formation and reclamation of tissue fluid

Pressure potential (hydrostatic potential) is the pressure exerted on a fluid by its surroundings, e.g. by **pumping action of heart** and **elastic recoil of arteries.**

Solute potential is the force of attraction towards water molecules caused by dissolved solutes, particularly **ions** and **plasma proteins**.

Net force driving fluid movement at any point
= pressure potential gradient – solute potential gradient

Venous end of capillaries: the PP gradient has fallen (1) as distance from pumping heart increases and (2) as volume of fluid in vessels falls. High concentration of plasma proteins means that blood solute potential is high.

SP gradient > PP gradient

Net movement of water from tissues to plasma.

During starvation, or following liver damage, the concentration of plasma proteins may fall. The SP gradient is reduced and tissue fluid is not reclaimed. Tissues swell, often causing a swollen abdomen (as in the protein deficiency disease, Kwashiorkor).

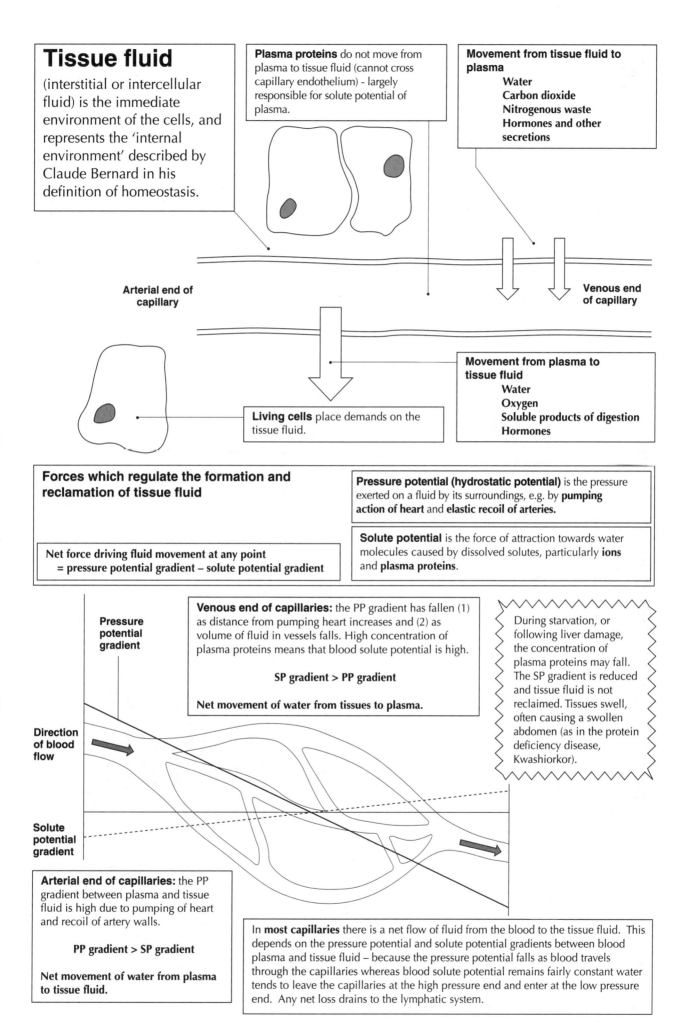

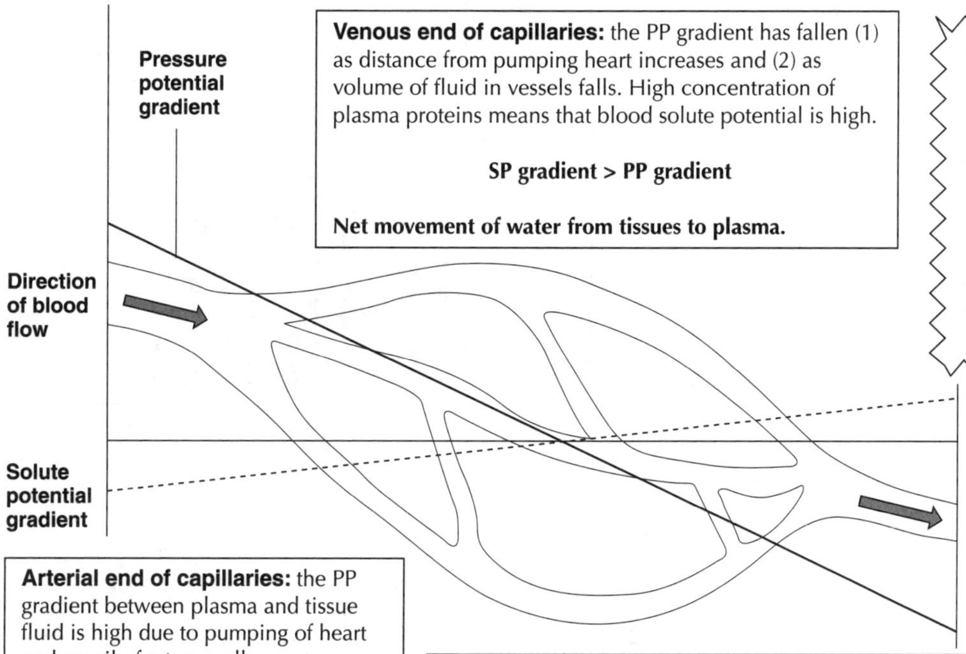

Arterial end of capillaries: the PP gradient between plasma and tissue fluid is high due to pumping of heart and recoil of artery walls.

PP gradient > SP gradient

Net movement of water from plasma to tissue fluid.

In **most capillaries** there is a net flow of fluid from the blood to the tissue fluid. This depends on the pressure potential and solute potential gradients between blood plasma and tissue fluid – because the pressure potential falls as blood travels through the capillaries whereas blood solute potential remains fairly constant water tends to leave the capillaries at the high pressure end and enter at the low pressure end. Any net loss drains to the lymphatic system.

Blood vessels transport blood to tissues and organs:

this is a **mass flow system** depending on **pressure generated by contraction of the ventricles.**

There is a higher pressure in the aorta because the muscular wall of the left ventricle is thicker than the wall of the right ventricle: contraction is therefore more powerful in the left ventricle.

Both sides of the heart contract together, so the **pattern** of contraction (systole) and relaxation (diastole) is identical.

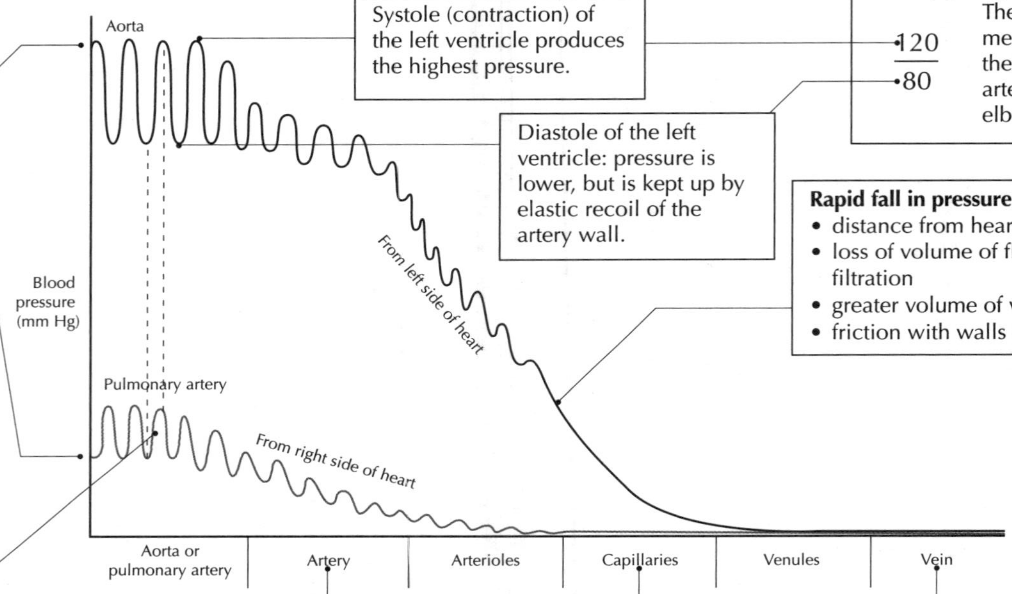

Systole (contraction) of the left ventricle produces the highest pressure.

Diastole of the left ventricle: pressure is lower, but is kept up by elastic recoil of the artery wall.

Normal values in a healthy young adult are: $\frac{120}{80}$ These are measured in the brachial artery at the elbow.

Rapid fall in pressure in capillaries:
- distance from heart
- loss of volume of fluid due to filtration
- greater volume of vessels
- friction with walls of vessels.

Hypertension (high blood pressure) is the most common circulatory disease. Normal blood pressure is $^{120}/_{80}$ – any value above $^{140}/_{90}$ indicates hypertension and $^{160}/_{95}$ is considered dangerous. Uncontrolled hypertension may damage heart (heart enlarges and demands more oxygen), brain (CVA, or stroke, which ruptures cerebral arteries supplying brain) and kidneys (glomerular arterioles are narrowed and deliver less blood). Treatment includes weight reduction in obese patients, restriction of Na^+ intake, increased Ca^{2-}/K^+ intake and stopping smoking. Vasolilators, beta-blockers and diuretics are widely prescribed.

Varicose veins are so dilated that the valves do not close to prevent backflow of blood. The veins lose their elasticity and become congested. Predisposing factors include age, heredity, obesity, gravity and compression by adjacent structures. Common sites are the legs (where poor circulation may lead to haemorrhage or ulceration of poorly-nourished skin), the rectal-anal junction (haemorrhoids may bleed and cause mild anaemia), the oesophagus (caused by liver cirrhosis or right-sided heart failure – may lead to haemorrhage and death) and the spermatic cord (may reduce spermatogenesis).

Structure and function of blood vessels

	Artery	Capillaries	Vein
Outer layer of collagen fibres	Yes	No	Yes
Middle layer of elastic fibres and fibres of smooth muscle	Thick	Absent	Thin
Inner layer: endothelium of pavement epithelium	Yes	Yes	Yes
Valves	No Transport of blood at high pressure, from heart to organs: strong and elastic to resist pressure changes	No Exchange of materials between blood and tissue fluid	Yes Return of blood at low pressure from organs to heart

These pressure changes can be felt as a **pulse** anywhere that an artery can be pressed against a bone (hard surface)
Most commonly:
- **radial pulse** – artery over end of radius at the wrist;
- **temporal pulse** – artery over temporal bone of skull (just in front of the top of the earlobe).

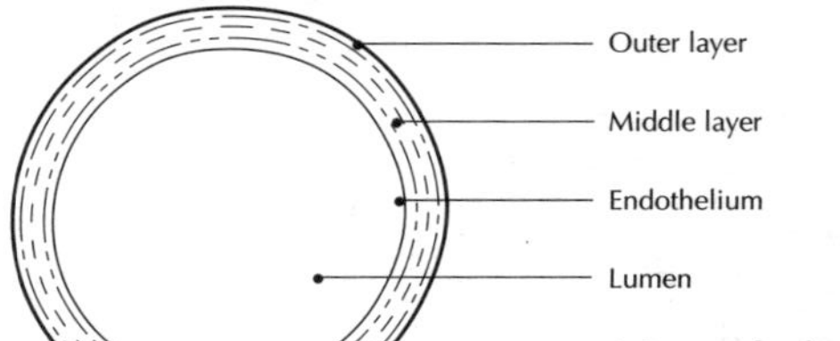

Artery and vein

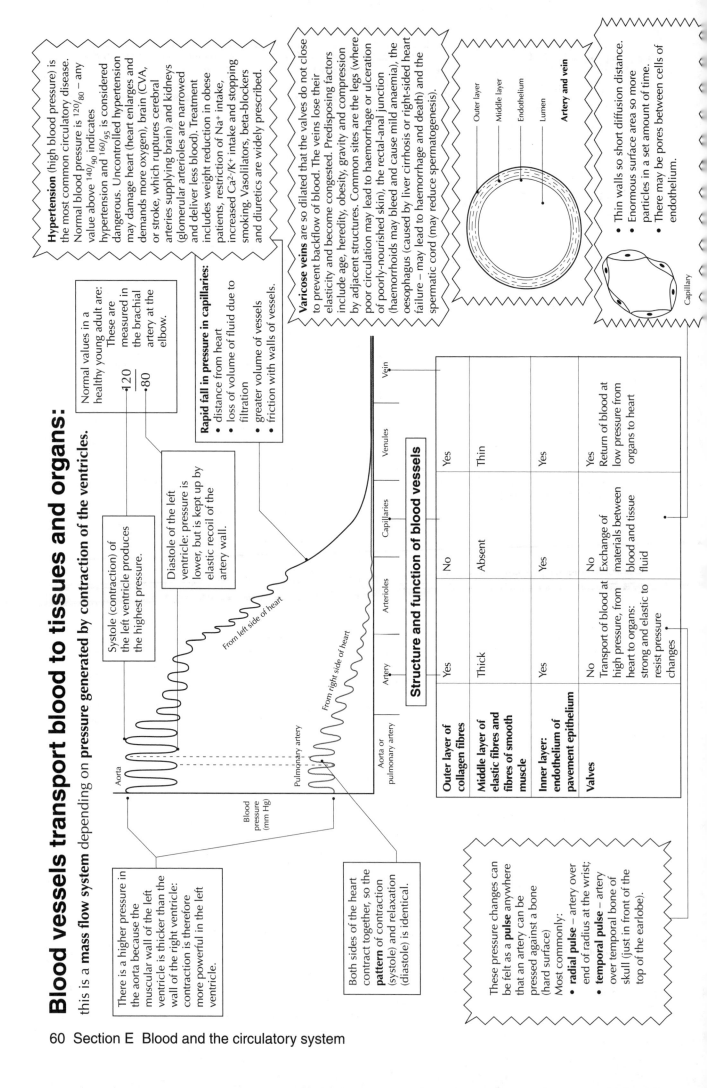

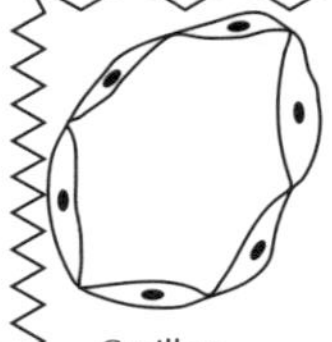

Capillary

- Thin walls so short diffusion distance.
- Enormous surface area so more particles in a set amount of time.
- There may be pores between cells of endothelium.

Mammalian double circulation comprises **pulmonary** (heart – lung – heart) and **systemic** (heart – rest of body – heart) **circuits**. The complete separation of the two circuits permits rapid, high-pressure distribution of oxygenated blood essential in active, endothermic animals. The circuits are named for the organ or system which they service – thus each kidney has a **renal** artery and vein. Each organ has an artery bringing oxygenated blood and nutrients, and a vein removing deoxygenated blood and waste.

Pulmonary artery: delivers deoxygenated blood to the lungs for reoxygenation.

Superior vena cava: carries deoxygenated blood back to the right atrium from the head and forelimbs. The venous return to the heart initiates the expansion which triggers the sino-atrial node to fire the impulse which generates the heart beat.

Inferior vena cava: returns deoxygenated blood at low pressure to the right atrium of the heart.

Hepatic vein: delivers blood with an optimum concentration of solutes (particularly glucose) from the liver to the general circulation.

Hepatic portal vein: transports blood with a very variable solute concentration from the site of solute uptake (the gut) to the site of storage or regulation (the liver).

Renal veins: return deoxygenated blood with a reduced concentration of urea and creatinine, and a regulated pH and Na^+/K^+ ratio, from the kidneys to the general circulation.

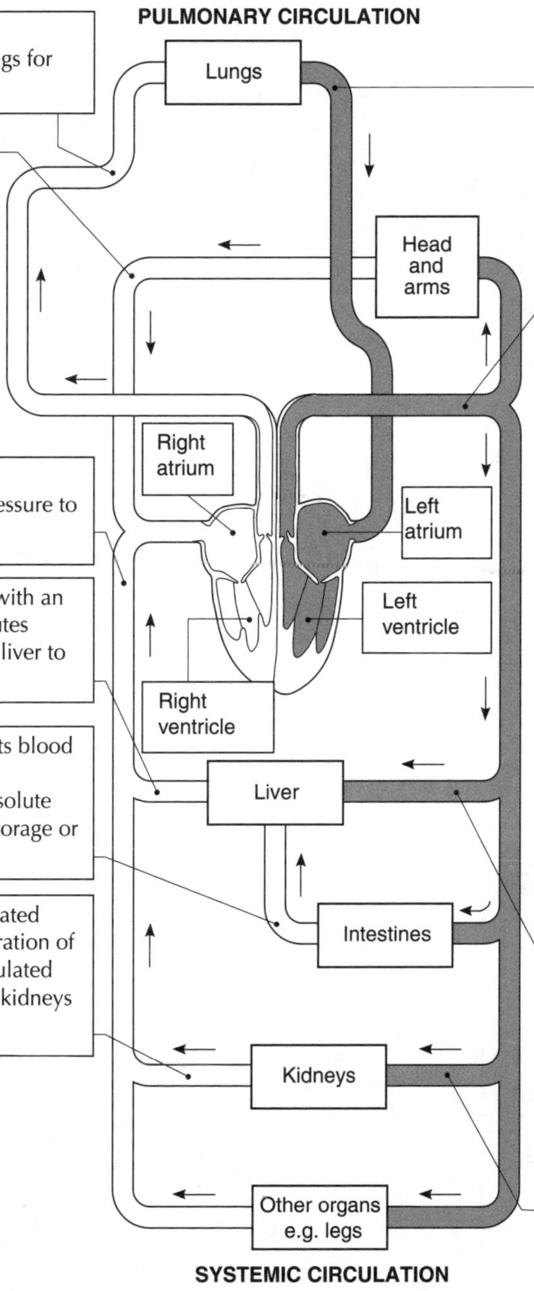

Pulmonary veins: return oxygenated blood at low pressure to the left atrium of the heart.

Aorta: the principal vessel which distributes oxygenated blood at high pressure to the systemic circulation.

The **flow of blood** is maintained in three ways:

1. **The pumping action of the heart:** the ventricles generate pressures great enough to drive blood through the arteries into the capillaries.

2. **Contraction of skeletal muscle:** the contraction of muscles during normal movements compress and relax the thin-walled veins causing pressure changes within them. Pocket valves in the veins ensure that this pressure directs the blood to the heart, without backflow.

3. **Inspiratory movements:** reducing thoracic pressure caused by chest and diaphragm movements during inspiration helps to draw blood back towards the heart.

Hepatic artery: delivers oxygenated blood to the liver - this organ is so active metabolically that oxygen demands are very high.

Renal artery: carries blood with high O_2 concentration and high concentrations of solutes such as urea, creatinine and Na^+.

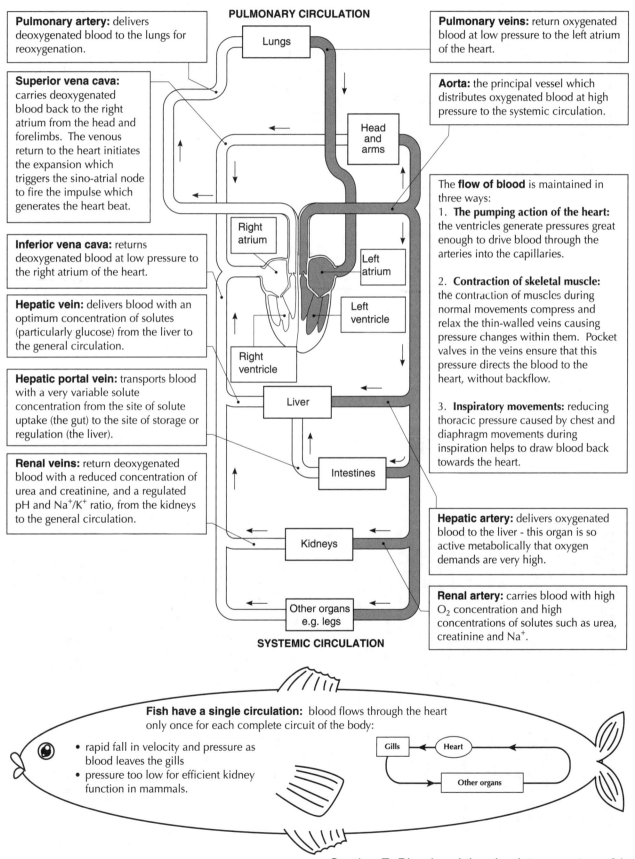

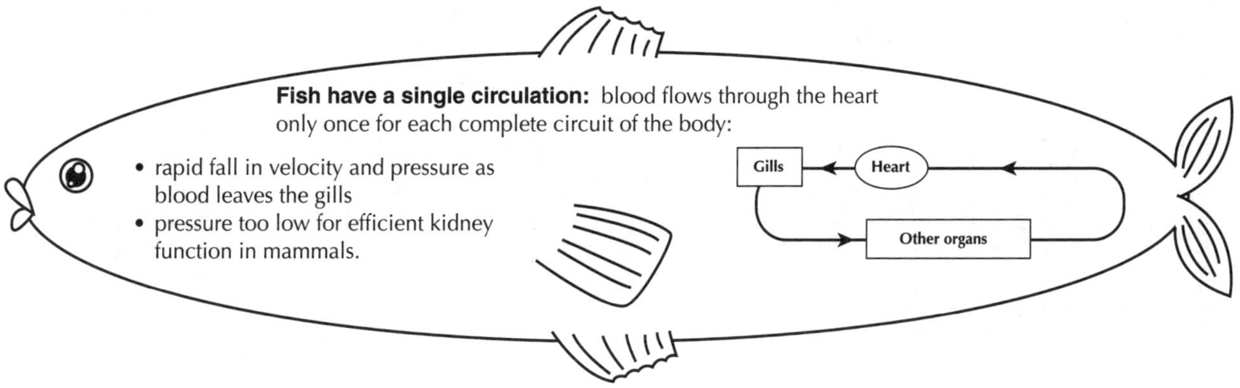

Fish have a single circulation: blood flows through the heart only once for each complete circuit of the body:

- rapid fall in velocity and pressure as blood leaves the gills
- pressure too low for efficient kidney function in mammals.

Problems associated with the immune response

Allergies occur when the immune system responds excessively to a common environmental challenge, such as house dust or pollen.

- may provoke sneezing, swollen and overactive mucous membranes, 'runny' eyes, itching and, in extreme cases, breathing difficulties
- 'allergic' individuals produce an excess of IgE antibodies as well as the more usual IgG type.
- binding of antigen to IgG on most cell surfaces causes release of **Histamine** and associated inflammation.

Hypersensitivity is an extreme and rapid allergic response. In the most severe cases **Anaphylaxis** may occur – this involves vasodilation and capillary leakage and the consequent lowering of blood pressure may even lead to death.

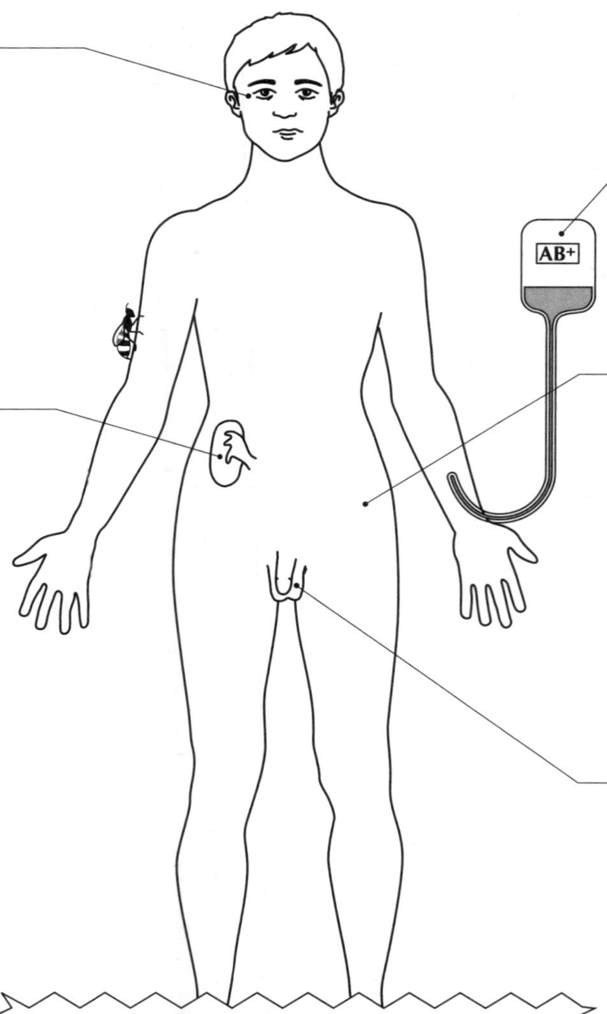

Blood transfusion may cause **agglutination** of red cells. Agglutinated blood may block capillaries, causing depletion of oxygen and glucose to respiring tissues. Agglutination is prevented by matching blood types between donor and recipient so that only **compatible** transfusions take place.

Autoimmunity arises when the 'immune' system fails to distinguish between 'self' and 'non-self' (foreign) cells, and attacks 'self', or body tissues.

- may affect one organ ('organ-specific' e.g. pernicious anaemia) or many ('organ non-specific' or 'multi-systemic' e.g. mixed connective tissue disease)
- causes generally not known, but may involve genetic, viral and environmental factors
- may be implicated in many conditions
  e.g. Rheumatoid arthritis
  Type I (insulin-dependent) diabetes mellitus
  Multiple sclerosis
  Haemolytic anaemia

Transplant rejection occurs when T-lymphocytes recognise, invade and destroy transplanted 'foreign' tissues.
T_{HELPER} cells recognise 'foreign' protein on tissue and secrete interleukin.

Immature $T_{CYTOTOXIC}$ cells recognise antigen and, under stimulation from IL-2, divide and mature into fully active $T_{CYTOTOXIC}$ cells.

Mature $T_{CYTOTOXIC}$ cells invade the tissue and destroy cells carrying the 'foreign' protein: tissue destruction causes loss of function of transplanted organ.

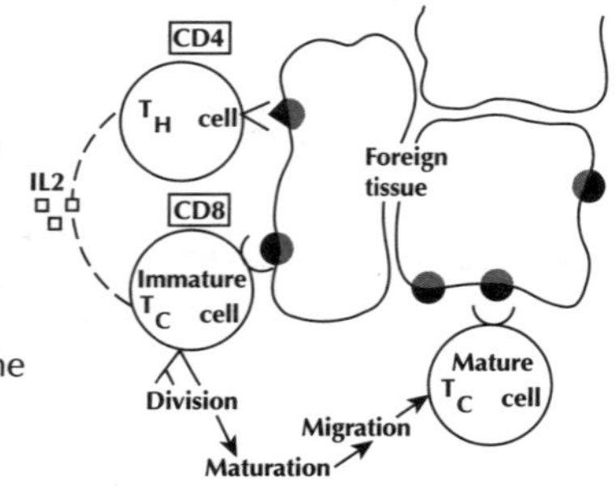

Precautions to minimise risk of rejection

- close tissue matching between donor and recipient
- irradiation of bone marrow and lymph tissue – reduces lymphocyte production and thus reduces rejection
- immunosuppression – cyclosporin B reduces T_C numbers but B lymphocytes unaffected. Thus humoral immunity is retained.

'Immune camouflage' is necessary for

- Sperm – hapoid so 'non-self' and head is 'hidden' within Sertoli cells of the seminiferous tubules.
- Foetus – carries some paternal antigens so must be protected from maternal immune surveillance by the placenta.

Xenotransplantation (organs transplanted between different species). e.g. some pig organs are the correct size for transplantation to humans, but would trigger a violent immune response unless they can be 'camouflaged' in some way.

Vaccination/immunisation deliberately provokes the immune system. Incorrect dosage or extreme sensitivity to antigen may lead to an allergic response.

Biological variation

- results from genetic and environmental factors
- the basis of evolution by natural selection.

Origins of variation

- Non-heritable (not passed on to offspring): mainly effect of environment e.g. sunburn in a light-skinned individual
- Heritable (e.g. skin colour in different races): results from genetic change, and is the more significant in evolution.

Mutation is any change in the structure or the amount of DNA in an organism.

- A **gene** or **point mutation** occurs at a single locus on a chromosome – most commonly by **deletion, addition** or **substitution** of a nucleotide base. Examples are sickle cell anaemia, phenylketonuria and cystic fibrosis.
- A **change in chromosome structure** occurs when a substantial portion of a chromosome is altered. For example, Cri-du-chat syndrome results from **deletion** of a part of human chromosome 5, and a form of white blood cell cancer follows **translocation** of a portion of chromosome 8 to chromosome 14.
- **Aneuploidy** (typically the **loss or gain of a single chromosome**) results from **non-disjunction** in which chromosomes fail to separate at anaphase of meiosis. The best known examples are Down's syndrome (extra chromosome 21), Klinefelter's syndrome (male with extra X chromosome) and Turner's syndrome (female with one fewer X chromosome).
- **Polyploidy** (the presence of additional **whole sets of chromosomes**) most commonly occurs when one or both gametes is diploid, forming a polyploid on fertilisation. Polyploidy is rare in animals, but there are many important examples in plants, e.g. bananas are triploid, and tetraploid tomatoes are larger and richer in vitamin C.

Sexual or genetic recombination is a most potent force in evolution, since it reshuffles genes into new combinations. It may involve

free assortment in gamete formation
crossing over during meiosis
random fusion during zygote formation.

Environment and variation
Important factors include:
- **diet** e.g. Ca^{2+} for bone development;
- **light** e.g. for development of flowers;
- **physical constraints** e.g. tight shoes!

Discontinuous variation occurs when a characteristic is either present or absent (the two extremes) and there are no intermediate forms.

Such variations do not give normal distribution curves but bar charts are often used to illustrate the distribution of a particular characteristic in a population.

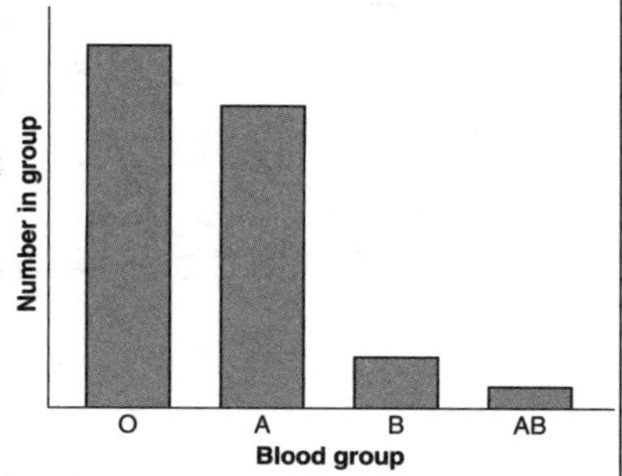

Examples are human blood groups in the ABO system (O, A, B or AB), basic fingerprint forms (loop, whorl or arch) and tongue-rolling (can or cannot).

A characteristic which shows discontinuous variation is normally controlled by a single gene – there may be two or more alleles of this gene.

Continuous variation occurs when there is a gradation between one extreme and the other of some given characteristic – all individuals exhibit the characteristic but to differing extents.

If a frequency distribution is plotted for such a characteristic a **normal distribution** is obtained.

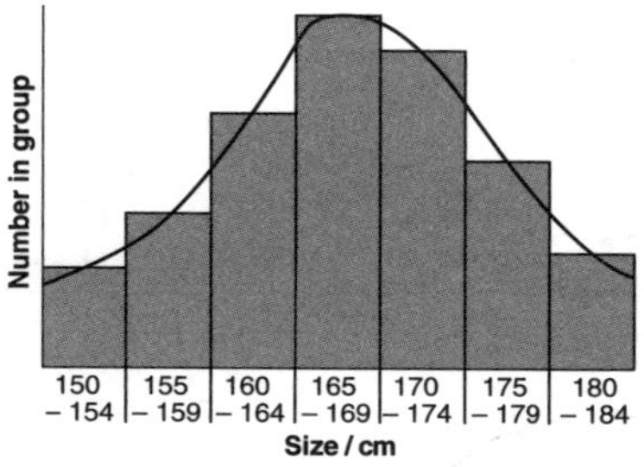

Typical examples are height, mass, handspan, or number of leaves on a plant.

Characteristics which show continuous variation are controlled by the **combined effect of a number of genes,** called **polygenes** (and are therefore **polygenic characteristics**) **or by the combined effects of genes and the environment.**

Phenotype = genotype + effects of environment.

Measuring variation in biology

Why take samples?

- Collecting all the values of measurable data would be very time consuming (and might not be easy – the tide could cover many limpets in a population!).
- Samples cause less damage to populations and to the environment.

N.B. The mean for your sample is very unlikely to be **exactly** the mean for the whole population of data …

Is the sample representative?

Collected data will only be valid for testing a hypothesis if the sample reflects the complete population of data: i.e. it is **representative**.

- **Randomly selected:** e.g. measurements may be made on individuals at selected points on a grid (could be real or imaginary) laid out over the sampling area. The coordinates of the points on the grid can be selected using random numbers generated by a calculator.
- **Large enough:** A running mean, when stable, suggests the sample, i.e. number of data values, is large enough.

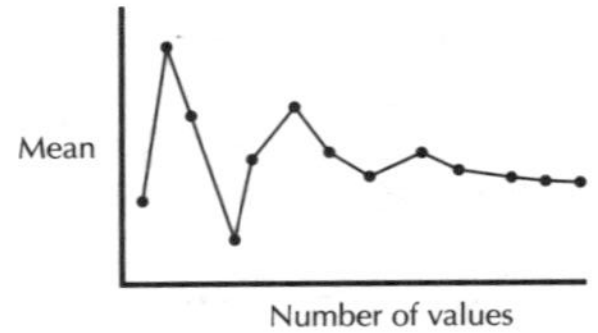

- **At least approximates to a normal distribution around the mean:** most continuous quantitative (i.e. measurable rather than countable) variables in biology do this.

Variability in continuous data is often plotted or displayed as a **frequency distribution**.

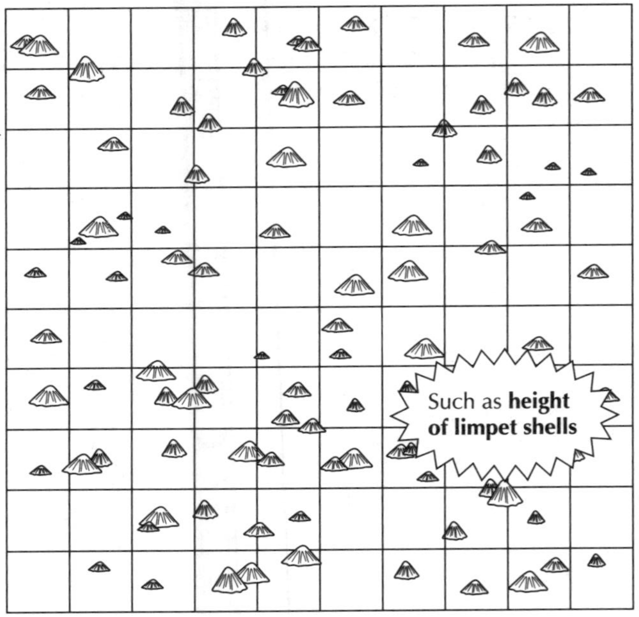

How confident can you be?

- A **Confidence Interval** (C.I.) gives a range of values likely to include a value for a measured variable.
- A **Confidence Level** is the probability that a value for a measured variable will fall within the C.I. In biology this is usually 95% – you could be 95% certain that any value for the measured variable falls within the C.I.

e.g. effect of temperature on activity of amylase

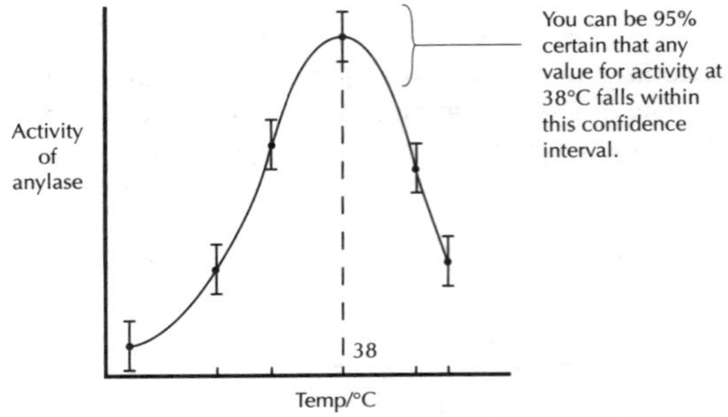

Standard deviation gives a good measure of variability (spread) of data: in a normally distributed set of data 95% of all values will be within 2 standard deviations of the mean.

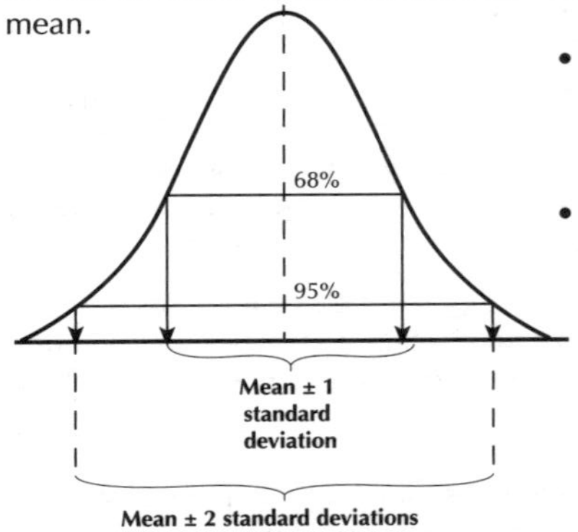

- Two data sets may have the same mean but different S.Ds.
- A smaller S.D. shows that data values are more closely bunched around the mean.

Mode: the most common data value.

Median: the middle value when data classes are presented in rank order.

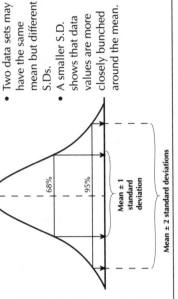

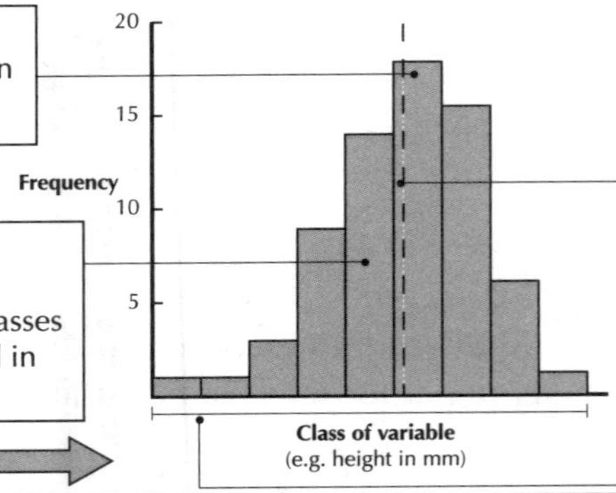

Mean: the average of all the data entries

- = $\frac{\text{sum of data entries}}{\text{total number of data entries}}$

N.B. a mean should not be calculated:

- from values that are already means themselves
- when the measurement scale is not linear (e.g. pH is a logarithmic scale, so 'mean pH' is inappropriate).

Range: the difference between smallest and largest data values. Gives a basic idea of spread of data, but …

Reduction of genetic variation

The Founder effect

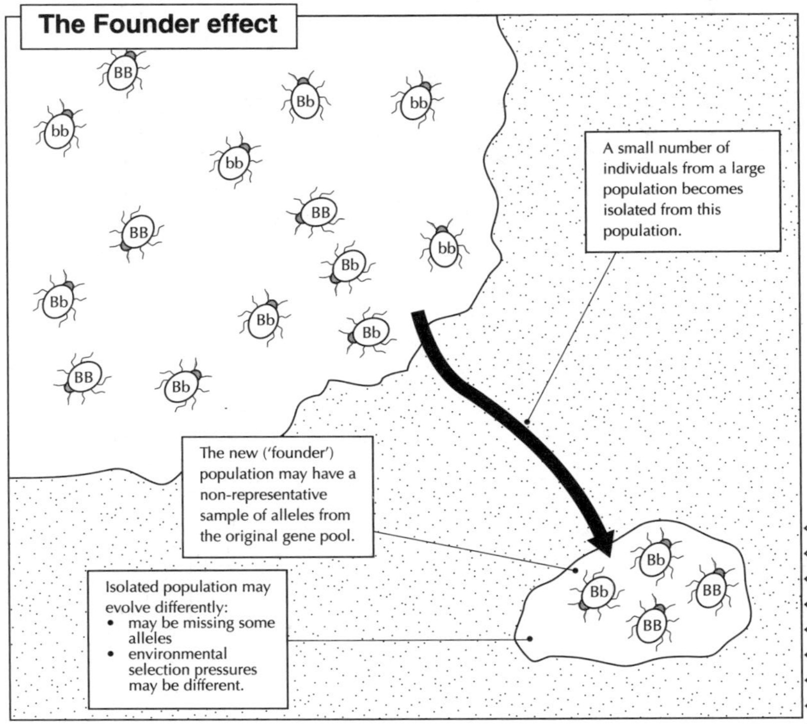

BB
Bb
bb
A small number of individuals from a large population becomes isolated from this population.
The new ('founder') population may have a non-representative sample of alleles from the original gene pool.
Isolated population may evolve differently:
• may be missing some alleles
• environmental selection pressures may be different.

Genetic bottlenecks

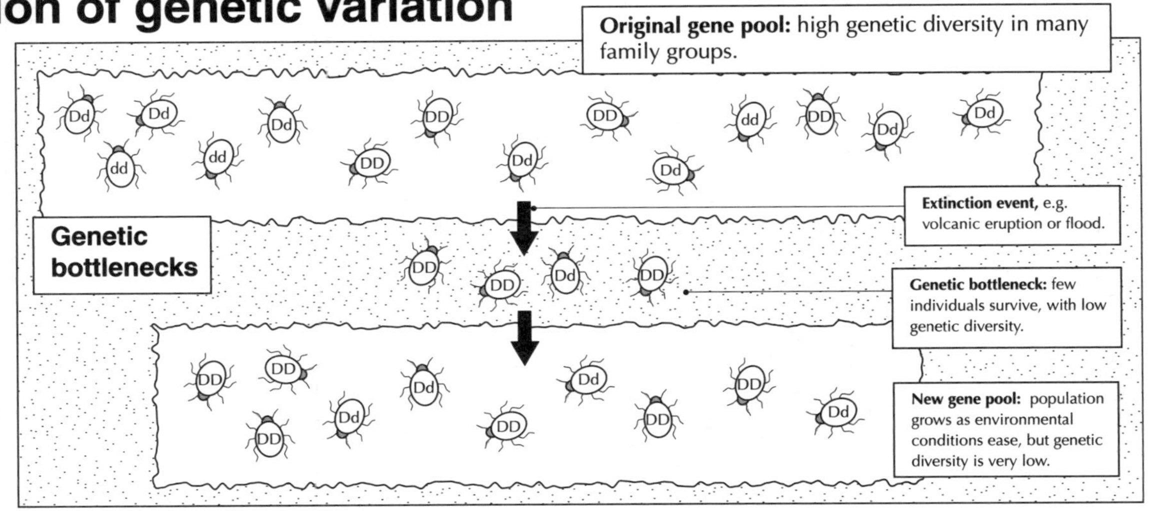

Original gene pool: high genetic diversity in many family groups.
DD
Dd
dd
Extinction event, e.g. volcanic eruption or flood.
Genetic bottleneck: few individuals survive, with low genetic diversity.
New gene pool: population grows as environmental conditions ease, but genetic diversity is very low.

Artificial selection

occurs when humans, rather than environmental factors, determine which genotypes will pass to successive generations.

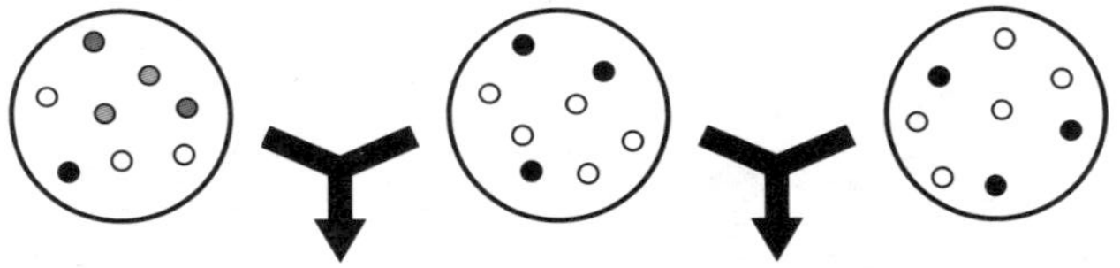

Outbreeding occurs when there is selective controlled reproduction between members of genetically distant populations (different strains or even, for plants, closely related but different species) e.g. introducing of disease resistance from wild sheep.

Inbreeding occurs when there is selective reproduction between closely related individuals, e.g. between offspring of same litter or between parent and child.

Tends to **keep desirable characteristics.**

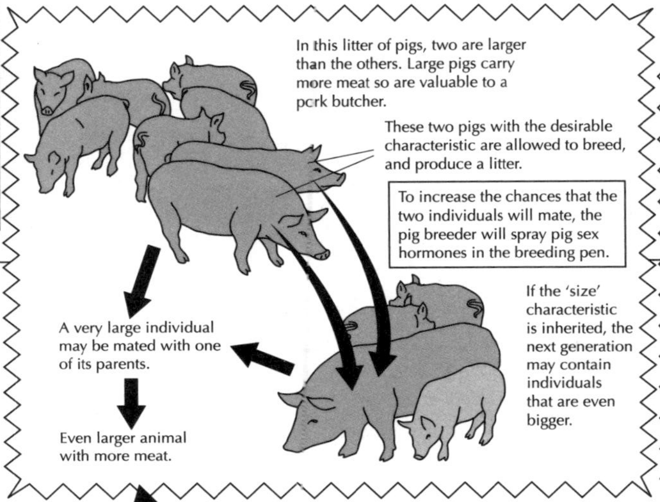

In this litter of pigs, two are larger than the others. Large pigs carry more meat so are valuable to a pork butcher.
These two pigs with the desirable characteristic are allowed to breed, and produce a litter.
To increase the chances that the two individuals will mate, the pig breeder will spray pig sex hormones in the breeding pen.
If the 'size' characteristic is inherited, the next generation may contain individuals that are even bigger.
A very large individual may be mated with one of its parents.
Even larger animal with more meat.

But! Inbreeding may cause:
- reduced fertility;
- lowered disease resistance as **genetic variation is reduced**.

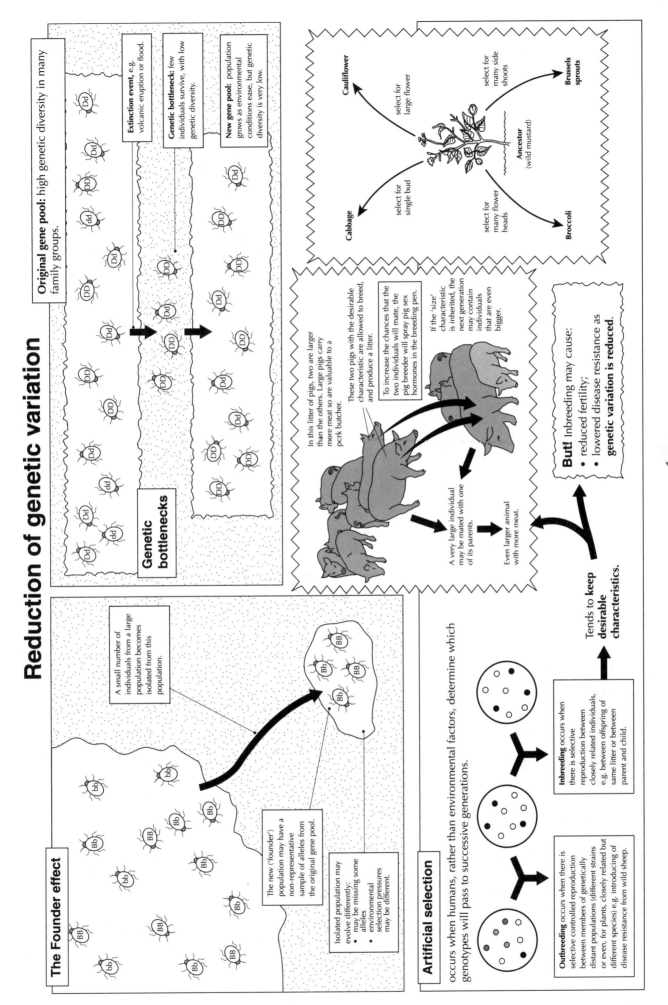

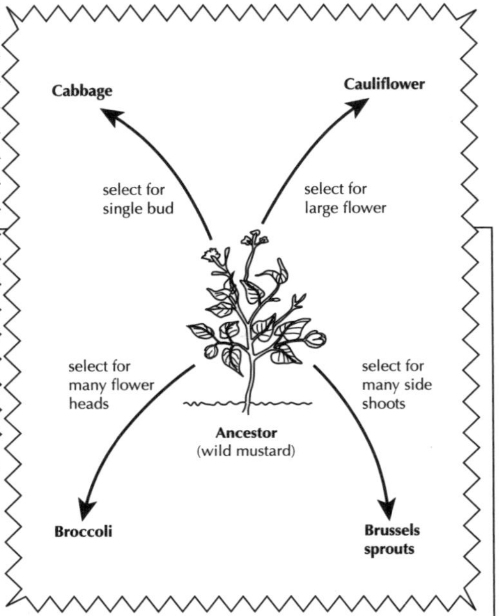

Cabbage
Cauliflower
select for single bud
select for large flower
select for many flower heads
select for many side shoots
Ancestor
(wild mustard)
Broccoli
Brussels sprouts

Genes control cell characteristics

DNA —Transcription→ RNA —Translation→ **Protein**
↓
Characteristics

DNA in nucleus contains the **genes** (specific sequences of bases which code for particular proteins), **regulators** (which switch genes 'on' or 'off') and many stretches with no apparent function (**junk** or **redundant DNA**).

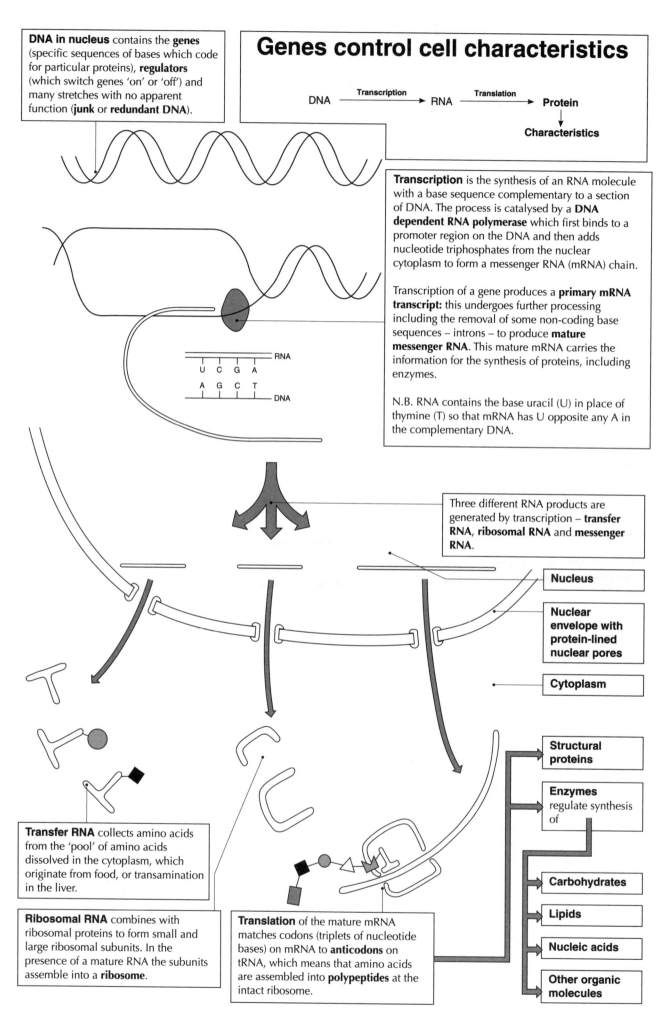

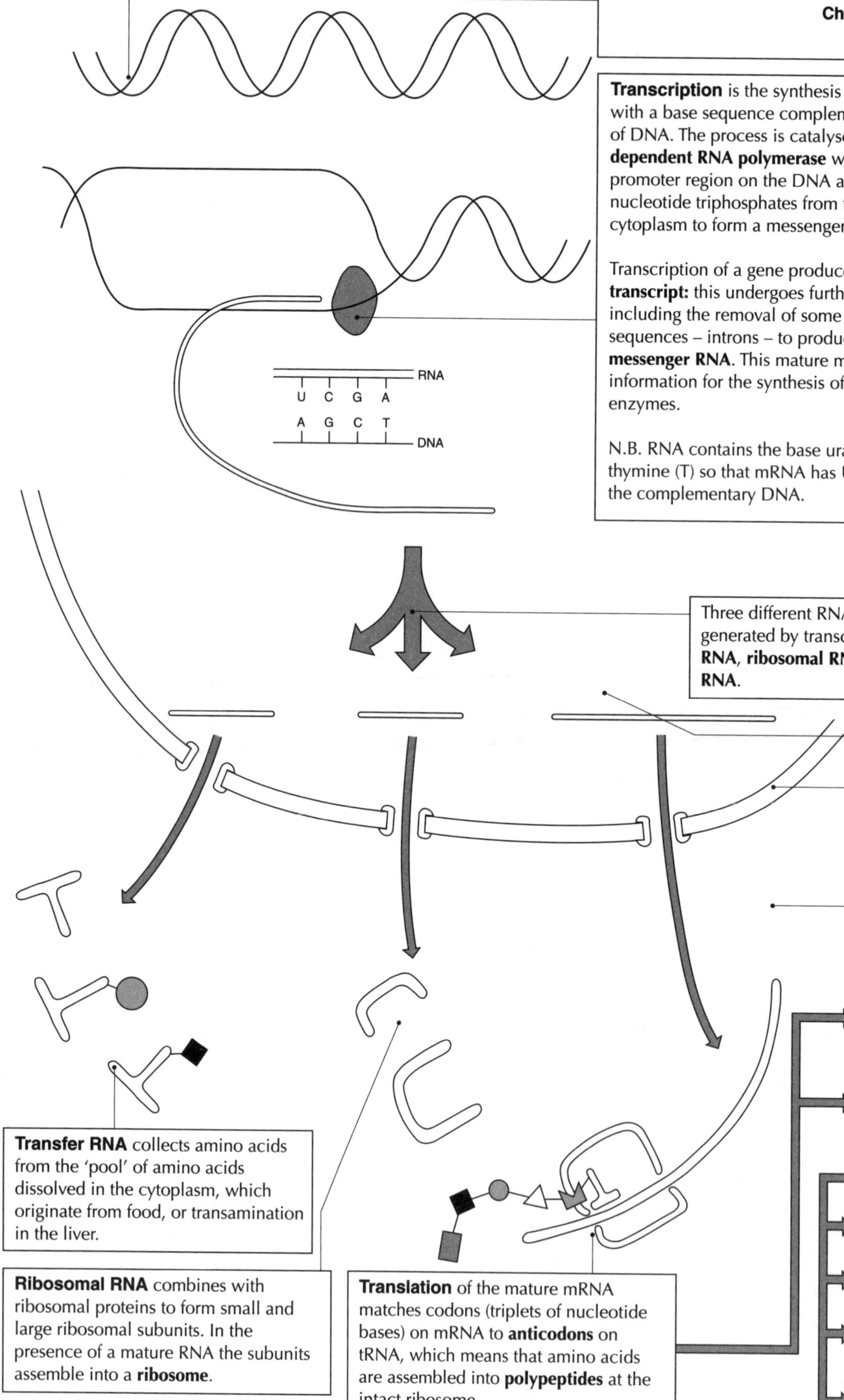

Transcription is the synthesis of an RNA molecule with a base sequence complementary to a section of DNA. The process is catalysed by a **DNA dependent RNA polymerase** which first binds to a promoter region on the DNA and then adds nucleotide triphosphates from the nuclear cytoplasm to form a messenger RNA (mRNA) chain.

Transcription of a gene produces a **primary mRNA transcript:** this undergoes further processing including the removal of some non-coding base sequences – introns – to produce **mature messenger RNA**. This mature mRNA carries the information for the synthesis of proteins, including enzymes.

N.B. RNA contains the base uracil (U) in place of thymine (T) so that mRNA has U opposite any A in the complementary DNA.

Three different RNA products are generated by transcription – **transfer RNA**, **ribosomal RNA** and **messenger RNA**.

Nucleus

Nuclear envelope with protein-lined nuclear pores

Cytoplasm

Transfer RNA collects amino acids from the 'pool' of amino acids dissolved in the cytoplasm, which originate from food, or transamination in the liver.

Ribosomal RNA combines with ribosomal proteins to form small and large ribosomal subunits. In the presence of a mature RNA the subunits assemble into a **ribosome**.

Translation of the mature mRNA matches codons (triplets of nucleotide bases) on mRNA to **anticodons** on tRNA, which means that amino acids are assembled into **polypeptides** at the intact ribosome.

Structural proteins

Enzymes regulate synthesis of

- **Carbohydrates**
- **Lipids**
- **Nucleic acids**
- **Other organic molecules**

Replication is DNA ⟶ DNA and occurs during **mitosis** and **meiosis**.

DON'T CONFUSE THESE PROCESSES!

Transcription is DNA ⟶ mRNA
Translation is mRNA ⟶ protein
and occur during **protein synthesis**.

The **genotype** of a cell or organism is the total DNA in its genes.

The **phenotype** of an organism is the total of its measurable characteristics.

phenotype results from **genotype + effects of the environment**.

Definitions in genetics

A **homologous pair** of chromosomes are the same length, have the centromere in the same position and carry the same genes (though not necessarily the same alleles of those genes) in identical positions. In most cases one of the pair is **paternal** (from the male parent), the other **maternal** (from the female parent).

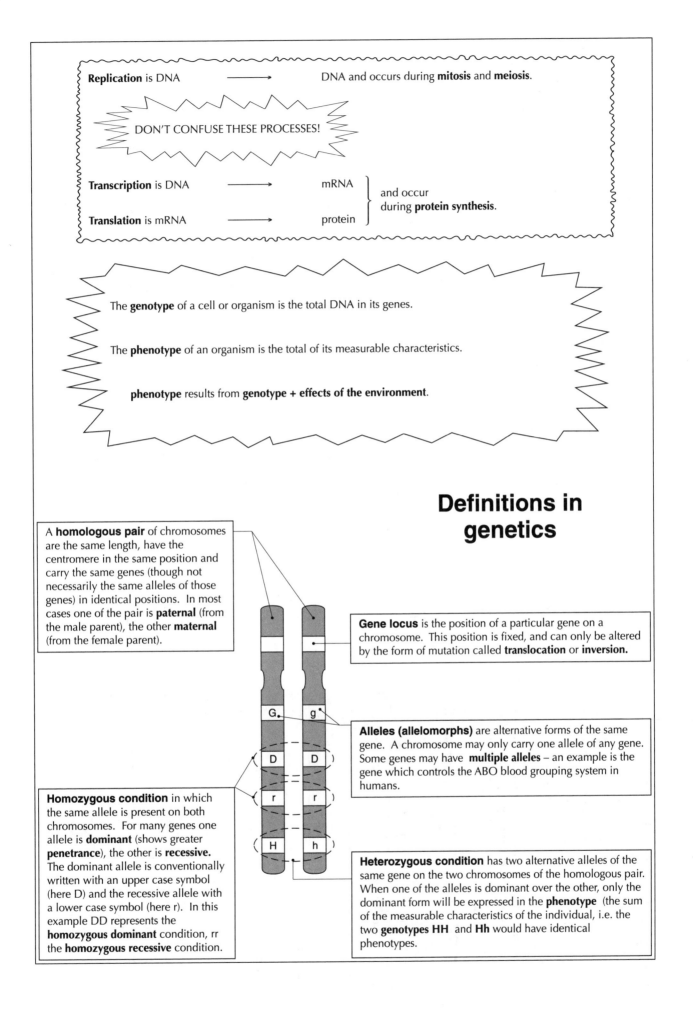

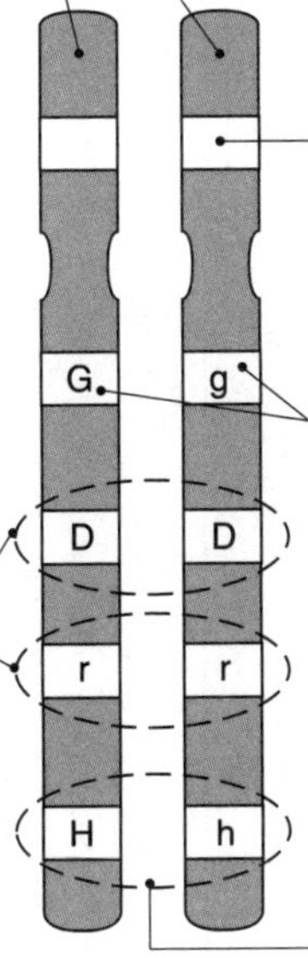

Gene locus is the position of a particular gene on a chromosome. This position is fixed, and can only be altered by the form of mutation called **translocation** or **inversion.**

Alleles (allelomorphs) are alternative forms of the same gene. A chromosome may only carry one allele of any gene. Some genes may have **multiple alleles** – an example is the gene which controls the ABO blood grouping system in humans.

Homozygous condition in which the same allele is present on both chromosomes. For many genes one allele is **dominant** (shows greater **penetrance**), the other is **recessive.** The dominant allele is conventionally written with an upper case symbol (here D) and the recessive allele with a lower case symbol (here r). In this example DD represents the **homozygous dominant** condition, rr the **homozygous recessive** condition.

Heterozygous condition has two alternative alleles of the same gene on the two chromosomes of the homologous pair. When one of the alleles is dominant over the other, only the dominant form will be expressed in the **phenotype** (the sum of the measurable characteristics of the individual, i.e. the two **genotypes HH** and **Hh** would have identical phenotypes.

Meiosis and variation

Meiosis separates chromosomes, halving the diploid number, and introduces variation to the haploid products.

During **prophase I** each replicated chromosome (comprising two chromatids) pairs with its **homologous partner,** i.e. the diploid number of chromosomes produces the haploid number of homologous pairs.

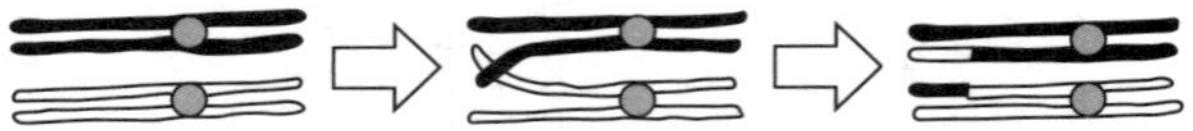

Crossing over occurs when all four chromatids are at **synapsis** (exactly aligned) – non-sister chromatids may cross over, break and reassemble so that **parental** gene combinations are replaced by **recombinants.** This is a major source of **genetic variation.**

At **anaphase I** and **telophase I** there is separation of **whole chromosomes** (i.e. of **pairs** of chromatids).

The products of meiosis I now contain the **haploid** (*n*) number of chromosomes, although each chromosome comprises two chromatids.

During the **second meiotic division** (metaphase II, anaphase II and telophase II) there is a modified mitosis which **separates the two sister chromatids of each chromosome.**

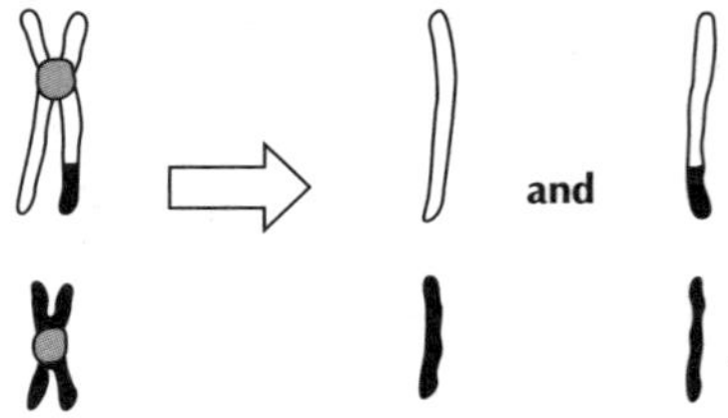

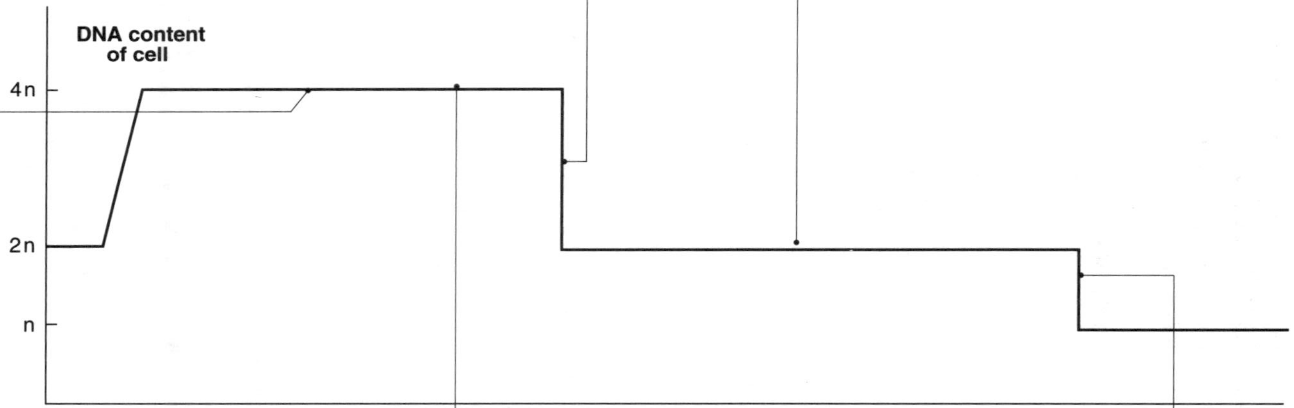

Importance of meiosis

1. It must occur in sexually reproducing organisms or the chromosome number would be doubled at fertilisation.

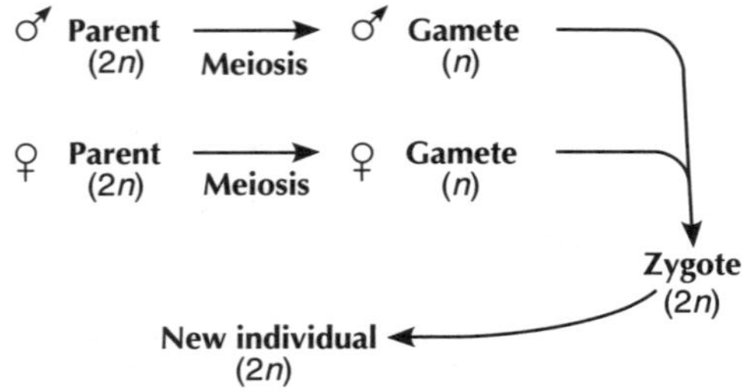

2. Crossing over, independent assortment and random fertilisation promote **genetic variation.** This provides new material for natural selection to work on during evolution.

Following prophase I, **independent assortment** can align the chromosomes in different ways on the **metaphase** plate.

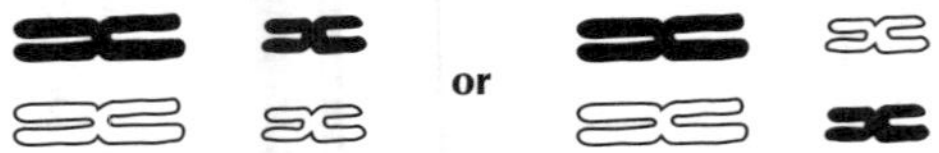

The number of possible combinations of chromosomes is great i.e. 2^n, where *n* is the number of homologous pairs. This is a second major source of **genetic variation** resulting from meiotic division.

At the end of telophase II **cytokinesis** produces daughter nuclei which have **half the number of chromosomes of the parent cell.**

i.e. Diploid (2*n*) ⟶ Haploid(*n*) Gametes

In addition, the unpaired chromosomes in the gametes may contain **new gene combinations** as a result of **crossing over** and **independent assortment.**

Further genetic variation results from the **random combination of gametes at fertilisation,** i.e. any male gamete may fuse with any female gamete.

Monohybrid inheritance

Phenylketonuria in humans is an example of **monohybrid inheritance.**

Phenylketonuria: results from a lack of the liver enzyme **phenylaline hydroxylase**: blood phenylalanine levels are raised, causing a number of effects.

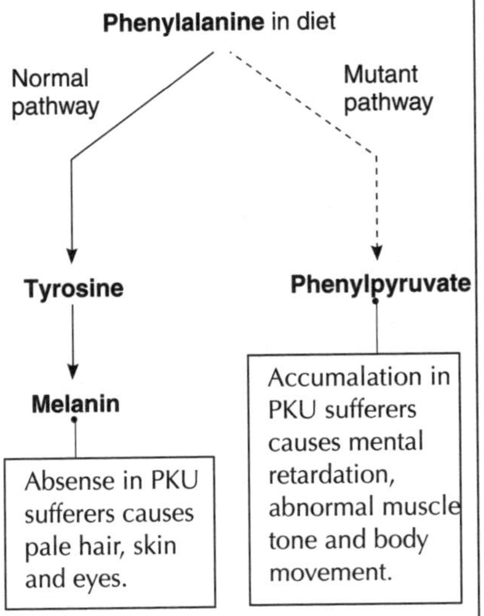

Early diagnosis advises a phenylalanine-restricted diet for children, markedly reducing PKU symptoms.

If one parent is homozygous normal, the other homozygous mutant:

The homozygous normal individual is represented as NN, the mutant individual as nn since in this case normal is dominant to mutant (with cystic fibrosis).

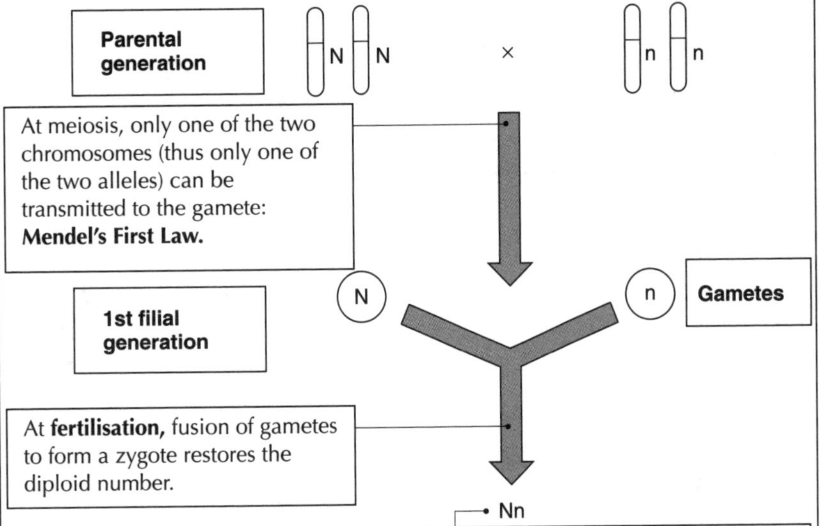

This individual is **genotypically** heterozygous, but **phenotypically** normal, i.e. a **carrier** of the recessive allele for cystic fibrosis.

If both parents are carriers (i.e. *heterozygous*)

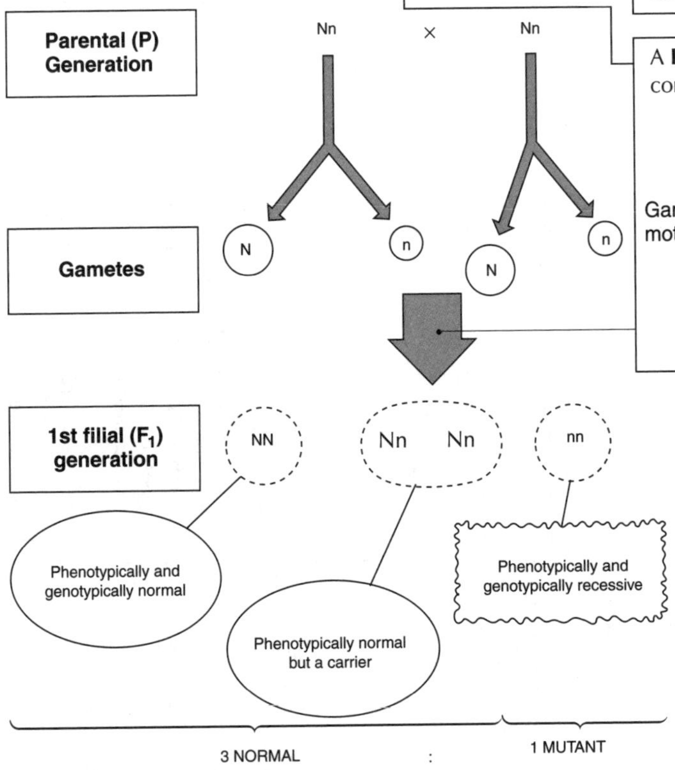

A **Punnett square** can be used to predict the possible combinations of alleles in the zygote.

	Gametes from father ♂ N	n
Gametes from mother ♀ N	NN	Nn
n	Nn	nn

N.B. the 3 : 1 ratio is only approximate unless the number of offspring is very large (unlikely in humans), because:

- Alleles may not be distributed between viable gametes in equal numbers;
- Fusion of gametes is completely random – it is a matter of chance whether one male gamete fuses with a particular female gamete.

Other significant examples of monohybrid inheritance in humans:

Albinism: autosomal recessive

Cystic fibrosis: autosomal recessive (the most common lethal allele in Caucasian populations)

Huntington's disease: autosomal dominant

Tay-Sach's disease: autosomal recessive (prevalent in Jews of Eastern European origin).

Mitosis and growth

The significance of mitosis is that it involves duplication of the genetic material and its equal distribution to each of two 'daughter' cells. **There is very little variation.**

Importance of mitosis

1. It is the process which provides the cells required for the **growth** of zygote into a functioning multicellular organism – this requires an increase in number of cells from one to 6 x 10^{13} in humans.
2. It supplies the cells to **repair** worn-out or damaged tissues. In the human the replacement of skin, gut and lung linings and blood cells requires about 1 x 10^{11} cells per day.
3. It maintains the chromosome number. Daughter cells have identical sets of chromosomes and so function harmoniously as part of the tissue, organ or organism.
4. **Asexual reproduction** provides offspring which are genetically identical to the parent – ideal when rapidly establishing a population. Mitosis provides the cells which make up the fragments of the parent body dispersed during this form of reproduction.

The cell cycle typically lasts from 8 to 24 h in humans – the nuclear division (mitosis) occupies about 10% of this time.

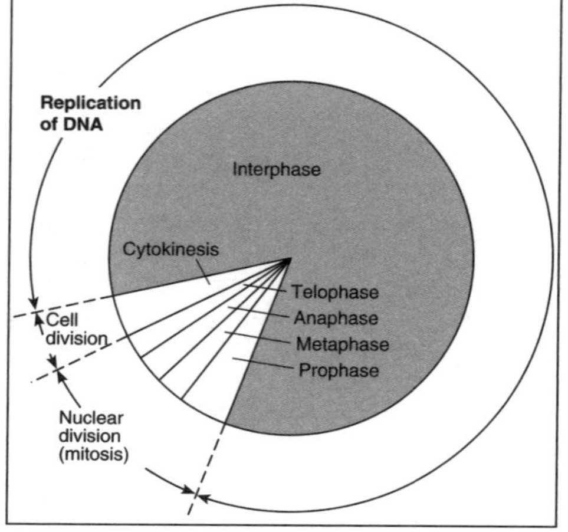

Interphase
DNA only visible as indistinct mass of **chromatin.** Nucleolus and nuclear membrane still intact. Centrioles lie close to one another. DNA replication occurs.

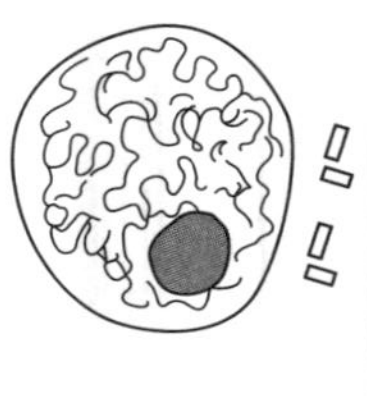

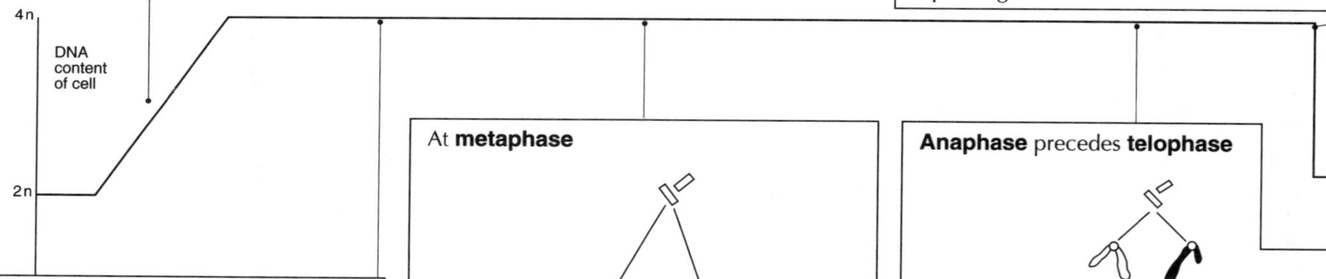

As **interphase** moves to **prophase**

DNA replicates. Each chromosome is now **two identical chromatids,** held together at the centromere. Nucleolus and nuclear membrane break down. **Centrioles** move to opposite poles, forming a **spindle** of **microtubules.**

The rate of replication may be extremely high – in humans 1 x 10^{11}m of DNA are produced per day (this is almost 700 miles per second!).

At **metaphase**

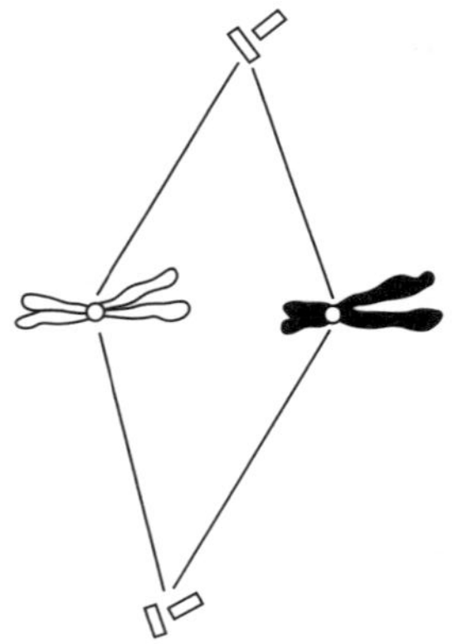

Chromosomes now attached to spindle at **kinetochore** on the centromere. The chromosomes are arranged in such a way that one chromatid from each pair lies on each side of the **equator.**

VINCRISTINE and VINBLASTINE are anti-cancer drugs which inhibit the formation of the spindle of microtubules.

Anaphase precedes **telophase**

Centromere divides and spindle fibres contract to pull **chromatids** to opposite poles. The early separation of the chromatids is **anaphase,** and the separation is complete (so that the chromatids are now **chromosomes**) when the spindle breaks down and the nuclear membrane reforms at **telophase.**

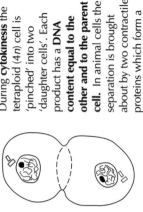

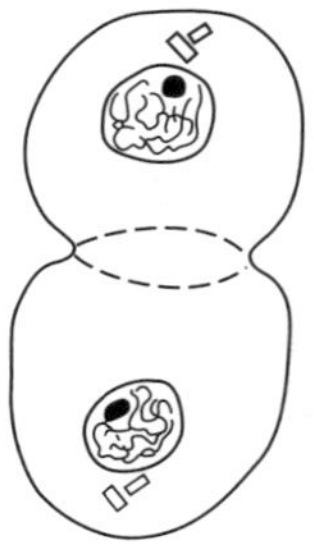

During **cytokinesis** the tetraploid (4*n*) cell is 'pinched' into two 'daughter cells'. Each product has a **DNA content equal to the other and to the parent cell.** In animal cells the separation is brought about by two contractile proteins which form a **cleavage furrow;** in plant cells a **cell plate** is laid down and covered with cellulose to form a separating **cell wall.**

Important sites of mitosis in humans include gut epithelium and bone marrow - these rapidly-dividing cells are susceptible to non-specific anti-cancer drugs and those patients undergoing treatment with these cytotoxic compounds often experience side effects which include mouth ulceration, irritation of gut and reduced blood cell production.

Cancer may result from uncontrolled cell division.

Normal healthy cell

In a healthy cell there are many genes concerned with the regulation of the cell cycle, i.e. with the control of cell division. These 'normal' genes may be called **proto-oncogenes.**

Proto-oncogenes may exert their effect via a number of different protein products.

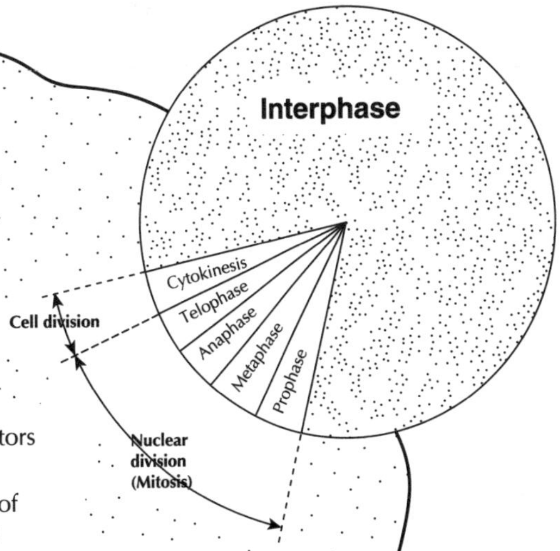

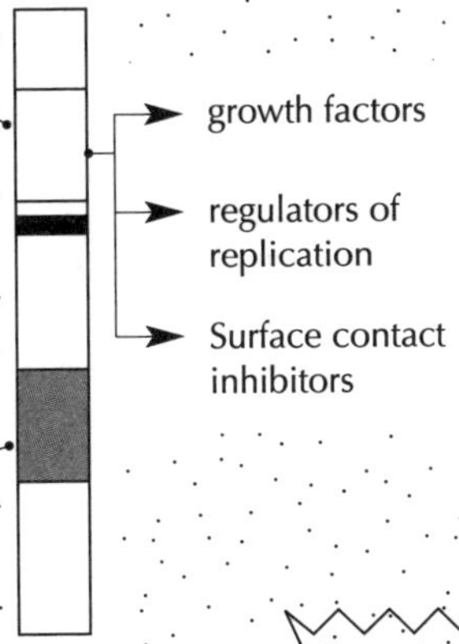

Tumour-suppressing genes: these normally prevent the over-expression of these proto-oncogenes.

Important anti-cancer drugs kill cancer cells or prevent cell division:

- **Alkylating agents** - inhibit DNA replication
- **Antimetabolites** - prevent normal cellular metabolism
- **Spindle inhibitors** - prevent separation of chromatids as no spindle is formed.

Factors promoting the proto-oncogene to oncogene conversion may include:

- ionising radiation, e.g. UV radiation and skin cancers;
- Carcinogens (cancer-causing chemicals), e.g. tars and lung cancer;
- Viruses, e.g. herpes simplex II may be linked to cervical cancer.

Conversion to a cancer cell is usually a **multi-step** process, involving a series of sequential changes to the genotype, which may depend upon a number of intracellular and extracellular factors acting over a very long period, but **inevitably involves a conversion of the normal proto-oncogene to the mutant oncogene.**

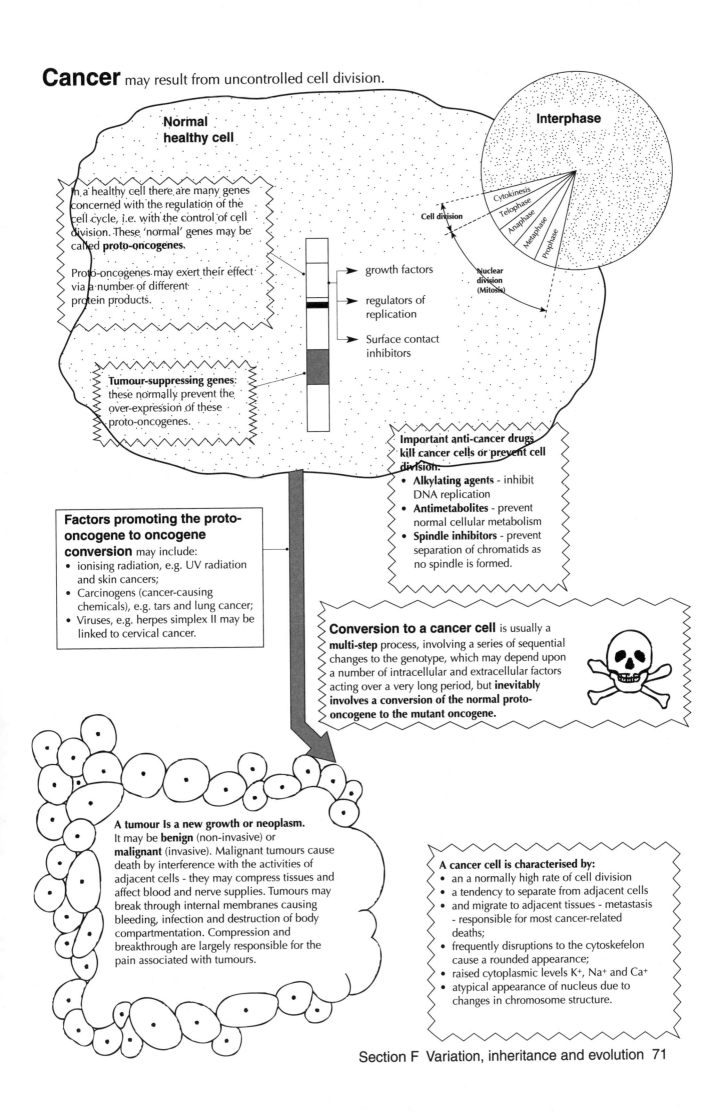

A tumour Is a new growth or neoplasm. It may be **benign** (non-invasive) or **malignant** (invasive). Malignant tumours cause death by interference with the activities of adjacent cells - they may compress tissues and affect blood and nerve supplies. Tumours may break through internal membranes causing bleeding, infection and destruction of body compartmentation. Compression and breakthrough are largely responsible for the pain associated with tumours.

A cancer cell is characterised by:

- an a normally high rate of cell division
- a tendency to separate from adjacent cells
- and migrate to adjacent tissues - metastasis - responsible for most cancer-related deaths;
- frequently disruptions to the cytoskefelon cause a rounded appearance;
- raised cytoplasmic levels K^+, Na^+ and Ca^+
- atypical appearance of nucleus due to changes in chromosome structure.

Gene mutation and sickle cell anaemia

Sickle cell anaemia is the result of a **single gene (point) mutation** and the resulting **error in protein synthesis.**

DNA carries the instructions for protein synthesis in the sequence of nucleotide bases in its structure.

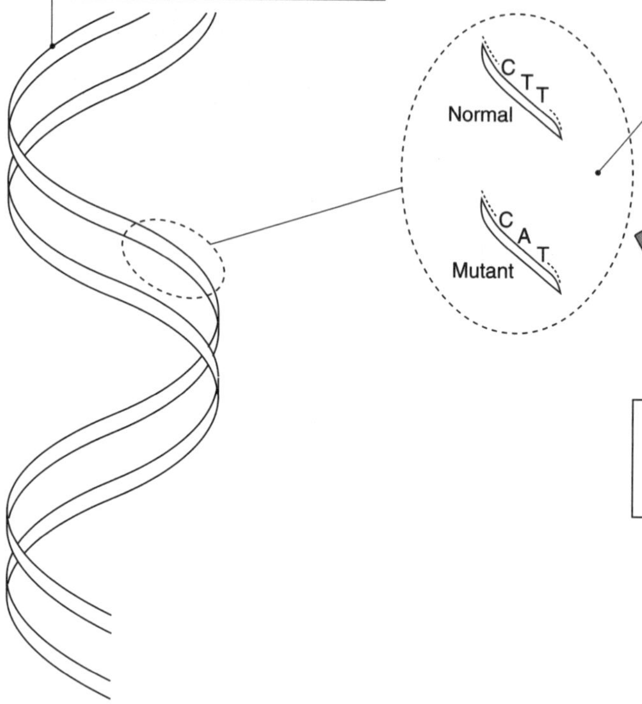

A **mutation** is a change in the DNA. In this case a **base substitution** has taken place – the base A (adenine) has replaced the base T (thymine) so that the triplet codeword CAT has replaced the normal triplet codeword CTT.

Messenger RNA **transcribed** from the mutant DNA will contain the codon GUA instead of the normal GAA.

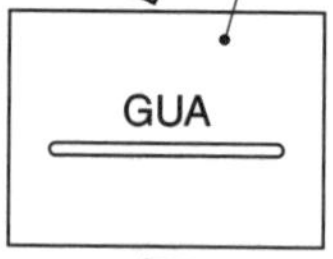

Translation of the 'mutant' mRNA will incorporate the amino acid **valine** (coded for by GUA) instead of **glutamate** (coded for by GAA).

The haemoglobin molecule has four protein chains – two α chains and two β chains. A short part of the amino acid sequence is shown.

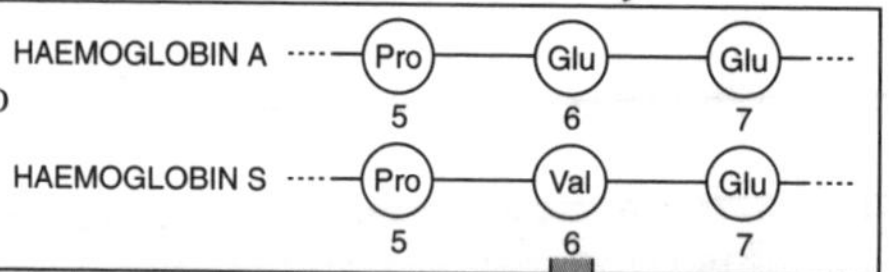

The 'mutant' protein, **haemoglobin S**, contains valine at position 6 of the β chain whereas the 'normal' protein, **heamoglobin A**, contains glutamate in the equivalent position.

The valine side chain is non-polar whereas the glutamate side chain is highly polar: the presence of valine makes haemoglobin S much less soluble than haemoglobin, causing 'sticky' fibres to form which distort the red blood cells.

Sickle cell anaemia

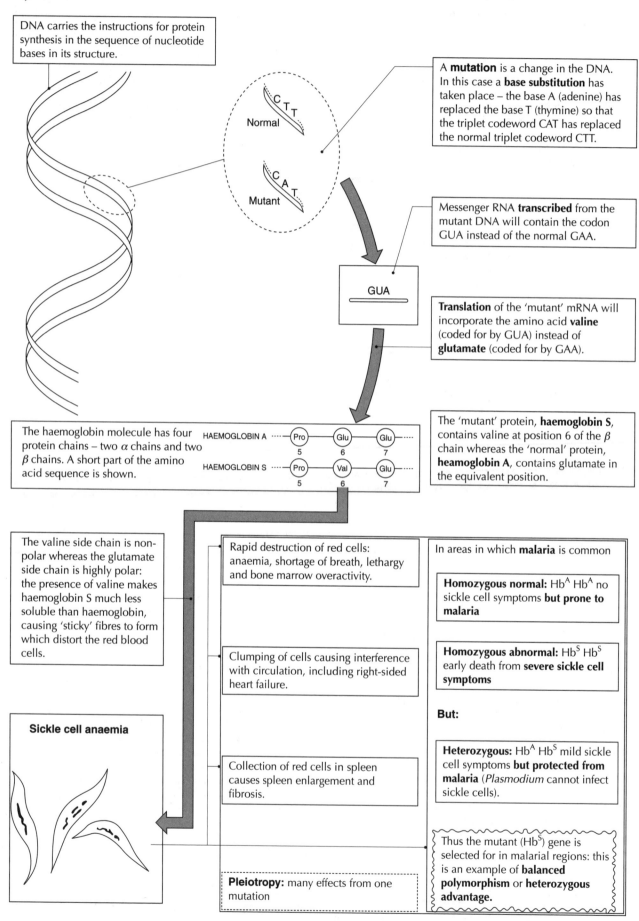

Rapid destruction of red cells: anaemia, shortage of breath, lethargy and bone marrow overactivity.

Clumping of cells causing interference with circulation, including right-sided heart failure.

Collection of red cells in spleen causes spleen enlargement and fibrosis.

Pleiotropy: many effects from one mutation

In areas in which **malaria** is common

Homozygous normal: $Hb^A Hb^A$ no sickle cell symptoms **but prone to malaria**

Homozygous abnormal: $Hb^S Hb^S$ early death from **severe sickle cell symptoms**

But:

Heterozygous: $Hb^A Hb^S$ mild sickle cell symptoms **but protected from malaria** (*Plasmodium* cannot infect sickle cells).

Thus the mutant (Hb^S) gene is selected for in malarial regions: this is an example of **balanced polymorphism** or **heterozygous advantage.**

Chromosome mutation and Down's syndrome

Down's syndrome (trisomy-21) is a chromosome mutation caused by **non-disjunction**.

In a normal meiotic division chromosomes are distributed equally between the gametes.

During **non-disjunction** there is an uneven distribution of the parental chromosomes at meiosis.

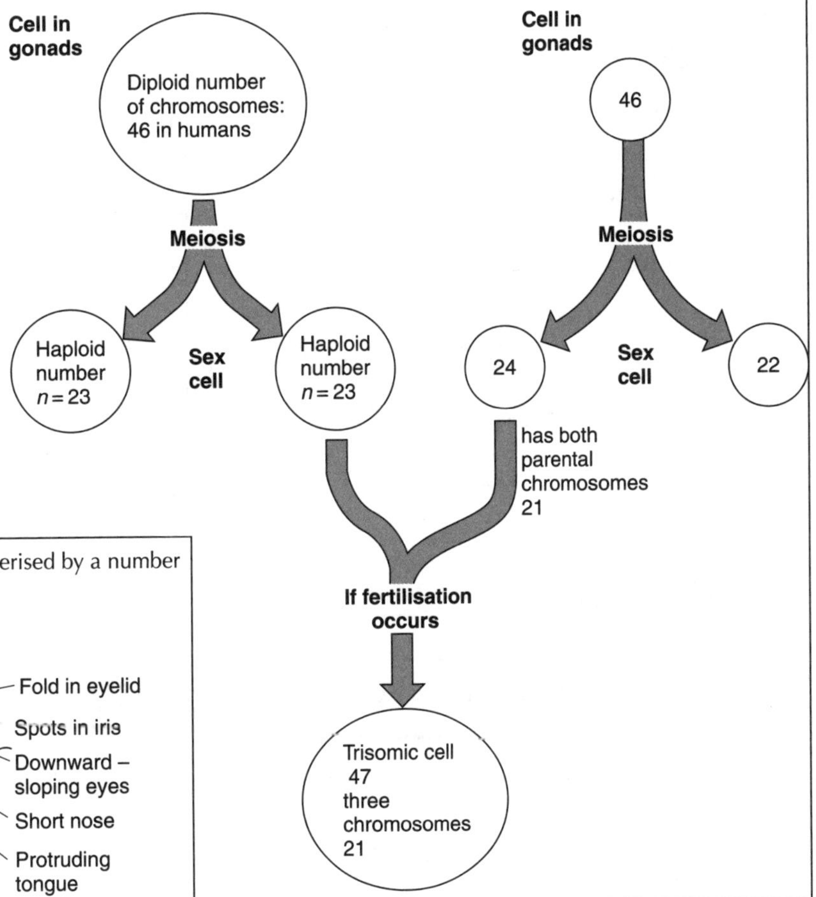

Down's syndrome (trisomy-21) is characterised by a number of distinctive physical features.

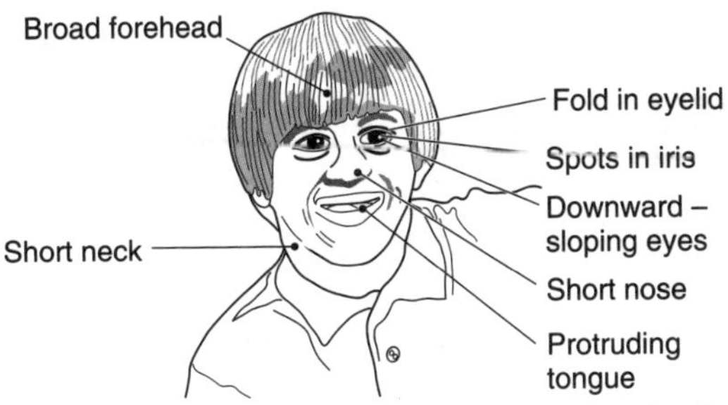

In addition there are congenital heart defects (30% of sufferers die before the age of 10) and mental retardation.

A **karyotype** is obtained by cutting out and rearranging photographic images of chromosomes stained during mitotic metaphase.

Karyotype of Down s Syndrome female

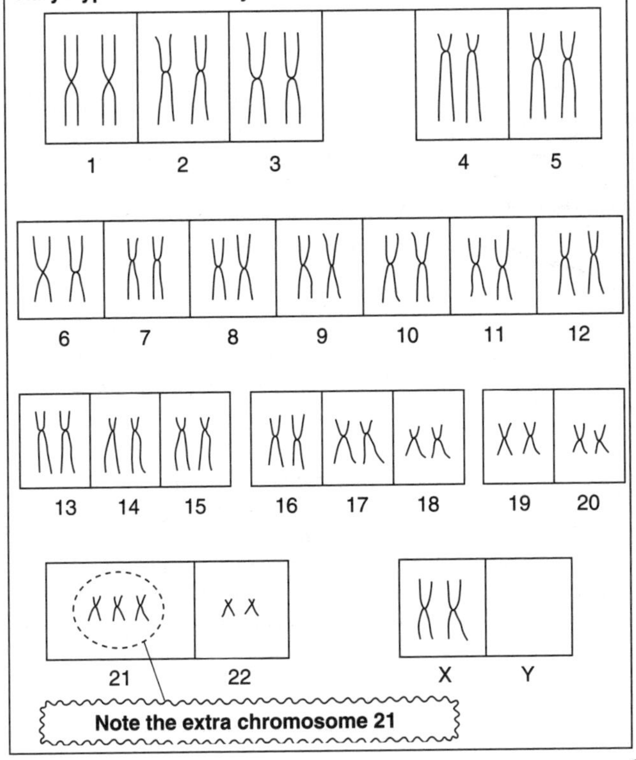

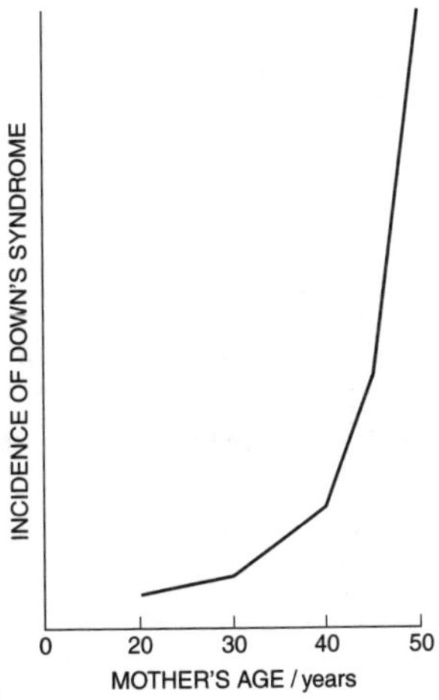

The non-disjunction is most usually the result of the failure of chromosomes to separate at anaphase I – the probability of this happening increases with the length of time the cell remains in prophase I. In the human female all meioses are initiated before puberty so there is an age-related incidence of Down's syndrome – the longer the oocyte takes to complete development the less accurate are the chromosome separations which follow.

Paternal non-disjunction accounts for only about 15% of cases of Down's syndrome.

Two other significant examples of non-disjunction:

Klinefelter's syndrome (XXY) caused by an **extra X chromosome** and resulting in a **sterile male** with **some breast development**.

Turner's syndrome (XO) caused by a **deleted X chromosome** and resulting in a **female** with **underdeveloped sexual characteristics**.

Codominance occurs when **both alleles of a gene express themselves equally in the phenotype**, e.g. flower colour in snapdragons.

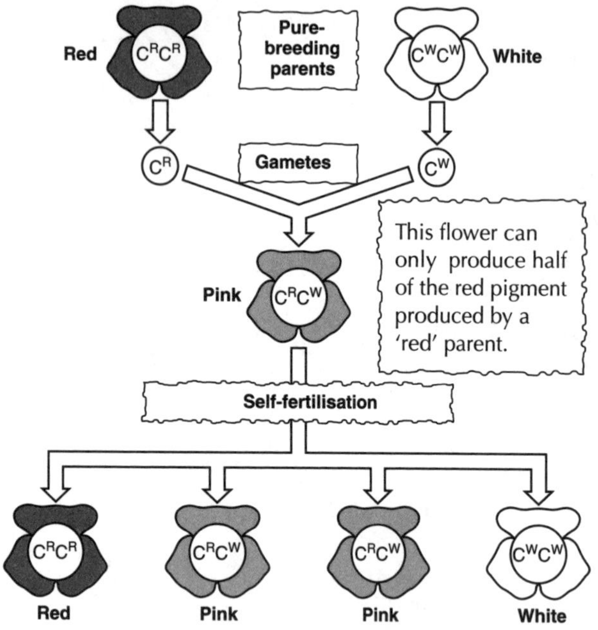

Multiple alleles exist when **a single gene has more than two alleles**. Even though there may be more than two alleles **only two may be present in any zygote**,

e.g. ABO blood group in humans – there are three different alleles of a single gene, I:

- I^A causes production of antigen A on red blood cell
- I^B causes production of antigen B on red blood cell
- I^O causes no antigen production.

I^A and I^B are **co-dominant** and I^O is **recessive** to both.

Consider a cross between heterozygous parents of blood groups A and B:

group A / group B	I^A	I^O
I^B	I^AI^B [AB]	I^BI^O [B]
I^O	I^AI^O [A]	I^OI^O [O]

This cross can produce any of the four possible phenotypes.

Other interactions of alleles and genes

Interaction between genes may provide new phenotypes, e.g. comb shape in poultry.

4 shapes determined by two interacting genes

Rose : R present P absent

Pea : R absent P present

Walnut : both R and P present

Single : neither R nor P present

Pure breeding parents	**Rose**	×	**Pea**
F_1	all walnut		
F_2	**9 Walnut** : **3 Rose** : **3 Pea** : **1 Single**		

Epistasis occurs **when the allele of one gene overrides or masks the effects of another gene**, e.g. coat colour in mice.

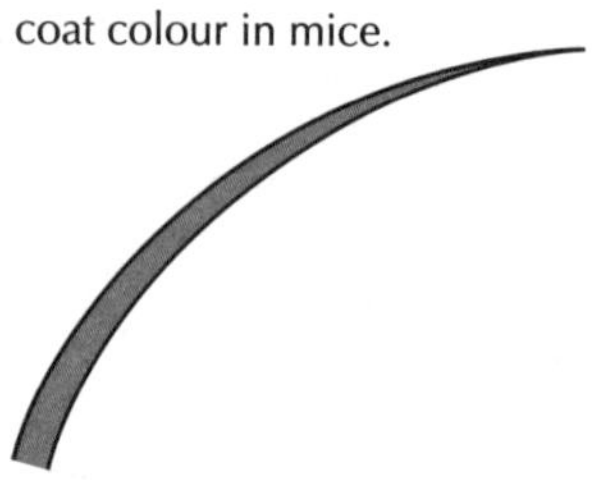

One gene determines presence or absence of colour:

A colour present
a colour absent

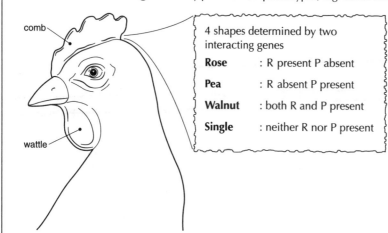

Second gene determines solid colour or banded colour:

B banded (agouti) colour
b solid colour

Gene one is **epistatic** to gene two – any homozygous aa mouse will be albino, whichever allele of gene two is present.

Don't be confused!
In **epistasis** one allele affects **an entirely different gene.**
In **dominance** one allele affects **an alternative allele of the same gene.**

Linkage between genes prevents free recombination of alleles.

Consider this dihybrid cross between two pure-breeding fruit flies

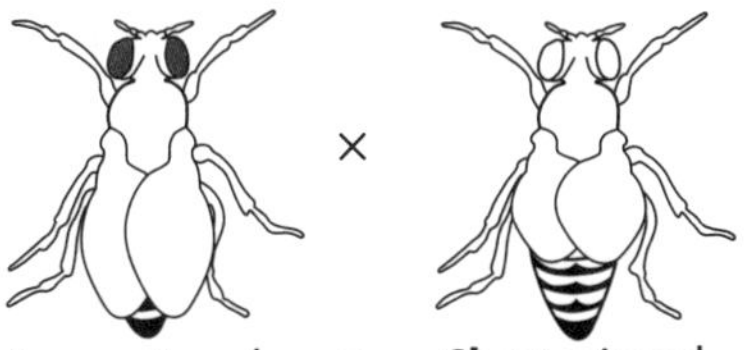

Mendel's 2nd law predicts		**Actual results obtained**	
Long-winged, **Red**-eyed	9	**Long**-winged, **Red**-eyed	12
Long-winged, **White**-eyed	3	**Long**-winged, **White**-eyed	1
Short-winged, **Red**-eyed	3	**Short**-winged, **Red**-eyed	1
Short-winged, **White**-eyed	1	**Short**-winged, **White**-eyed	2

In the F_2 generation

i.e. far more of the F_2 offspring have the **parental combination of characterisitics** than predicted

WHY?

The genes for **wing length** and **eye colour** are **linked**. This means that **they are located on the same chromosome** and thus tend to **pass into gametes together**.

i.e. **parental** chromosomes can be represented as

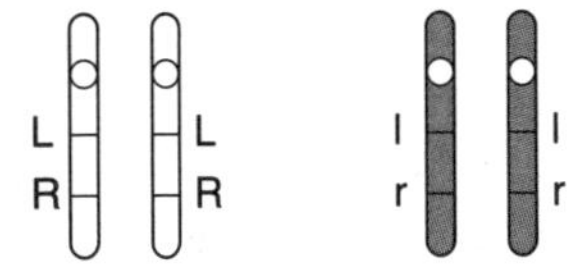

producing gametes

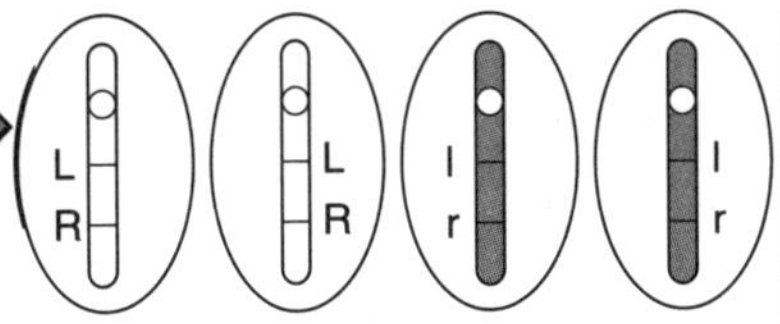

which produce F_1 individuals with

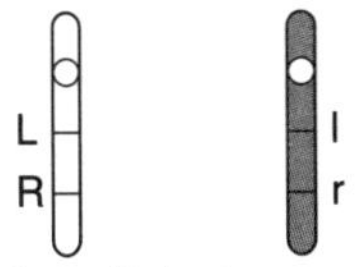

and since LR are linked, as are lr, the F_2 offspring will tend to be

from ♀ \ from ♂	LR	lr
LR	LONG, RED	LONG, RED
lr	LONG, RED	SHORT, WHITE

That is, the alleles tend to remain in the original parental combinations and so parental phenotypes predominate.

BUT

How can linked alleles be separated?

During meiosis, homologous chromosomes pair up to form bivalents and replicate to form tetrads of chromatids

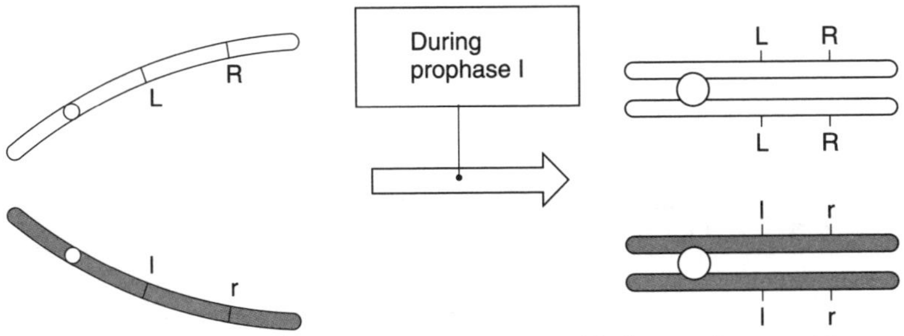

As the chromatids lie alongside one another it is possible for **crossing over** (the exchange of genetic material between adjacent members of a homologous pair) to occur.

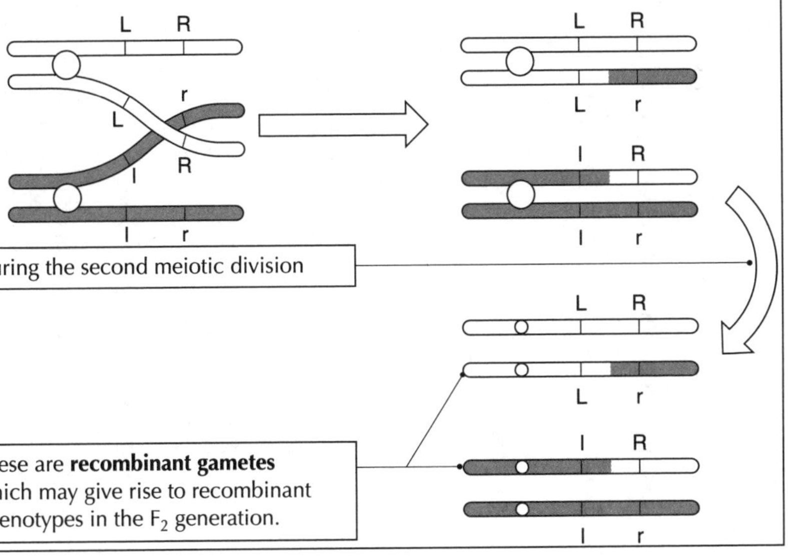

During the second meiotic division

These are **recombinant gametes** which may give rise to recombinant phenotypes in the F_2 generation.

One well-known example of **linkage in humans** involves the genes for **ABO blood groups** and the **nail-patella syndrome**.

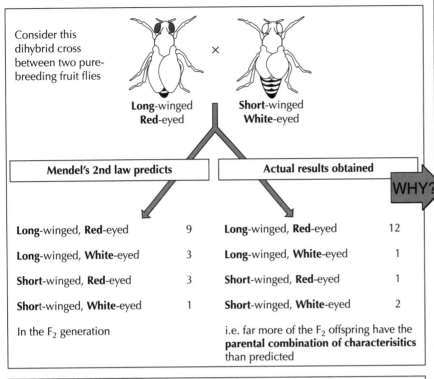

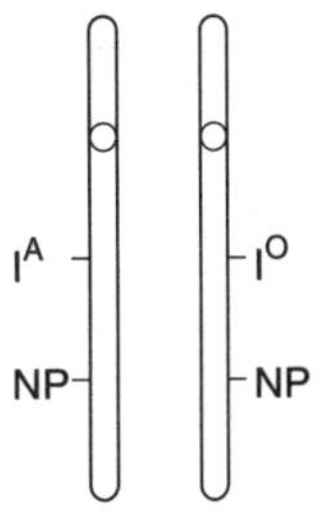

The NP allele is a dominant one: people with this syndrome have small, discoloured nails, and the patella is missing or small and pushed to one side.

Most people with N-P syndrome belong to either A or O blood group.

Sex linkage and the inheritance of sex

Any genes on the X chromosomes will be inherited by both sexes, but whereas the male can only receive **one** of the alleles (he will be XY, and therefore must be **homozygous** for the X-linked allele) the female will be XX and thus may be either **homozygous** or **heterozygous.** This gives females a tremendous genetic advantage since any recessive lethal allele will not be expressed in the heterozygote.

For example, **haemophilia** is an X-linked condition.

Normal gene = H, mutant gene = h

Parents X^HX^h female, carrier × X^HY male, normal

i.e. **both parents have normal phenotype …**

Gametes (X^H) (X^h) (X^H) (Y)

F_1 offspring

♂ Gametes / ♀ Gametes	X^H	Y
X^H	X^HX^H	X^HY
X^h	X^HX^h	X^hY

ie. normal, female X^HX^H; carrier, female X^HX^h; normal, male X^HY; haemophiliac, male X^hY

… but may have a haemophiliac son.

Other significant X-linked conditions include **Duchenne muscular dystrophy, red-green colour blindness** and **coat colour in cats** (where the alleles for ginger (G) and black (g) produce tortoiseshell in the heterozygote X^GX^g: thus there should, in theory, be **no male tortoiseshell cats!**

XX
XY

Any genes carried on the Y chromosome will be received by **all** the male offspring – there is little space for other than sex genes, but one well-known example concerns **webbed toes.**

XY^W — only Y chromosome carries W gene × X X

Gametes (X) (Y^W) (X) (X)

Offspring XY^W XY^W (males with webbed toes) X X X X (normal females)

In mammals sex is determined by two chromosomes which are very different to one another. These are the **heterosomes** – the male is **heterogametic** (XY: can produce both X and Y gametes) and the female is **homogametic** (can only produce X gametes).

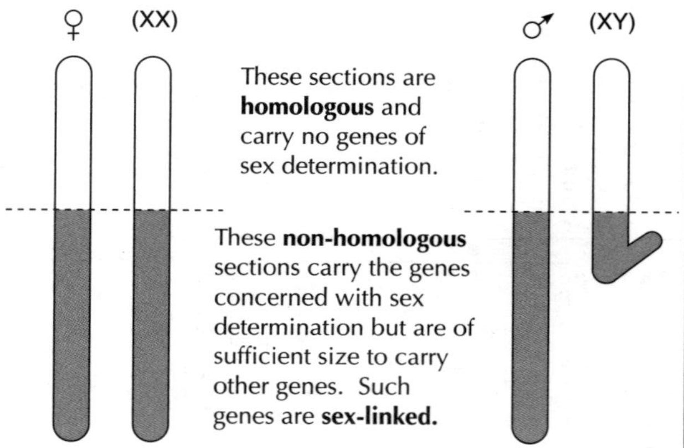

♀ (XX)
♂ (XY)
These sections are **homologous** and carry no genes of sex determination.
These **non-homologous** sections carry the genes concerned with sex determination but are of sufficient size to carry other genes. Such genes are **sex-linked.**

Inheritance of sex is a special form of Mendelian segregation.

♂ XY × ♀ XX

Gametes (X) (Y) (X) (X)

F_1 generation: sex of offspring can be determined from a Punnett square.

♂ Gametes / ♀ Gametes	X	Y
X	XX (female)	XY (male)
X	XX (female)	XY (male)

Theoretically there should be a 1 : 1 ratio of male : female offspring. In humans various factors can upset the ratio – the Y sperm tend to have greater mobility; the XY zygote and embryo is more delicate than the XX embryo. The balance is just about maintained.

Darwin's theory of natural selection

depends on:

- overproduction
- a struggle for existence
- variation within a species and
- survival of the fittest.

Overproduction

Plants and animals in Nature produce more offspring than can possibly survive, yet the population remains relatively constant. There must be many deaths in Nature.

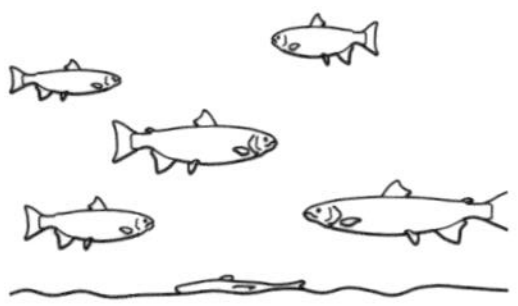

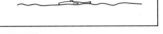

Struggle for existence

Overproduction of this type leads to **competition** – for food, shelter and breeding sites, for example. There is thus a **struggle for existence.** Those factors in the environment for which competition occurs represent **selection pressures.**

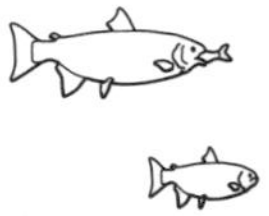

Variation

Much variation is of the **continuous type**, i.e. a range of phenotypes exists between two extremes. The range of phenotypes within the environment will typically show a **normal distribution.**

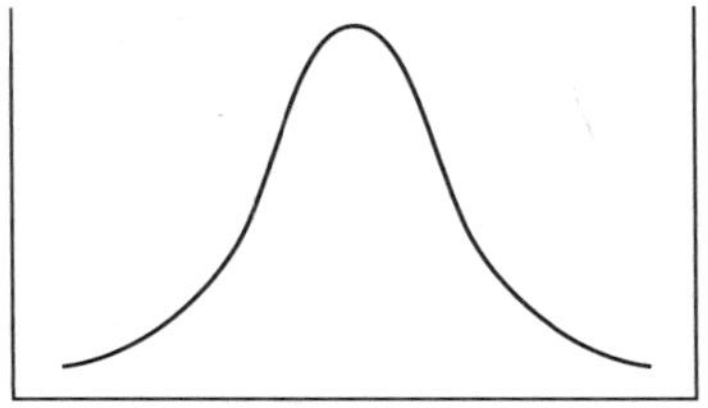

Within a population of individuals there may be considerable **variation** in genotype and thus in phenotype.

Lamarck has an alternative proposal!
Darwin believed:

VARIATION —environment **selects**→ ADAPTATION

Lamarck believed:
Environment **causes** VARIATION

This is called **inheritance of acquired characteristics.**

Survival of the fittest

Variation means that some individuals possess characteristics which would be advantageous in the struggle for existence (and some would be the opposite, of course).

Those possessing the best combination of characteristics would be more competitive in the struggle for existence: they would be more 'fit' to cope with the selection pressures imposed by the environment. This is **natural selection** and promotes **survival of the fittest.**

If variation is **heritable** (i.e. caused by an alteration in genotype) new generations will tend to contain a higher proportion of individuals suited to survival.

Stabilising selection favours intermediate phenotypic classes and operates against extreme forms – there is thus a **decrease** in the frequency of alleles representing the extreme forms.

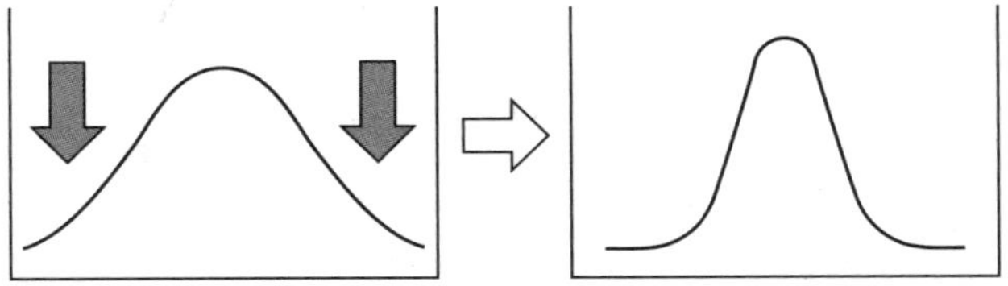

Stabilising selection operates when the phenotype corresponds with optimal environmental conditions, and competition is not severe. It is probable that this form of selection has favoured heterozygotes for **sickle cell anaemia** in an environment in which **malaria** is common, and also works against **extremes of birth weight** in humans.

Directional selection favours one phenotype at one extreme of the range of variation. It moves the phenotype towards a new optimum environment; then stabilising selection takes over. There is a change in the allele frequencies corresponding to the new phenotype.

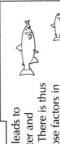

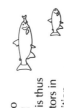

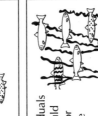

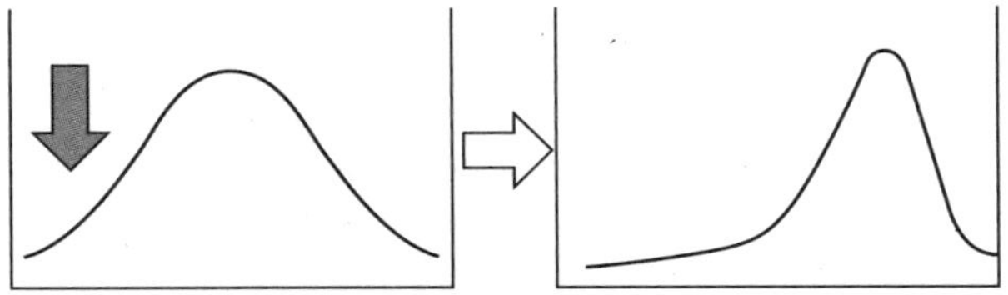

Directional selection has occurred in the case of the peppered moth, *Biston betularia*, where the dark form was favoured in the sooty suburban environments of Britain during the industrial revolution: **industrial melanism.** Another significant example is the development of **antibiotic resistance** in populations of bacteria – mutant genes confer an advantage in the presence of an antibiotic.

Reproductive isolation and speciation

A species is:

'the lowest taxonomic group';

'a group of organisms which can interbreed and produce fertile offspring';

'a group of organisms which share the same ecological niche';

... and members of the same species have a very high proportion of their DNA in common.

Basic process:

Speciation is the end product of evolution

Variation within a population

Evolution by natural selection.

Adaptation

Reproductive isolation: prevents mixing of alleles from different populations.

New species

Mechanical isolation takes place when the reproductive structures are physically incompatible,
e.g. a Great Dane will not mate with a Chihuahua, and some flower species cannot be entered by the same pollinating insect.

Sympatric (= 'same country') **speciation** occurs when populations are reproductively isolated within the same environment.

Behavioural isolation takes place when two different species or populations evolve courtship displays which are essential for successful mating,
e.g. Peahen is only stimulated to mate by peacock's visual courtship display; fireflies' flight patterns and flash displays prevent interspecific mating.

BLEEP

FLASH

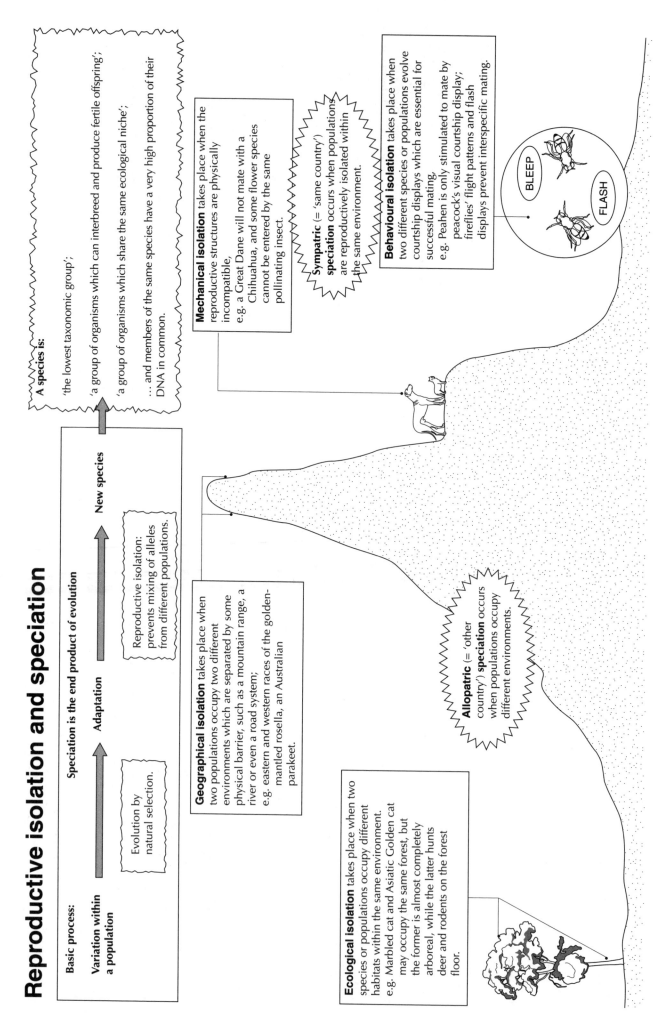

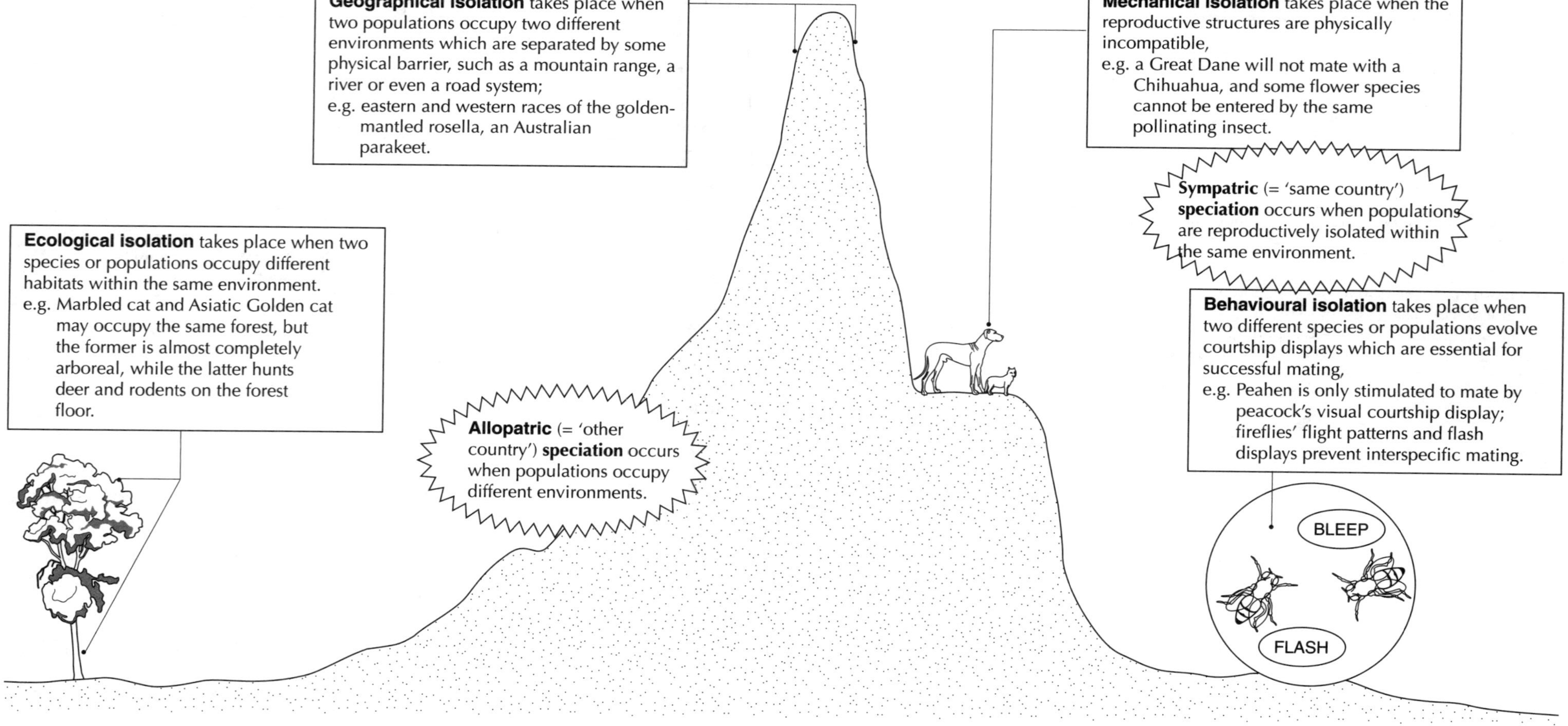

Geographical isolation takes place when two populations occupy two different environments which are separated by some physical barrier, such as a mountain range, a river or even a road system;
e.g. eastern and western races of the golden-mantled rosella, an Australian parakeet.

Allopatric (= 'other country') **speciation** occurs when populations occupy different environments.

Ecological isolation takes place when two species or populations occupy different habitats within the same environment.
e.g. Marbled cat and Asiatic Golden cat may occupy the same forest, but the former is almost completely arboreal, while the latter hunts deer and rodents on the forest floor.

Questions on evolution

Is there any evidence?

Is there a mechanism?

Is evolution still occurring?

Fossils
- Remains of dead animals and plants which pressure, mineralisation and poor conditions for decomposition have turned to stone.
- Sedimentary rocks have been formed at the bottom of ancient seas. Oldest rocks (and fossils) are found in the deepest layers.
- Collection of fossils from different layers gives a time-limit for structural adaptations.
- Important examples are evolution of horse (limbs and teeth), development of the inner ear, links between monocot/dicot plants … and **humans**.

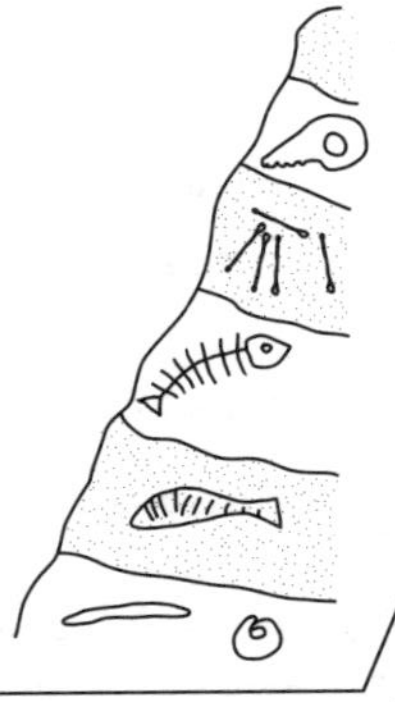

Proteins
- These are composed of long chains of amino acids.
- Some are found very widely through animal and plant groups.
- The number of amino acid changes between same protein from different species generally shows how closely they are related in evolutionary terms.
- Important examples are haemoglobin, cytochrome C and a hormone similar to prolactin.

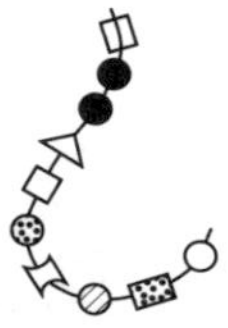

DNA

- polynucleotide chain
- has 'non-coding' sections
- comparison of base sequences in non-coding regions provides a genetic clock for the time of divergence of similar species.

Darwin's theory: natural selection

Overproduction — **Competition** — **Variation** — **Survival of the fittest**

Many examples

Darwin's Finches: selection for beak shape.

Grasses on spoil heaps: selection for metal tolerance.

Peppered moths: selection for camouflage against predators.

This can be bad news for humans!

e.g. Pesticide resistance insects:
- mutations/reproduction **generates** variation;
- pesticide **selects** the pesticide resistance;
- a similar process selects drug-resistant strains of bacteria.

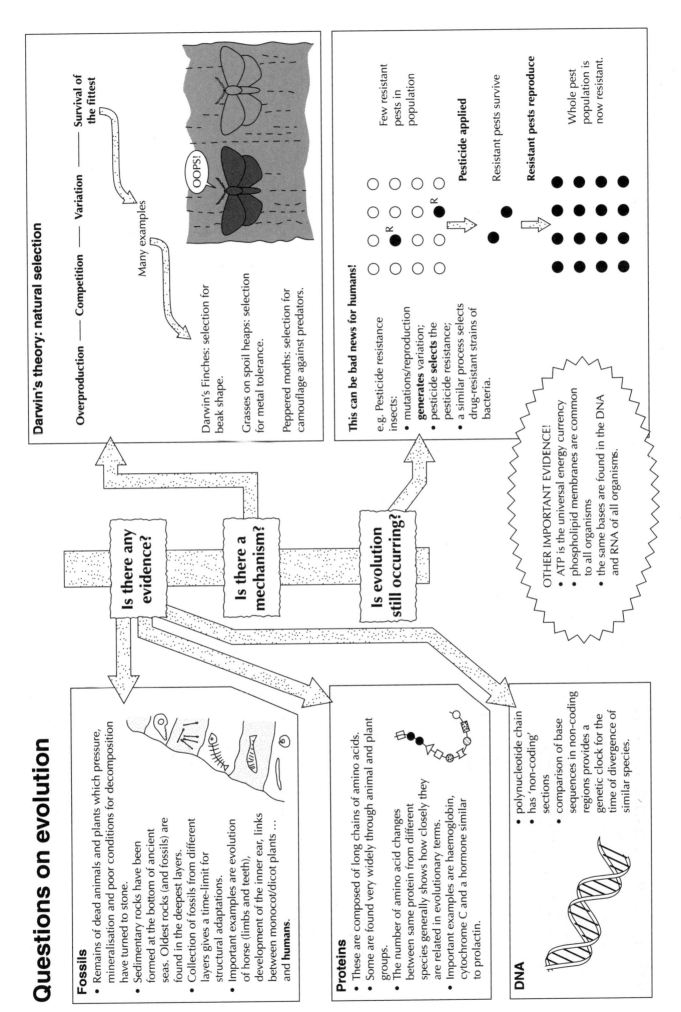

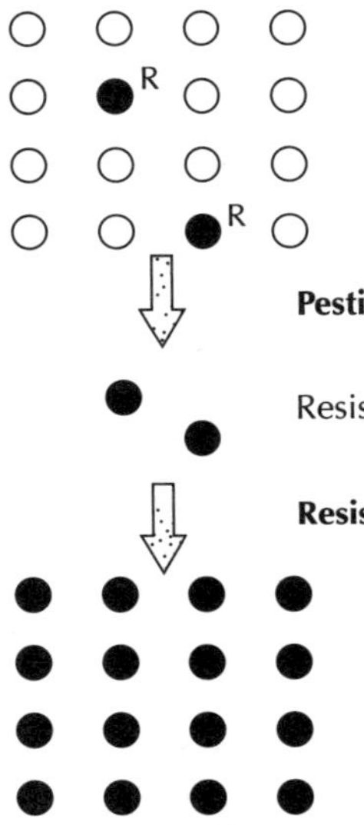

Few resistant pests in population

Pesticide applied

Resistant pests survive

Resistant pests reproduce

Whole pest population is now resistant.

OTHER IMPORTANT EVIDENCE!
- ATP is the universal energy currency
- phospholipid membranes are common to all organisms
- the same bases are found in the DNA and RNA of all organisms.

Artificial selection occurs when humans, rather than environmental factors, determine which genotypes will pass to successive generations.

Polyploidy and plant breeding

Polyploids contain **multiple sets of chromosomes** (chromosome multiplication can be induced by treatment with **colchicine** during mitosis – this inhibits spindle formation and prevents chromatid separation).

Autopolyploids (all chromosomes from the **same** species) e.g. all **bananas** are **triploid** – they are infertile and contain no seeds. Most **potatoes** are **tetraploid** – cells are bigger and tubers are larger. Cultivated **strawberries** are **octoploid.**

Allopolyploidy (sets of chromosomes from **different** species) is possible if the two species have a chromosome complement similar in number and shape. This might allow plant breeders to **combine the beneficial characteristics of more than one species.**

The evolution of **bread wheat** is an important example.

Wild wheat: has brittle ears which fall off on harvesting.

↓ **Selective breeding**

Einkorn wheat: non-brittle but low yielding.

↓ **Polyploidy** with *Agropyron* grass

Emmer wheat: high yielding but difficult to separate seed during threshing.

↓ **Polyploidy** with *Aegilops* grass

Bread wheat: high yielding with easily separated 'naked' seeds.

Inbreeding and outbreeding

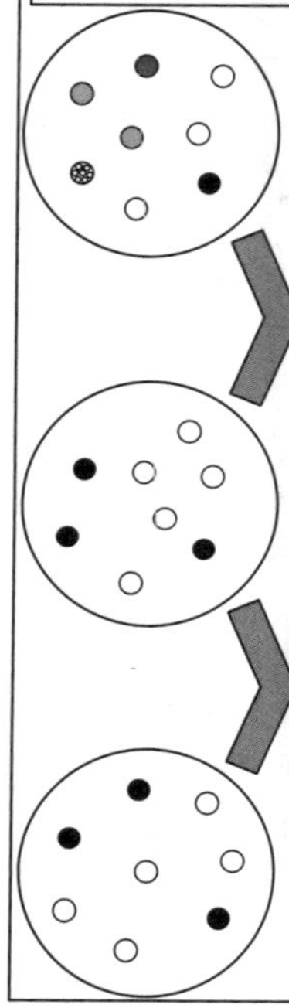

Outbreeding occurs when there is selective controlled reproduction between members of genetically distant populations (different strains or even, for plants, closely related but different species).

→ Tends to **introduce new and superior phenotypes:** the progeny are known as **hybrids** and the development of improved characteristics is called **heterosis** or **hybrid vigour.**

e.g. introduction of disease resistance from wild sheep to domestic strains;
combination of shorter-stemmed 'wild' wheat and heavy yielding 'domestic' wheat.

This may result from increased numbers of dominant alleles or from new opportunities for gene interaction.

Inbreeding occurs when there is selective reproduction between closely related individuals, e.g. between offspring of same litter or between parent and child.

→ Tends to **maintain desirable characteristics**

e.g. uniform height in maize (easier mechanical harvesting);
maximum oil content of linseed (more economical extraction);
milk production by Jersey cows (high cream content).

But it may cause **reduced fertility** and **lowered disease resistance** as genetic variation is reduced.

Thus inbreeding is not favoured by animal breeders.

Techniques with animals are less well advanced than those with plants because:

- animals have a longer generation time and few offspring;
- more food will be made available from improved plants;
- there are many ethical problems which limit genetic experiments with animals.

Two important animal techniques are:

Artificial insemination: allows sperm from a male with desirable characteristics to fertilise a number of female animals.

Embryo transplantation: allows the use of **surrogate mothers** (thus increasing number of offspring) and **cloning** (production of many identical animals with the desired characteristics).

Selection of dairy cattle

- Identify parents that produce offspring with high milk yield
- Father can now inseminate many daughters.

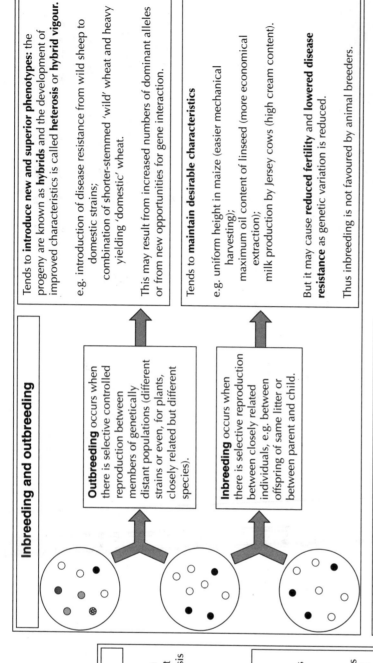

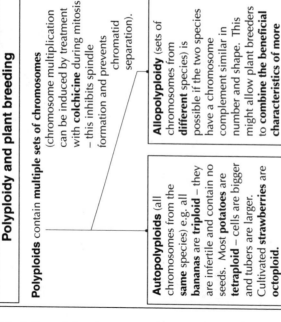

Desirable features:
- high yield of milk;
- high butterfat content;
- docile during milking;
- teat placement on udders.

The Chi-squared (χ^2) test

Analysis of data consisting of discrete (discontinuous) variables

The *expected* ratio of individuals which **show** the dark allele to those which show the light allele is 3:1

In this sample:

	Observed (O)	Expected (E)
Dark	70	81
Light	30	27

If this represents, for example, the results of a monohybrid cross between two heterozygous parents and .

Apply the equation

$$\chi^2 = \sum \frac{(O - E)^2}{E}$$

Category	**O**	**E**	**(O–E)**	**(O–E)²**	**(O–E)²/E**
Dark	78	81	3	9	0.11
Light	30	27	3	9	0.33

$\chi^2 = 0.44$

Substitute χ^2 in a **probability table**: it is necessary to know the number of **degrees of freedom** (= number of possible clauses – 1 : here 2–1 = 1 d.o.f.)

Probability (p) = 0.50

What does this mean?
p = 0.50 means that there is a 0.5 probability (50%) that *any differences between observed and expected results are due solely to chance* : any p > 5% suggests that any deviations from expected results are *not significant.*

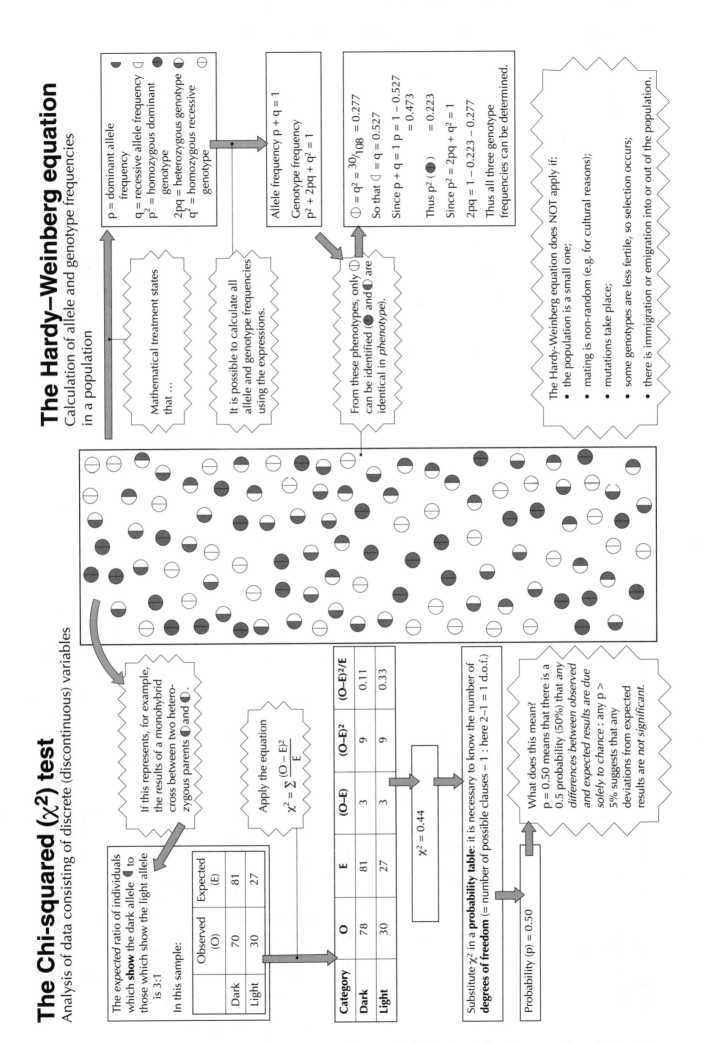

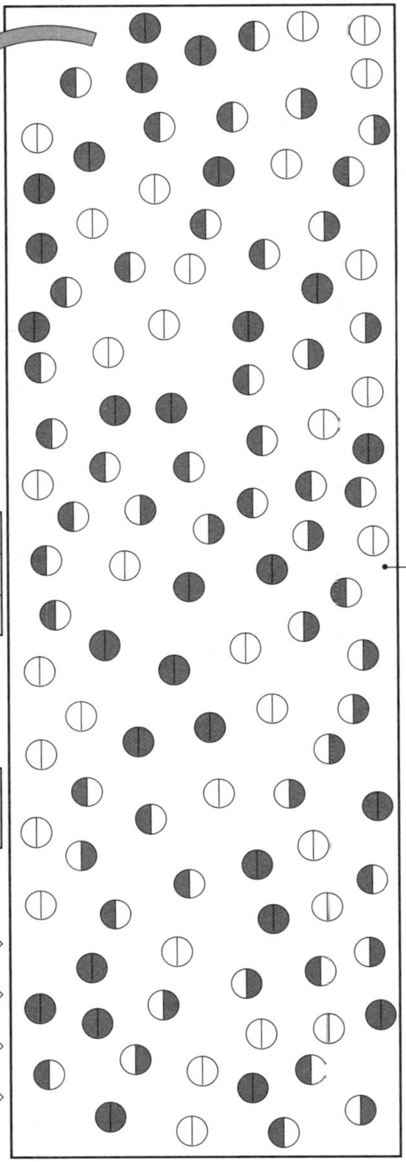

The Hardy–Weinberg equation

Calculation of allele and genotype frequencies in a population

Mathematical treatment states that …

p = dominant allele frequency
q = recessive allele frequency
p^2 = homozygous dominant genotype
2pq = heterozygous genotype
q^2 = homozygous recessive genotype

It is possible to calculate all allele and genotype frequencies using the expressions.

Allele frequency p + q = 1

Genotype frequency
$p^2 + 2pq + q^2 = 1$

From these phenotypes, only can be identified (and are identical in *phenotype*).

$= q^2 = {}^{30}/_{108} = 0.277$

So that $= q = 0.527$

Since p + q = 1 p = 1 – 0.527 = 0.473

Thus p^2 ( ) = 0.223

Since $p^2 = 2pq + q^2 = 1$

2pq = 1 – 0.223 – 0.277

Thus all three genotype frequencies can be determined.

The Hardy-Weinberg equation does NOT apply if:

- the population is a small one;
- mating is non-random (e.g. for cultural reasons);
- mutations take place;
- some genotypes are less fertile, so selection occurs;
- there is immigration or emigration into or out of the population.

DNA (deoxyribonucleic acid) is the genetic material

The Watson–Crick model for DNA suggests that the molecule is a double helix of two complementary, polynucleotide chains.

Nucleotides are the subunits of nucleic acids, including DNA. Each of these subunits is made up of:

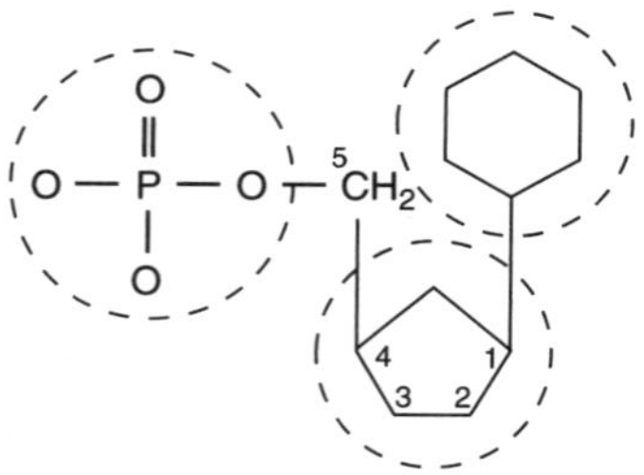

Note that the phosphate group is bonded to the C_5 atom of the pentose sugar.

There are **four different nucleotides** in a DNA molecule; they differ only in the organic (nitrogen) base present.

There are two different **pyrimidine (single ring)** bases, called **cytosine (C)** and **thymine (T)**

and

two different **purine (double ring)** bases called **adenine (A)** and **guanine (G)**.

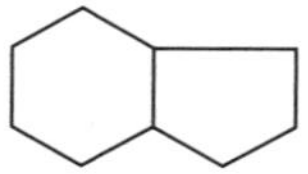

The different dimensions of the purine and pyrimidine bases is extremely important in the formation of the double-stranded DNA molecule.

There are **ten base pairs** for each pitch of the double helix.

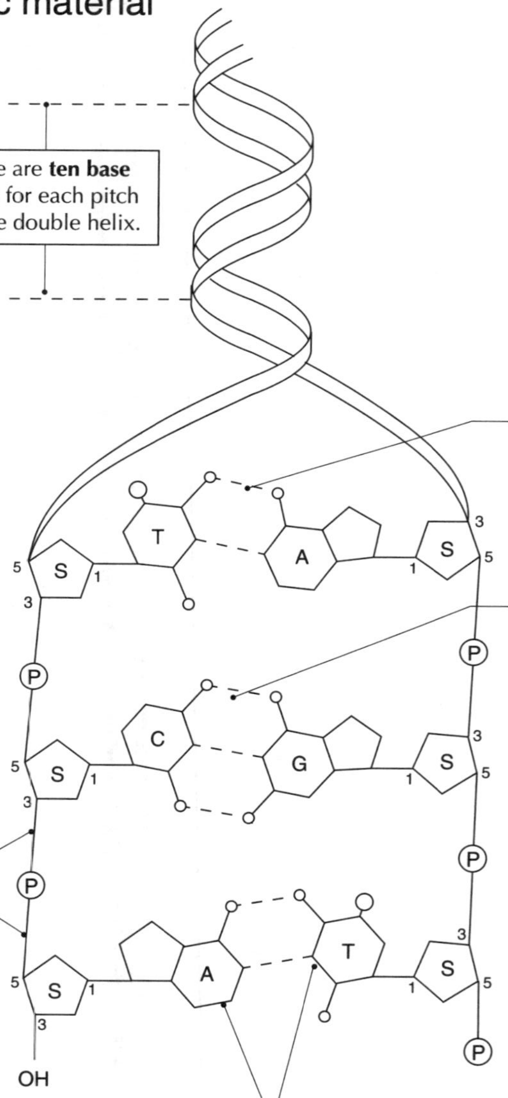

Nucleotides are linked to form a **polynucleotide** by the formation of 3′ 5′ **phosphodiester links** (phosphate bridges) in which a phosphate group forms a bridge between the C_3 of one sugar molecule and the C_5 of the next sugar molecule.

The chains are complementary: because of base pairing, the base sequence on one of the chains automatically dictates the base sequence on the other.

Requirements for a genetic material

- Accurate replication during cell growth and duplication.
- Stable structure to reduce the risk of mutation.
- Must have the potential to carry biological information.
- It must offer a way of transferring information into cell characteristics.

Base pairing in DNA was proposed to explain how two polynucleotide chains could be held together by hydrogen bonds. To accommodate the measured dimensions of the molecule each base pair comprises **one purine - one pyrimidine**.

The double helix is most stable, that is the greatest number of hydrogen bonds is formed, when the base pairs

A :::::::: T (two hydrogen bonds)

and G :::::::: C (three hydrogen bonds)

are formed. These are **complementary base pairs**.

Note that in order to form and maintain this number of ◯ hydrogen bonds the nucleotides are inverted with respect to one another so that the phosphate groups (here shown as P) face in **opposite directions**.

DNA structure is suited to function

- It is very stable: nucleotides are linked by **covalent** bonds.
- It carries coded information: the **order of the nucleotide bases** can be used to control production of other molecules.
- It can be replicated: **specific base pairing** means that DNA can be copied when cells divide.
- It is compact: **folding** of the molecule means a great deal of information can be packed into a small volume.

DNA replication and chromosomes

The double helix is unwound and the base-pairs are separated by the enzyme **DNA helicase.**

Nucleotides are located opposite their complementary base (A–T and G–C) on the **DNA template strand,** the hydrogen bonds form and the nucleotides are linked covalently by the enzyme **DNA polymerase.**

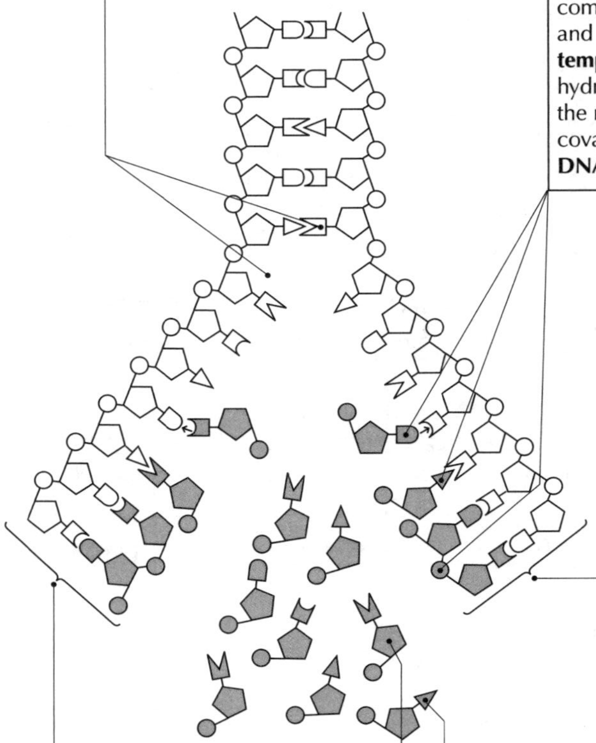

The two daughter DNA strands are synthesised in slightly different ways – one is made as a continuous strand, the other as a series of short strands joined together by the enzyme **DNA ligase.**

Each replica DNA double helix is a hybrid of **one parent** and **one daughter** strand: this is called **semi-conservative replication** because half of the original (parent) DNA has been conserved in each new DNA molecule.

Supply of free nucleotides for assembly into new DNA molecule alongside unwinding template.

In prokaryotes, e.g. bacteria:
- the DNA is not enclosed in a nucleus;
- the DNA is 'naked' – it has no protein scaffold;
- the DNA forms circles – some (the plasmids) can be very small.

Nucleolus is the site of manufacture of ribosomal subunits. Within the nucleolus are **nucleolar organisers** which contain multiple copies of the genes which are transcribed to ribosomal RNA.

The nucleolus breaks down in preparation for nuclear division, and is reassembled at the end of telophase.

Nuclear Pore can regulate the entry (e.g. ribosomal proteins, nucleotides) and exit (e.g., ribosomal subunits, messenger RNA) of molecules to and from the nucleus. There is a highly organised arrangement of proteins around the nuclear pore to carry out this controlled transport.

Nuclear envelope is a double membrane.

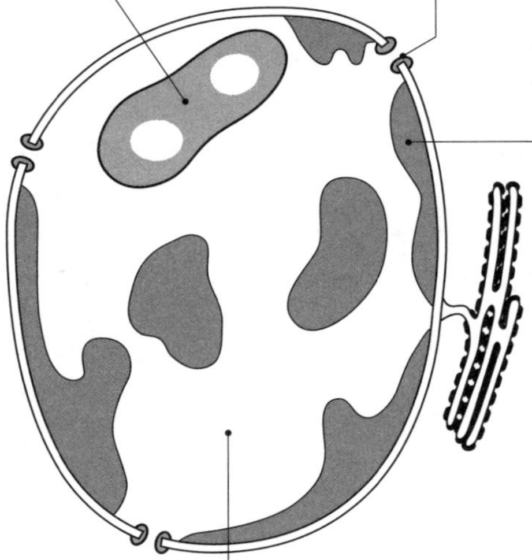

Chromatin is the genetic material containing the coded information for protein synthesis in the cell. It is made up of DNA bound to basic proteins called **histones**. The protein acts like a scaffold (support) for the DNA.

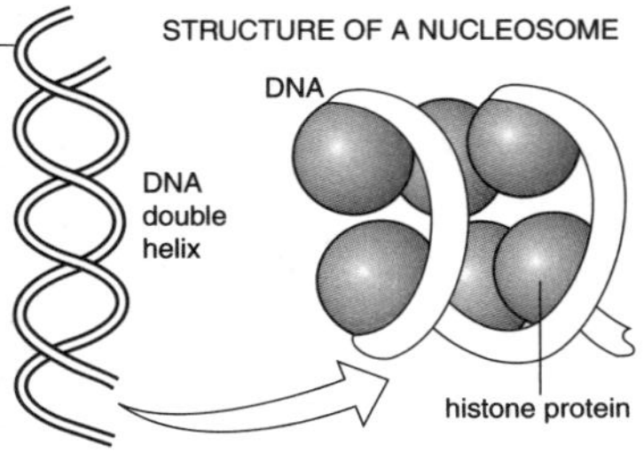

During nuclear division the chromatin condenses to form the **chromosomes**, and the chromatin containing DNA which is being 'expressed' (transcribed into mRNA) becomes visible as more loosely coiled threads.

Nucleoplasm, the cytoplasm of the nucleus, contains a variety of solutes, including ATP (energy source) and other nucleoside triphosphates which are the raw materials for DNA replication. The enzyme complex which regulates the replication of DNA **(DNA polymerase)** is also found here, as are ribosomal proteins awaiting assembly with ribosomal RNA into ribosomal subunits.

Centromere and kinetochores attach chromosome to nuclear spindle and control separation of chromatids at anaphase.

Sister chromatids

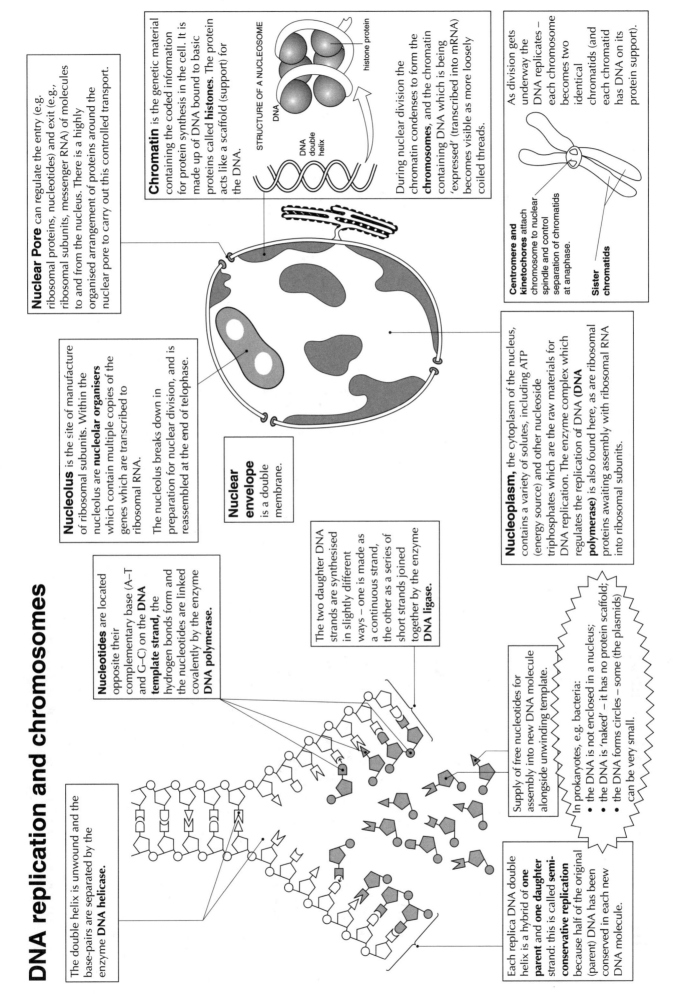

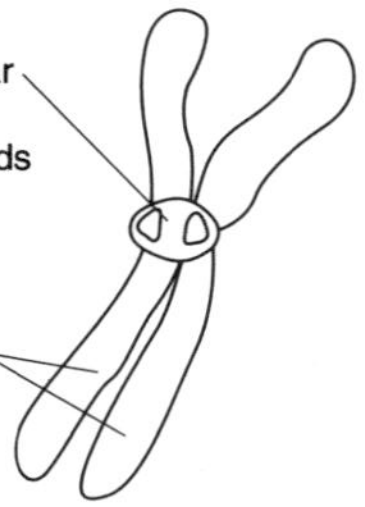

As division gets underway the DNA replicates – each chromosome becomes two identical chromatids (and each chromatid has DNA on its protein support).

Experiments on DNA function

Griffith's experiment: DNA is the genetic material

Griffith used two strains of **Pneumococcus:**

S
Smooth strain has capsule and causes pneumonia
R
Rough strain has no capsule and is harmless

BOTH STRAINS CAN BE KILLED BY HEAT TREATMENT.

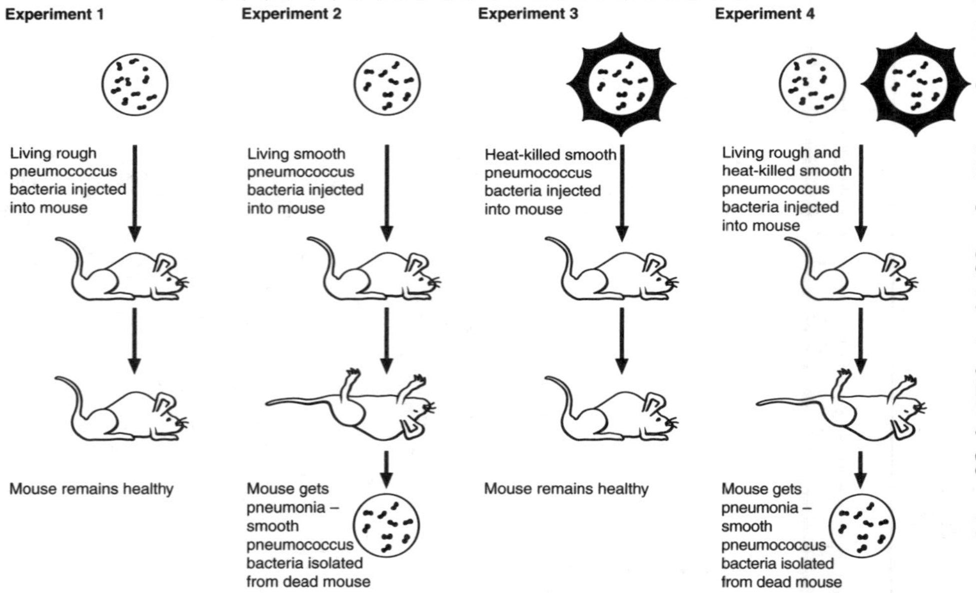
Experiment 1
Living rough pneumococcus bacteria injected into mouse
Mouse remains healthy
Experiment 2
Living smooth pneumococcus bacteria injected into mouse
Mouse gets pneumonia – smooth pneumococcus bacteria isolated from dead mouse
Experiment 3
Heat-killed smooth pneumococcus bacteria injected into mouse
Mouse remains healthy
Experiment 4
Living rough and heat-killed smooth pneumococcus bacteria injected into mouse
Mouse gets pneumonia – smooth pneumococcus bacteria isolated from dead mouse

Interpretation	**Interpretation**	**Interpretation**	**Interpretation**
Rough pneumococcus bacteria are not infective.	Smooth pneumococcus bacteria are infective.	Smooth pneumococcus bacteria that are killed by heat are not infective.	Non-infective rough bacteria have been transformed into smooth bacteria as a result of being mixed with heat-killed smooth bacteria.

The **transforming principle** discovered by Griffith:

- is **not** affected by **protease;**
- **is** destroyed by **DNAase;**

i.e. **the transforming principle is DNA.**

Genetic engineering shows that transfer of DNA alters characteristics of organisms.

This is the most compelling evidence that DNA is the genetic material.

Meselsohn and Stahl demonstrate that DNA replication is semi-conservative

- Used *Escherichia coli*, a harmless gut bacterium.
- Grew colonies of *E. coli* with NH_4Cl as the nitrogen source for DNA synthesis.
- Were able to use **density gradient centrifugation** to identify DNA.
- A 'heavy' (not radioactive) isotope of nitrogen, ^{15}N, is available.

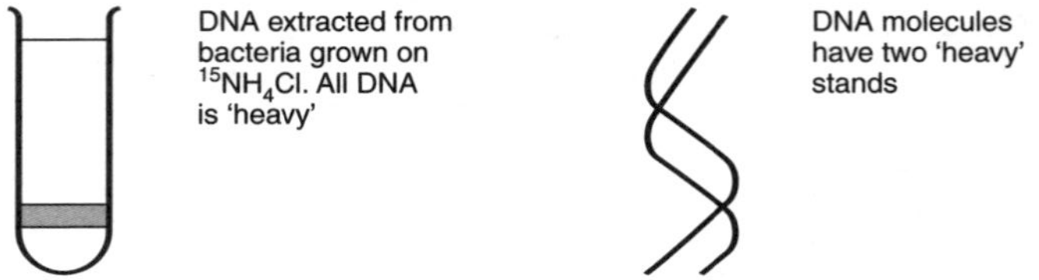
DNA extracted from bacteria grown on $^{15}NH_4Cl$. All DNA is 'heavy'
DNA molecules have two 'heavy' stands

Bacteria grown on $^{15}NH_4Cl$ for many generations are transferred to $^{14}NH_4Cl$ for several generations.

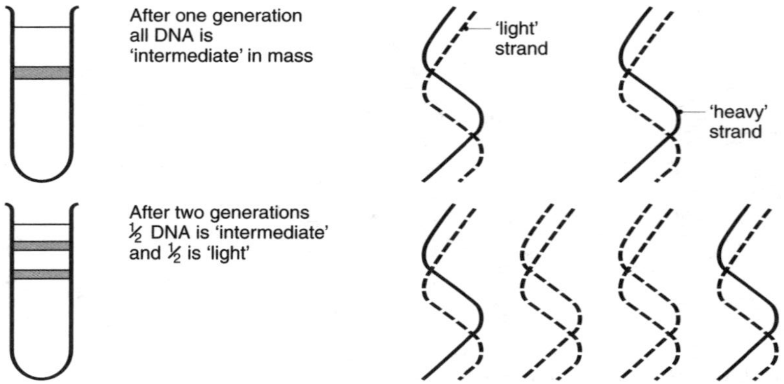
After one generation all DNA is 'intermediate' in mass
'light' strand
'heavy' strand
After two generations ½ DNA is 'intermediate' and ½ is 'light'

- The presence of 'intermediate' DNA after growth in $^{14}NH_4Cl$ supports the **semi-conservative principle.**
- **Conservative replication** would produce two bands, one 'heavy' and one 'light', after one generation in $^{14}NH_4Cl$.

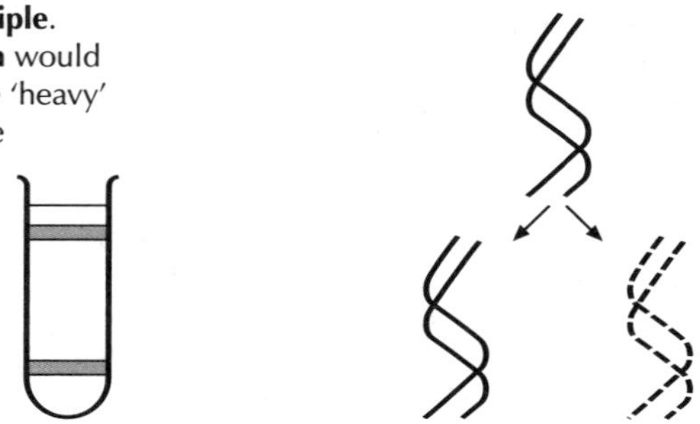

Embryo development: homeobox genes and apoptosis

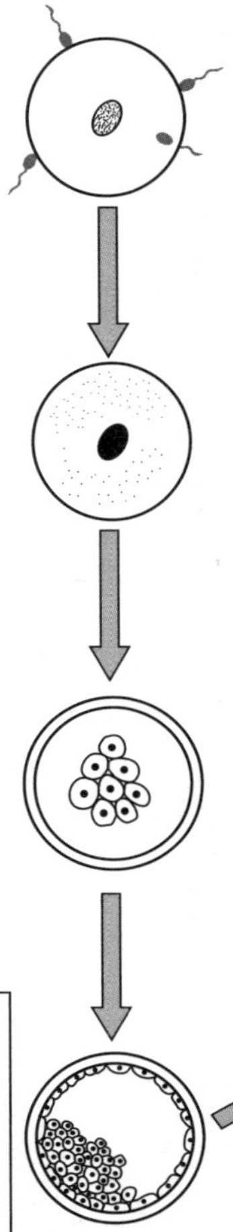

Fertilisation: fusion of male and female gametes forms a **zygote.**

Chemical gradients are set up inside the zygote.

Mitosis produces a **morula**, a ball of small, identical cells.

Chemical gradients begin to influence cells. Cells 'know' their position, and only switch on the genes they need for their particular function.

Not all of the genome is 'read' in any one cell type.

Adult organism has:
- front (**ventral**) and back (**dorsal**) surfaces;
- top (**anterior**) and bottom (**posterior**) ends;
- may develop a head (**cephalisation**);
- could have paired limbs.

It goes on and on!
- Between 50 and 70 million cells per day die due to apoptosis.
- In a teenager, the entire body mass is 'turned over' in a year.

Homeobox genes
- guide the development of shape and form in embryos;
- all contain the same sequence of 180 nucleotides somewhere along their length;
- the common sequence (the **homeobox**) codes for proteins that can bind to DNA;
- control which sections of DNA (i.e. which genes) can be transcribed, so determine which proteins will be produced in each cell;
- are **transcription factors** – genes which control other genes.

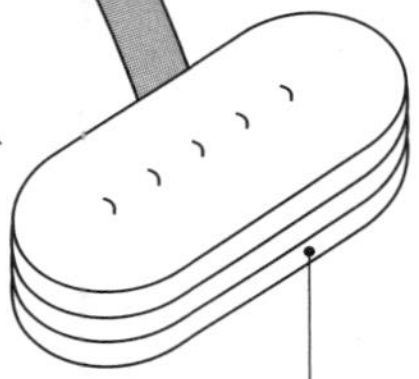

Simple embryo has three germ layers (**ecto-, meso-** and **endoderm**) which provide the basis for all tissues and organs.

How does it happen?

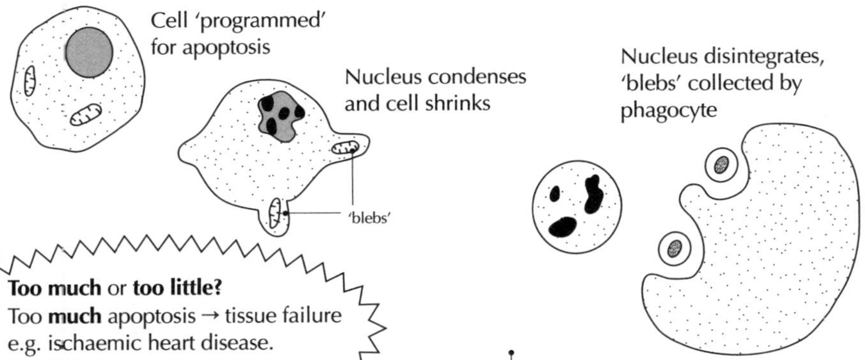
Cell 'programmed' for apoptosis
Nucleus condenses and cell shrinks
'blebs'
Nucleus disintegrates, 'blebs' collected by phagocyte

Too much or **too little?**
Too **much** apoptosis → tissue failure e.g. ischaemic heart disease.
Too **little** apoptosis → cancers.

Apoptosis
A form of **programmed cell death** in multicellular organisms:
- This is significant in both plant and animal tissue development.
- Tissue development involves

cell
↓ mitosis
ball of cells
↓ differentiation
tissue with correct **function**
↓ 'pruning' by apoptosis
tissue with correct **function** and **shape**.

- Selective apoptosis is controlled by gradients of signalling molecules – presence allows cells to multiply, absence allows them to be destroyed by apoptosis.
- Can remove useless structures, e.g. tadpole tail at metamorphosis.
- Can 'trim' structures to shape, e.g. 'paddle' limbs to hands and feet.

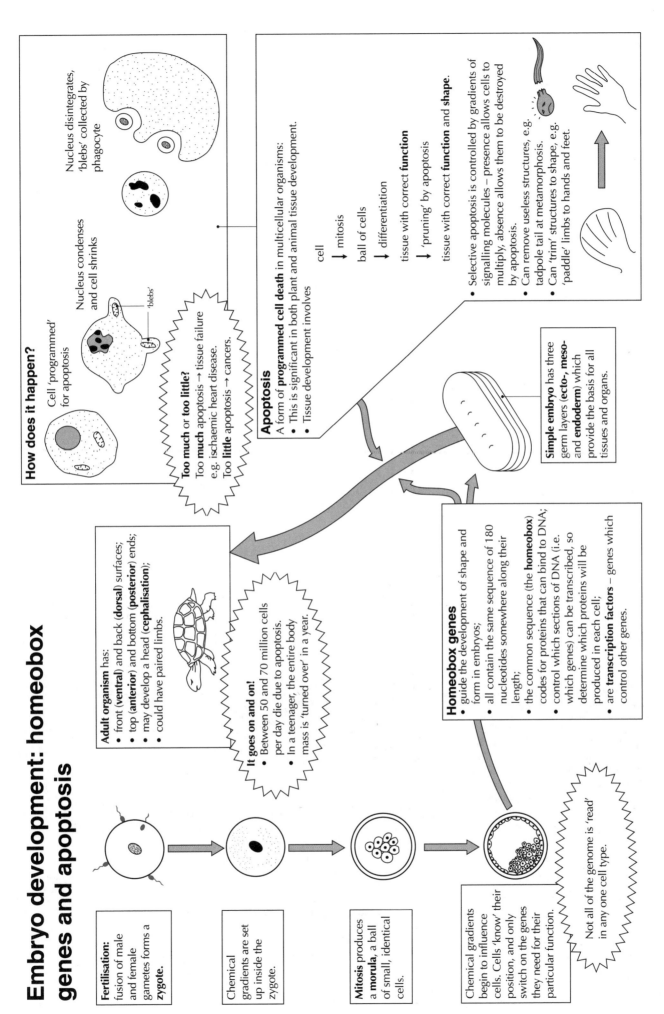

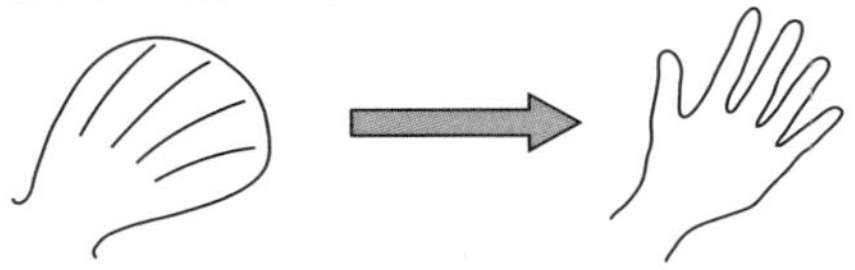

Nucleic acids: RNA **Ribonucleic acid** has a number of functions in protein synthesis.

RNA is composed of nucleotides, but differs from DNA in that ...

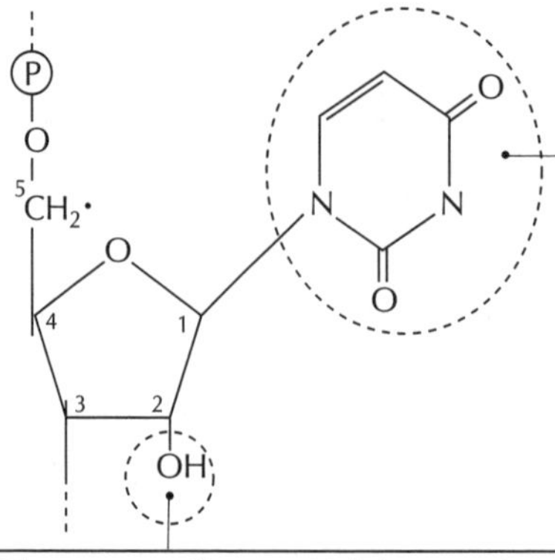

The pyrimidine base, **uracil**, replaces the base **thymine**. When intramolecular (RNA-RNA) or intermolecular t-RNA--m-RNA or DNA-RNA) base pairing occurs uracil hydrogen bonds to adenine.

The pentose **ribose**, which has an -OH group at the 2′ carbon atom, replaces **2′-deoxyribose**.

... and RNA is usually single-stranded, although it may take on complex structures depending on its function.

Messenger RNA is single-stranded RNA formed by **transcription** of a single strand of double-stranded DNA molecule.

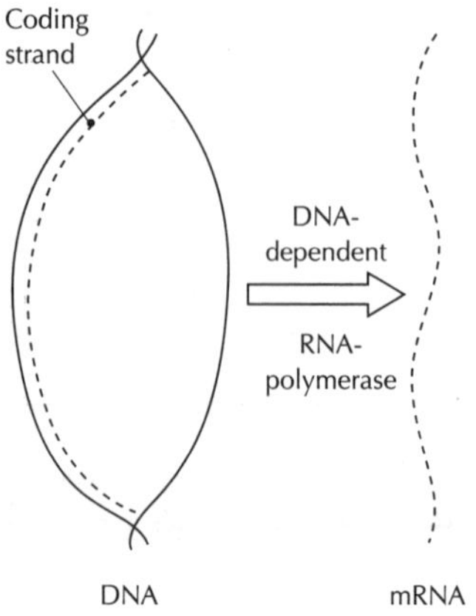

Transcription occurs in the nucleus, and allows the flow of genetic information (in the form of triplets of nucleotide bases) from the nucleus to the sites of protein synthesis at the ribosomes in the cytoplasm.

RNA is always used as a **primer** in **replication of DNA** and ...

... RNA may possibly catalyse its own preperation for translation (i.e. **ribozyme** is an RNA-enzyme).

Ribosomal RNA is single-stranded but folded into a complex series of shapes which form aggregates with ribonuclear proteins in the **ribosomes**. the rRNA molecules are synthesized in the nucleolus, and the many ribosomal subunits are transported via nuclear pores to be assembled in the cytoplasm. The ribosomes are the sites of the translation stage of protein synthesis.

Retroviruses such as HIV have RNA as their genetic material. In order to insert their genome into the host cell (typically a T-lymphocyte) these viruses use the enzyme **reverse transcriptase** to convert the viral RNA template into DNA.

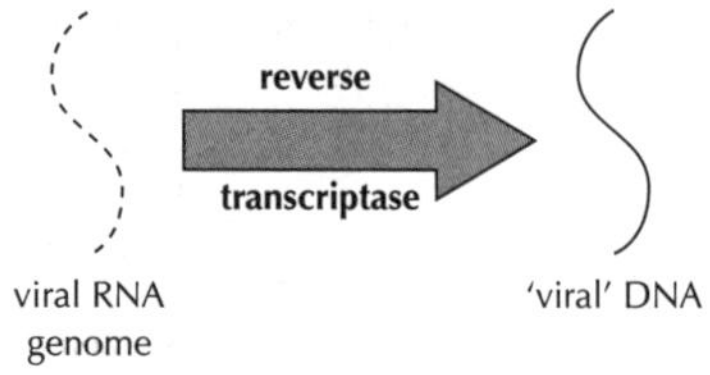

The drug **zidovudine** (formerly AZT) blocks this reaction and thus prevents the viral genome being inserted into host DNA.

Transfer DNA is single-stranded, but has a complex three-dimensional shape formed by intramolecular hydrogen bonding.

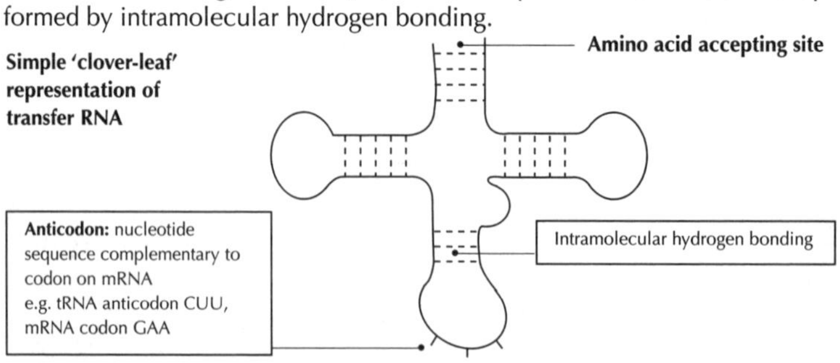

Transfer RNA 'reads' the genetic code in that it is able to pick up specific amino acids and deliver them to the ribosomes where the amino acids may be assembled into proteins according to the instructions carried on messenger RNA. This assembly process is called **translation.**

e.g.

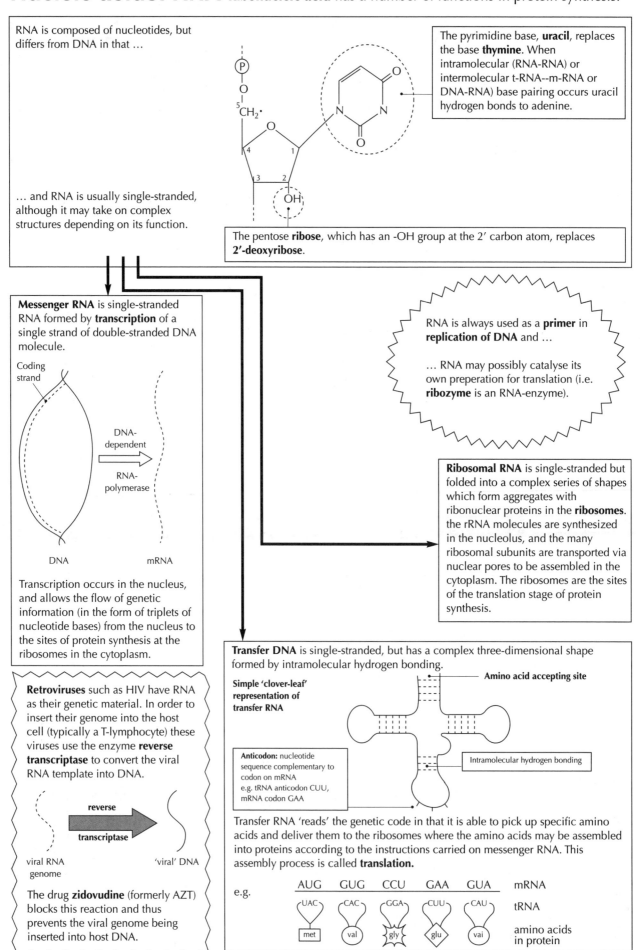

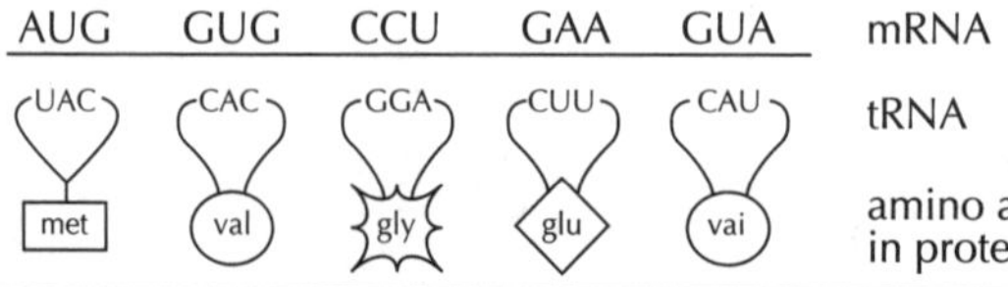

Plant and animal cloning

Cloning - the production of genetically identical individuals - has many applications.

Remove MERISTEM (growing point) from plant with desirable characteristics.

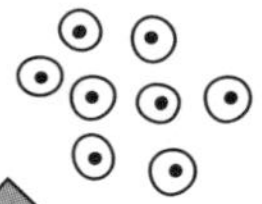

The tissue is sterilised (with sodium hypochlorite solution), broken up into cells, has cell walls removed (with the enzyme cellulase) to produce PROTOPLASTS.

These can be treated with plant hormones to make them divide.

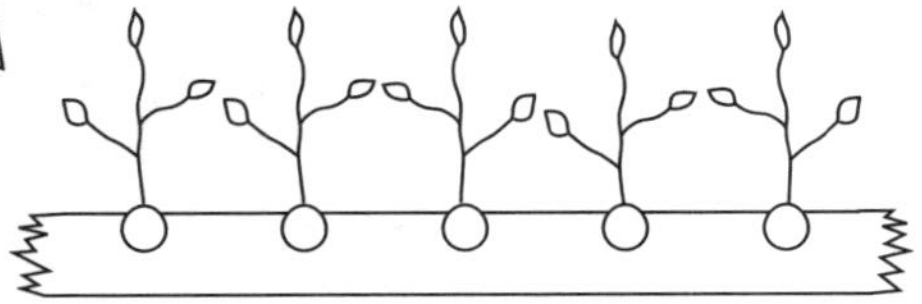

Plantlets can grow on a nutrient medium.

Tissue cloning:
allows growth of many identical individuals in a small space - ideal for transport, for example rubber 'trees' can be started in Britain then transferred to Malaysia for planting.

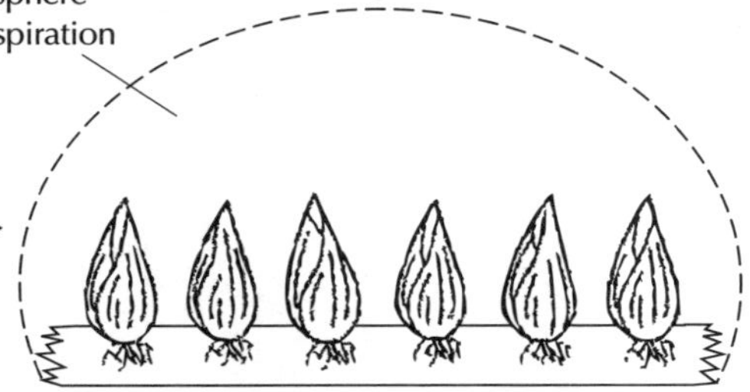

transparent cover allows entry of light but excludes fungi and bacteria

humid atmosphere reduces transpiration

nutrient medium containing rooting and growth hormones

Many ornamental house plants are produced in this way.

Vegetative propogation:
Fragments of plants can be seperated from the parent and induced to grow in a propogator. These fragments, for example root or leaf cuttings, are genetically identical and if kept under ideal conditions will grow by mitosis into plants **identical to each other and to the parent.**

'In vitro' (test tube) fertilisation

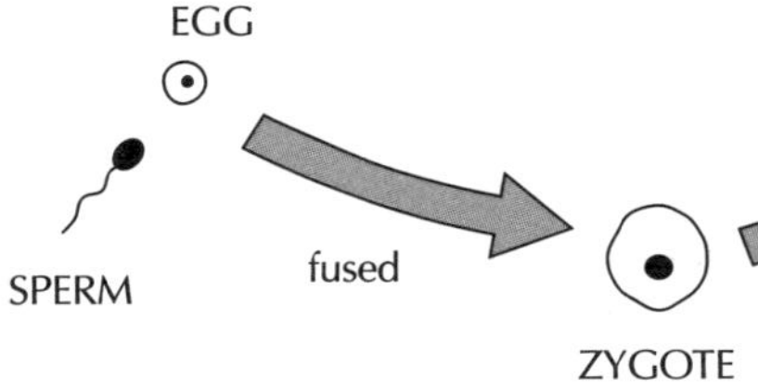

BALL OF CELLS kept in laboratory (embryo)

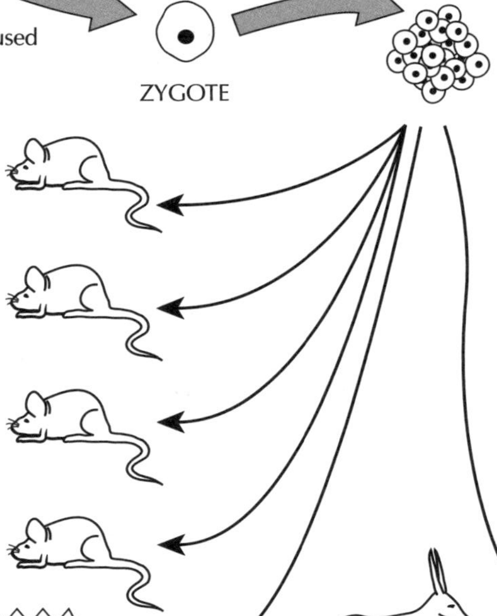

The ball of cells can be divided to produce several identical embryos which can be returned to the womb e.g. identical mice for use in medical research (this makes the test animal into a 'fixed variable').

Screening for inherited disease: a ball of human cells can be divided and each'embryo' tested for defective genes - only a 'normal' embryo will then be returned to the mother's womb to develop into a foetus.

Surrogacy in conservation

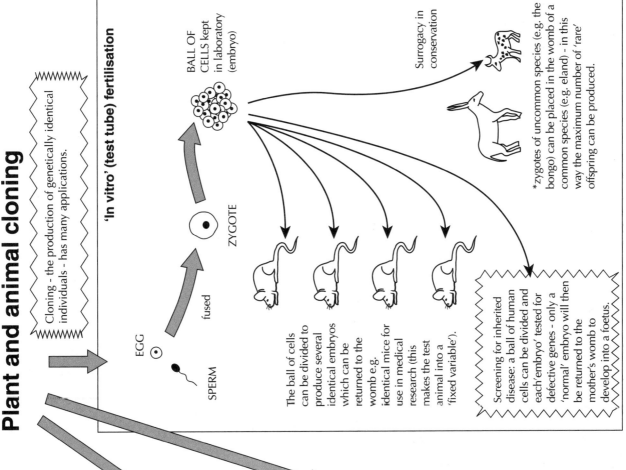

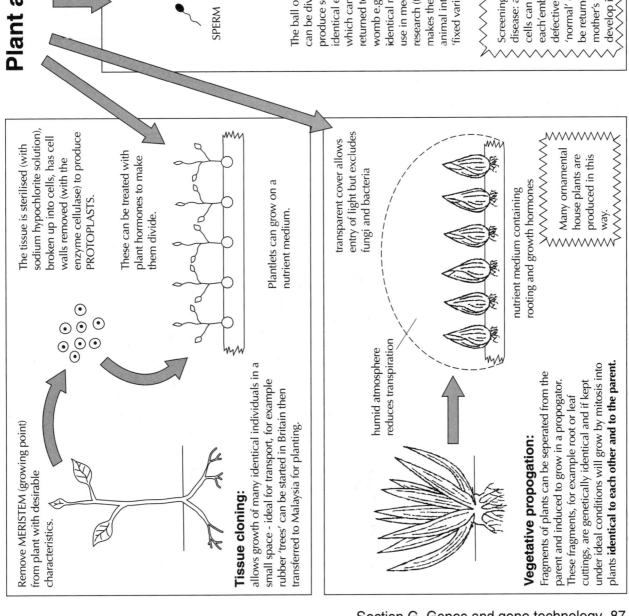

*zygotes of uncommon species (e.g. the bongo) can be placed in the womb of a common species (e.g. eland) - in this way the maximum number of 'rare' offspring can be produced.

The Genetic code

In the DNA molecule each 'code letter' (nucleotide base) **always** pairs with a specific partner i.e.

A············T
and G············C

This is complementary base pairing

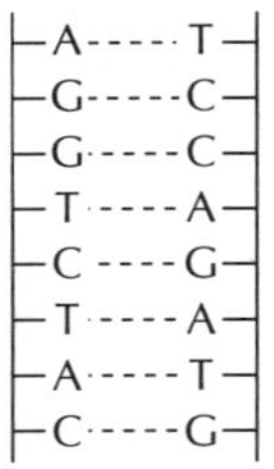

The **genetic information in DNA** is carried as a sequence of 'codewords'. Each 'codeword' on the DNA is made up of **three** bases and each 'codeword' corresponds to a **single amino acid in a protein.**

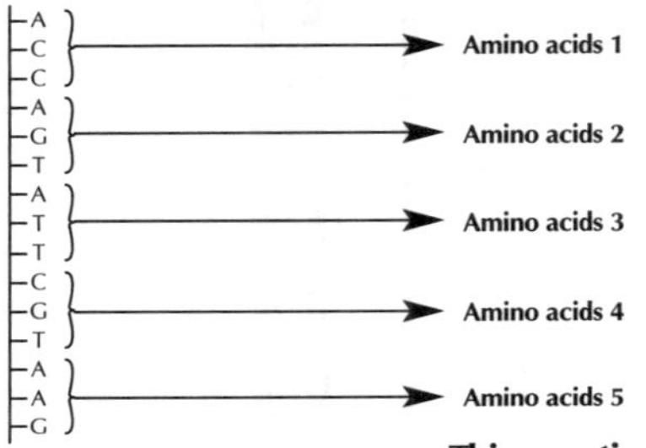

This genetic code is

Triplet:

- There are four possible DNA bases (A, C, G and T).
- There are twenty common amino acids to be 'coded for'.

If 1 base coded for 1 amino acid: only FOUR (4^1) possibilities.
2 bases coded for 1 amino acid: only SIXTEEN (4^2) possibilities.
3 bases coded for 1 amino acid: SIXTY FOUR (4^3) possibilities.

The triplet code was confirmed by **frame shift mutations**:
* remove 1 or 2 bases from DNA sequence → **no functional protein**;
* remove 3 bases (i.e. 1 'word') → **protein with one missing amino acid**.

Non-overlapping:

Each individual DNA base contributes to the code for a single amino acid:

e.g. 9 bases: G A T A C C G A C

3 triplets = 3 amino acids

not

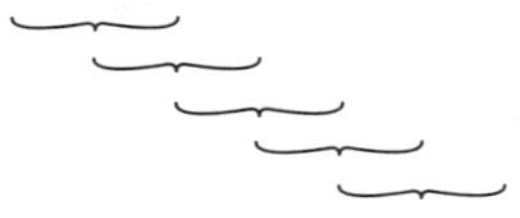

etc.

Good point: Substitution of one base for another has little effect.
Bad point: A long sequence of DNA is needed to code for a protein.

Degenerate:

- 64 triplets for 20 amino acids;
- there is spare 'capacity' – some amino acids are coded for by more than one triplet.

Good point! – a mutation which substitutes one base for another may not alter the amino acid coded for and so may not alter the protein produced.

Universal:

all organisms share the same code (strong evidence for a common evolutionary origin).

Alterations in phenotype may result from gene mutations:

gene mutations occur when some part of the 'code letter' sequence in the DNA is altered. As a result a defective protein, or no protein at all, may be made.

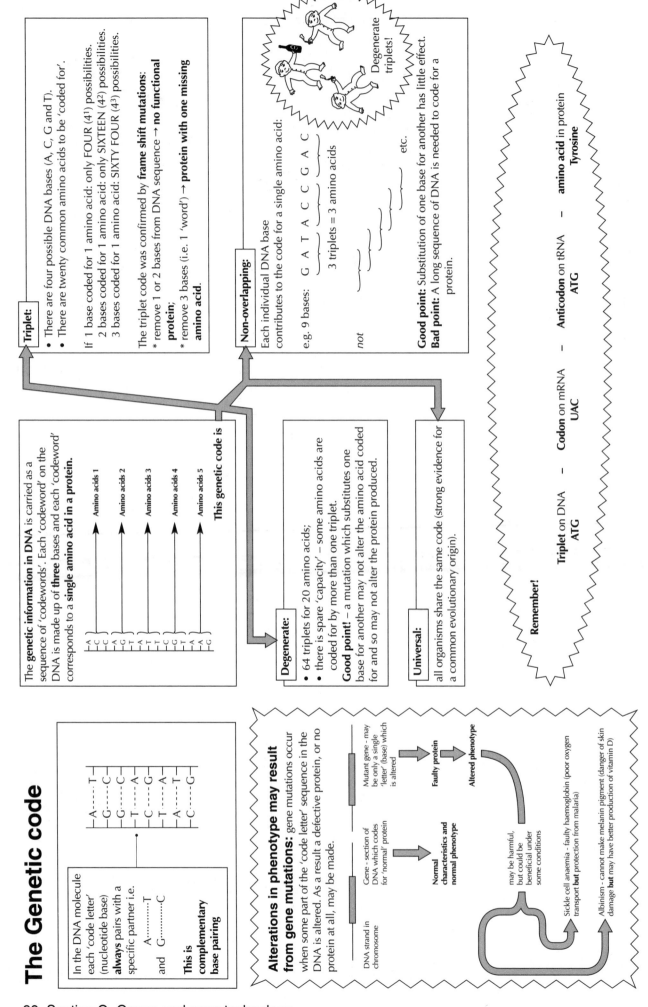

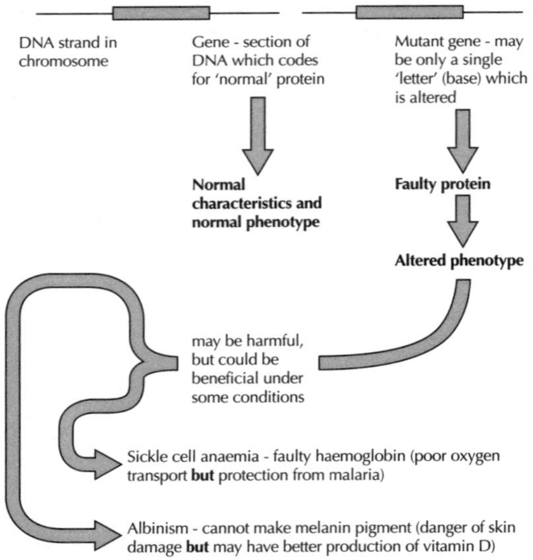

Remember!

Triplet on DNA	–	**Codon** on mRNA	–	**Anticodon** on tRNA	–	**amino acid** in protein
ATG		**UAC**		**ATG**		**Tyrosine**

Translation of messenger RNA

9. The ribosome reaches a **terminator (stop) codon, UAG** for example, which codes for a **releasing factor** rather than another aminoacyl tRNA.

Termination

10. Hydrolysis of tRNA to amino acid now releases completed peptide.

11. The components of the translation complex are disassembled.

Translation of messenger RNA involves initiation, elongation and termination.

8. The steps described in (5) are repeated and the peptide grows in length by two amino acids.

7. Ribosome A site is now located over a new codon which may then bind to a new aminoacyl tRNA (carrying tryptophan in this case).

Elongation

6. The unbound tRNA may now leave the P site, and the ribosome moves along the mRNA by one codon.

5. A **peptide bond** is formed between the two adjacent amino acids, and at the same time the covalent bond between the amino acid in the P site and its tRNA molecule is hydrolysed.

Hydrolysis of bond

TRANSCRIPTION
DNA
RNA POLYMERASE transcribes DNA to mRNA
Transcription from 3' to 5' end of DNA.

A single-strand messenger RNA molecule is produced complementary to one strand of the DNA molecule.

1. **Initiation** takes place when a tRNA molecule carrying the amino acid methionine associates with the **small ribosomal subunit.** N.B. The methionine is often removed after translation so that not every protein has methionine as the first amino acid in its primary sequence.

Initiation

2. Initiation complex locates the initiation codon, AUG, at the 5' end of the mRNA molecule.

3. The ribosome is completed by the addition of the **large ribosomal subunit.** This has two 'sites' – P site (binds to growing peptide chain) and A site (binds to the incoming aminoacyl tRNA)

P SITE
A SITE

4. A second amino acid (leucine in this case) is delivered to the A site located adjacent to the correct codon on the mRNA.

Leucine-tRNA

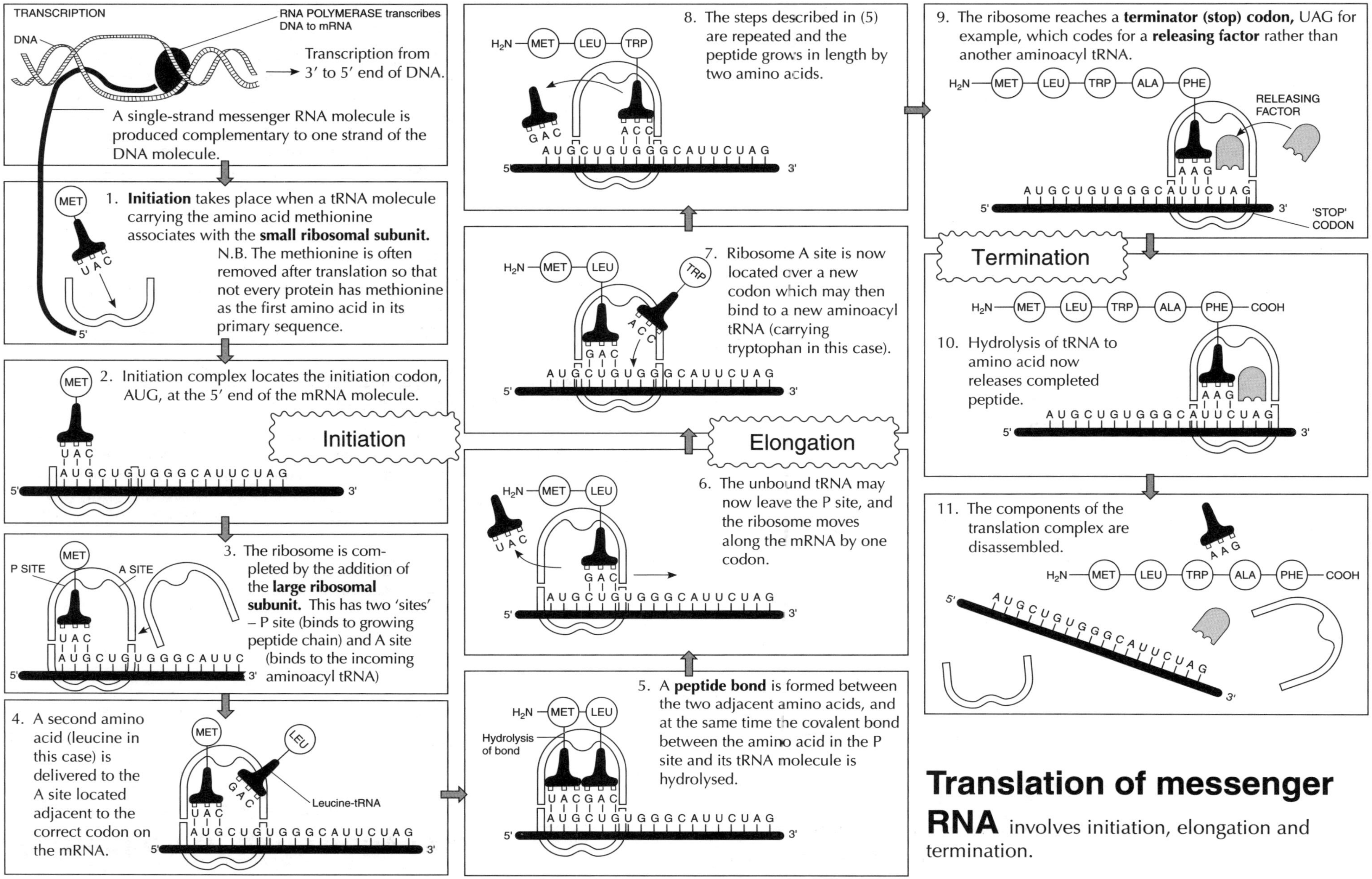

Regulation of gene activity:

control of transcription of DNA to mRNA.

Jacob and Monod: experimented on the gut bacterium, *Escherichia coli.*

They measured the activity of enzymes concerned with the metabolism of lactose.

β-galactosidase – hydrolyses lactose ⟶ glucose + galactose

β-galactosidase permease
thiogalactoside transacetylase } uptake and metabolism of lactose

1. If no lactose in growth medium : <10 molecules of each enzyme present.
2. When lactose available in medium: 3–5000 molecules of each enzyme, **and all three enzymes synthesised at the same rate**.

- lactose is an **inducer**
- the three enzymes are **inducible**.

3. Some mutants of *E. coli* make large amounts of all three enzymes even in the absence of lactose: these are **constitutive** mutants and Jacob and Monod proposed that the mutation had occurred in a **regulator gene**.
4. The regulator controls the synthesis of a **signal molecule** which 'switches' off the action of the **structural genes** which code for the synthesis of the three enzyme molecules.
5. Other *E. coli* mutants are unable to receive this signal – the **repressor** – these were found to have a mutation very close to the structural gene for β-galactosidase. The mutated region is called the **operator gene**.

Short interfering RNA (siRNA) is involved in **post-transcriptional gene suppression!** A short double-stranded RNA molecule can destroy a mRNA molecule, so no translation to protein is possible. This process is **very specific** so different **siRNA** molecules can be used to prevent expression of different proteins. This could be useful in preventing formation of 'mutant' proteins e.g. in management of Huntington's disease.

The Operon hypothesis

a. Gene order in the lac operon (the operon for the lactose enzyme genes)

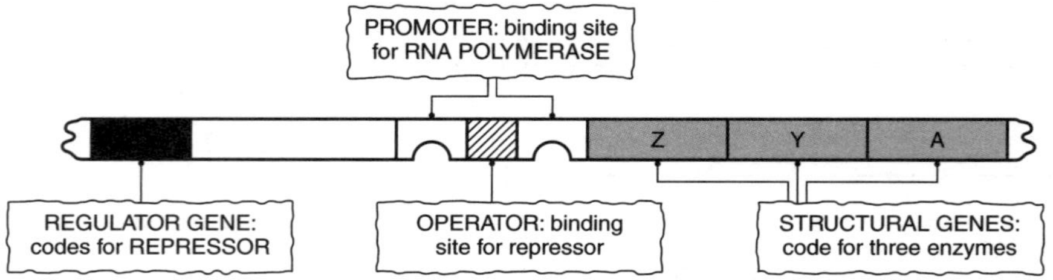
PROMOTER: binding site for RNA POLYMERASE
Z
Y
A
REGULATOR GENE: codes for REPRESSOR
OPERATOR: binding site for repressor
STRUCTURAL GENES: code for three enzymes

b. Inducer absent: no lactose in medium

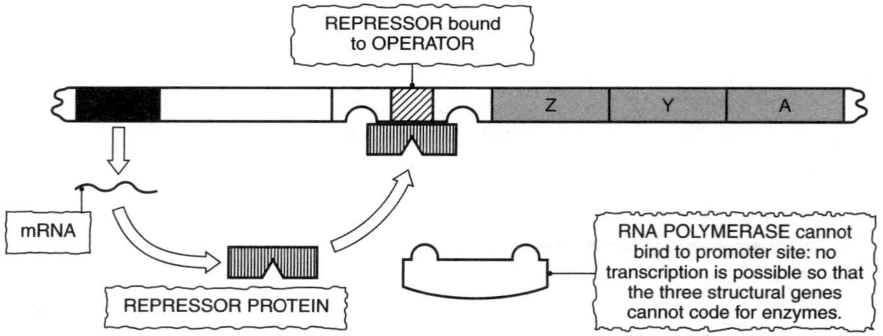
REPRESSOR bound to OPERATOR
Z
Y
A
mRNA
REPRESSOR PROTEIN
RNA POLYMERASE cannot bind to promoter site: no transcription is possible so that the three structural genes cannot code for enzymes.

c. Inducer present: lactose in medium

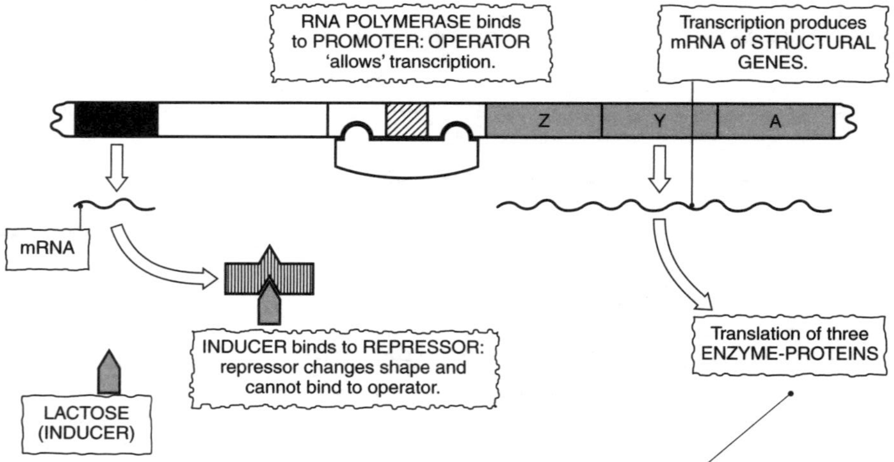
RNA POLYMERASE binds to PROMOTER: OPERATOR 'allows' transcription.
Transcription produces mRNA of STRUCTURAL GENES.
Z
Y
A
mRNA
INDUCER binds to REPRESSOR: repressor changes shape and cannot bind to operator.
LACTOSE (INDUCER)
Translation of three ENZYME-PROTEINS

Gene cloning

involves **recombination**, **transformation** and **selection**.

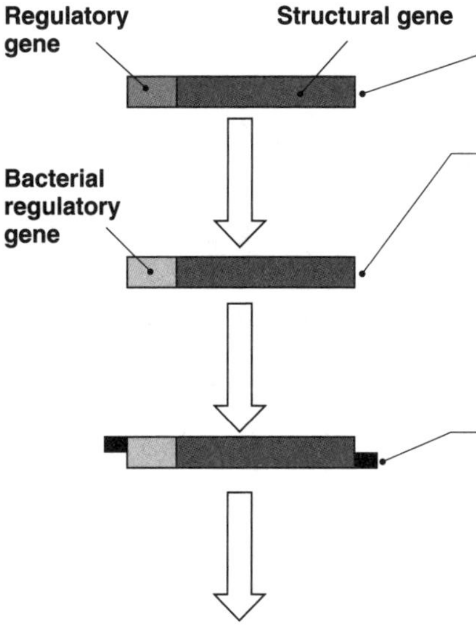

Regulatory gene
Structural gene
Bacterial regulatory gene

Donor DNA, e.g. a human gene. This may have been extracted from a **genome library**, been manufactured from messenger RNA using **reverse transcriptase** or, very rarely, synthesised in an **automatic polynucleotide synthesizer**.

Human regulatory gene has been replaced by a **bacterial regulator**. This will ensure a high rate of transcription and of the synthesis of gene product in a bacterial system.

'Sticky ends' have been added to the donor DNA. These are short nucleotide sequences (3–6 bases long) which will locate complementary sequences in the recipient plasmid DNA. For example, the 'sticky end' AGCT must be added to locate the complementary 'sticky end' TGCA on the opened recipient plasmid.

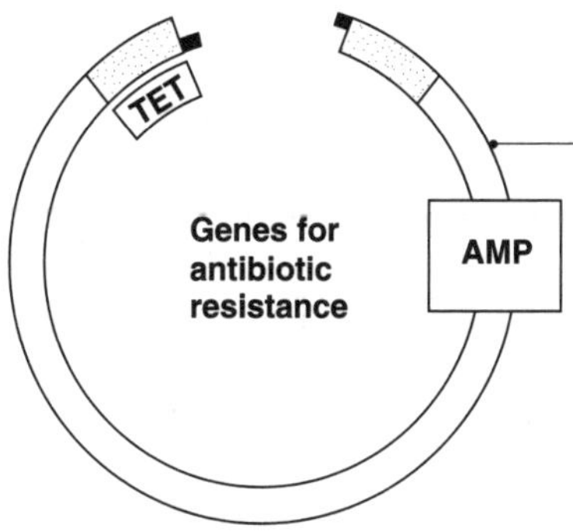

TET
Genes for antibiotic resistance
AMP

Recipient DNA, typically a bacterial plasmid, has been 'opened' at the sequence complementary to the donor DNA sticky ends by a specific **restriction endonuclease**. The recipient plasmid also contains two genes which confer resistance to antibiotics (**tetracycline** and **ampicillin**). This is vital for the selection process which will occur later.

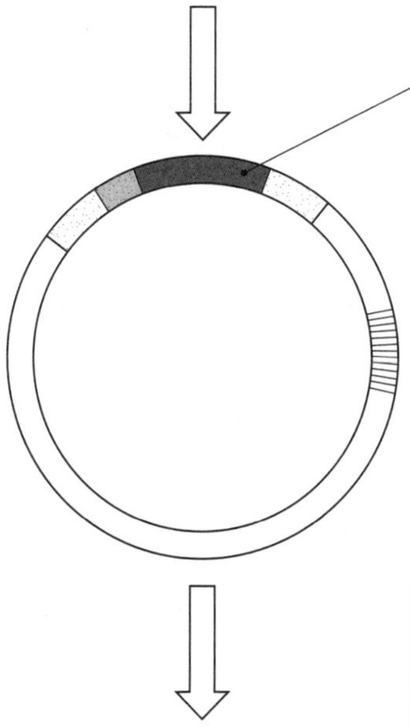

The donor DNA is now attached to the open plasmid using the enzyme **DNA ligase**.

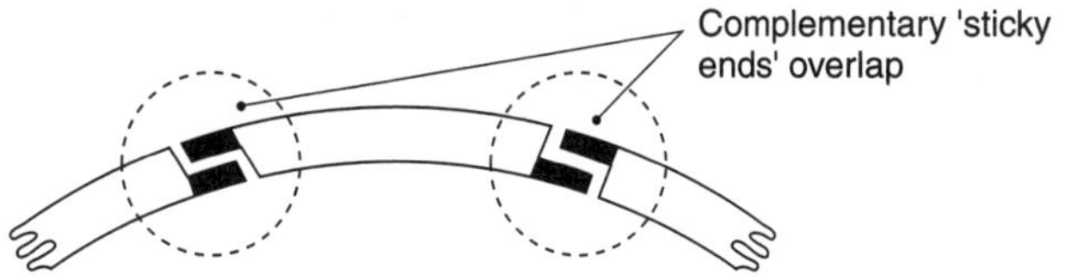

Complementary 'sticky ends' overlap

The resulting **recombinant DNA** typically carries a human gene within a plasmid: it is then called a **plasmid vector**.

The recipient bacterial cell, typically a non-pathogenic species such as *E. coli*, is treated with Ca^{2+} ions to make it 'leaky'. Such a cell may take up the plasmid vector. The incubation mixture contains few plasmids per cell to encourage uptake of single plasmids, but usually only about 1% of the *E.coli* cells will be transformed in this way.

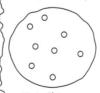

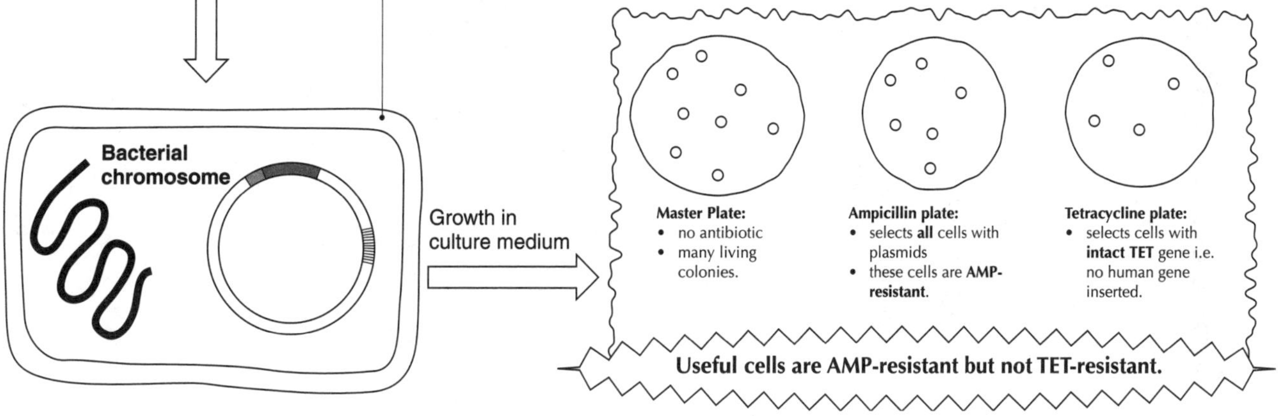

Bacterial chromosome
Growth in culture medium
Master Plate:
• no antibiotic
• many living colonies.
Ampicillin plate:
• selects all cells with plasmids
• these cells are AMP-resistant.
Tetracycline plate:
• selects cells with intact TET gene i.e. no human gene inserted.
Useful cells are AMP-resistant but not TET-resistant.

Genetic engineering (recombinant DNA technology)

depends on enzymes and culture of micro-organisms.

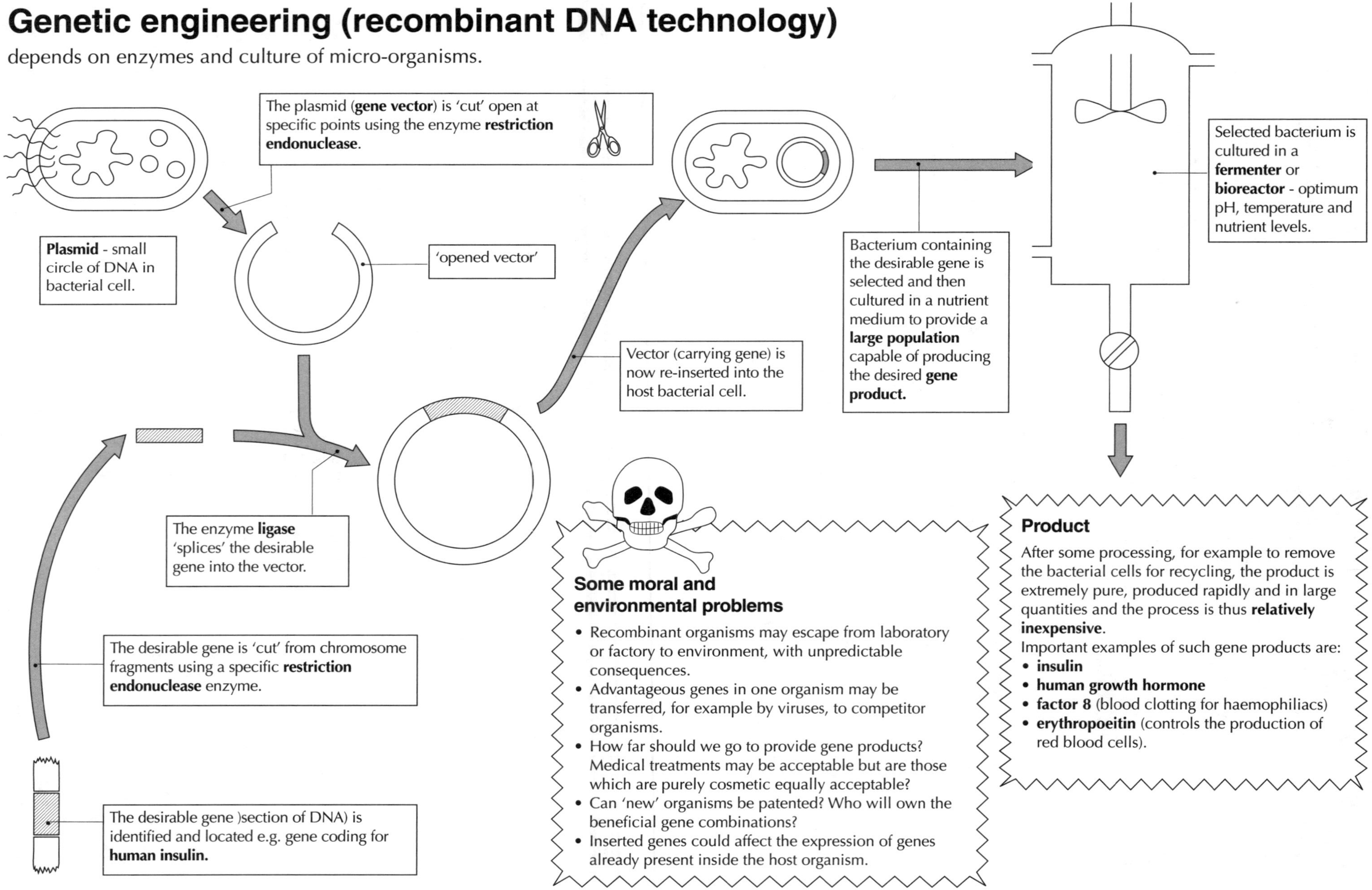
The plasmid (**gene vector**) is 'cut' open at specific points using the enzyme **restriction endonuclease**.
Plasmid - small circle of DNA in bacterial cell.
'opened vector'
The enzyme **ligase** 'splices' the desirable gene into the vector.
The desirable gene is 'cut' from chromosome fragments using a specific **restriction endonuclease** enzyme.
The desirable gene)section of DNA) is identified and located e.g. gene coding for **human insulin.**
Vector (carrying gene) is now re-inserted into the host bacterial cell.
Bacterium containing the desirable gene is selected and then cultured in a nutrient medium to provide a **large population** capable of producing the desired **gene product.**
Selected bacterium is cultured in a **fermenter** or **bioreactor** - optimum pH, temperature and nutrient levels.

Some moral and environmental problems

- Recombinant organisms may escape from laboratory or factory to environment, with unpredictable consequences.
- Advantageous genes in one organism may be transferred, for example by viruses, to competitor organisms.
- How far should we go to provide gene products? Medical treatments may be acceptable but are those which are purely cosmetic equally acceptable?
- Can 'new' organisms be patented? Who will own the beneficial gene combinations?
- Inserted genes could affect the expression of genes already present inside the host organism.

Product

After some processing, for example to remove the bacterial cells for recycling, the product is extremely pure, produced rapidly and in large quantities and the process is thus **relatively inexpensive**.

Important examples of such gene products are:

- **insulin**
- **human growth hormone**
- **factor 8** (blood clotting for haemophiliacs)
- **erythropoeitin** (controls the production of red blood cells).

Gene transfer

can promote desirable characteristics.

Resistance to pests and herbicides

A gene is inserted into the plant which enables it to make **insecticidal crystal protein** (ICP) which affects the gut of the caterpillars so that they cannot feed and eventually die.

The crop plant has a gene transferred into it which makes it resistant to herbicides. The field of growing crop can then be sprayed with the herbicide which will selectively kill the 'weeds' since they do not possess the 'resistance' gene.

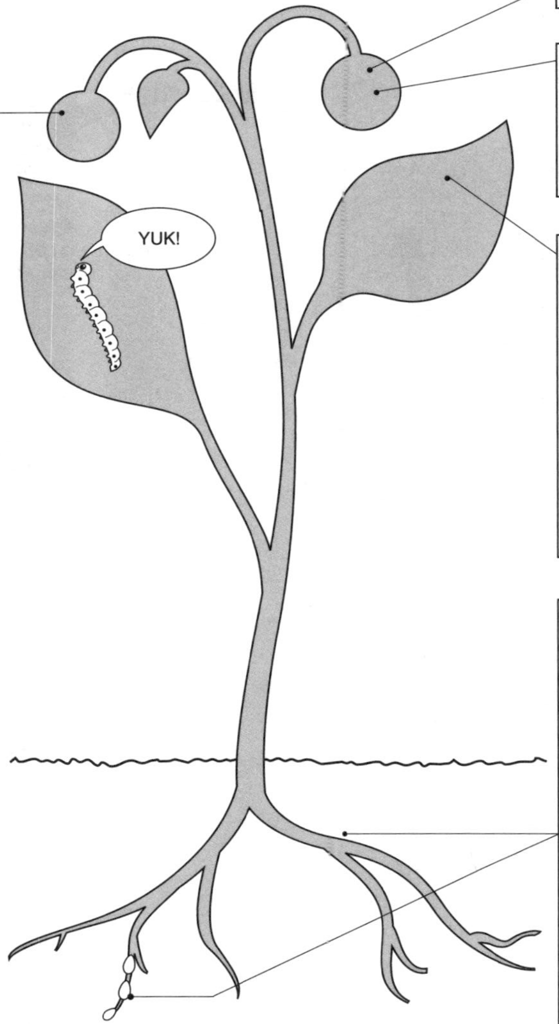

Improved shelf life: many fruits are wasted because they deteriorate before they can be sold/eaten. A gene has been introduced into tomatoes which inhibits the enzymes causing deterioration – the new '**flavr-savr**' tomatoes last for several weeks.

Transfer of nutrients: rice has been genetically modified to produce large quantities of vitamin A. This vitamin is deficient in the diet of many children in less-developed countries: the 'Golden Rice' may prevent the symptoms of vitamin A deficiency (which can lead to blindness).

Transferring genes with *Agrobacterium tumefaciens*

Introduce desired gene into T1 plasmid.

Return plasmid to bacterium.

Bacterium infects plant – plant produces a tumour (crown gall). Each cell contains the plasmid with the desired gene.

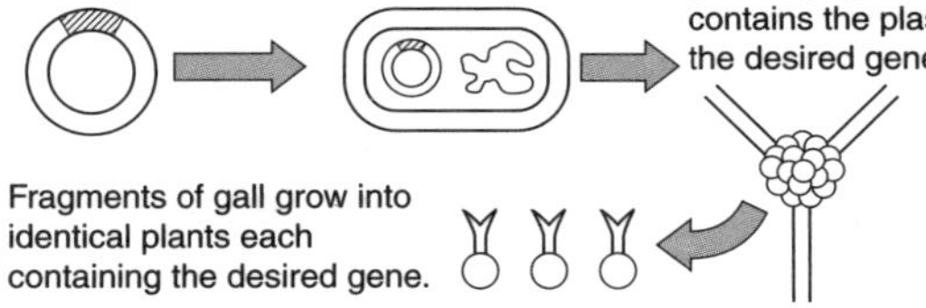

Fragments of gall grow into identical plants each containing the desired gene.

Gene transfer is also important in animals

TRANSGENIC ANIMALS

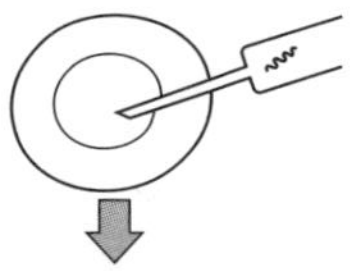

DNA containing desired gene can be introduced into nucleus using a fine pipette.

Cells are cultured, implanted into female animal which gives birth to transgenic animal.

Animal releases protein made from desirable gene in its milk. Factor 8 (essential for blood clotting) is made in this way.

Nitrogen fixation: this process involves the reduction of nitrogen gas from the atmosphere into a form suitable for conversion into amino acids, nucleotides and other essential compounds.

N_2 (from atmosphere) → NH_4^+ (ammonium ions) → amino acids and other compounds

this key step is controlled by enzymes coded for by 12 genes called the Nif genes.

Most plants cannot fix nitrogen but it is hoped that gene transfer might either:

- insert these genes directly into a plant;
- make a plant more susceptible to the formation of root nodules with nitrogen-fixing bacteria of the genus *Rhizobium*.

This could:

- produce cereal crops which also manufacture large amounts of protein;
- reduce the demand for nitrogenous fertilisers.

Producing spare body parts!
Genes for human cell surface proteins have been transferred to pigs! These animals have organs, such as the heart and kidneys, with **human** antigens on their surface.
It may be possible to use these organs as **xenotransplants** (transplants between different species) as the human immune system may not recognise them, and they may not be rejected.

Enzymes and genetic engineering

Restriction endonucleases recognise specific palindromic nucleotide sequences in DNA and cut both strands of the double helix at those points. In this example the endonuclease called HindIII recognises the four base sequence AGCT.

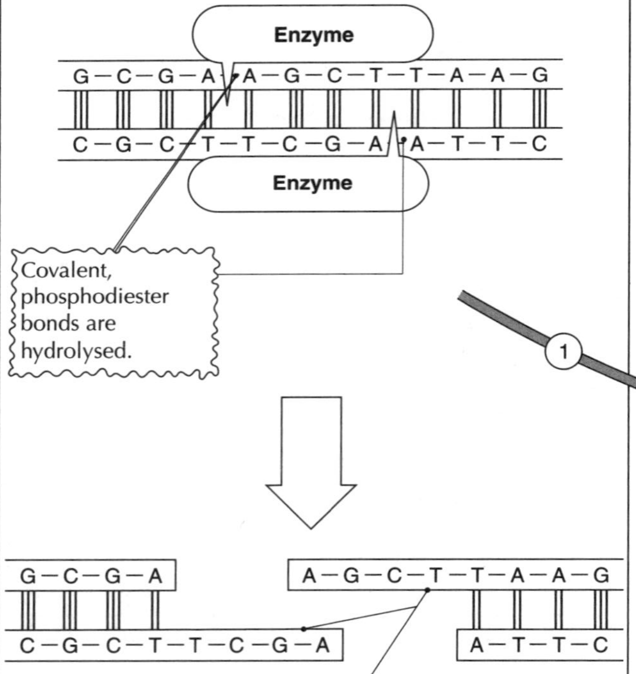

These sections of single stranded DNA are called '**sticky ends**' because they have the potential to form hydrogen bonds with, and hence 'stick to', complementary bases at the end of another DNA single strand.

DNA ligases are enzymes which can rejoin the complementary **sticky ends** of fragments of DNA. N.B. These fragments need not have been from the **same piece of double helical DNA,** although in order to have complementary sticky ends they will be produced by the **same restriction endonuclease.**

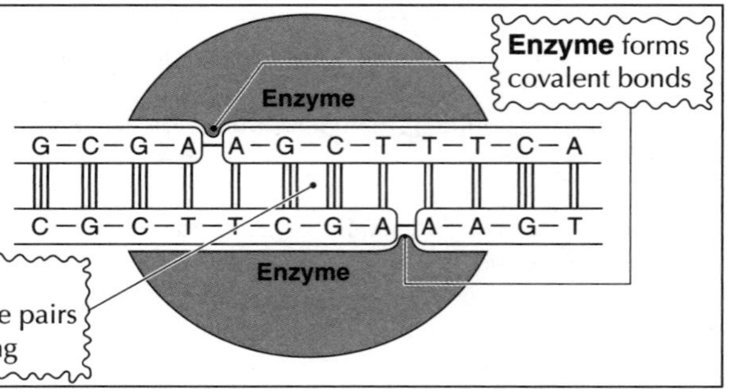

A **molecular vector** is a piece of DNA which can be manipulated to carry a 'foreign' gene into a host cell.

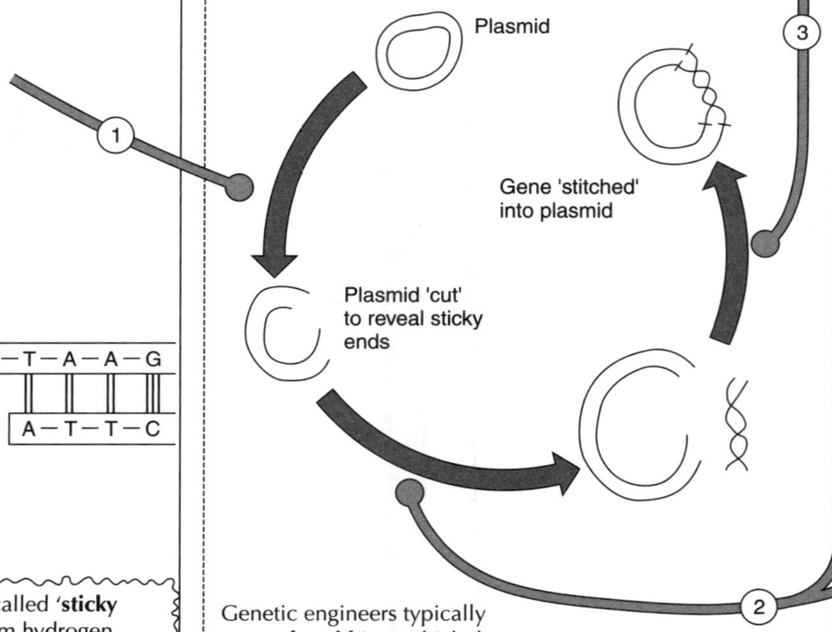

Genetic engineers typically use a **plasmid** into which they stitch a foreign gene to form **recombinant DNA.**

Reverse transcriptase is an enzyme which can synthesize DNA using an RNA molecule as a template.

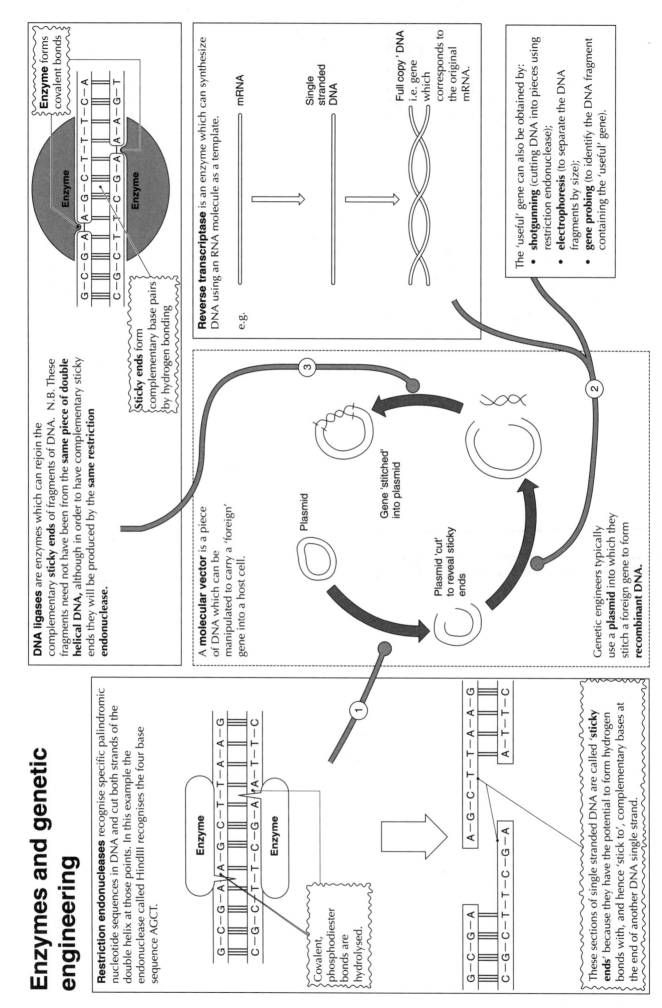

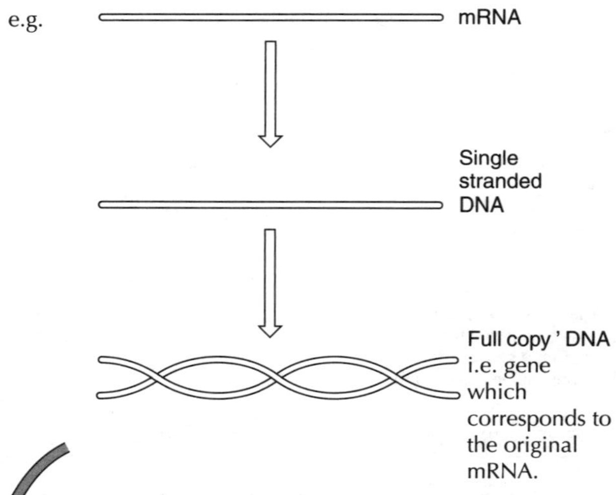

The 'useful' gene can also be obtained by:

- **shotgunning** (cutting DNA into pieces using restriction endonuclease);
- **electrophoresis** (to separate the DNA fragments by size);
- **gene probing** (to identify the DNA fragment containing the 'useful' gene).

The sequencing of DNA

involves a number of other techniques:

- cutting DNA with **restriction enzymes;**
- separation of fragments by **electrophoresis;**
- copying by **polymerase chain reaction.**

The technique depends on using specially modified nucleotides to halt DNA synthesis: the deoxyribose has a second oxygen atom removed so that it cannot form phosphate bridges.

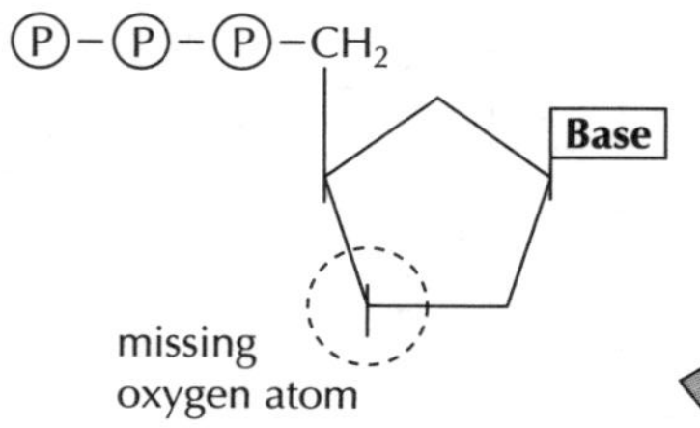

The basic method:
Four test tubes are labelled T, G, A and C. Into each test tube is added:

- the DNA to be sequenced
- a primer (radioactively-labelled, so that it can be 'seen' on the gel at a later stage)
- the four normal nucleotides
- the enzyme DNA polymerase
- the appropriate modified (dideoxy) nucleotide e.g. T* into tube T.

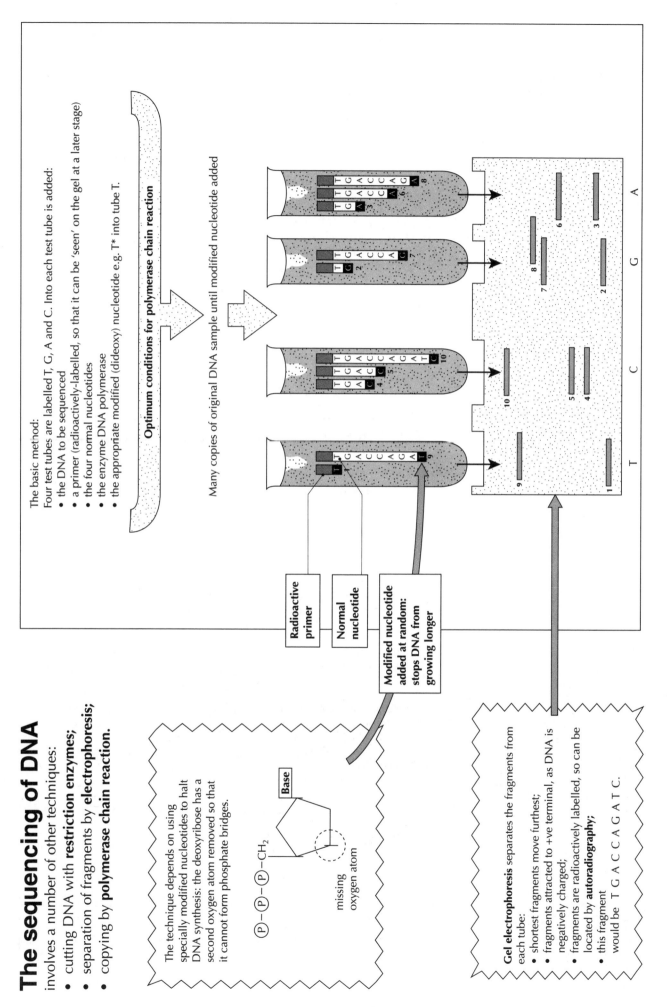

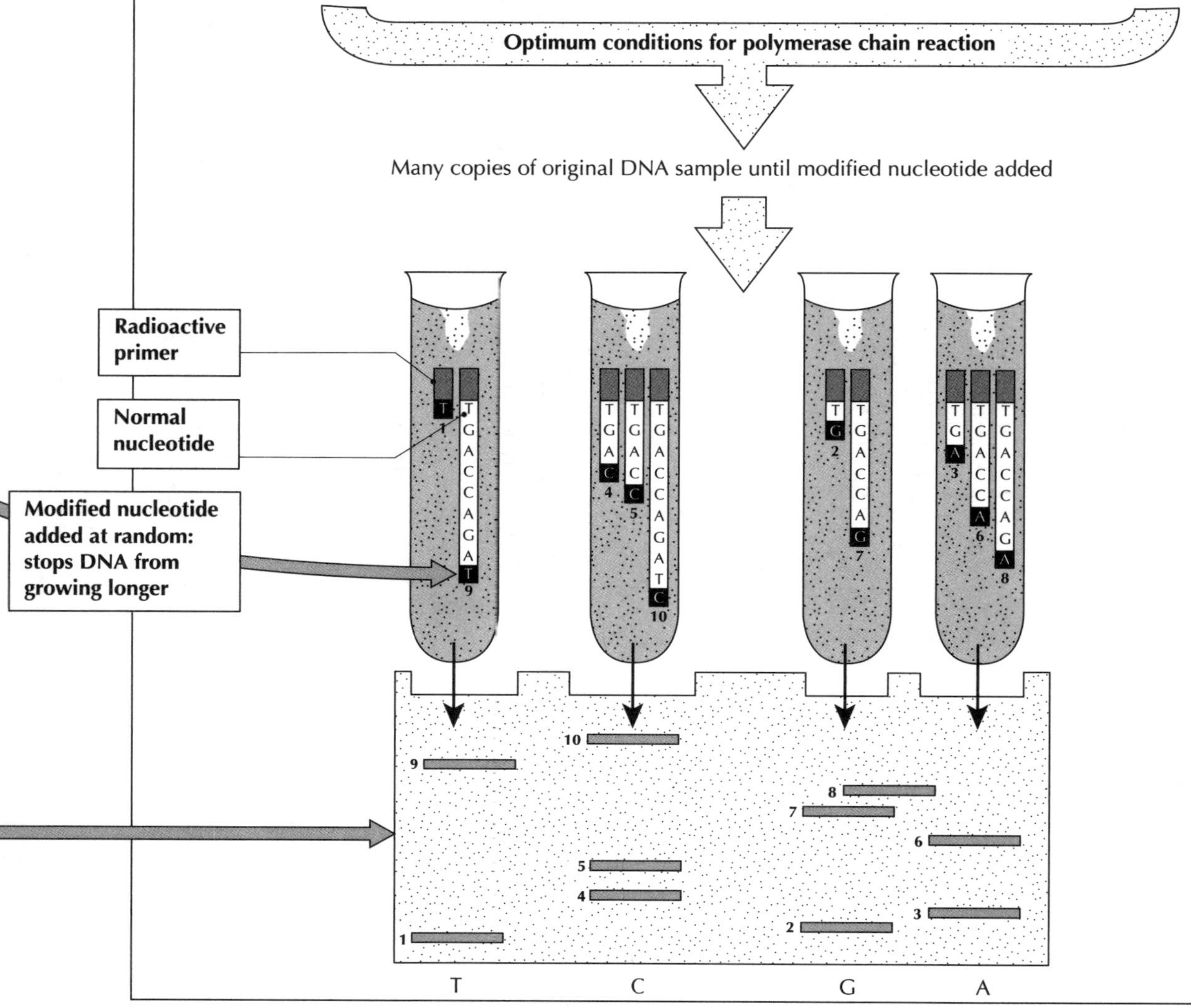

Gel electrophoresis separates the fragments from each tube:

- shortest fragments move furthest;
- fragments attracted to +ve terminal, as DNA is negatively charged;
- fragments are radioactively labelled, so can be located by **autoradiography;**
- this fragment would be T G A C C A G A T C.

Genetic profiling: minisatellites and probes.

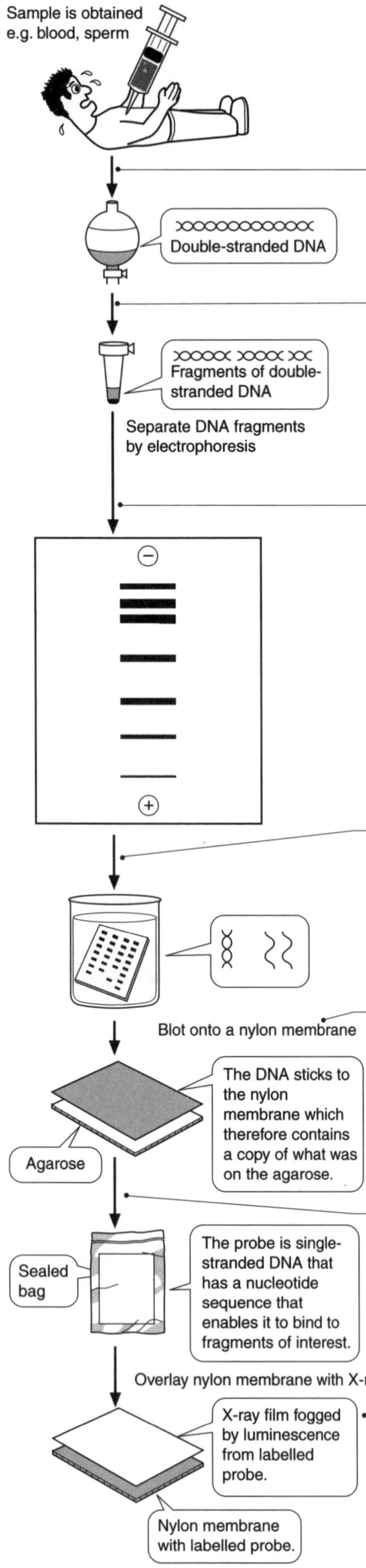

Sample is obtained
e.g. blood, sperm
Double-stranded DNA
Fragments of double-stranded DNA
Separate DNA fragments by electrophoresis
Blot onto a nylon membrane
The DNA sticks to the nylon membrane which therefore contains a copy of what was on the agarose.
Agarose
Sealed bag
The probe is single-stranded DNA that has a nucleotide sequence that enables it to bind to fragments of interest.
Overlay nylon membrane with X-ray film
X-ray film fogged by luminescence from labelled probe.
Nylon membrane with labelled probe.

Extraction of DNA: shake tissue sample with water-saturated phenol and chloroform at pH 7.8. Protein is precipitated and DNA enters aqueous layer.

Digestion of DNA: using bacterial restriction endonucleases (**restriction enzymes**) cut long DNA molecules into fragments 1000–20 000 nucleotides in length. These fragments contain unique non-coding regions (minisatellites).

Electrophoresis: fragments of DNA are separated according to size – the sample of DNA fragments is placed in a well cut into an agarose gel and an electric current is applied (20 hours, room temperature, pH 7.0). The gel acts as a sieve so that the smaller fragments of DNA pass more quickly towards the anode (at this pH all of the DNA fragments are negatively charged).

Denaturation: the double-stranded DNA fragments are immersed in alkali to convert them to single strands (so that 'probes' can bind to them later).

Blotting: a thin nylon sheet is laid over the gel and blotting paper is laid on top of this – the buffer is drawn up by capillary action, the DNA fragments are dragged along and stick to the nylon membrane.

Hybridisation: incubation of the nylon membrane with a single-stranded DNA PROBE – the probe will only bind to a single-stranded DNA which contains a complementary nucleotide sequence.

Disclosure: the bound probes are located using 'fogging' of X-ray film (either the probes are radioactive, or in a less hazardous procedure chemicals which are luminescent are used.)

Comparison of DNA profiles: a sample from a suspected rapist, arsonist or murderer can be compared with a sample obtained from the victim. **Multi-locus probes** can establish identity to 1 in 10^6 individuals, but require large samples of pure DNA. **Single locus probes** are less rigorous (unless several, typically 4 or 5, are used sequentially) but can be applied to tiny samples.

Problems of contamination

Forensic samples are rarely pure:

- contain DNA from vagina/anus of rape victims;
- DNA from fungi and bacteria is usually present;
- DNA decays quickly so that restriction sites may be lost – may give too few or too many fragments;
- dyes from clothes may affect specificity of restriction enzymes;
- environmental contaminants may bind to DNA fragments and affect their rate of movement under electrophoresis.

Applications of profiling:

- Paternity disputes
- Apprehension of criminals e.g. rapists
- Identification of wild animal family trees – important in breeding programmes.

Minisatellites are short regions of DNA, with unique repeating nucleotide sequences, found within the non-coding regions (**the introns**) of the chromosomal DNA. These non-coding regions, which are not responsible for the expression of proteins, are very variable and thus ideally suited to **genetic profiling**.

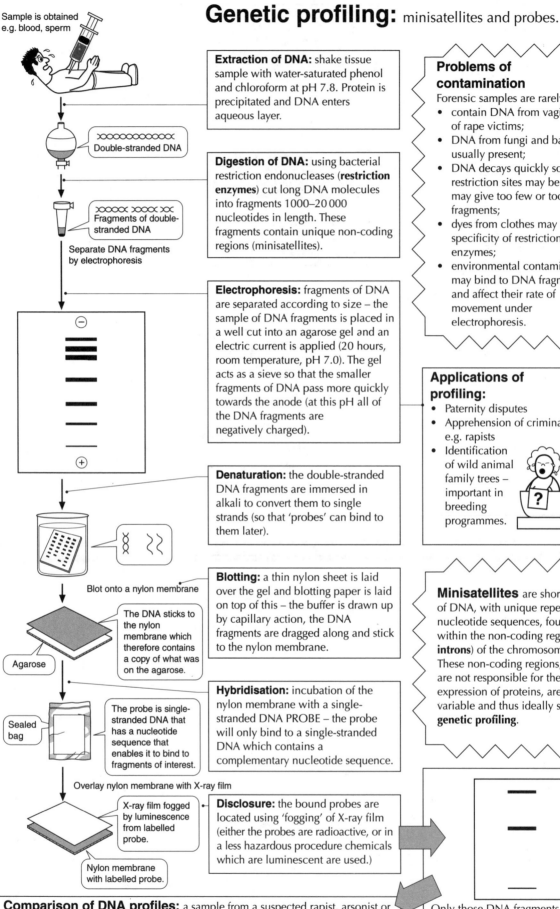

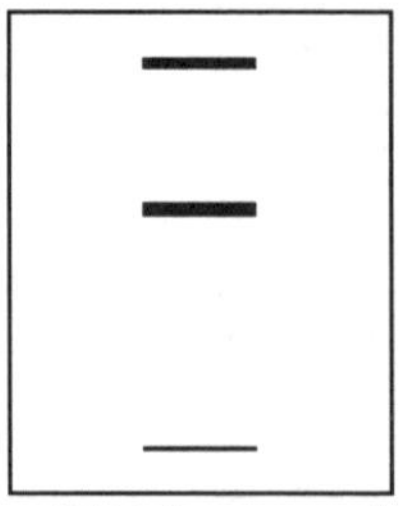

Only those DNA fragments that bound the labelled probe show up on the X-ray film : the resulting image is a **DNA profile**, commonly called a **DNA fingerprint**.

The polymerase chain reaction (PCR)

makes DNA amplification possible.

Original target DNA is typical double-stranded DNA with two complementary and anti-parallel polynucleotide chains.

Heating to 93 °C breaks the hydrogen bonds which hold the two strands of the DNA double helix together: the DNA is '**melted**'.

Cooling to 55 °C allows hydrogen bonds to reform **but** if this is carried out in the presence of excess 'primer' DNA the hydrogen bonds form between 'primer' (---) and one original strand rather than between the two original complementary strands of the DNA double helix. Thus two 'primer-DNA' hybrids are formed, by the process of **annealing**.

DNA polymerase resynthesises double-stranded DNA, using the short 'primers' introduced at the previous stage as the starting point for nucleotide condensation.
The reaction occurs rapidly at 72 °C; the DNA polymerase is a thermostable form isolated from a 'hot springs' bacterium, *Thermophilus aquaticus*.

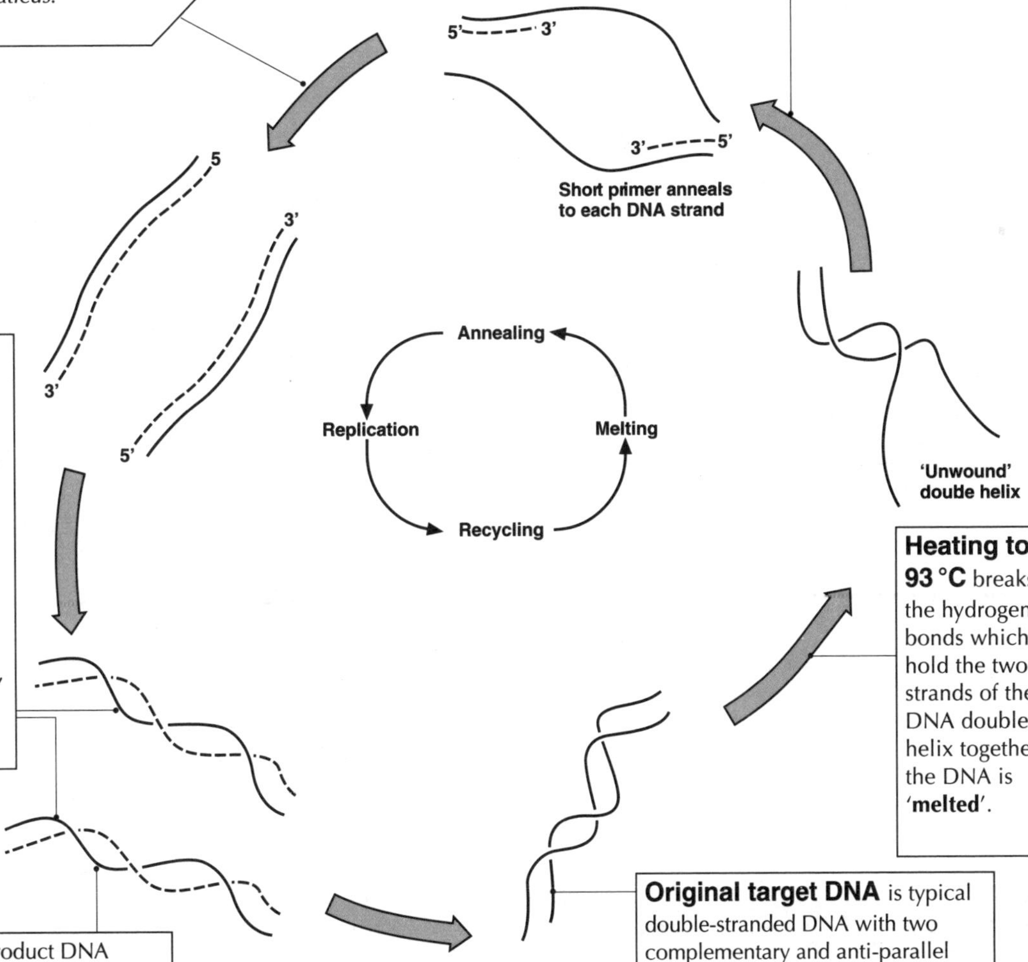

Completion of replication results in two identical copies of the original DNA. The process can be made **selective** by choosing a primer which only binds with a particular region of the DNA. Only the DNA'downstream' of the primer will be replicated.

Each of the two product DNA double helices can now act as a target for another cycle of reactions: *n* cycles produce 2^n copies. Even allowing for some operating inefficiencies 20 cycles will produce 1000 000 copies of an original DNA sample.

DNA amplification has many applications

- **Pre-natal diagnosis:** DNA from a single cell removed from an embryo can be amplified, screened for inherited disorders such as cystic fibrosis, and the results used to advise on possible re-implantation of the embryos.
- **Historical studies:** DNA from skeletal and mummified material has been amplified and analysed to provide 'genetic history', e.g. the presence/absence of certain mutations.
- **Forensic analysis:** a minute sample of sperm can provide enough DNA to be amplified and generate a genetic fingerprint which might be vital in rape or assault cases. There is still some debate about the acceptability of PCR evidence in the courts.
- **Wildlife conservation:** DNA from a single hair can be amplified and analysed to provide information about the genetic relationships between individual members of rare species. This may be useful in determining the best strategy for captive breeding schemes.
- **Dinosaur production:** if you believe **Jurassic Park!**

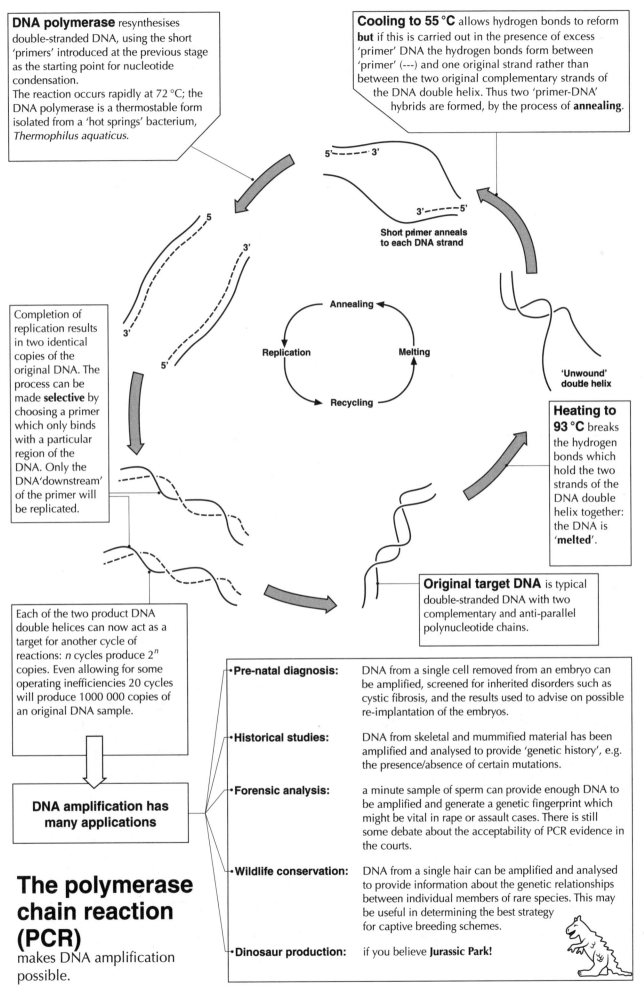

Cystic fibrosis is caused by faulty ion transport, and is the most common inherited fatal disease in Caucasian populations.

Screening for CF involves using a **gene probe** for identification of the defective gene. The CFTR gene is located on human chromosome 7, but direct screening for mutant forms is complicated since there are many mutations of this gene, although it is possible for the most common (70%) of the mutation. This is called ΔF508 since it results in the omission of one phenylalanine molecule from position 508 in the protein chain. The direct screening is usually carried out on cells from a buccal smear, on white blood cells, or on chorionic villus/amniotic fluid cells if the test is pre-natal.

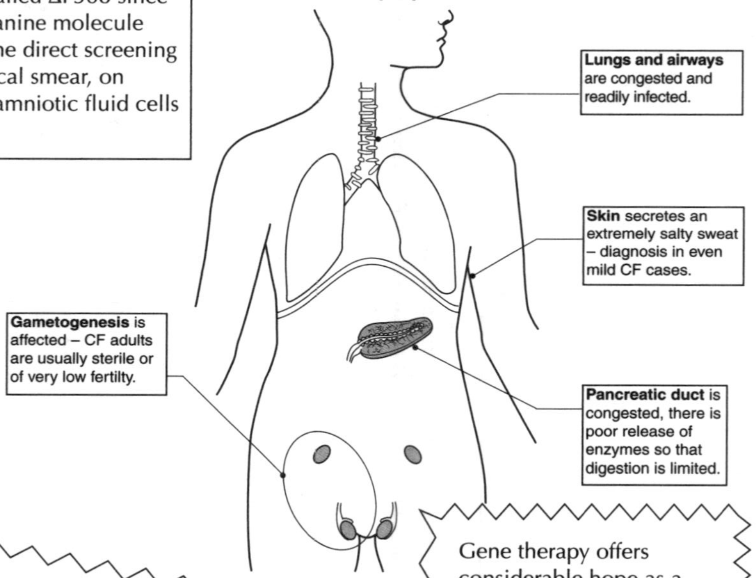

Lungs and airways are congested and readily infected.
Skin secretes an extremely salty sweat – diagnosis in even mild CF cases.
Pancreatic duct is congested, there is poor release of enzymes so that digestion is limited.
Gametogenesis is affected – CF adults are usually sterile or of very low fertility.

How a gene probe (DNA probe) works

A gene probe is a short section of DNA labelled with a tag (usually a radioactive label or a fluorescent tag).

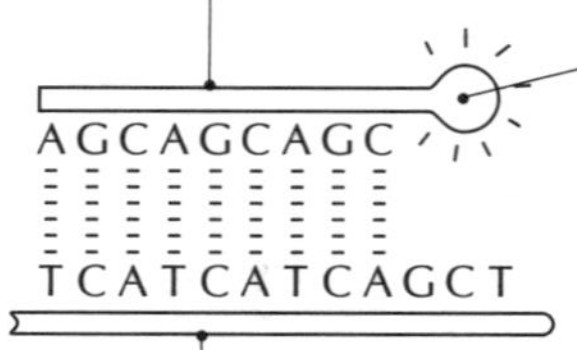

AGCAGCAGC
TCATCATCAGCT

Tag: radioactive – is located on a photographic plate; fluorescent – is located by uv light.

Target DNA - has a base sequence complementary to the probe. Often has a 'tandem repeat' sequence.

Target is heated to separate strands, mixed with probe, cooled to anneal probe with target.

Patients suffering from CF have an ionic imbalance across the membranes of epithelial cells. This imbalance is caused by the absence of the normal form of a protein – **Cystic Fibrosis Transmembrane Regulator (CFTR)** – which controls the passage of chloride ions across these cell membranes.

Normal cell: chloride ions flow out, are followed by water, and lining of passages are kept moist with a free-moving layer of mucus.

CF cell: chloride movement is restricted, water does not enter the passages which become clogged with a drier, more viscous mucus.

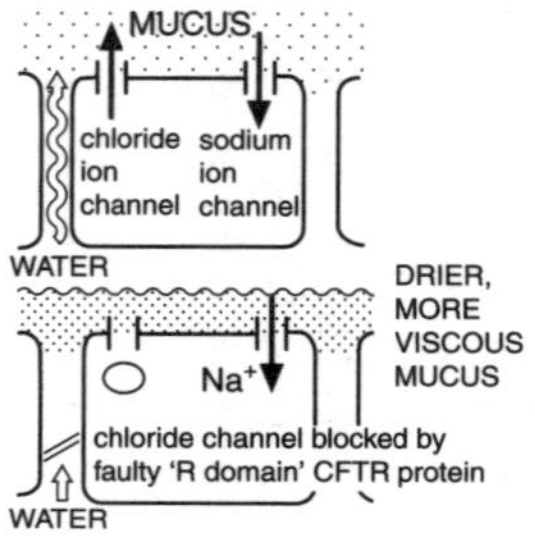

MUCUS
chloride ion channel
sodium ion channel
WATER
DRIER, MORE VISCOUS MUCUS
Na⁺
chloride channel blocked by faulty 'R domain' CFTR protein
WATER

The presence of this mucus restricts the movement of materials along body passages, for example, breathing is difficult as airflow is restricted through the respiratory tree.

The situation is made worse because long fibres of DNA are released from the epithilial cells and from white blood cells in the locality. These fibres make the mucus even more viscous and difficult to move – physiotherapy involving lung percussion is only partially successful.

The thick, stagnant mucus offers a nutrient source for bacteria. These invade rapidly and cause severe lung infections – most CF sufferers have died of lung infections by their mid-twenties. One common bacterial coliniser is *Pseudomonas aeruginosa* – this bacterium has a thick capsule which further restricts the flow of mucus. CF sufferers try to restrict infection with a wide antibiotic regime.

Gene therapy offers considerable hope as a cure for CF. This therapy would introduce healthy CFTR genes into defective epithelial cells which could then synthesise functional protein resulting in a normal flow of chloride ions and water across these membranes. Success depends on identification of the faulty CFTR gene, large scale synthesis of replacement 'normal' CFTR gene, and development of vector systems to transfer the healthy gene into the DNA of target epithelial cells.

The basic principle is:

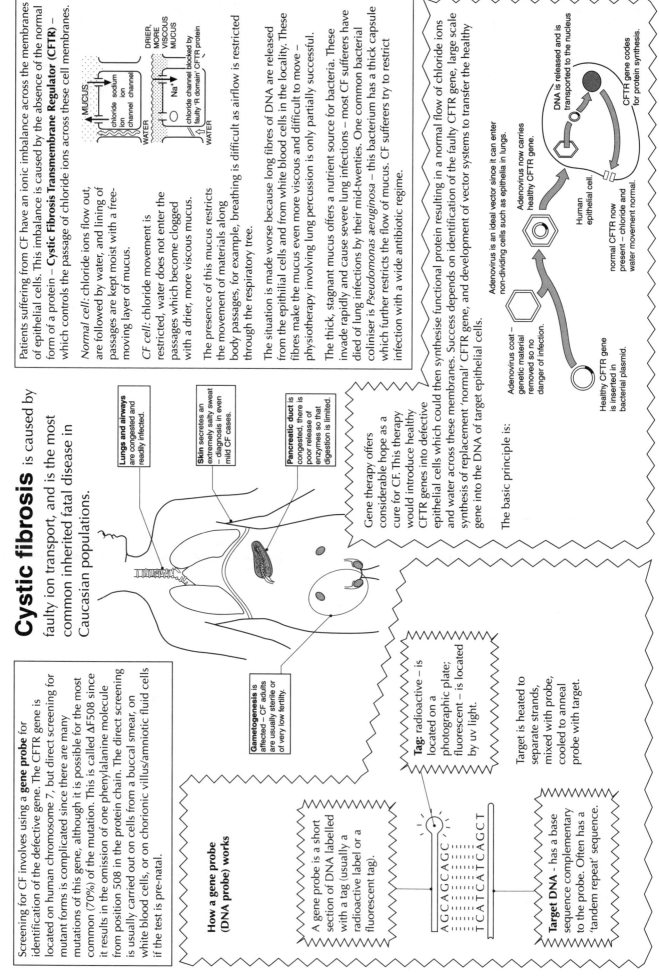

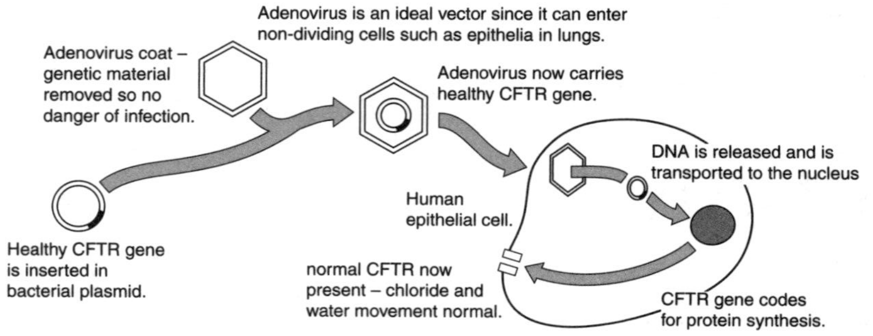

Adenovirus is an ideal vector since it can enter non-dividing cells such as epithelia in lungs.
Adenovirus coat – genetic material removed so no danger of infection.
Adenovirus now carries healthy CFTR gene.
DNA is released and is transported to the nucleus
Human epithelial cell.
Healthy CFTR gene is inserted in bacterial plasmid.
normal CFTR now present – chloride and water movement normal.
CFTR gene codes for protein synthesis.

Section H Biodiversity and conservation

Biodiversity can be considered at different levels

- **Habitats:** how many different possibilities of food, shelter and breeding opportunities are offered by an environment.
- **Species:** how many species, and how many individuals of each species, can be supported by a habitat.
- **Genetic:** how many alleles of how many genes are present in a particular population.

Biodiversity (Biological diversity)

can describe:

- the **numbers** of species present;
- the **range of types** of species in a particular habitat.

Why is it important?

- Extinction of one species may affect many other members of an ecosystem.
- It provides great resource of food and medicines.
- Many people gain great pleasure from the study of wildlife and wild places.

Biodiversity can be reduced

Deforestation:

- direct loss of species e.g. by cutting down hardwood trees;
- reduction in soil fertility, as trees may contain 90% of the nutrients in a forest;
- erosion, reducing possibilities for establishment of food chains;
- alterations to water cycle as transpiration is reduced.

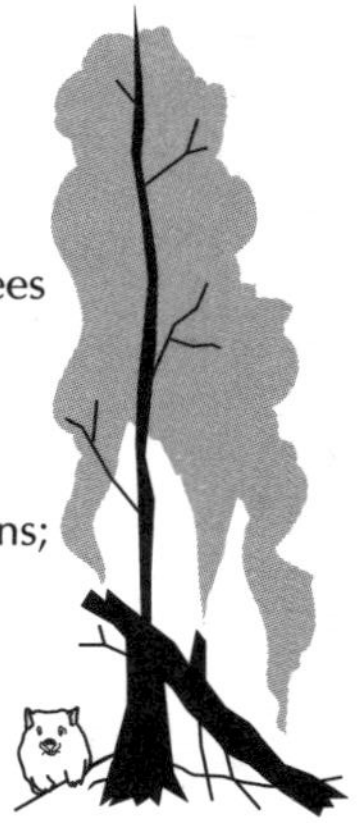

Agriculture:

- promotes species useful as crops;
- pesticides may directly remove species;
- removes hedgerows to increase efficiency;
- over-use of fertilisers may cause eutrophication of nearby ponds and lakes;

Pollution:

- greenhouse gases may → climate change;
- heavy metals may accumulate in predators;
- acid gases may affect forests and lakes.

The Simpson species diversity index

- a **single** number, the **diversity index**
- includes **species richness** (number of species present), and
- **abundance** (the number of each individual species)

$$D = \frac{N(N-1)}{\sum n(n-1)}$$

- D: a single number, the diversity index
- N: the total number of individuals of all species
- n: the total number of individuals in a particular species
- $\sum$: the 'sum of'

Weaknesses:

- makes no allowance for size – a whale is counted the same as a shrimp;
- hard to calculate for plants – many different stems may belong to the same individual.

Biodiversity can be maintained

- **Conservation** by management of habitats **in situ**
- **Zoos** and **captive breeding programmes**
- **Botanic gardens** and **seed banks**

Taxonomy is the science of classification

HIERARCHICAL CLASSIFICATION OF THE LION

On moving down the hierarchy of groups, note that there are …

MORE SIMILARITIES and FEWER DIFFERENCES between the members

- KINGDOM (Animalia) — Other phyla
  - PHYLUM (Chordata) — Other classes
    - CLASS (Mammalia) — Other orders
      - ORDER (Carnivora) — Other families
        - FAMILY (Felidae) — Other genera
          - GENUS (Panthera) — Other species
            - SPECIES: Panthera leo

All animals are ingestive heterotrophs.

All chordates have a notochord (→backbone).

All mammals have fur and mammary glands.

All carnivores have well-developed carnassial (flesh-cutting) teeth.

All felidae have retractable claws.

All Panthera (big cats) can roar but cannot purr.

All lions can mate and produce fertile offspring with other lions.

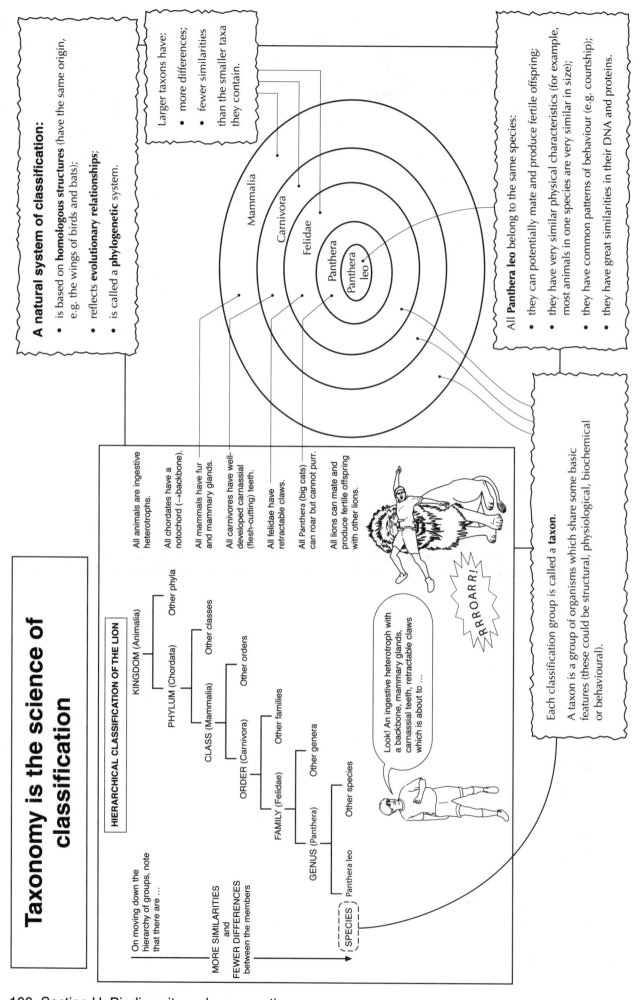

Look! An ingestive heterotroph with a backbone, mammary glands, carnassial teeth, retractable claws which is about to …
RRROARR!

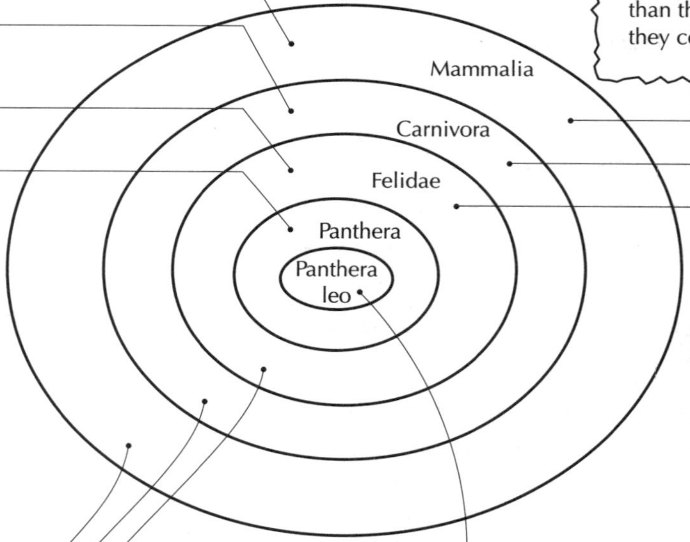
Mammalia
Carnivora
Felidae
Panthera
Panthera leo

A natural system of classification:

- is based on **homologous structures** (have the same origin, e.g. the wings of birds and bats);
- reflects **evolutionary relationships**;
- is called a **phylogenetic** system.

Larger taxons have:

- more differences;
- fewer similarities

than the smaller taxa they contain.

Each classification group is called a **taxon**.

A taxon is a group of organisms which share some basic features (these could be structural, physiological, biochemical or behavioural).

All **Panthera leo** belong to the same species:

- they can potentially mate and produce fertile offspring;
- they have very similar physical characteristics (for example, most animals in one species are very similar in size);
- they have common patterns of behaviour (e.g. courtship);
- they have great similarities in their DNA and proteins.

Naming the organisms

THE BINOMIAL SYSTEM OF NOMENCLATURE

- provided by Linnaeus
- successful since
  (a) each species has a unique name
  (b) shows which species are closely related e.g. *Panthera leo* and *Panthera tigris*
- incorporates both species and genus into specific name
  e.g. *Homo sapiens*
  (i) Underline or *italicise*
  (ii) Capital letter for genus, lower case for species.
- the 'naming' of organisms (usually after they have been placed into 'groups' or taxons)
- needed to aid communication between groups of scientists so 'name' must be unambiguous and easily understood
- usually in Latin since
  (a) original scientific language
  (b) universally accepted

What about viruses?

- made of a protein coat surrounding RNA or DNA;
- can only reproduce inside living cells.

So could be:

- a collection of molecules, organised but not alive;
- a highly evolved parasite which originated as a prokaryote.

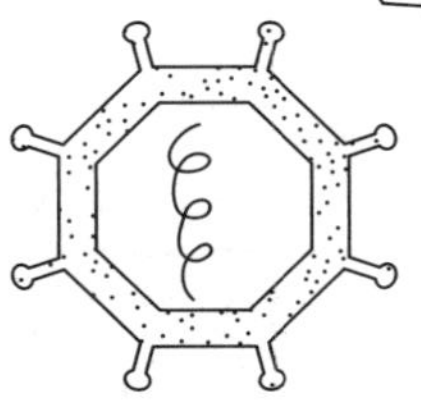

The five kingdom system of classification

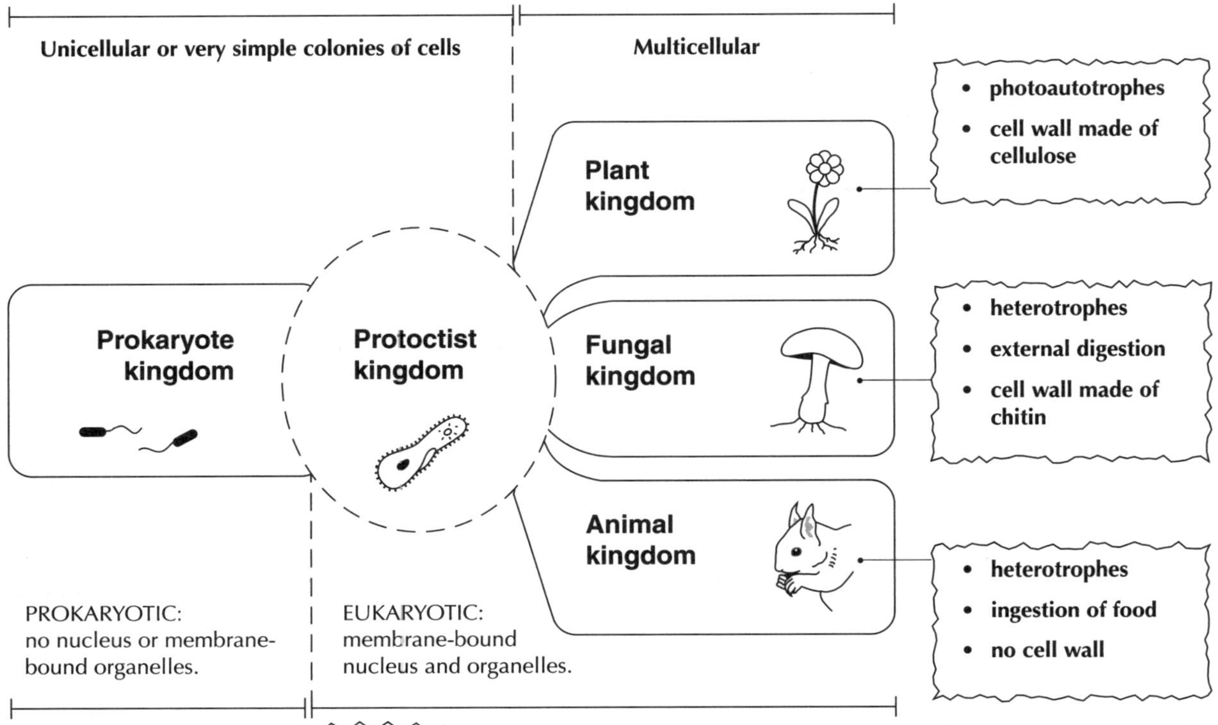
Unicellular or very simple colonies of cells
Multicellular
Prokaryote kingdom
Protoctist kingdom
Plant kingdom
Fungal kingdom
Animal kingdom
• photoautotrophes
• cell wall made of cellulose
• heterotrophes
• external digestion
• cell wall made of chitin
• heterotrophes
• ingestion of food
• no cell wall
PROKARYOTIC: no nucleus or membrane-bound organelles.
EUKARYOTIC: membrane-bound nucleus and organelles.

Sponges are another exception!

- made of many cells;
- cells may be specialised, e.g. for feeding or reproduction;
- all cells are enclosed in a form of skeleton.

The three domain system

- uses genetic data to divide the prokanjotes into two groups
- two groups are the **bacteria** and the **archaea** ('ancient ones')
- all other organisms are placed into one group, the **eukarya** (including animals and plants).

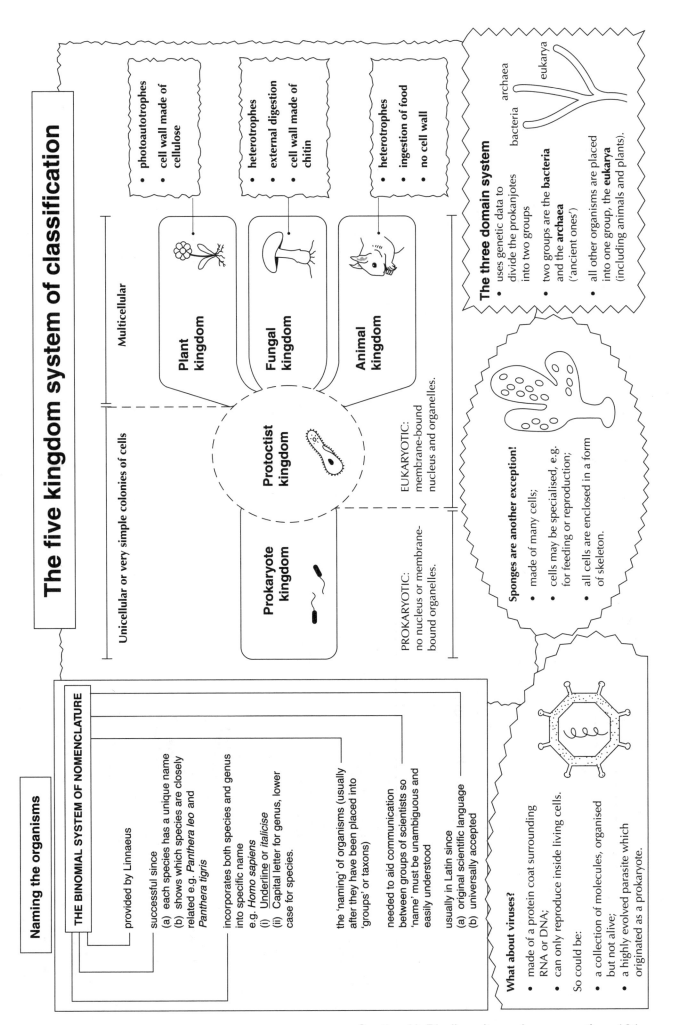

bacteria
archaea
eukarya

Keys and classification

A key enables identification of an organism by observation of its characteristics. Close observation allows a series of questions (the branch points in this key) to be answered, eventually leading to the organism being studied.

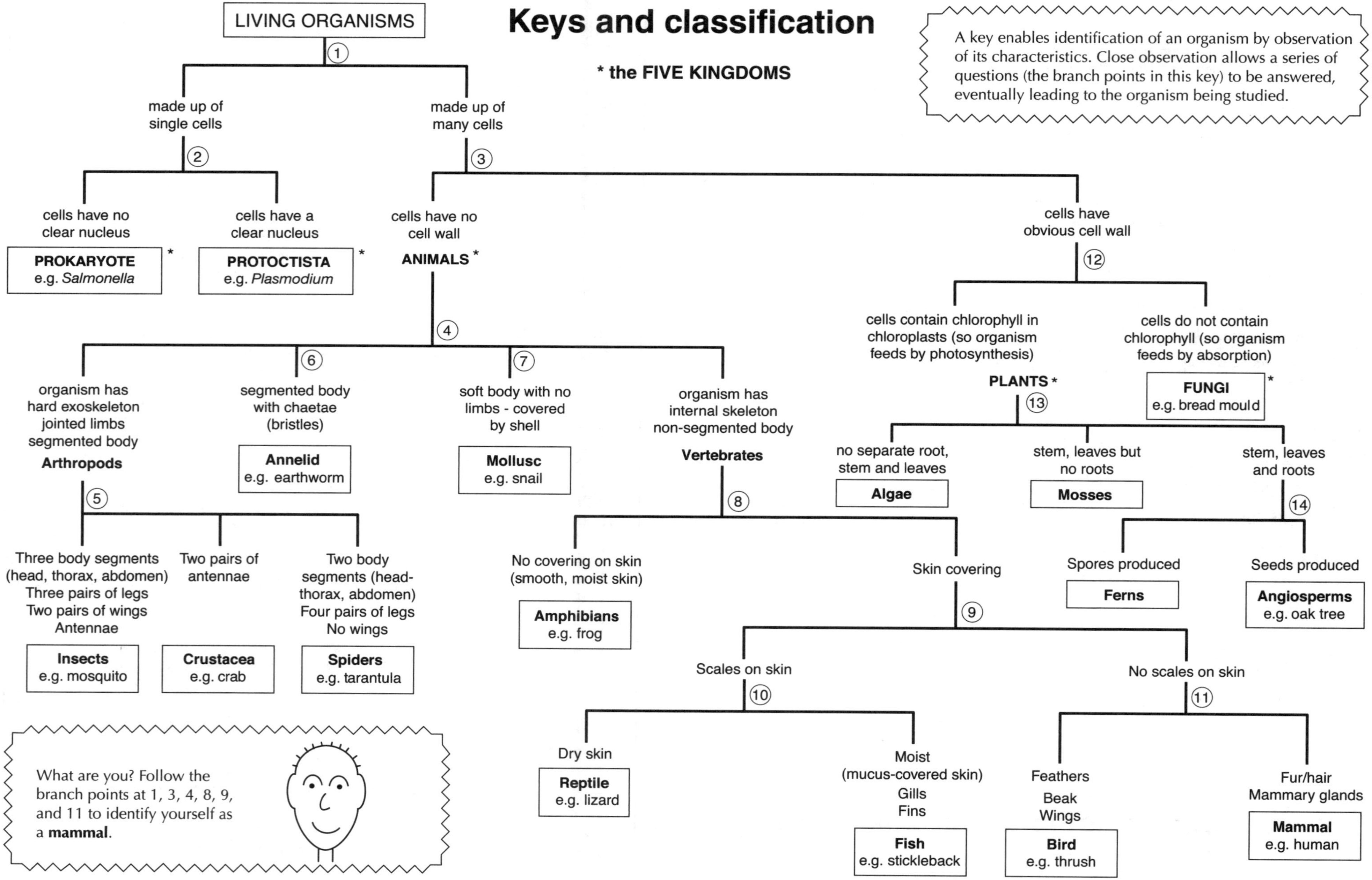

What are you? Follow the branch points at 1, 3, 4, 8, 9, and 11 to identify yourself as a **mammal**.

Modern techniques in taxonomy include protein and DNA sequencing

Haemoglobin

- Haemoglobin is a protein with two α and two β chains.
- The amino acid sequence in the β chains can be analysed.
- A table of the number of amino acid differences between human β chain and the haemoglobin of other species can be prepared.

Human β-chain	0
Chimpanzee	0
Gorilla	1
Gibbon	2
Rhesus monkey	8
Dog	15
Mouse	27
Grey kangaroo	38
Chicken	45
Frog	67

This number is inversely proportional to the closeness of the evolutionary relationship e.g. humans are more closely related to chimps than to rhesus monkeys.

Other important proteins which have been used to provide similar data are:

- **cytochrome c** (part of the electron transfer chain);
- short sections of **fibrinogen** (a blood clotting protein).

Immunological comparison of proteins

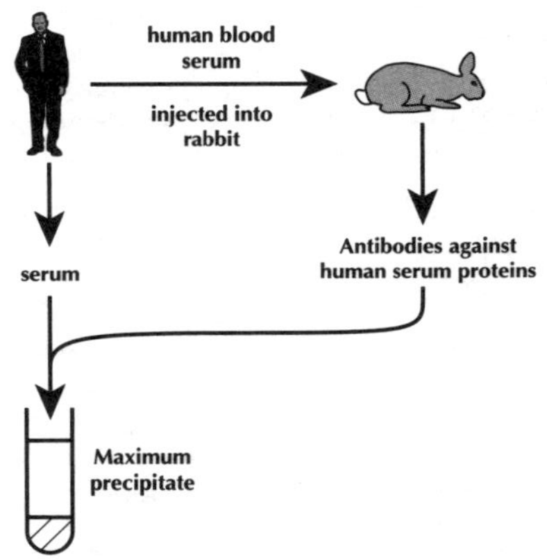

- Rabbit antibodies can be mixed with serum from other species.
- The amount of precipitate shows how closely the other species is related to humans.

Base sequences in DNA

- DNA is extracted from the organisms to be compared.
- The samples of DNA undergo amplification by PCR.
- The PCR (Polymerase Chain Reaction) products are separated by gel electrophoresis.
- The arrangement of bases in the DNA products is determined by Sanger sequencing.

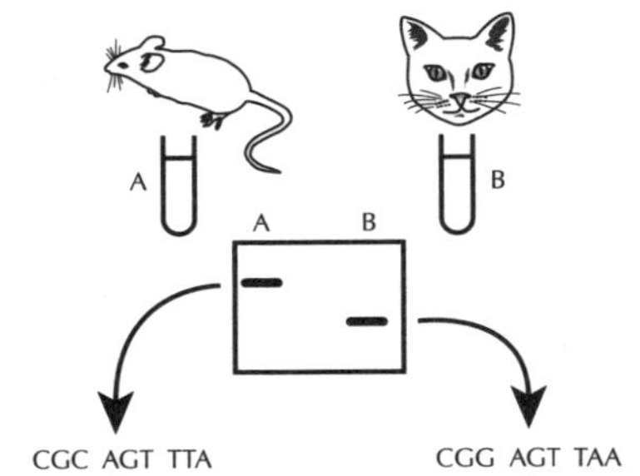

Many sequences compared

Computer analysis

Phylogenetic tree

e.g. similarities between monocot- and dicot- plants

classification of orchids

evolutionary relationships of seaweeds.

DNA-DNA hybridisation allows comparison of individual genes or the total genome from two species …

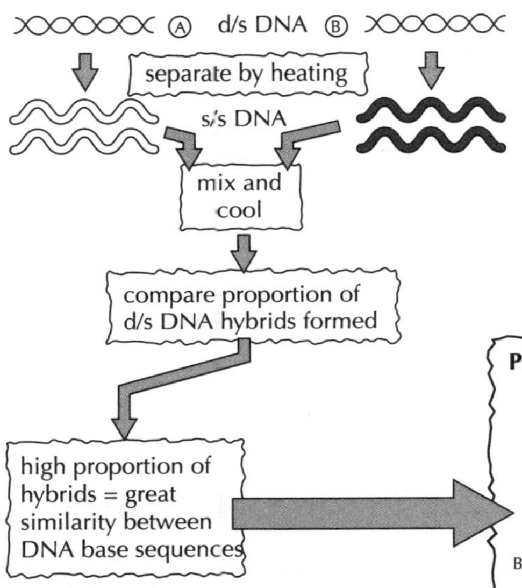

… and depends on:

- base-pairing rules A–T G–C;
- weakness of hydrogen bonds linking the two strands in DNA;
- techniques which can separate double-stranded from single-stranded DNA.

Phylogenetic tree of hominoids

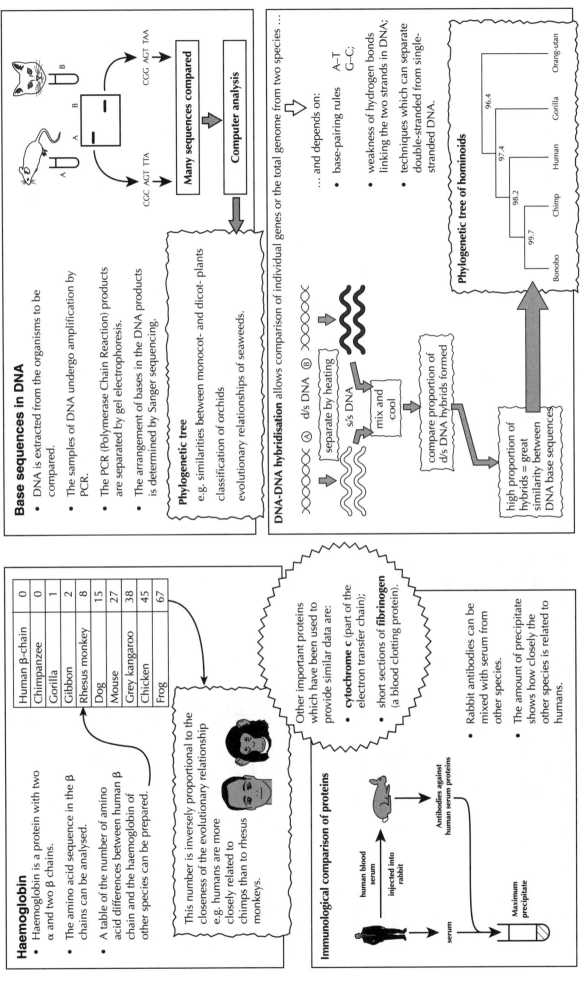

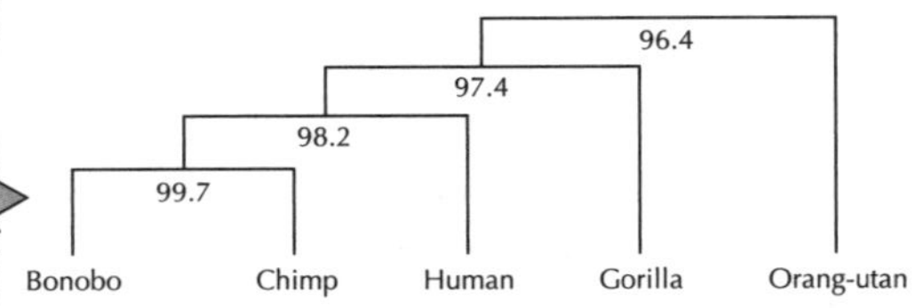

Courtship is behaviour used to attract and keep a mate

Environment and hormones affect behaviour

Photoperiod = number of hours of light/24h. Increases in Spring in the northern hemisphere.

increasing photoperiod in spring → **detected in brain** → **stimulation of pituitary gland** → **release of gonadotrophins** → **primary sex organs** → **sex hormones**

The ovary and testis

For example, **oestrogen** and **testosterone:** Steroids synthesised from cholesterol

sex hormones → **development of gametes**; **breeding colaboration**; **protection of territory**

breeding colaboration → **Courtship behaviour**; **protection of territory** → **Courtship behaviour**

Pair bonding is important

- co-operation in choice of **breeding site** (e.g. nest)
- defence of **territory** – must include breeding site, shelter and adequate supply of food
- **rearing of young** – may include male bringing food to female.

Sexual selection

- ♀♀'s choose biggest ♂ with best courtship display
- most obvious where one ♂ has harem of ♀♀ e.g. elephant seal.

Courtship behaviour

- allows animals to **approach closely**, and limit aggression (♀ spiders may eat the ♂ if courtship fails);
- makes sure that animals **recognise their own species** (inter-species mating rarely produces fertile offspring);
- ensures **choice of strong, healthy mate** (most 'suitable' genes to the next generation);
- **synchronises reproductive behaviour** (gametes viable at the same time).

depends on signals

Human lips have an extremely high density of touch receptors!

Touch: some species of spider use a method in which the male plucks a strand of the female's web – the frequency of plucking is species-specific.

Sound is widely used, for example:

- birdsong – males will call for a female – bitterns 'boom';
- stridulation – crickets rub the edges of their wings together to produce a species-specific 'chirping' sound;
- croaking – used by many amphibians.

N.B. This can be very dangerous – the ♂ is also very obvious to predators!

Scent: species-specific sex attractants (examples of **pheromones**):

- many moths – females attract males who detect chemical stimuli on antennae;
- mammals – civet cats produce **civetone**, and musk deer produce **muscone** from modified anal glands.

Visual: very commonly studied (because we can see them!)

- red belly of ♂ stickleback;
- colour changes over body surface of cuttlefish;
- waving of enlarged claw by fiddler crabs.

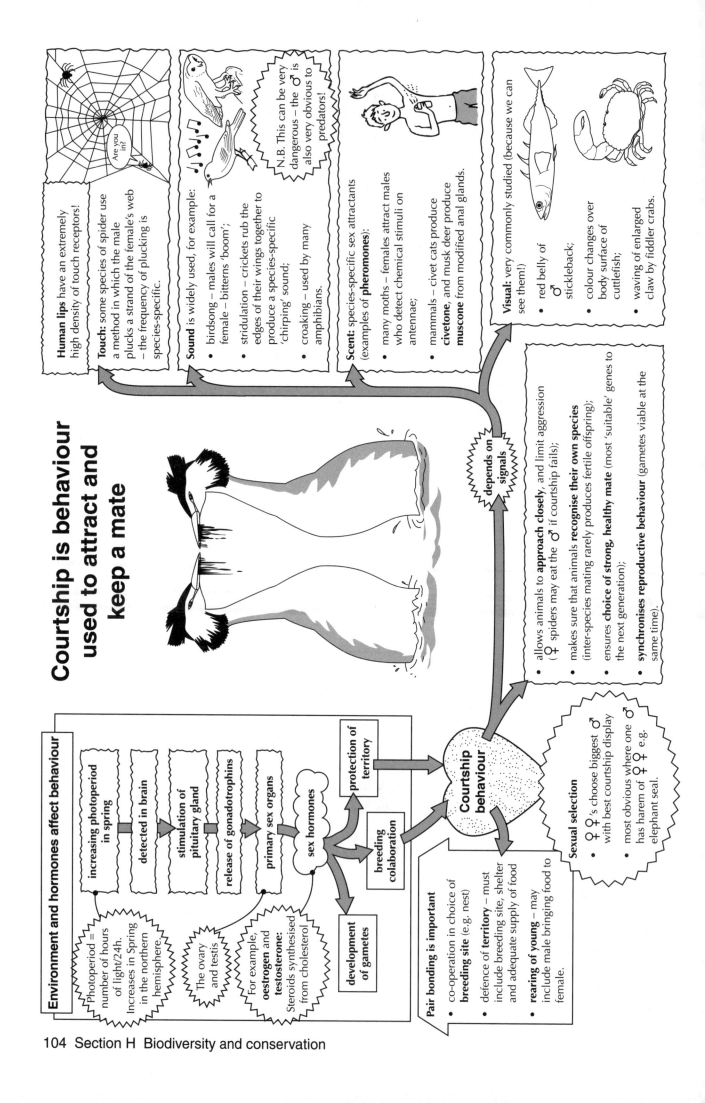

The greenhouse effect

could lead to global climate change.

Origins of greenhouse gases

Photosynthesis in forests and grasslands removes carbon dioxide (CO_2) from the atmosphere.

Car exhaust emissions contain much CO_2 – released to the atmosphere.

Combustion of fossil fuels by industrial plants releases large amounts of CO_2.

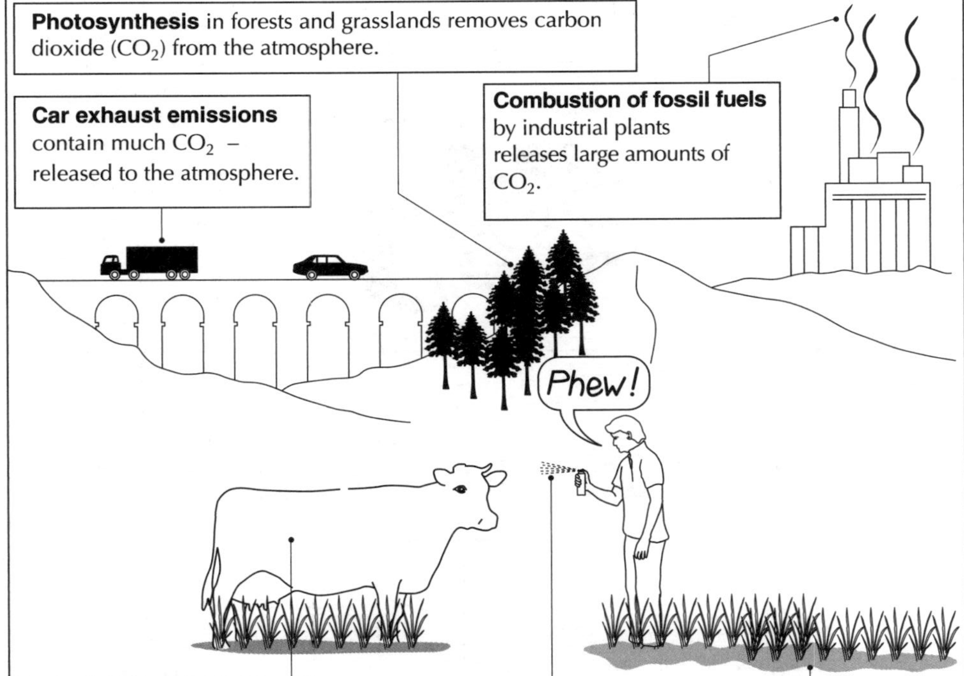

Ruminant fermentation produces **methane** (CH_4) which cattle release into the atmosphere. Intensive cattle ranching increases CH_4 release at the expense of CO_2 uptake by photosynthesis.

Aerosol propellants contain **chlorofluorocarbons** (CFCs) which are 10^5 x worse than carbon dioxide as greenhouse gases.

Anaerobic fermentation in swamps and paddy fields produces CH_4. Inorganic fertilisers cause release of nitric oxide (NO).

All living organisms release carbon dioxide by respiration – the additional **greenhouse gases** contributed by humans **(anthropogenic contributions)** include methane and CFCs in addition to greater quantities of carbon dioxide.

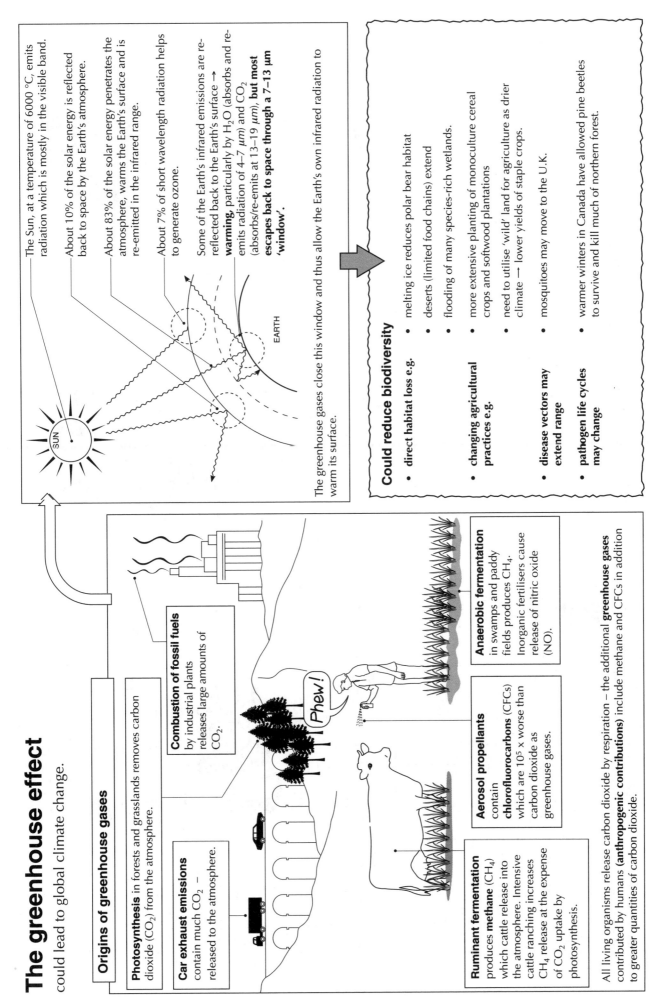

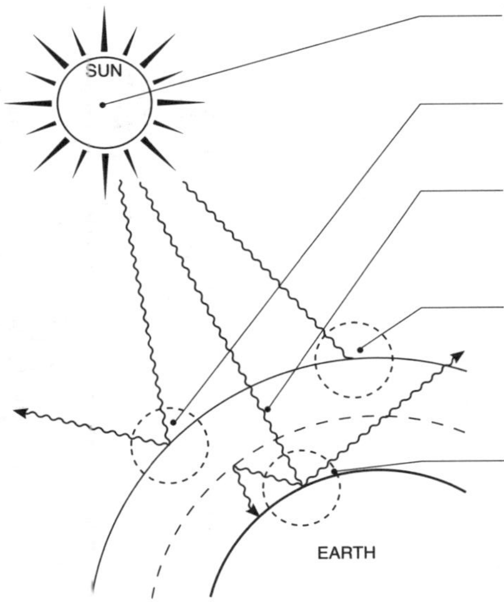

The Sun, at a temperature of 6000 °C, emits radiation which is mostly in the visible band.

About 10% of the solar energy is reflected back to space by the Earth's atmosphere.

About 83% of the solar energy penetrates the atmosphere, warms the Earth's surface and is re-emitted in the infrared range.

About 7% of short wavelength radiation helps to generate ozone.

Some of the Earth's infrared emissions are re-reflected back to the Earth's surface → **warming**, particularly by H_2O (absorbs and re-emits radiation of 4–7 μm) and CO_2 (absorbs/re-emits at 13–19 μm), **but most escapes back to space through a 7–13 µm 'window'.**

The greenhouse gases close this window and thus allow the Earth's own infrared radiation to warm its surface.

Could reduce biodiversity

- **direct habitat loss e.g.**
  - melting ice reduces polar bear habitat
  - deserts (limited food chains) extend
  - flooding of many species-rich wetlands.
- **changing agricultural practices e.g.**
  - more extensive planting of monoculture cereal crops and softwood plantations
  - need to utilise 'wild' land for agriculture as drier climate → lower yields of staple crops.
- **disease vectors may extend range**
  - mosquitoes may move to the U.K.
- **pathogen life cycles may change**
  - warmer winters in Canada have allowed pine beetles to survive and kill much of northern forest.

Conservation at many levels is needed to maintain biodiversity

The Royal Society for the Protection of Birds is primarily concerned with national issues (e.g. preservation of Caledonian pine forest in Scotland) but plays a large part in international conservation by:

- monitoring trade in birds;
- advising on the management of overseas nature reserves.

RSPB

The RSPB was founded in Didsbury, Manchester by a group of women trying to protect the Great Crested Grebe from hunters collecting feathers to decorate women's hats!

GREENPEACE runs a series of non-violent, direct action campaigns such as **Protection of Marine Mammals** and the **Regulation of the Disposal of Toxic Waste.**

COUNCIL FOR THE PROTECTION OF RURAL ENGLAND (C.P.R.E.) is a national body which has particular concerns for local issues e.g. **the planning of roads through sensitive areas** such as ancient woodlands. They take advice from scientists who conduct **environmental impact assessments.**

OUTSTANDING SCIENTIFIC AND POLITICAL CONTRIBUTIONS:

Diane Fossey: Gorillas

Jane Goodall: Chimpanzees

Gerald Durrell: Captive breeding of Golden Lion Tamarin

John Aspinall: Tigers

International + **National** + **Local** + **Individuals**

CONVENTION ON INTERNATIONAL TRADE IN ENDANGERED SPECIES (C.I.T.E.S.) monitors trade in wildlife and wildlife products and tries to impose controls where necessary.

CITES

A licence like this is necessary for the import of any wildlife product

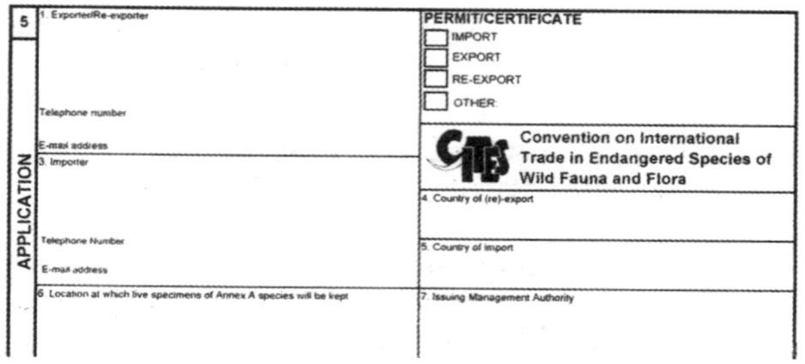

WORLDWIDE FUND FOR NATURE (WWF) attempts to maintain **habitats** and thus the biodiversity they contain. May utilise individual **flagship species** to draw attention to and raise funding for the need to conserve habitats. These species are usually attractive mammals like the panda shown in the WWF logo.

WWF

INTERNATIONAL UNION FOR THE CONSERVATION OF NATURE AND NATURAL RESOURCES (I.U.C.N.) tries to co-ordinate action and monitor success on species survival. The I.U.C.N. publishes the Red Data Books which list in detail the animal and plant species threatened with extinction.

The I.U.C.N. has identified which of the world's 2500 species of amphibians are critically endangered (CR), endangered (EN), and vulnerable (VU), so that conservation work can be channelled in the right direction.

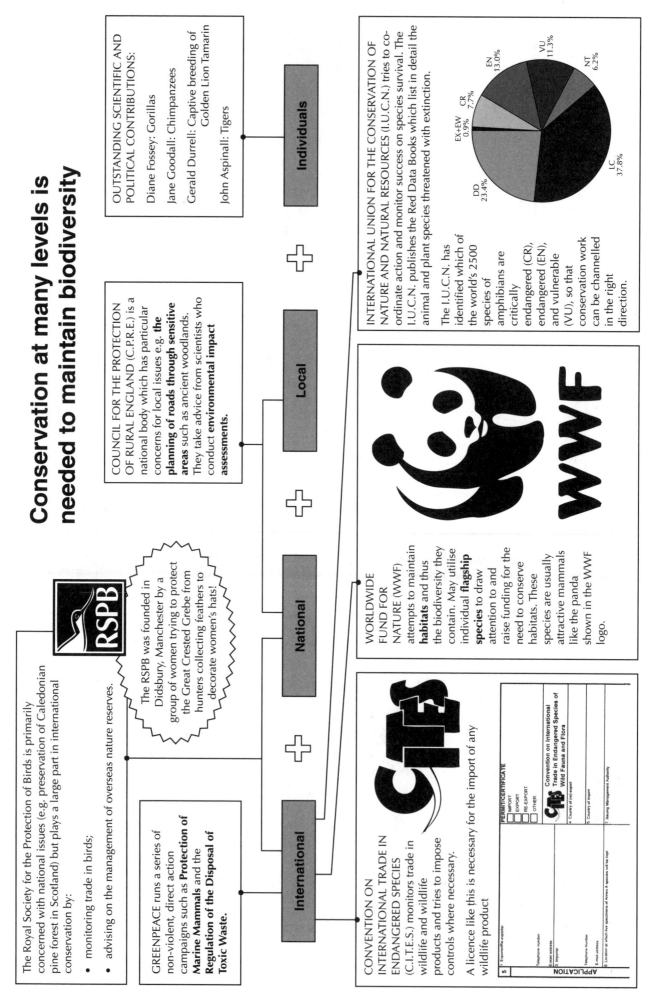

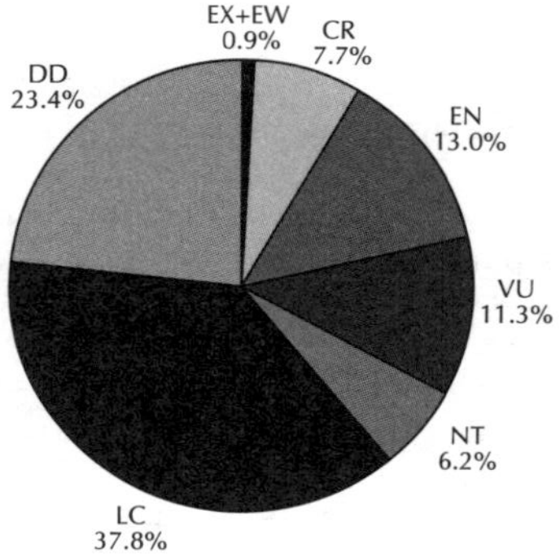

Madagascar is a biodiversity 'hotspot': protection requires international cooperation

The problems:

Deforestation: typical slash and burn to provide **crop-growing** space, wood-cutting to provide **building materials**, **fuel** and forest products such as **charcoal**.

This has produced the typical devastating problems of deforestation, erosion of land by wind and water (suspended solids cause enormous damage to machinery in irrigation projects), loss of habitat, particularly breeding sites for lemurs such as the Indris and changes in both water and nitrogen cycling.

Madagascan forest is unique – 90% of the animals found there are found nowhere else on Earth – but now covers less than 20% of its original area.

Mining: rich deposits of ilminite, a titanium ore, have been discovered but their exploitation would lead to the loss of a unique area of littoral forest on the south-east coast.

Commercial fishing: as with elsewhere in the world, highly efficient fishing techniques (used by non-Malagasy fishermen, especially the Japanese fleet) are decimating fish stocks and threatening the livelihood of local fishermen.

Loss of forest cover

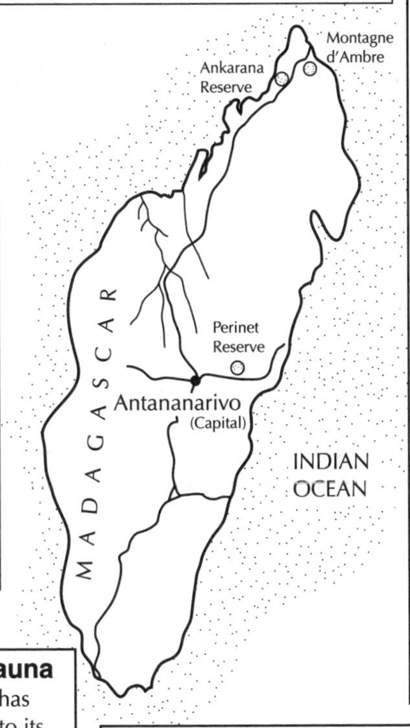

Suggested solutions

Replanting of vegetables, hardwoods, 'charcoal' thorns, particularly in a tree nursery close to the Montagne d'Ambre National Park, is being funded by the WWF in return for the farmers themselves acting as wardens. 'Wildlife corridors' are being planted to link some of the remaining areas of woodland.

Debt swapping is a scheme initiated in the United States. Bankers in developed countries agree to wipe out the debts of the Third World country by allowing the debts to be purchased at the London Bank at a fraction of their face value and in the currency of the debtor nation. The 'debt' (which would probably never have been paid without this scheme) is then used to finance conservation work. One species to benefit from this is the Angonoka, the ploughshare tortoise, which is being rescued by work at a captive breeding centre managed by the Jersey Wildlife Preservation Trust in the north-west of the island.

Madagascar has a unique flora and fauna

Madagascar is an island continent and as such has evolved a range of plants and animals adapted to its own conditions. As many as 90% of the species are endemic to Madagascar. Animal species include lemurs such as the Golden-crowned Sifaka (may be only 100 remaining), the ploughshare tortoise (slowly increasing in number, but still fewer than 100), the fosa (a mongoose-like big 'cat'), a range of chameleons (all feared by the Malagasy as 'human spirits not yet at rest'), whilst the 10 000 plant species include 1000 orchids, numerous pitcher plants and the Madagascan periwinkle (which contains the most effective known treatment for childhood leukaemia).

Pitcher Plant

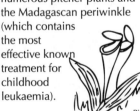

Chameleon

The involvement and cooperation of international organisations

Many conservation projects in Madagascar have failed, but there is optimism that a cooperative effort may yet stem the tide. Some organisations involved in conservation work in Madagascar are Earthwatch (monitoring lemur populations) and the Jersey Wildlife Preservation Trust, but the most significant operation is between:

The **World Bank** which

- oversees the debt-swapping scheme
- controls the operations of the Rio-Tinto Zinc mining subsidiary – no investment without reassurance on ecological restoration
- directly finances some of the work of the WWF.

The **Worldwide Fund for Nature** which co-ordinates many of the individual projects being carried out on the island, using its own funds and support from the World Bank and from

UNISAID (the United States Agency for International Development), which has been particularly heavily involved in financing the Montagne d'Ambre conservation project.

The particular cause for pessimism:

- The enormous destruction of forest has been so **rapid** – most of the losses have occurred within the last 70 years.
- The local people are the major cause of the destruction – they believe that the spirits of their ancestors will help to save them.

Madagascar is seen as a conservation 'flagship'. If an area of such outstanding beauty and wildlife significance cannot be saved, there may be little hope for less spectacular parts of the world.

Forestry can involve in situ' conservation of wildlife.

'In situ' means 'on site' conservation and is the process of conserving an endangered species **in its natural habitat.**

The habitat must be managed
- ensure food/nesting sites for endangered species
- may need to control predators or competitors (e.g. grey squirrel)
- should have the same population density as in equivalent 'natural' habitats.

'In situ' is better: endangered population has the genotype well-suited to this habitat!

Population must be large enough to allow enough genetic diversity so that endangered species can adapt and evolve over time.

DECIDUOUS, BROADLEAVED SPECIES:
- planted along edges of commercial woodlands improve the amenity value/appearance of the woodland;
- support very many more species of insect, and thus contribute more food to insectivorous species;
- native species usually fruit more easily providing an additional food source;
- can be planted to provide corridors for movement between suitable habitats for species with specific feeding or nesting requirements.

STACKS OF 'BRASH': the cuttings from commercial timber, or from roots extracted during forestry operations provide:
- corridors for wildlife;
- habitats for many insects;
- calling/nesting sites for song birds;
- windbreaks for the re-establishment of planted trees;
- lying-up sites for nocturnal mammals.

OPEN SPACE: clearing in woodland offer:
- 'lighter' habitat which encourages growth of an herbaceous layer;
- higher temperatures which may encourage basking reptiles and flying butterflies;
- an 'edge' effect – the greatest number of species is found along the edges of woodland.

STANDS OF DIFFERENT AGES: at different stages of development trees may provide cover, food or breeding sites (rarely all three at the same time).
Thus a variety of ages of tree has optimum value for wildlife. The maximum benefit is derived from both mixed age and mixed species stands.

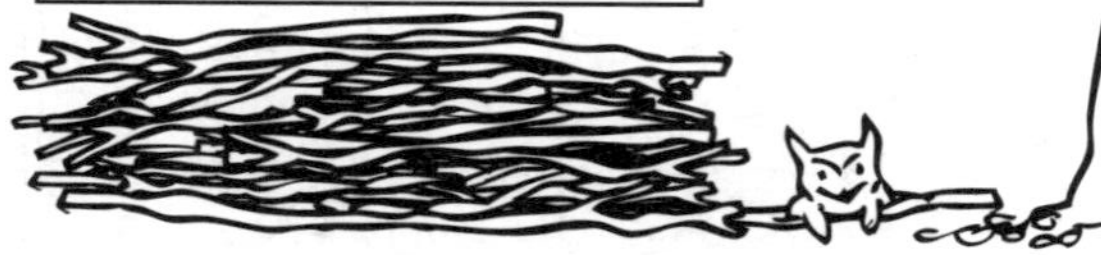

LEAF LITTER: the fallen leaves and fruits from broad-leaved species decomposes more quickly and produces a less acidic environment than the litter of 'needles' from coniferous species. This permits the germination and growth of a more biodiverse community of herbaceous plants and fungi.

DEADWOOD: fallen trees, or even some deliberately felled, are left to rot naturally.
- This provides a habitat for fungi, mosses and ferns as well as for insects and their larvae which may be significant food sources for insectivorous birds and mammals.
- Nutrients are naturally returned to the soil.

Ex-situ ('off site') conservation
- involves removing part of a population from a habitat where it is threatened to a new location where it can be protected:

could be

colony relocation

(e.g. some bat species to man-made 'caves'): very difficult to recreate original environment (e.g. climate, other species present).

human care

botanic gardens
- seed banks may keep viable seeds for 100+ years
- plants may be induced to breed and produce seed by controlling environmental conditions.

zoos
- captive breeding
- education of public
- revenue generation (10% of world population visits a zoo each year!)
- academic research
- keep stud books and gene banks.

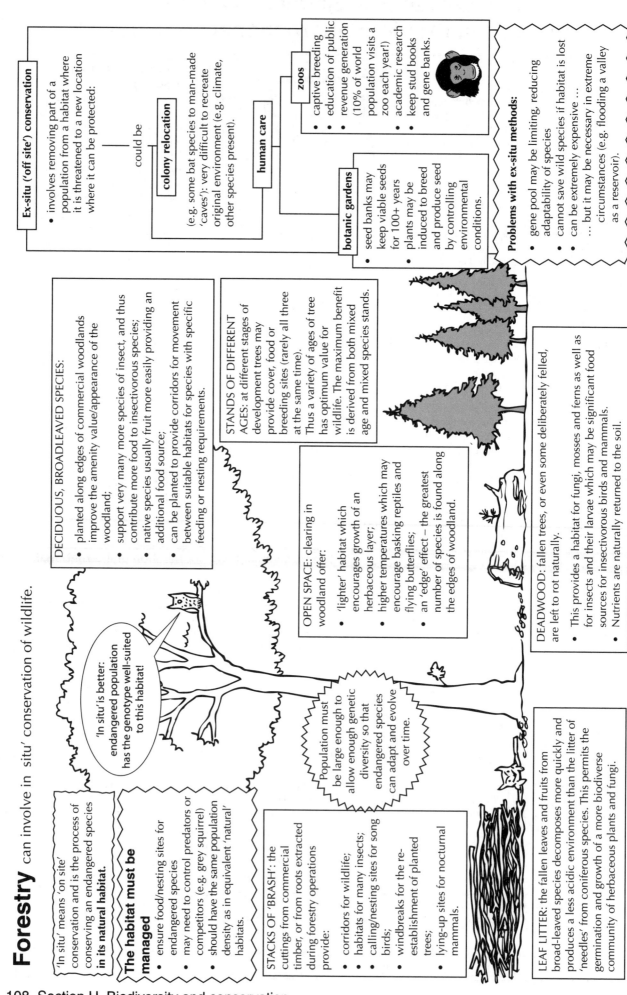

Problems with ex-situ methods:
- gene pool may be limiting, reducing adaptability of species
- cannot save wild species if habitat is lost
- can be extremely expensive …
  … but it may be necessary in extreme circumstances (e.g. flooding a valley as a reservoir).

Conservation of species

There are several possible strategies:

Preservation: involves keeping some part of the environment **without any change**. Only possible if the area can be fenced off/protected.

Reclamation: involves the restoration of **damaged** habitats. Often applies to the recovery of former industrial sites, such as mineworkings.

Creation: involves the production of **new** habitats. Only really possible with small areas e.g. digging a garden pond, or planting of a new forest.

all involve management of a habitat

Pressures on a habitat

- Humans have a significant **biotic** impact on the environment

Temporary when humans were **nomadic**: environment has periods of recovery

Permanent as humans became **cultivators** and **settlers**

- Use of tools and domestication of animals → more efficient agriculture/support for larger population.
- Greater demands for shelter, agricultural land and fuel → greater rate of deforestation.
- Development of fossil fuels → more use of machines/greater 'cropping'/larger populations.
- Pollution as greater use of fossil fuels/pesticides/fertilisers and development of nuclear power.

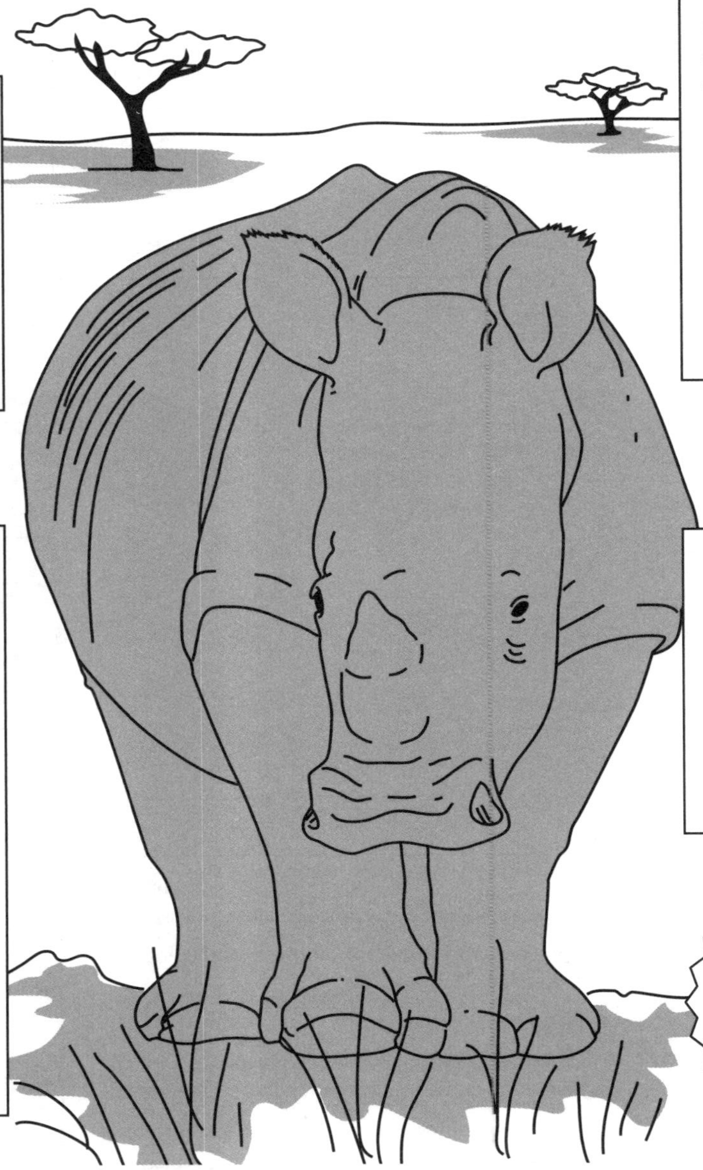

Direct protection may be necessary

Even when a suitable **habitat** (able to provide **food**, **shelter** and **breeding sites**) is available, individual species may be at direct risk from humans.

- Rhinoceros may be hunted for their horns, mistakenly believed to have medicinal or aphrodisiac properties.
- Elephant may be hunted for their ivory tusks.
- Primates (e.g. chimpanzees) and other species may be hunted as 'bush meat'.
- Butterflies, molluscs (shells) and plants may be 'collected'.
- Parrots, primates and fish may be collected for the pet trade.

One useful technique involves ...

Flagship species

Large, attractive and 'cuddly' species attract funding from agencies and donations from the public protection for less attractive species (e.g. beetles and worms) and live in the same habitat.

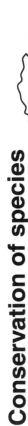

Management is a compromise!

- Maintenance of a particular habitat (e.g. chalk hillside for wildflowers/butterflies) often means **halting succession**.
- The requirements for wildlife must be balanced by human demands for **resources** (e.g. mining for uranium), **recreation** (e.g. diving around coral reefs) and **agricultural land**.

Galapagos islands: conservation issues

Alien invaders!

Introduced species arrived when sailors (and pirates) visited the islands for food and fresh water.

Goats: eat all vegetation – reduce food available to tortoises and land iguanas.

Pigs: omnivorous scavengers which also destroy delicate ecosystems as they dig for plant roots.

Rats: a particular problem as they eat eggs of ground-nesting birds (e.g. boobies).

Cats: direct predators on ground-nesting species.

A flagship species: attractive, so attracts financial support used for benefit of whole habitat.

Galapagos Penguin

Endangered by:

- El Nino
- Overfishing of prey species
- Feral cats
- Becoming trapped in fishing nets.

Humans cause problems!

- Ecotourism – generates money **but** increases pollution
- Increasing conversion of land to agricultural use removes 'wild' habitat and diverts fresh water
- Use of wild species as a food resource (the Beagle took more than 100 tortoises!)
- Oil spillage – disastrous for marine iguanas and seals
- Overfishing – nets also drown dolphins, iguanas and penguins.

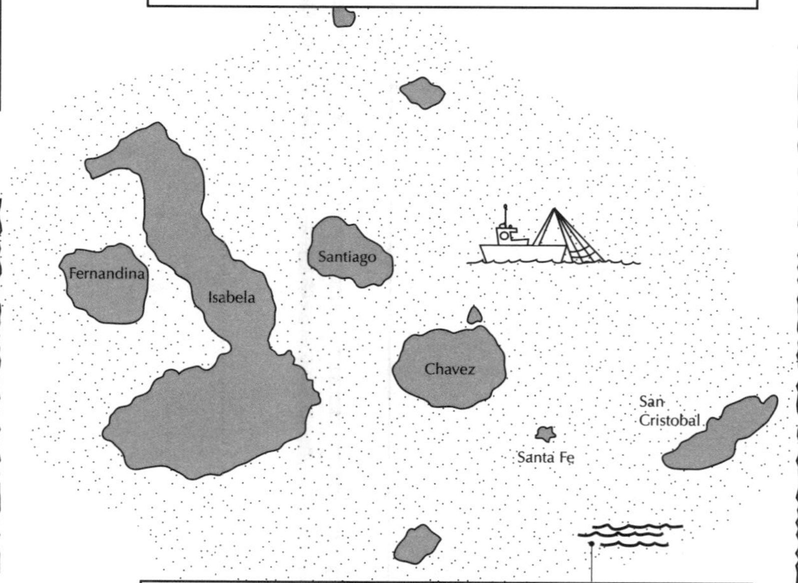

Fernandina
Isabela
Santiago
Chavez
Santa Fe
San Cristobal

El Nino:

- follows warm ocean conditions off the west coast of Ecuador and Peru
- changes to sea temperatures and wind patterns affect rainfall – becomes very heavy in these regions
- rising sea temperatures reduce nutrient content of water, affecting ocean food chains.

But humans are conservationists too!

- removal of introduced species e.g. Isabela is now free of goats
- reintroduction of species – Espanola tortoises (there are plenty) have been moved to Pinta (where there's only one left!)
- increasing landing fees to limit tourism.

Galapagos: ideas on natural selection

Charles Darwin, on the voyage of the Beagle, made many observations that contributed to his ideas on Natural Selection. These included the **Adaptive Radiation** of the **Galapagos finches**

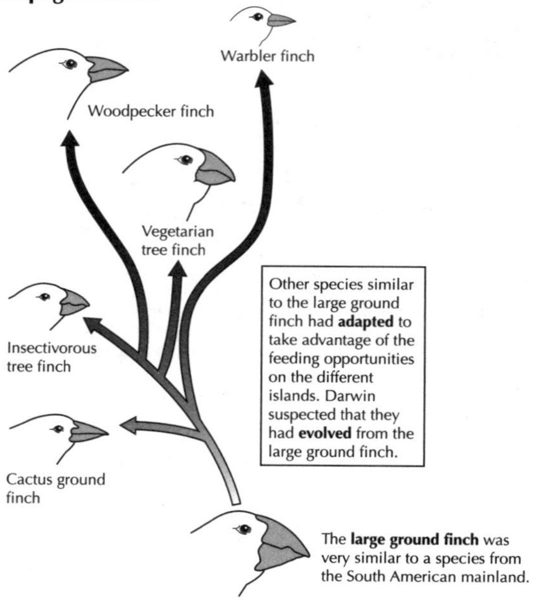

Warbler finch
Woodpecker finch
Vegetarian tree finch
Insectivorous tree finch
Cactus ground finch
Other species similar to the large ground finch had **adapted** to take advantage of the feeding opportunities on the different islands. Darwin suspected that they had **evolved** from the large ground finch.
The **large ground finch** was very similar to a species from the South American mainland.

Finches of the Galapagos islands show adaptations to the food resources on the different islands.

Medicines are drugs used in the management of disease: many medicines are produced by plants.

Functions of medicines

- Cure diseases caused by bacteria: **antibiotics**
- Relieve symptoms, such as pain or high temperatures: **analgesics**
- Prevent the development of disease: **anti-hypertensives**
- Substitute for endogenous (found in the body) substances e.g. **insulin, oral contraceptives.**

The juice of the Silphium plant was used as an oral contraceptive in ancient Greece. It was used so widely it became extinct!

I feel better already!

Vincristine:

- anti-cancer drug (inhibits spindle formation)
- isolated from Madagascan Periwinkle – this plant almost became extinct during mining operations in Madagascar

Biodiversity must be maintained

The loss of any plant, fungal or bacterial species could mean:

- potential medicines are not discovered;
- possible 'factories' for genetic modification may be lost.

Opiates:

- include heroin and morphine;
- important for relief or severe pain;
- obtained from seed pods of opium poppy (*Papaver somniferum*).

Belladonna:

- used to treat some intestinal disorders;
- as Atropine, it relaxes the radial muscles of the iris to increase pupil diameter;
- extracted from flowers and leaves of Deadly Nightshade (*Atropa belladonna*).

Digoxin and Digitoxin:

- increases the strength of contraction of cardiac muscle, so important in treatment of heart failure;
- **Digitoxin** is more powerful (may be fatal!) but **Digoxin** lasts for a shorter time with fewer side effects;
- produced from the leaves of Foxglove (*Digitalis purpurea*).

Aspirin (Acetylsalicylic acid):

- mild analgesic, used to treat headache, period pains and joint discomfort;
- important preventative role in reducing the risk of blood clots, as it lowers the 'stickiness' of platelets;
- over use may lead to gastric bleeding – especially dangerous to sufferers from stomach ulcers;
- originally isolated from the bark of members of willow family (*Salix* sp.).

Vaccines and antibodies

- Some plants have been engineered to deliver medicines in edible structures.
- Examples include vaccines in potato tubers, antibodies in tomato fruits and steroid hormones in tropical Brassicas.

Antibiotics

- Anti-bacterial compounds, effective against many bacterial diseases including wound infections, meningitis and some gut disorders;
- Original compounds were isolated from molds (e.g. *Penicillium* sp.) – modern versions are made by laboratory synthesis.

Genetic modification:

Bacteria may be modified by the insertion of genetic material from other organisms – the genetic material must be identified and isolated from the original organism. The modified bacterium may now produce a valuable medicine – insulin and human growth hormone are made in this way.

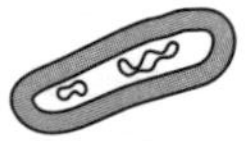

Leaf structure is adapted for photosynthesis and for gas exchange.

Phototropism is a growth response which allows shoots to grow towards the light to allow optimum illumination of the leaves.

Large leaf surface area is held perpendicular to the light source (and kept there by a 'tracking' system).

Leaves are thin so that there are few cell layers to absorb light before it is received by the photosynthetic cells.

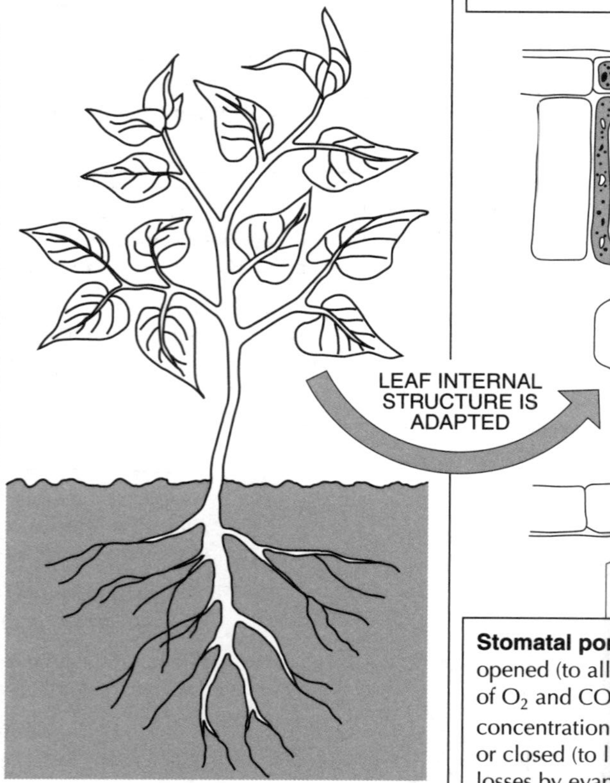

Cuticle: is made up of a waxy compound called **cutin** and is secreted by the epidermis. It is the cuticle which reduces water loss by evaporation, not the epidermis itself. May be considerably thickened in xerophytes.

Upper epidermis: is one or two cells thick – protects against **water loss** (either by cuticle – see left – or by epidermal hairs which trap moisture and reflect light) and against **invasion by pathogens** so that the moist inner leaf is quite sterile. Transparent to visible light.

Palisade mesophyll: is the major site of photosynthesis – cells packed vertically with many chloroplasts which may move by cytoplasmic streaming to optimum position within the cell for light absorption and subsequent photosynthesis.

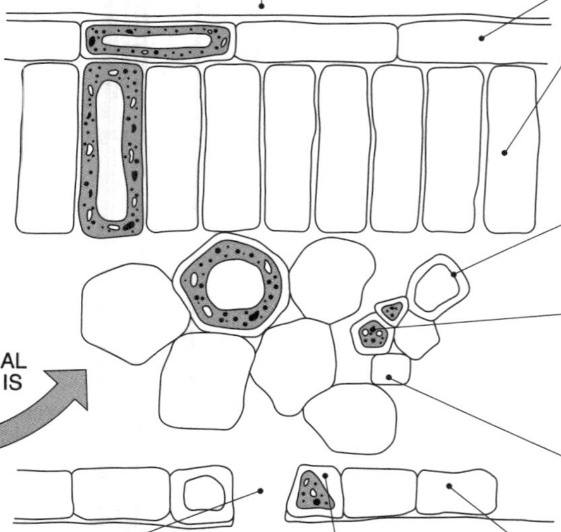

Xylem vessel: transports water and mineral salts to the leaves. Heavily lignified cell walls help to maintain extension of the leaf blade.

Phloem sieve tube: removes products of photosynthesis (principally sucrose) and may import other organic solutes such as amino acids/amides and help to redistribute ions such as phosphate.

Spongy mesophyll: irregularly shaped cells which fit together loosely to leave large air spaces which permit diffusion of gases through leaves. There is a great deal of water loss by evaporation from the surface of these cells.

Lower epidermis: similar protective functions to upper epidermis. Cuticle usually thinner (less light → lower temperature → lower evaporation rate).

Stomatal pore can be opened (to allow diffusion of O_2 and CO_2 down concentration gradients) or closed (to limit water losses by evaporation to a drier atmosphere).

Guard cell: has chloroplasts and membrane proteins to permit pumping of K^+ ions to drive osmotic movement of water. Uneven thickening of cellulose cell wall permits opening/closure of stomatal pore as turgidity changes.

Leaf mosaic is the arrangement of leaves in a pattern which minimises overlapping/shading but maximises leaf exposure to light.

Shoot system holds leaves in optimum position for illumination and CO_2 uptake.

Etiolation causes rapid elongation of internodes kept in darkness to extend shoot.

Epidermis, mesophyll, phloem, and xylem are examples of **tissues**.

Each of them is made up of cells
- with similar structure
- with the same functions
- which have a common origin.

Xerophytes are plants adapted for water conservation: they show many **xeromorphic adaptations**.

CACTI

are adapted to desert conditions.

Flowering is controlled by rainfall so that flower petals, which lose water by evaporation, are very short-lived. Flowering, and fruit formation, may occupy only a few days of the short rainy season.

Stem has numerous water-conservation features:

- Ridged structure which leads to 'self-shading' – less of the plant is exposed to sunlight at any one time.
- Sunken stomata – these trap a moist pocket close to the surface and so reduce water potential gradient between inside and outside of stem.

Altered stomatal rhythm keeps stomata closed during hottest periods of the day.

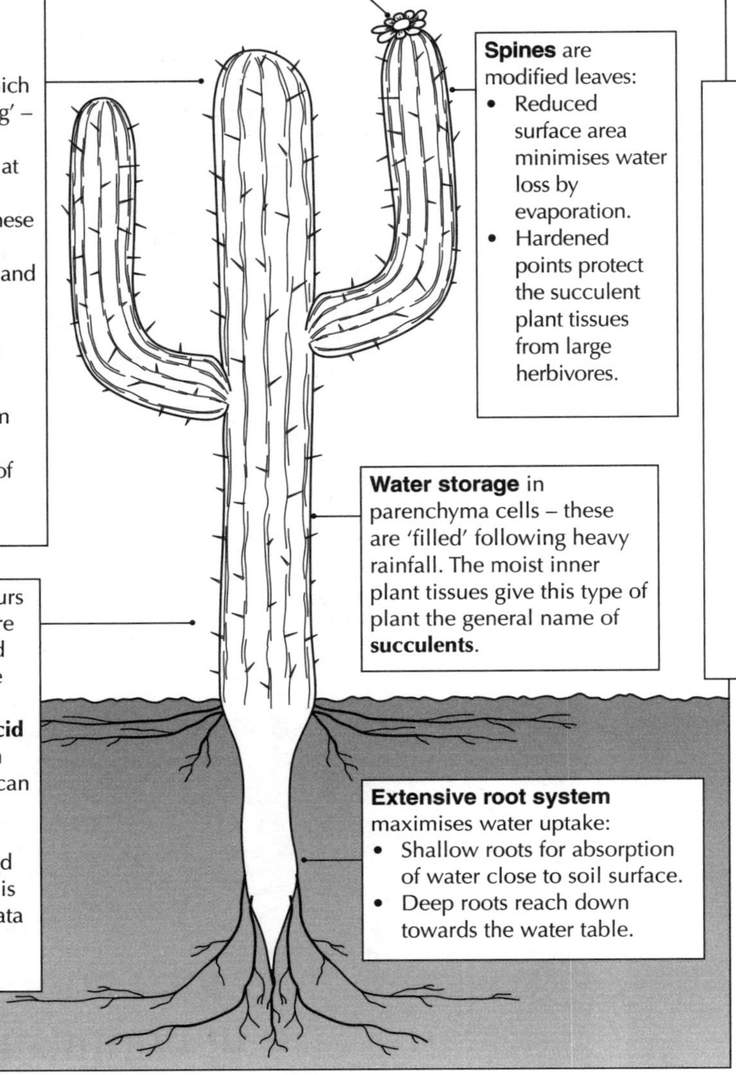

Spines are modified leaves:

- Reduced surface area minimises water loss by evaporation.
- Hardened points protect the succulent plant tissues from large herbivores.

Water storage in parenchyma cells – these are 'filled' following heavy rainfall. The moist inner plant tissues give this type of plant the general name of **succulents**.

Photosynthesis occurs in stem cells (leaves are reduced to spines) and may use an alternative biochemical pathway called **crassulacean acid metabolism (CAM)**. In CAM carbon dioxide can be absorbed at night, stored as an organic acid, and then released during the day. Water is conserved since stomata need only be open during the night.

Extensive root system maximises water uptake:

- Shallow roots for absorption of water close to soil surface.
- Deep roots reach down towards the water table.

Marram grass is adapted to drying conditions of the seashore.

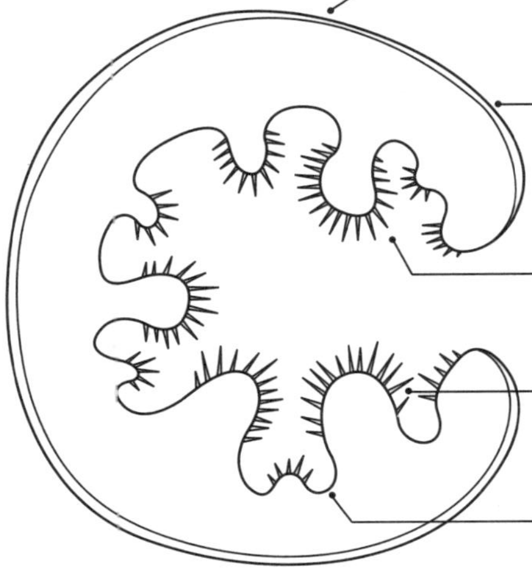

Thick waxy cuticle reduces cuticular transpiration.

Absence of stomata on upper epidermis reduces stomatal transpiration.

Leaf rolling traps humid layer close to lower epidermis. Caused by **hinge cells** which become flaccid when water is in short supply.

Leaf hairs limit air movement over leaf surface.

Sunken stomata keep humid air close to stomatal pore and so reduce water potential gradient.

Growth form

Rosette pattern has overlapping leaves which shade one another to limit evaporation.

Prostrate form has leaves close to ground – minimises surface to evaporation and reduces effects of air currents.

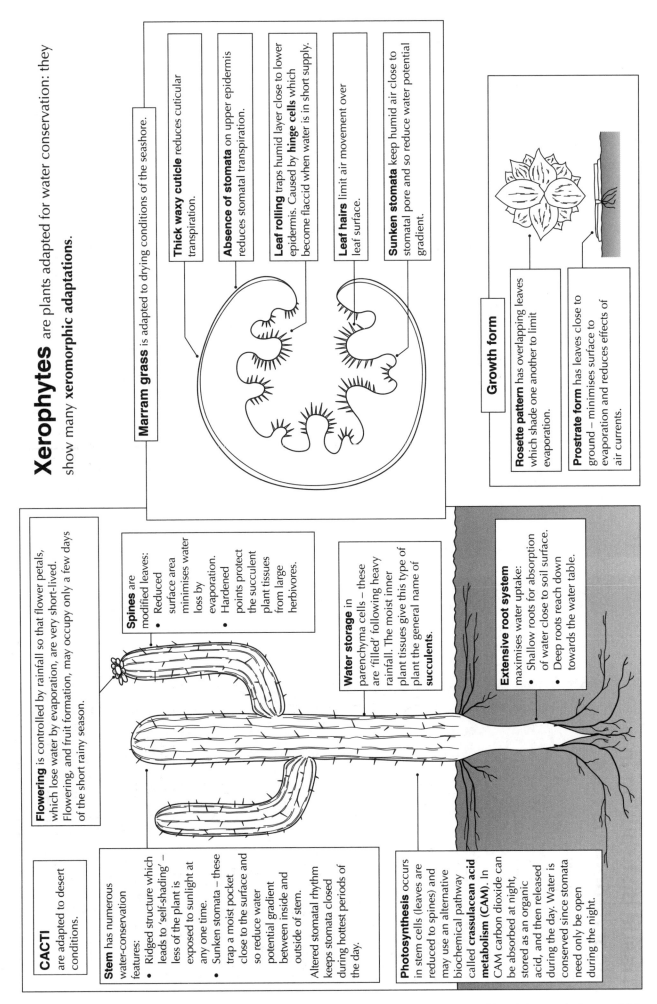

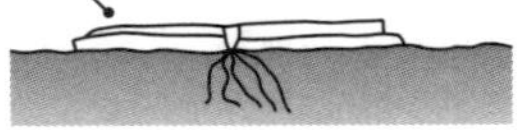

Tissue distribution in a dicotyledonous root

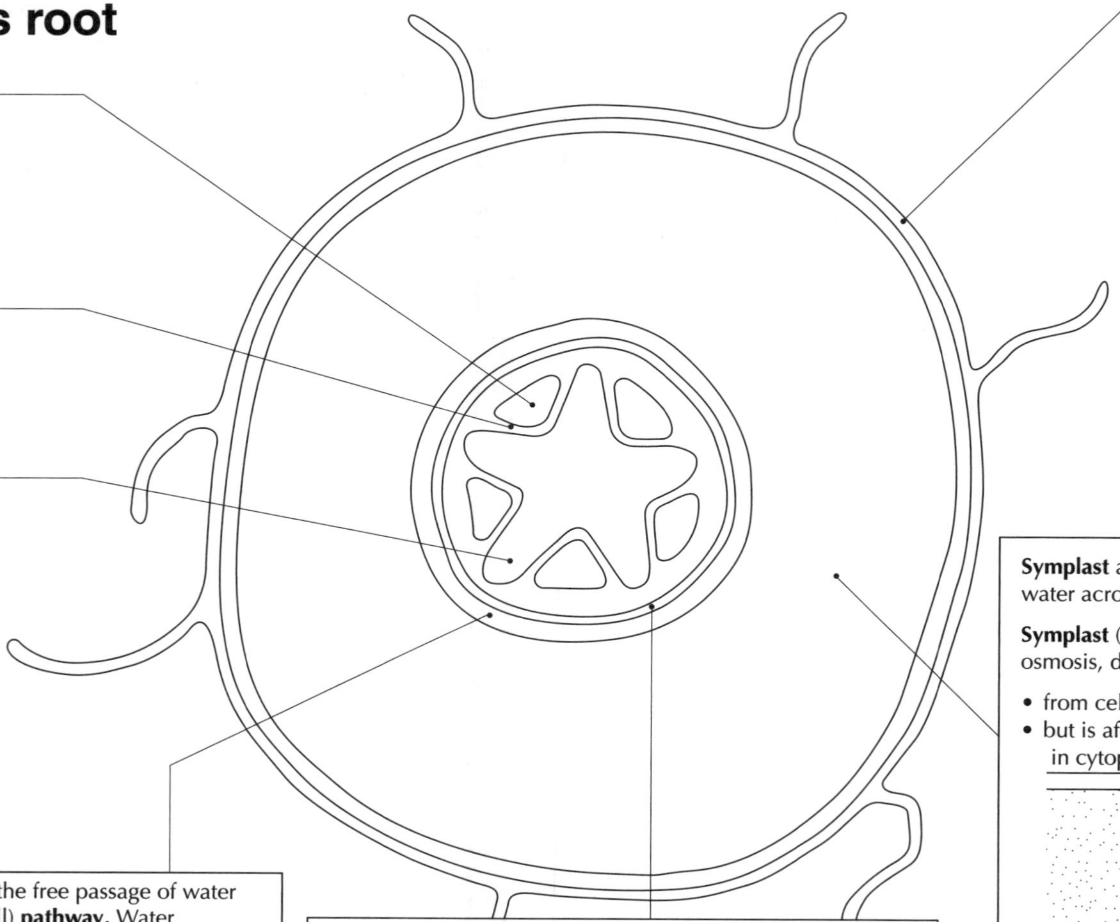

Phloem: transport of organic solutes. These solutes are osmotically active so phloem is situated between 'arms' of xylem enabling absorbed water to enter xylem without crossing phloem.

Cambium: present in some dicotyledonous species – meristematic and divides to produce new phloem and xylem cells.

Xylem: transport of water and ions. Less lignified than in stem – less need for resistance to lateral stresses. Arranged in a series of 'arms' – in this case a **pentarch** (five armed) arrangement.

The phloem, cambium and xylem make up the **stele** – central position surrounded by softer tissues is the optimum arrangement to resist 'upward pulling' forces.

Endodermis: acts as a barrier to the free passage of water and ions via the **apoplast** (cell wall) **pathway.** Water and ions must therefore cross the membrane of the endodermal cell, and here some selection of ions for absorption may take place.

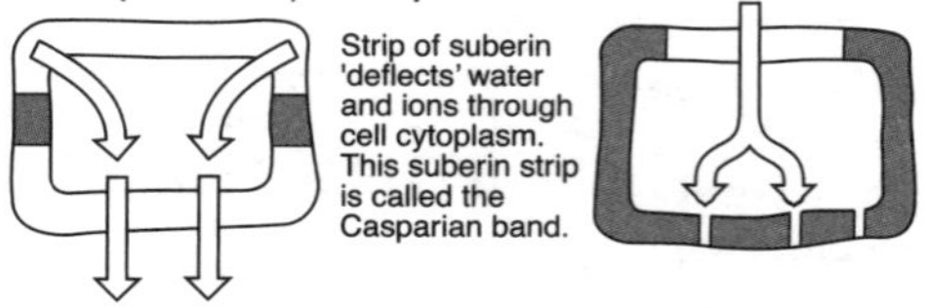
Strip of suberin 'deflects' water and ions through cell cytoplasm. This suberin strip is called the Casparian band.
Heavily suberized wall in older endodermal cell.
Plasmodesmata permit limited onward flow.

Cortex: composed of parenchyma. May store starch and can be infected by *Rhizobium* species to form root nodules in leguminous plants.

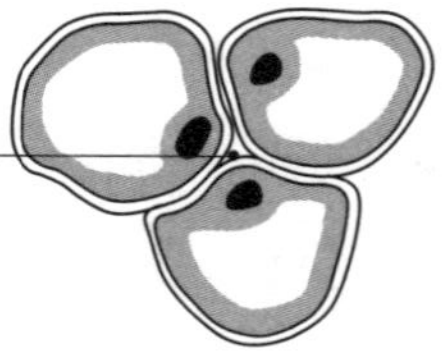
Intercellular air space allows diffusion of gases through the root.

Piliferous layer is an epidermis modified to produce root hairs. The root hairs are outgrowths of a **single cell.** Layer is only one cell thick, has no protective function (hence no cuticle) and is soon lost by abrasion – hence root hairs only present on youngest part of root. Root hairs provide increased surface area for uptake of water and ions in solution.

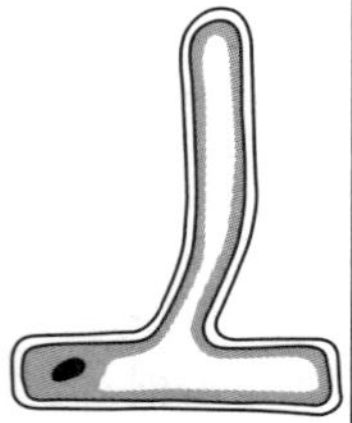

Symplast and **apoplast** are pathways for the passage of water across the living tissues of the root.

Symplast ('inside the cytoplasm'). Water moves by osmosis, down a water potential gradient

- from cell-to-cell via plasmodesmata
- but is affected by dissolved solutes in cytoplasm.

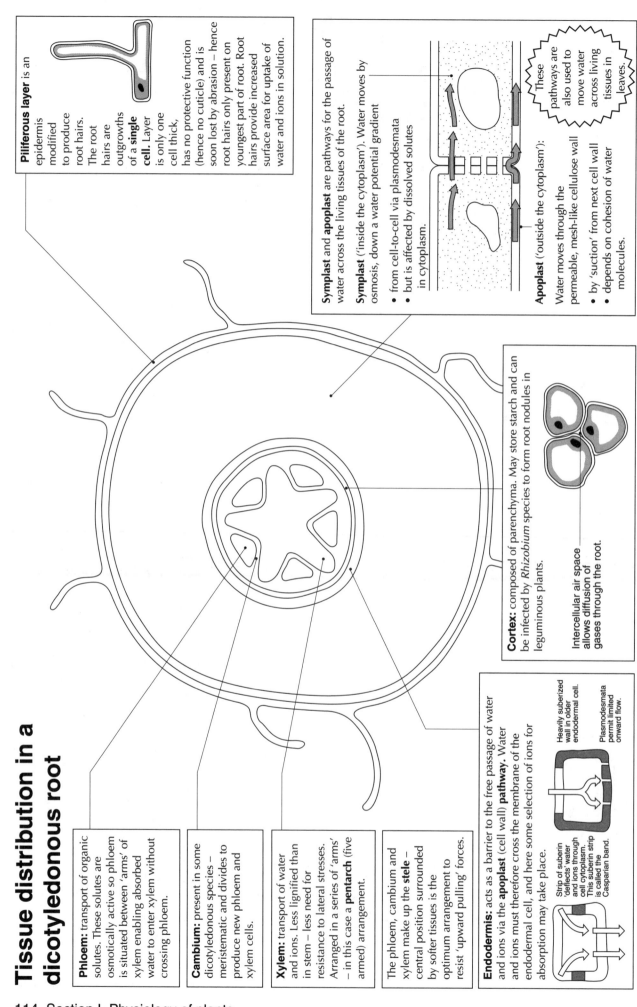

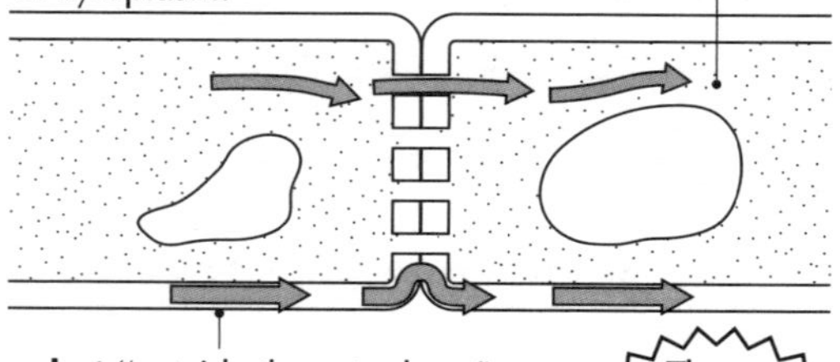

Apoplast ('outside the cytoplasm'):

Water moves through the permeable, mesh-like cellulose wall

- by 'suction' from next cell wall
- depends on cohesion of water molecules.

These pathways are also used to move water across living tissues in leaves.

Tissue distribution in a herbaceous stem

The tissue location provides **mechanical support** and **transport** of water and solutes.

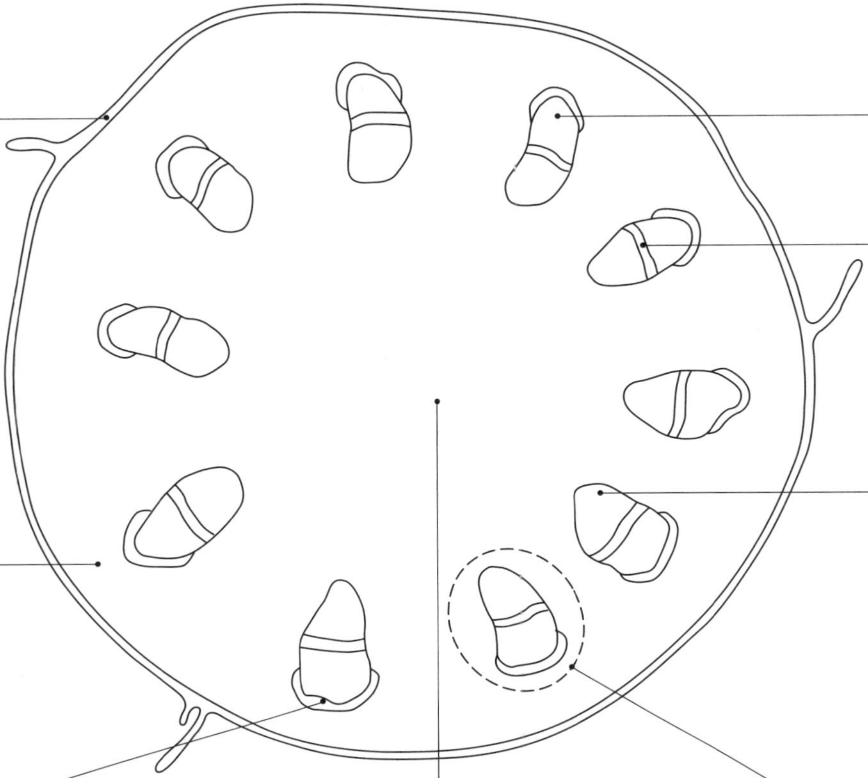

Epidermis: protects the moist inner tissues of the stem against drying out and invasion by pathogens. Epidermis also contributes to support by holding in the turgid parenchyma cells of the cortex.

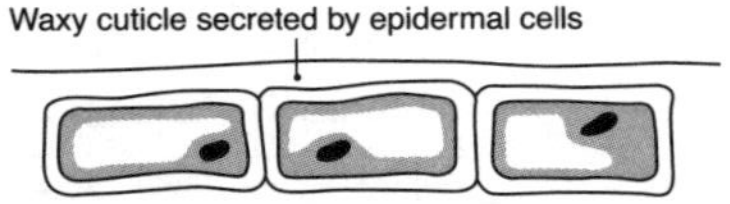

The epidermis is usually one cell thick but may have extended unicellular or multicellular hairs which (a) reduce air movements and water loss and (b) reflect light and prevent overheating.

Collenchyma: provide support, especially in young stems as they are living cells and thus can expand as the stem grows.

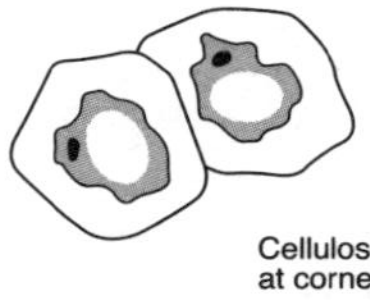

Sclerenchyma: are elongated into supporting fibres. Have heavily lignified secondary cell walls and thus have no living contents.

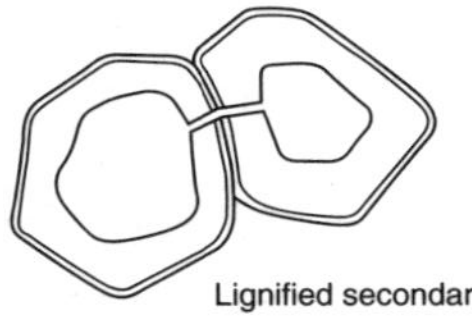

Parenchyma: is the basic unspecialised packing tissue of the plant body. Cells may store food (commonly starch) and may provide support when turgid. Intracellular air spaces aid diffusion of gases through plant body. Forms the **pith (inner cortex).**

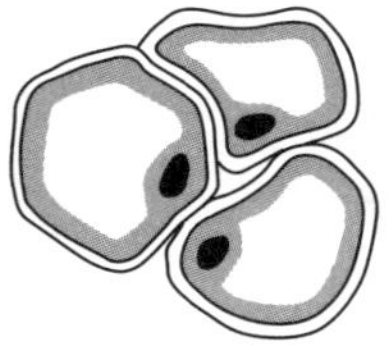

Phloem: transport of organic solutes such as sucrose, amino acids. Some redistribution of ions.

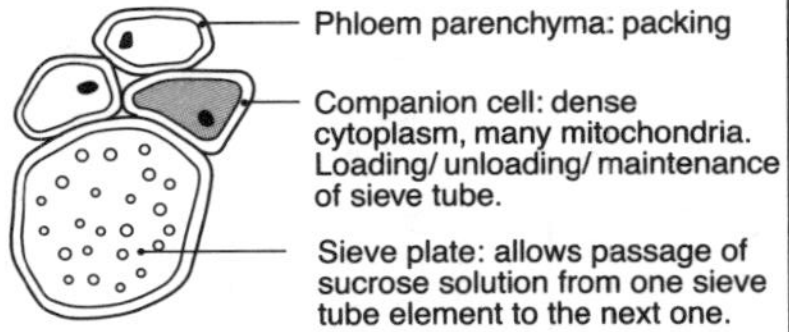

Cambium: a lateral meristem. Simple, non-specialised cells divide and then differentiate into phloem (to outside) and xylem (to inside).

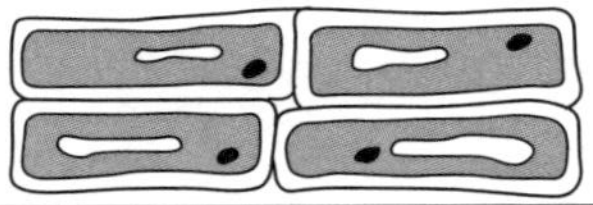

Xylem: transport of water and ions. Lignified cells offer support.

Protoxylem: incomplete lignification – may have cytoplasm

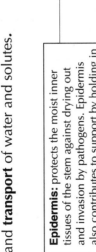

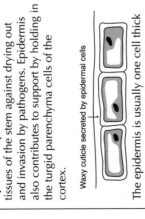

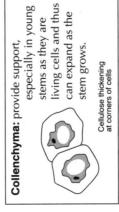

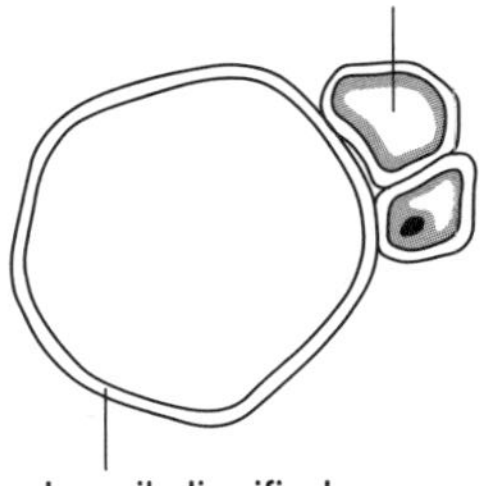

Metaxylem: heavily lignified secondary cell wall – no living contents

Phloem
Intra-fascicular cambium
Xylem
} together make up a **vascular bundle**

Cohesion–tension theory of transpiration

Follow the sequence ①→⑨, which begins with **evaporation of water from the surface of the leaf.**

① **Water evaporates** from the surface of epidermal cells.

② **Water diffuses** from sub-stomatal chamber into the atmosphere: the rate of movement will be influenced by any factor which changes the water potential gradient between this chamber and the atmosphere.

③ **Water evaporates** from the cellulose cell walls of cells of the palisade and spongy mesophyll into the sub-stomatal chamber.

④ **Water moves through leaf tissues** via **apoplast** (cell wall), **symplast** (cytoplasmic) and **vacuolar** pathways.

⑤ **Water leaves the xylem** via vessels with very little lignification (i.e. with freely permeable cellulose cell walls). Moves into apoplast due to **cohesion** and into symplast/vacuoles by **osmosis** along a **water potential gradient.**

⑥ **Water is drawn up xylem** since transpiration causes a state of **tension** or **negative pressure** in the xylem vessels. The replacement molecules of water form a continuous stream due to **cohesion** (attractive forces between water molecules caused by hydrogen bonding) and **adhesion** (attractive forces between molecules of water and the inside surfaces of the lignified xylem vessels).

⑦ **Water enters xylem** through poorly lignified cell walls due to **cohesion** (from apoplast) and **osmosis** along a **water potential gradient** from the symplast/vacuolar pathways.

⑧ **Casparian band** (impregnation of cell wall suberin) of the **endodermis** prevents water movement via apoplast pathway – all water movement via **symplast/vacuolar** pathways, from endodermis to pericycle.

⑨ **Water uptake** occurs by **osmosis** along a **water potential gradient** from the soil solution primarily, but not completely, into **root hair cells.**

Root pressure:

- ions are pumped by active transport (so dependent on respiration for ATP)
- water follows by osmosis
- can 'push' water a few cm up the stem.

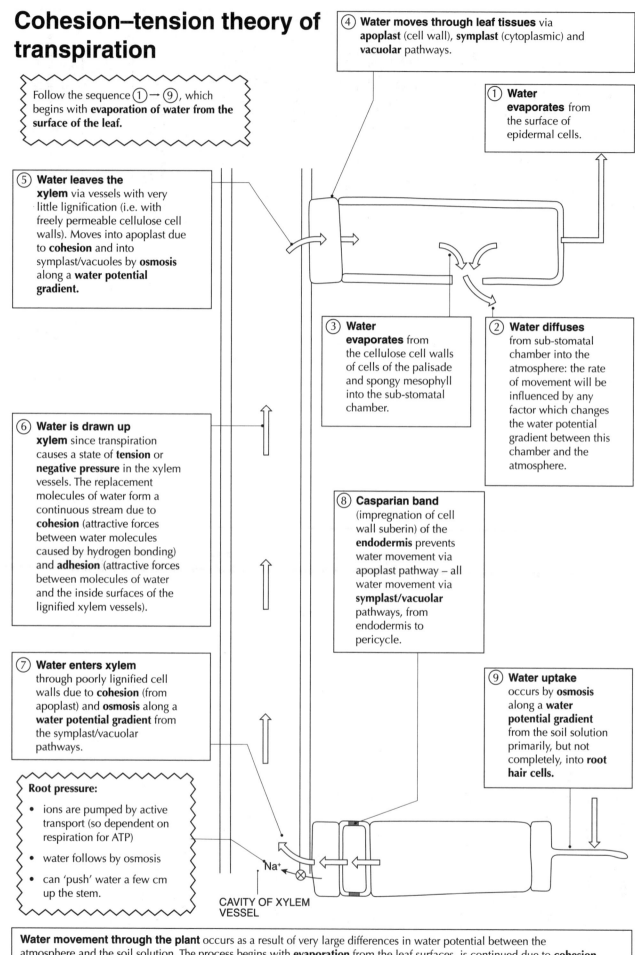

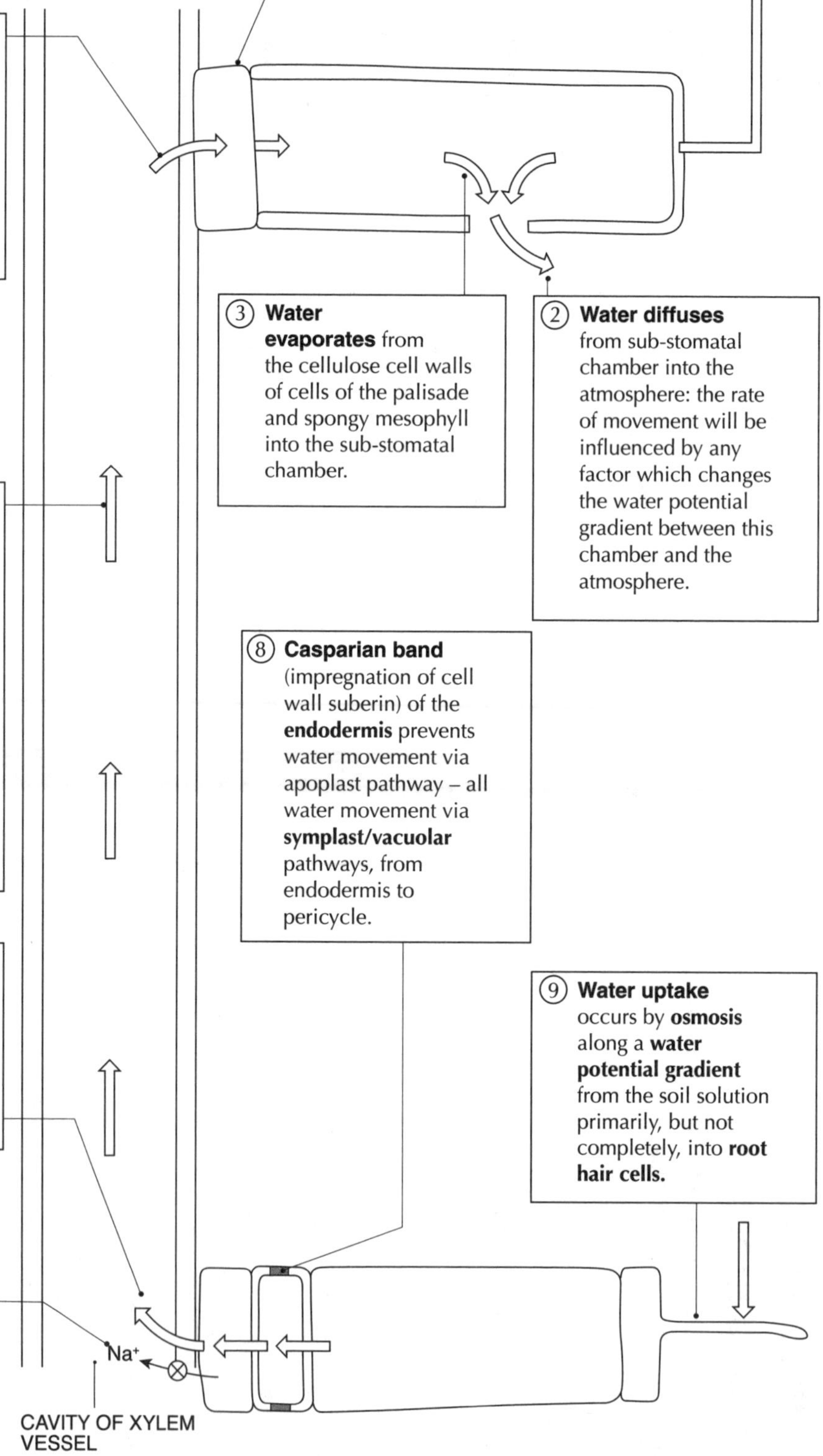

Water movement through the plant occurs as a result of very large differences in water potential between the atmosphere and the soil solution. The process begins with **evaporation** from the leaf surfaces, is continued due to **cohesion** between water molecules and **tension** in the xylem vessels and is completed by **osmosis** from the soil solution.

The bubble potometer

measures **water uptake** (= water loss by transpiration + water consumption for cell expansion and photosynthesis).

Water uptake

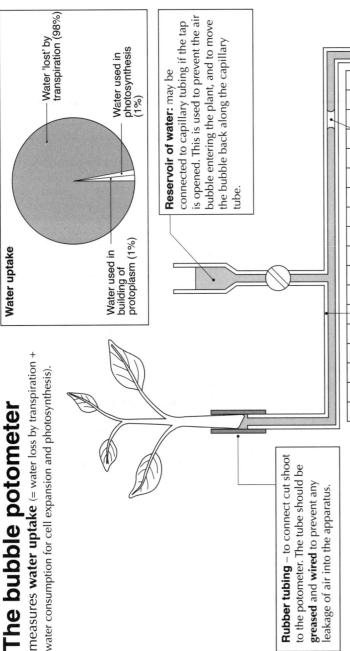

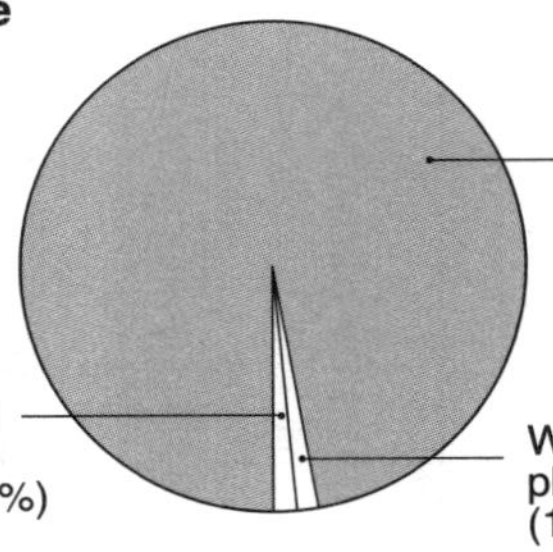

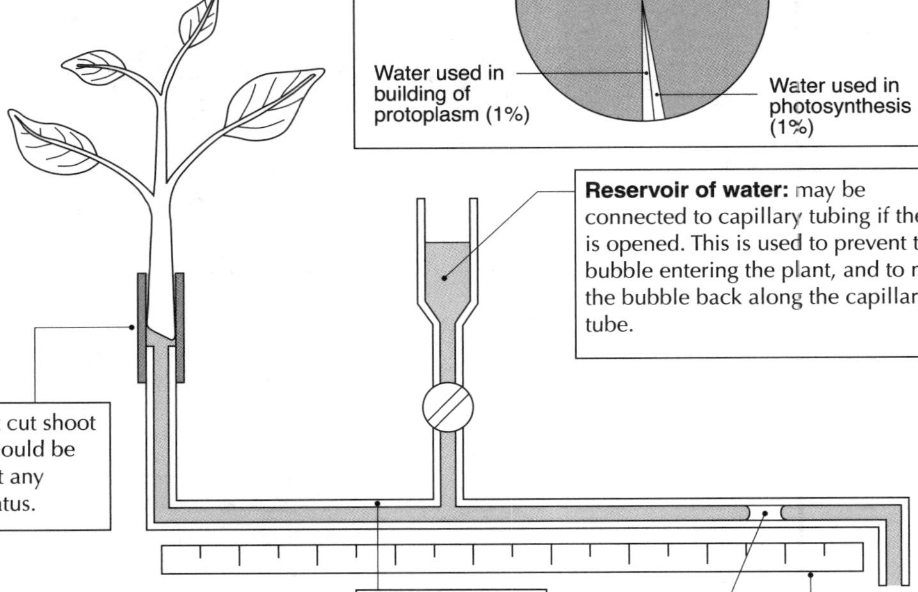

Rubber tubing – to connect cut shoot to the potometer. The tube should be **greased** and **wired** to prevent any leakage of air into the apparatus.

Reservoir of water: may be connected to capillary tubing if the tap is opened. This is used to prevent the air bubble entering the plant, and to move the bubble back along the capillary tube.

Capillary tube: must be kept horizontal to prevent the bubble moving due to its density compared with water.

Air bubble: inserted by removal of tube end from beaker of water. Movement corresponds to water uptake by the cut shoot.

Graduated scale: permits direct reading of bubble movement/water uptake.

Atmometer control: The atmometer is an instrument which can measure evaporation from a non-living surface. When subjected to the same conditions as a potometer the changes in the rate of evaporation from a plant and from a purely physical system can be compared – for example, a reduction in light intensity will show a decrease in water loss **only from a potometer** (due to stomatal closure). The atmometer control indicates when the potometer is acting as a free evaporator and when it is affected by physiological factors such as photosynthesis and stomatal closure.

This porous pot replaces the cut shoot.

Procedure

1. The leafy shoot must be cut **under water,** the apparatus must be filled **under water** and the shoot fixed to the potometer **under water** to prevent air locks in the system.
2. Allow plant to equilibrate (5 min) before introduction of air bubble. Take at least three readings of rate of bubble movement, and use reservoir to return bubble to zero on each occasion. Calculate mean of readings. Record air temperature.
3. Scale can be calibrated by introducing a known mass of mercury into the capillary tubing and using $\rho=m/v$ (ρ for mercury is known, m can be measured, thus v corresponding to a measured distance of bubble movement can be determined).
4. Rate of water uptake per unit area of leaves can be calculated by measurement of leaf area.

External factors affecting transpiration

Light intensity: use bench lamp (with water bath to act as heat filter) to increase light intensity. To simulate 'darkness' enclose shoot in black polythene bag.

Humidity: enclose shoot in clear plastic bag to **increase** relative humidity of atmosphere – include water absorbant such as calcium chloride to **decrease** relative humidity.

Wind: use small electric fan with 'cool' control to mimic air movements whilst avoiding effects of temperature changes.

May also determine relative importance of upper surface/lower surface/stem/petiole in water loss by smearing with vaseline (acts like a waxy cuticle) as appropriate.

N.B. It is sometimes difficult to change only one condition at a time, e.g. enclosure in a black bag to eliminate light will also increase the relative humidity of the atmosphere.

Mass flow theory – transport in the phloem: movements of water and sucrose generate a gradient of hydrostatic pressure which drives the translocation of organic solutes from leaf (the 'source' of solutes) to root (the 'sink' for solutes).

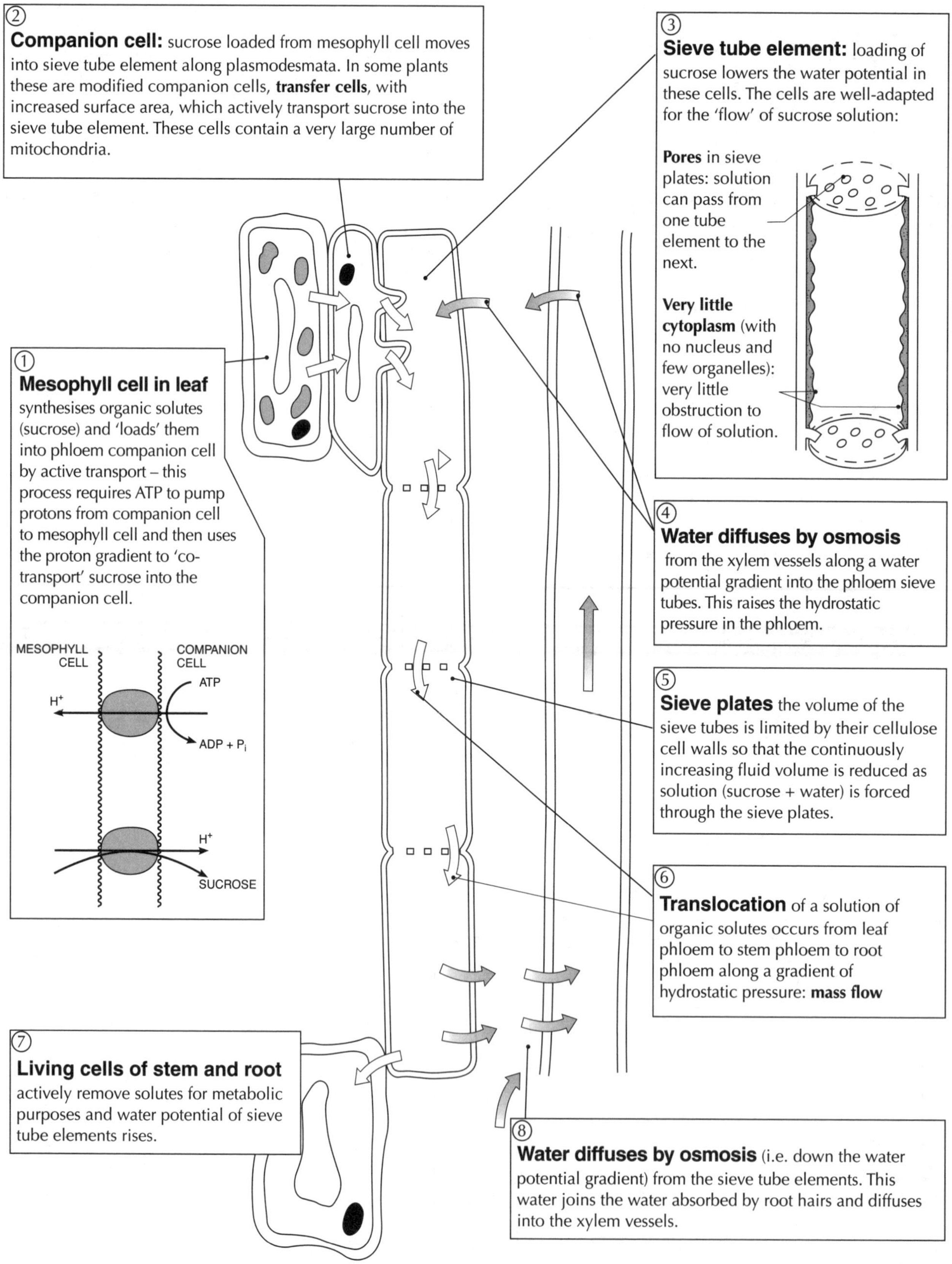

② Companion cell: sucrose loaded from mesophyll cell moves into sieve tube element along plasmodesmata. In some plants these are modified companion cells, transfer cells, with increased surface area, which actively transport sucrose into the sieve tube element. These cells contain a very large number of mitochondria.
③ Sieve tube element: loading of sucrose lowers the water potential in these cells. The cells are well-adapted for the 'flow' of sucrose solution:
Pores in sieve plates: solution can pass from one tube element to the next.
Very little cytoplasm (with no nucleus and few organelles): very little obstruction to flow of solution.
① Mesophyll cell in leaf synthesises organic solutes (sucrose) and 'loads' them into phloem companion cell by active transport – this process requires ATP to pump protons from companion cell to mesophyll cell and then uses the proton gradient to 'co-transport' sucrose into the companion cell.
MESOPHYLL CELL
COMPANION CELL
ATP
H+
ADP + Pi
H+
SUCROSE
④ Water diffuses by osmosis from the xylem vessels along a water potential gradient into the phloem sieve tubes. This raises the hydrostatic pressure in the phloem.
⑤ Sieve plates the volume of the sieve tubes is limited by their cellulose cell walls so that the continuously increasing fluid volume is reduced as solution (sucrose + water) is forced through the sieve plates.
⑥ Translocation of a solution of organic solutes occurs from leaf phloem to stem phloem to root phloem along a gradient of hydrostatic pressure: mass flow
⑦ Living cells of stem and root actively remove solutes for metabolic purposes and water potential of sieve tube elements rises.
⑧ Water diffuses by osmosis (i.e. down the water potential gradient) from the sieve tube elements. This water joins the water absorbed by root hairs and diffuses into the xylem vessels.

Chloroplasts: sites of photosynthesis contain photosensitive pigments.

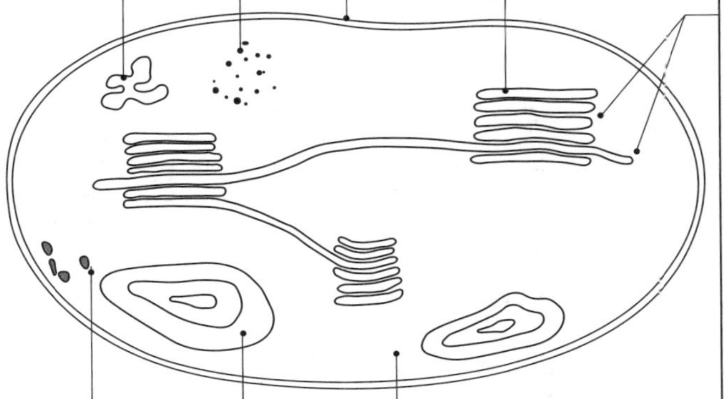

Chloroplast envelope: is permeable to glucose, ions such as Fe^{2+} and Mg^{2+}, oxygen and carbon dioxide.

Grana are stacks of **thylakoid membrane:** the grana enclose a space which is the hydrogen reservoir used in **chemiosmotic synthesis of ATP.**

Thylakoid membranes are the site of the **light-dependent reactions** of photosynthesis. They have on their surface, or embedded within them, the necessary molecules of **chlorophyll** and **accessory pigments,** together with **electron transport systems** and **enzymes.** The light absorbing molecules are arranged in **photosystems.**

Starch granule is the insoluble storage carbohydrate product of photosynthesis.

The stroma (chloroplast matrix) contains the enzymes which catalyse the reactions of the **light–independent reactions** of photosynthesis, such as **ribulose bisphosphate carboxylase.**

$$CO_2 + RuBP \longrightarrow 2\ PGA$$

and the **starch synthase complex**

$$(glucose)_n + glucose\text{–}1\text{–}\textcircled{P} \longrightarrow \text{"starch"}$$

Chromatography is a technique for separating the components of a mixture. The mixture is passed over a **stationary phase**, commonly of paper or a matrix supported in a glass column, dissolved in a solvent or **mobile phase**. The molecules in the mixture are separated according to their relative affinities to (their tendency to 'stick' to) the mobile and stationary phases.

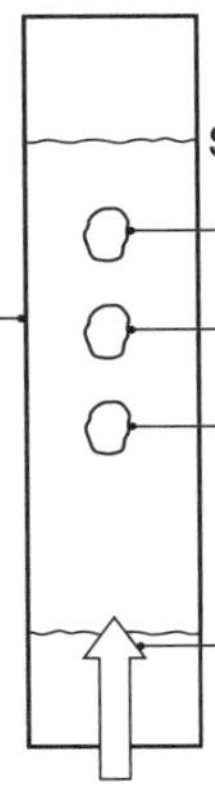

In paper chromatography molecules in the mixture can be identified by their R_f (**retardation factor**) values:

$$R_f = \frac{\text{distance travelled by compound}}{\text{distance travelled by solvent-front}}$$

Chromatography reveals that 'chlorophyll' is a mixture of pigments

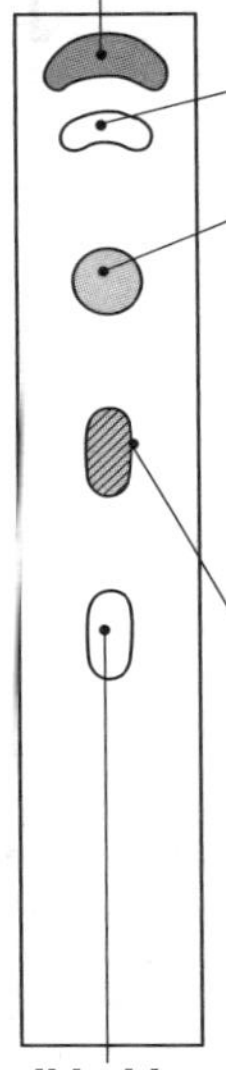

ß-carotene (orange)

Phaeophytin (grey): is a breakdown product of chlorophyll.

Xanthophyll (yellow)

ß-carotene and **xanthophyll** are both **carotenoids.** They are **accessory pigments** which absorb strongly in the blue-violet range and transfer absorbed energy to the **primary pigments** in the photosynthetic **reaction centre.**

Chlorophyll a is the **primary pigment** which uses absorbed light energy to drive the chemical reactions of photosynthesis (production of ATP and $NADPH_2$).

Chlorophyll b (blue-green) is an **accessory pigment.**

Seaweeds are well-adapted!

They contain a brown pigment (fucoxanthin), which absorbs blue and green light. These wavelengths penetrate the seawater much more efficiently than red light.

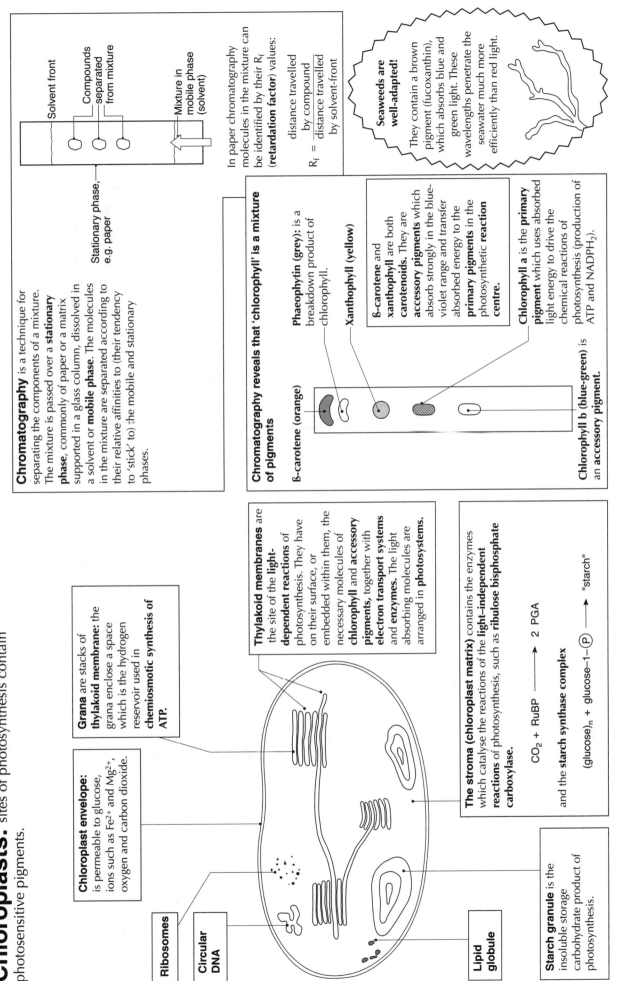

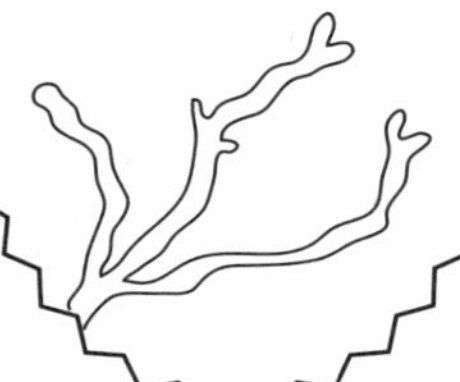

Human nutrition depends on food chains

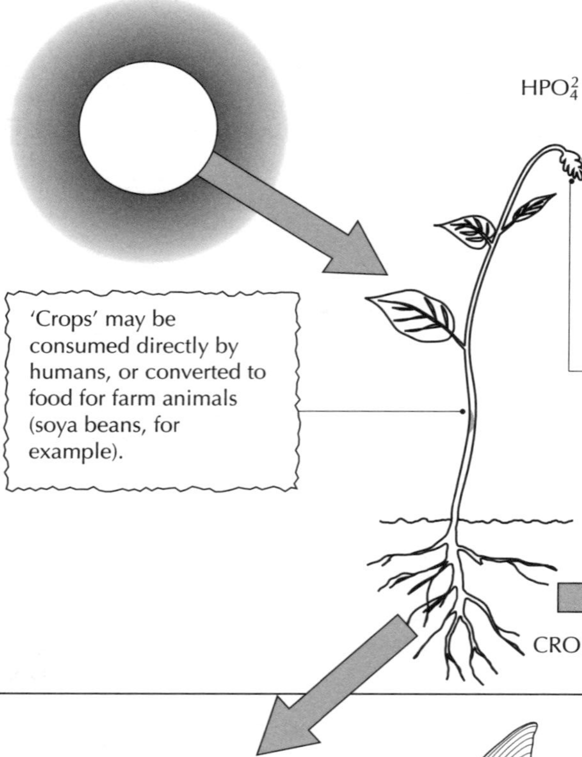

'Crops' may be consumed directly by humans, or converted to food for farm animals (soya beans, for example).

Fertilisers improve the mineral nutrient content of the soil

Mineral element	Absorbed as	Biological function
Nitrogen	Nitrate (NO_3^-)	Synthesis of amino acids, proteins, nucleic acids, vitamins and hormones
Phosphorus	Phosphate	Production of ATP, nucleic acids and phospholipids
Potassium	Potassium (K^+)	Cofactor for photosynthesis, and needed to operate stomatal guard cells

Pesticides reduce competition for/between humans and other organisms:

- **Herbicides** kill other plants, especially grasses in cereal crops.
- **Insecticides** kill insects, such as aphids, locusts, beetles and caterpillars.
- **Fungicides** kill fungi, especially rust diseases on cereal crops.
- **Rodenticides** kill rodents, such as mice and rats, which can 'steal' from food stores.

There can be problems with agricultural chemicals!

- Fertilisers can leach into bodies of water, causing **eutrophication.**
- Pesticides can accumulate in food chains, particularly affecting top predators: **bioaccumulation.**
- Pesticides can select organisms not killed by the chemical: **pest resistance.**

This 'extra' energy transfer loses 90-95% of stored energy in respiration. Human feeding on animals is **very** inefficient.

Fish are important food animals too!

- Salmon species popular – high conversion ratio and high sale price!
- High protein food available ('trash' fish): often has artificial colourant added to it.

Pesticides improve yield of animals

Insecticide kills lice on fish.
Insecticide can control disease-vectors for cattle.
Fungicide controls fungal infections of gills and scales in farmed fish.

Animal husbandry can improve productivity

Strain of animal: **selective breeding** for animals with a high **conversion ratio** (i.e. most efficient transfer of food intake to body mass)

Reduce movement/keep warm: less energy consumption and so more efficient 'conversion'.

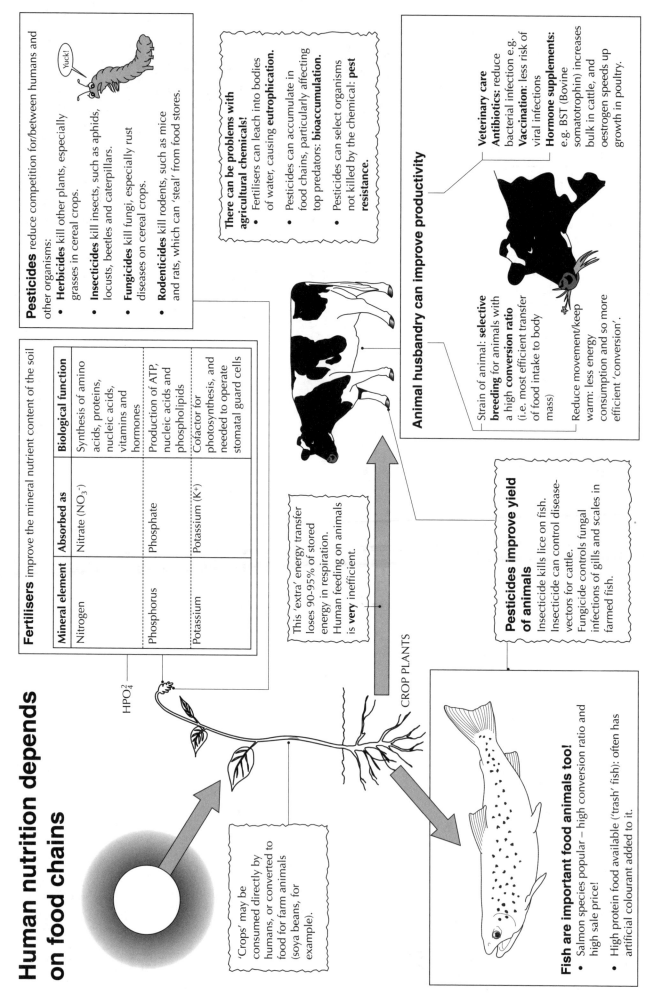

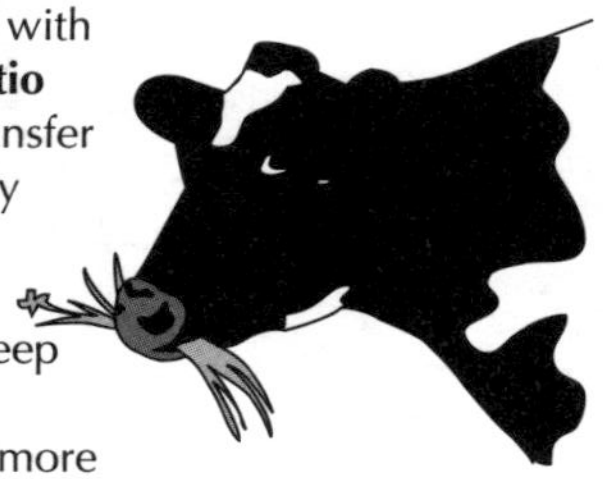

Veterinary care
Antibiotics: reduce bacterial infection e.g.
Vaccination: less risk of viral infections
Hormone supplements: e.g. BST (Bovine somatotrophin) increases bulk in cattle, and oestrogen speeds up growth in poultry.

Microbes and food: I

Bread, wine and mycoprotein

If a culture of yeast is supplied with glucose and water, at a temperature of around 28 °C it will reproduce by **budding**.

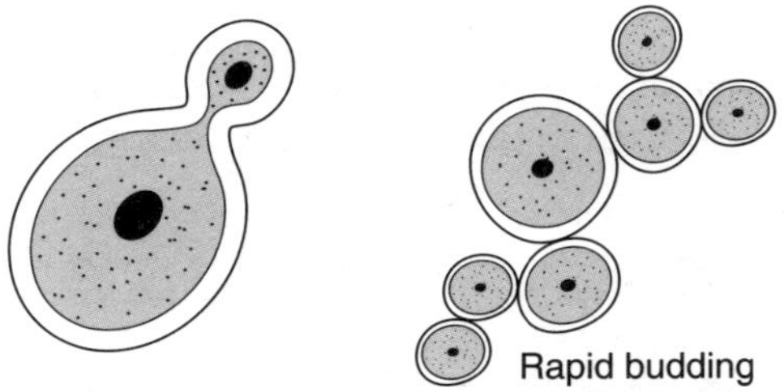

Bud forming on parent cell.

Rapid budding produces chains of cells.

Different strains of the yeast *Saccharomyces cerevisiae* are specialised for **baking** and for different forms of **brewing**.

Wine making from grape sugar

GRAPES → CRUSHED → 'MUST' → FILTERED → JUICE → CLOUDY SOLUTION → CLEAR SOLUTION →

May **add sulphite** or **pasteurise** to stop action of bacteria (or wine may turn to vinegar!)

Some skin remains in red wine: colour and healthy antioxidants.

Adsorbents (e.g. crushed shells) may be added to remove suspended particles.

Fermentation may include conversion of **malic acid** to **lactic acid** by *Lactobacillus*.

GLUCOSE —ANAEROBIC CONDITIONS→ ENERGY + CARBON DIOXIDE + ETHANOL

N.B. Alcohol, like lactic acid, is a poison. It eventually kills the yeast cells which produce it, and does the same to human cells if taken in too large a quantity!

There are GOOD and BAD points

GOOD:
- Microbes will often ferment waste products from other industries.
- Microbes can be grown under very controlled conditions, reducing risk of contamination.
- Microbes can be genetically-engineered to work with great efficiency.

BAD:
- There is some risk of harmful by-products.
- Microbes may mutate and become less efficient.

Carbon dioxide is particularly important in bread making

AMYLOSE and AMYLOPECTIN —α and β amylase→ MALTOSE + GLUCOSE

Yeast + sugar warmed together = RAISING AGENT

FLOUR, SALT, WARM WATER → DOUGH —FERMENTATION→ RISEN DOUGH —BAKING→

(Carbon dioxide 'bubbles' cause the dough to swell.)

BAKING • kills yeast • evaporates alcohol.

Mixing (kneading) dough creates a protein framework which traps CO_2 → texture of bread. Framework is created when **gluten** molecules form **disulphide bridge** cross-links.

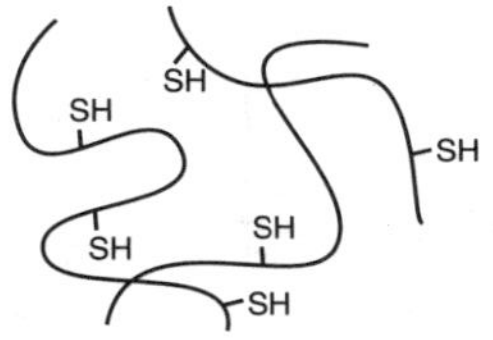

FLOUR IMPROVERS e.g. **ascorbic acid** speed up this process by oxidising –SH groups.

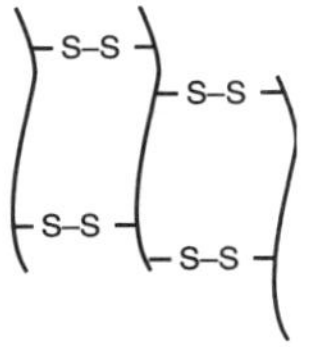

Mycoprotein

Mycoprotein is the bodies of a filamentous fungus.

Glucose syrup (carbon source for energy and organic molecules)

Ammonia (nitrogen source → amino acids)

Mineral salts

Choline (stimulates growth of long fibres)

Biotin (vitamin required for respiration)

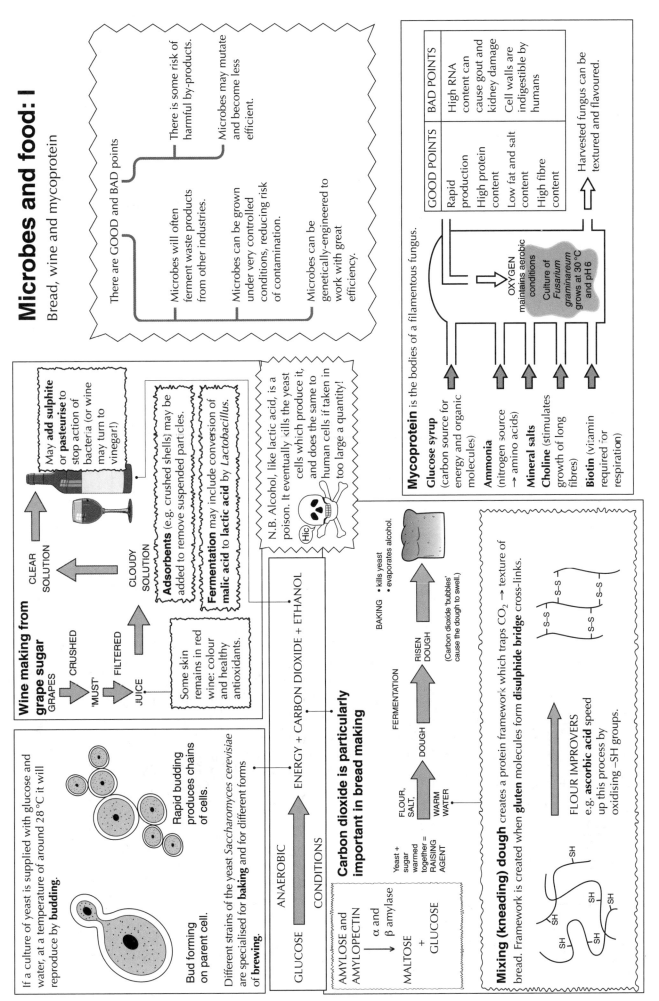

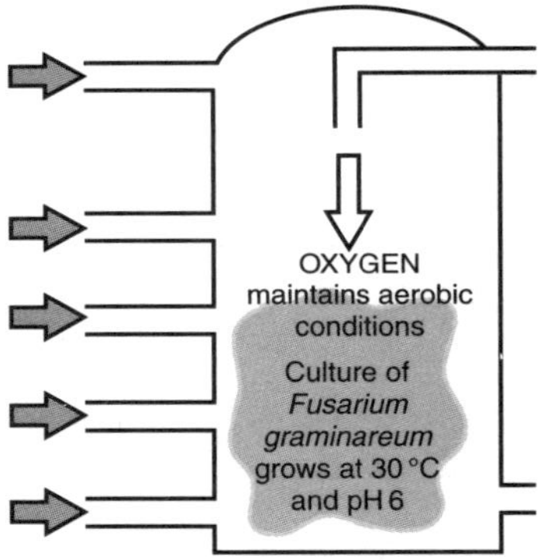

GOOD POINTS	BAD POINTS
Rapid production High protein content Low fat and salt content High fibre content	High RNA content can cause gout and kidney damage Cell walls are indigestible by humans

Harvested fungus can be textured and flavoured.

Microbes and food: II

Milk

Why does milk go sour?

Fresh milk from a healthy cow contains very few bacteria.

Infection from:
- surface of udder;
- hands of milker;
- milking machine.

At factory may now contain 10^{10} bacteria /dm³.

Lactobacilli

LACTOSE —lactase→ GLUCOSE + GALACTOSE

↓

LACTIC ACID

↓

fall in pH

↓

eventually denatures enzymes and stops process

Other bacteria → (some pathogenic)

spoilage

H_2S, CO_2, amines, ethanol → **'bad smell'**

As pH ↓, casein salts dissociate so casein (milk protein) becomes less soluble = **curds**.

Whey contains soluble albumin and globulin.

Resazurin: how fresh is milk?

Resazurin is a pH indicator so is affected by lactic acid:

FRESH MILK (pH 6.5) —souring→ pH 4.0
PURPLISH-BLUE ——→ RED-PINK

YOGHURT: this is milk which has been slightly 'soured' by the excretion of lactic acid from bacterial cultures.

→ 'STANDARD' MILK → 'MIXED WHOLE MILK' → Safe Milk → 'SET' YOGHURT with sharp, acidic taste →

Lasts about 2 weeks at refrigerator temperature (4°C).

Standardisation of content of fat, vitamins A and D, calcium, and protein. May require **fortification** (addition of nutrients) and extra preservatives.

Homogenisation shaking/stirring to distribute fat evenly.

Pasteurisation destroys pathogenic bacteria. Usually 'flash' process – 72°C for 15s – since milk can be stored for 2 to 3 days.

*Add enzyme **penicillinase** to minimise penicillin inhibition of bacteria (antibiotics are used in the treatment of **mastitis**, an udder infection).

Inoculation with starter culture
Mutualism between *Streptococcus thermophilus* and *Lactobacillus bulgaricus* (*S. thermophilus* → formic acid → *L. bulgaricus*; *L. bulgaricus* → amino acids → *S. thermophilus*)

Convert
LACTOSE + H_2O → LACTIC ACID
and secrete vit. B_3 + folic acid

Chilling: inhibition of microbial activity – to 5°C. Add fruit pulp (colour, taste), gums and stabilisers (texture) and, sometimes, preservatives.

Cheese: in its simplest form this is condensed yoghurt, formed by draining off liquid whey – it is sometimes called **lactic cheese**.

Steps 1, 2, and 3 as above (i.e. produce 'safe' milk), then

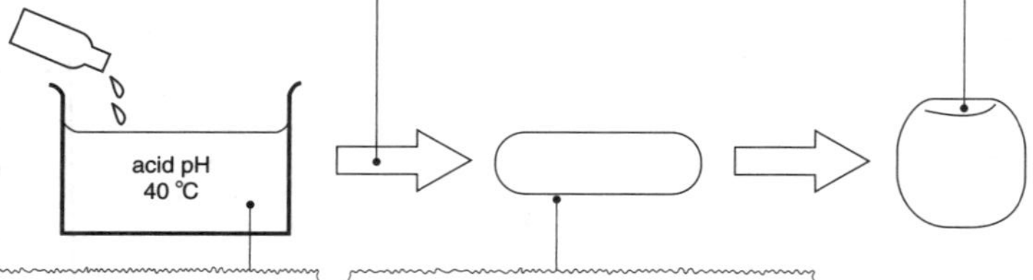

Addition of rennet (enzyme extract)
Chymosin (rennin) + pepsin
* formerly from calves' stomachs, now genetically engineered.

Enzyme action
Casein (milk protein) → insoluble '**curd**' which surrounds the soft inner **whey** = '**soft curd**'.

Pressing
removes the watery **whey** e.g. 87% H_2O → 40% H_2O in cheddar: increases resistance to microbial spoilage.

Salting
immersion in brine or rubbing with salt – affects activity of ripening bacteria. Varies for different cheese types.

Ripening: bacteria growing on curd
Proteins → amino acids → **compounds giving taste and aroma**
Fats → fatty acids →

'**Hard**' cheeses e.g. cheddar/Edam retain much protein and fat.
'**Soft**' cheeses e.g. Camembert – much more digestion to other compounds.

CAMEMBERT 'les pieds de Dieu'

'Cheesy' smell of feet is caused by microbes acting on skin fats and proteins!

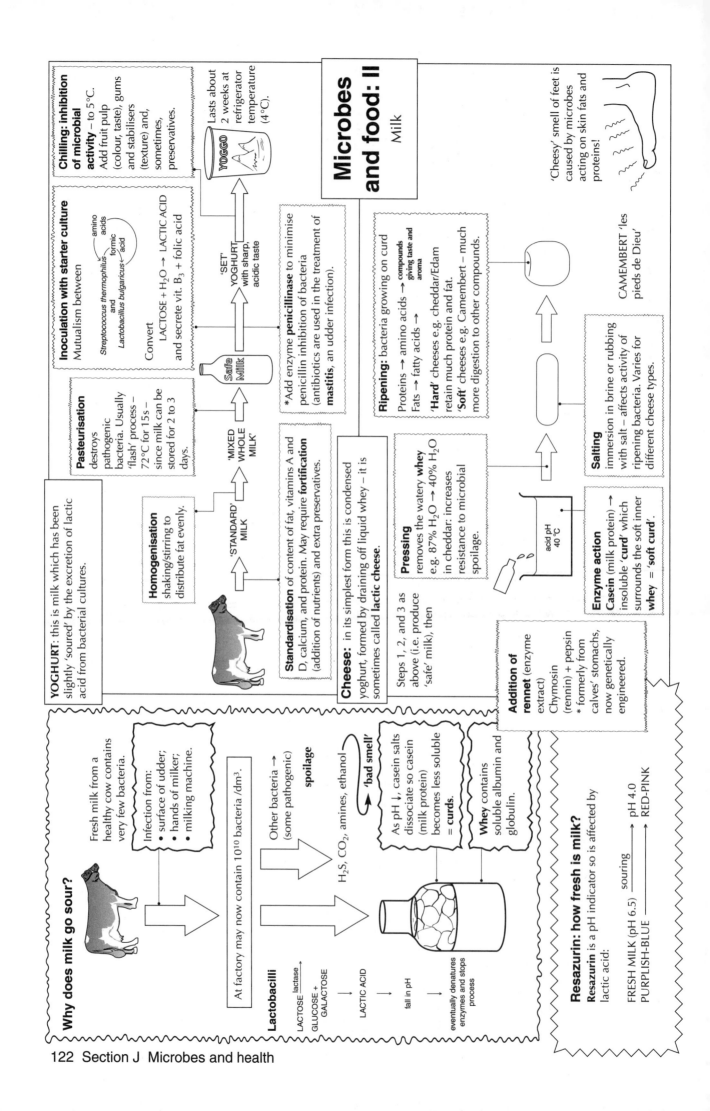

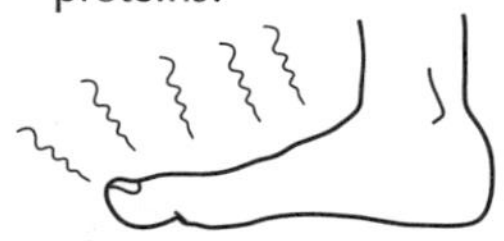

Food spoilage is caused by enzymes – it can be dangerous.

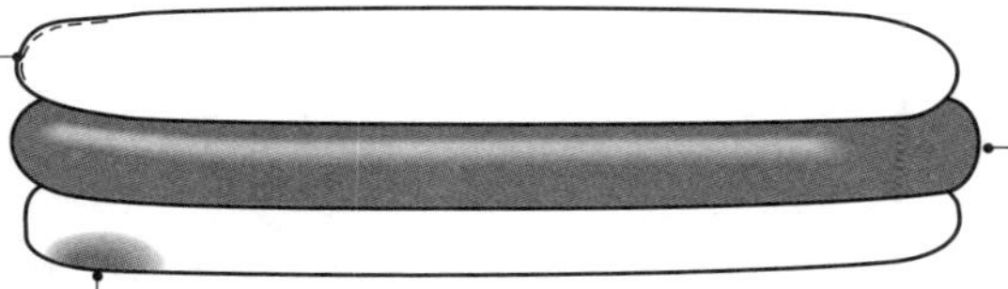

Autolysis: this is the deterioration of food caused by enzymes within the food itself.

- Begins on death of animal or harvesting of crop
- Benefits: meat tenderisation and ripening of fruit
- Drawbacks: fat oxidation → rancid flavours and tastes
  'browning' of cut surfaces caused by action of polyphenol oxidase
- Natural protection by outer layers e.g. skin of apple/shell of nut.

Microbial spoilage : this is caused by opportunistic organisms which can exploit a newly available food source – effects are cyclic

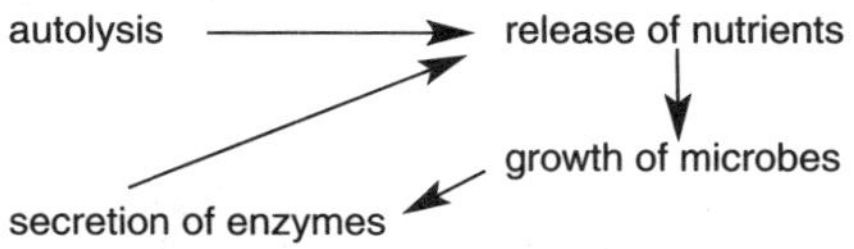

When food Is spoiled

- Appearance is less attractive
- Surface dries out – reduced palatability
- Reduction in food content
- Release of **enterotoxins** (see opposite).

N.B. Partial spoilage brings some benefits.
e.g. ripening of cheese and maturation of yoghurt.

Genetic engineers have now produced an 'inhibited polyphenol oxidase' to prevent cut potatoes from browning and softening.

£

- higher proportion of crop is useable
- greater financial benefits for crisp and processed potato manufacturers.

Some specific spoilage organisms

Food	Damage	Organisms
Wine, beer	souring due to ethanol oxidised to acetic acid	*Acetobacter*
Peanuts, cereals, fruit, bread, dried foods	aflatoxin produced causing food poisoning, mycelial growth through food	*Aspergillus spp*
Pasteurised milk	'bitty milk' from lecithin breakdown	*Bacillus cereus*
bread, fruit, vegetables	visible mycelial growth	*Rhizopus*

Aflatoxin is one of the most dangerous of toxins

– haemorrhage due to leaky capillaries
– liver infection leading to hepatitis
– cancer of the liver, kidneys and colon

The fungus requires high O_2 concentration / moisture / carbohydrate source: **control by careful harvesting at high [CO_2]**

Food poisoning may result from ingested pathogens or from the release of enterotoxins onto the food source.

Pathogens then multiply within the body, causing, in particular, diarrhoea and vomiting:

e.g. Hepatitis A in shell fish
Listeria in soft cheeses and pâté
Bacillus cereus in cooked rice which is stored before use.

Enterotoxins include **Botulin** from *Clostridium botulinum:*

- One of the most lethal toxins known – 1 g might kill 100 000 humans
- Only produced under anaerobic conditions
- Paralysis, e.g. of heart due to nerve damage.

Salmonella food poisoning
(*S.typhimurium* most likely)

Intensive rearing of chickens makes it easy for *Salmonella* bacteria to spread from chicken gut to faeces to floor to chicken carcasses – then in water/on hands/on work surfaces.

68 °C : cooking to this temperature kills *Salmonella* in meat

The bacterium is not very virulent – about 10 000 000 are needed to initiate an infection. Once in the gut the organism multiplies under ideal conditions of temperature/food availability and releases a TOXIN – this leads to a severe inflammatory response in the gut lining and fever, pain, vomiting and diarrhoea.

Treatment: antibiotics are not very effective because they do not easily reach the gut lining which is damaged.

Replace body fluids: to minimise dehydrating effects of diarrhoea.

Control by good food hygiene: washing work surfaces cooking food thoroughly care when handling food.

Food preservation

involves inhibition of microbial activity.

High Osmolarity – salt or sugar can be used to generate high solute concentrations in foods. Such conditions inhibit the growth of micro-organisms which inevitably lose water by exosmosis.

Freezing – most fresh foods contain over 60% water. If water is frozen it cannot be used by micro-organisms.

There are **positive** points
very little loss of nutritive value – 'quick frozen' food may have more nutrients than food which is 'fresh' but takes 2–3 days to reach point of sale

... but some **negative** points too
expansion of water → cell damage → 'mush' (e.g. strawberries)
'drip' from frozen food → loss of soluble nutrients (e.g. vitamin C) on thawing.

Irradiation – although sterilisation by radiation is permitted for medical supplies and drugs it is not permitted for food preservation in the UK.

^{60}cobalt, ^{137}caesium → γ-radiation (dose required to kill microbes is usually greater than does permitted for humans, but should decay before consumption)

... but
- May be some induced radiation
- Very expensive, so use tends to be confined to high-cost foods e.g. prawns
- some loss of vitamins A, B, C and E.

N.B. not yet allowed in UK (although currently the only effective method for reducing *Salmonella* in frozen meat).

Dehydration – one of the most efficient methods involves **freeze drying**: freeze rapidly then rewarm under reduced pressure → ice sublimation → **porous structure**

Benefits: better for rehydration
allied to an N_2-containing atmosphere offers 2–3 years storage.

Drawback: any fats present are easily oxidised (**rancidity**) because of open structure.

Chemical Preservatives – include sulphur dioxide and sulphites, benzoic acid and benzoates and nitrates/nitrites.

- Typically act as **anti-oxidants** and inhibit auto-oxidation of fats
- Removal of O_2 gas and ↓ pH both limit growth of microbes
- SO_2/SO_2^-: widely used in sausages/dried fruits/soft drinks
- Benzoates: soft drinks
- NO_3^-/NO_2^-: meats e.g. ham/bacon are cured this way; some danger from carcinogenic **nitrosamines** (NO_2^- + $-NH_2$ in food → $-NONH_2$).

Before vegetable foods are dehydrated (or, indeed, stored in any other way) they are **blanched** in boiling water or steam. This denatures enzymes such as catalase and ascorbic acid oxidase and improves both appearance and storage life of the food.

Canning – heat sterilisation kills micro-organisms and their spores. If the sterilised food is to be stored for long periods, it must be sterilised in a sealed container which will prevent recolonisation by new microbes.

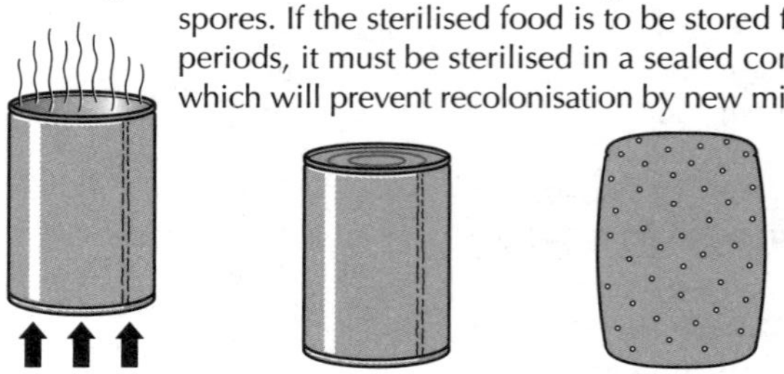

Heat – steam drives out air.

Can closed – falling volume should suck ends inwards.

'Bulge' – evidence of a 'blown' can due to gas released by respiring microbes.

Cans of food must be cooled in water – any gap in the sealing might allow entry of microbes as water would be drawn inwards as the can's contents contracted. Cooling waters are almost always chlorinated to prevent accidental reintroduction of microbes.

The **botulinum cook** – the use of pressure cookers allows boiling water to exceed 100 °C. The modern HTST (high temperature, short time) technique allows a temperature of 121 °C for three minutes. These conditions will even kill the spores of *Clostridium botulinum* – this organism produces a toxin – **botulin** – which is so toxic that it is estimated that 500 g would kill almost the whole population of the UK.

Control of pH – microbes may grow more slowly under acidic conditions.

Yoghurt
Lactic acid is produced by lactose fermentation.

Pickles
Preserved by added vinegar (3% aqueous solution of ethanoic acid).

A balanced human diet contains

fat, protein, carbohydrate, vitamins, minerals, water and fibre **in the correct proportions.**

An adequate diet provides sufficient **energy** for the performance of metabolic work, although the 'energy food' is in unspecified form.

A balanced diet provides all dietary requirements **in the correct proportions.** Ideally this would be $^1/_7$ **fat,** $^1/_7$ **protein** and $^5/_7$ **carbohydrate.**

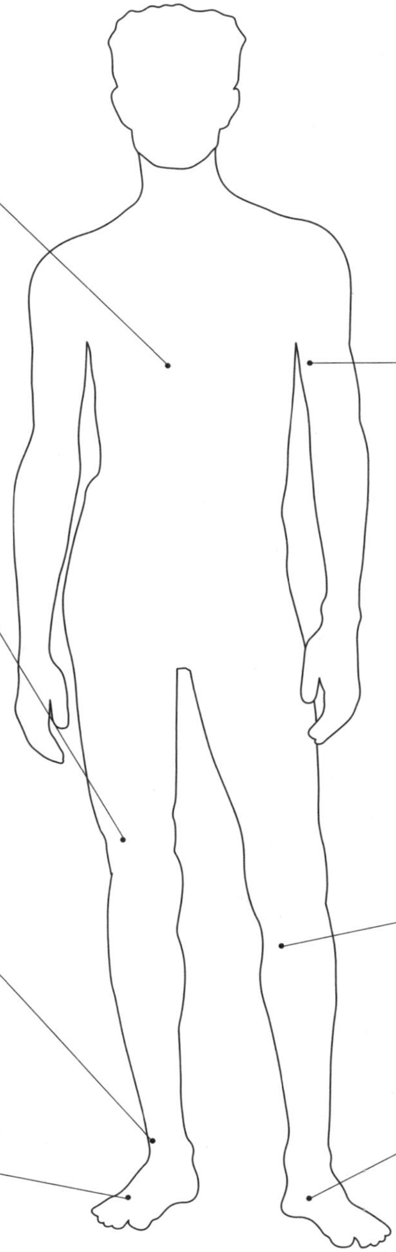

Carbohydrates

Mainly as a **respiratory substrate,** i.e. to be oxidized to release **energy** for active transport, synthesis of macromolecules, cell division and muscle contraction.

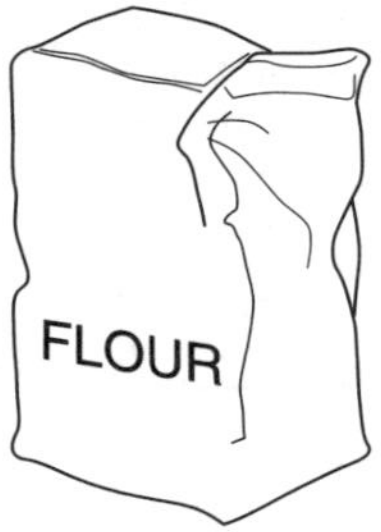

Common sources: rice, potatoes, wheat and other cereal grains, i.e. as **starch,** and as refined sugar, **sucrose,** in food sweetenings and preservatives.

Digested in duodenum and ileum and absorbed as **glucose.**

Lipids

Highly reduced and therefore can be oxidised to release **energy.** Also important in **cell membranes** and as a component of **steroid hormones.**

Common sources: meat and animal foods are rich in **saturated fats** and **cholesterol,** plant sources such as sunflower and soya are rich in **unsaturated fats.**

Digested in duodenum and ileum and absorbed as **fatty acids and glycerol.**

Vitamins have no common structure or function but are essential in small amounts to use other dietary components efficiently. **Fat-soluble vitamins** (e.g. A, D and E) are ingested with fatty foods and **water-soluble vitamins** (B group, C) are common in fruits and vegetables.

Fibre (originally known as **roughage**) is mainly cellulose from plant cell walls and is common in fresh vegetables and cereals. It **may** provide some energy but mainly serves to aid faeces formation, prevent constipation and ensure the continued health of the muscles of large intestine.

Proteins are **building blocks** for growth and repair of many body tissues (e.g. myosin in muscle, collagen in connective tissues), as **enzymes,** as **transport systems** (e.g. haemoglobin), as **hormones** (e.g. insulin) and as **antibodies.**

Common source: meat, fish, eggs and legumes/pulses. Must contain eight **essential amino acids** since humans are not able to synthesise them. Animal sources generally contain more of the essential amino acids.

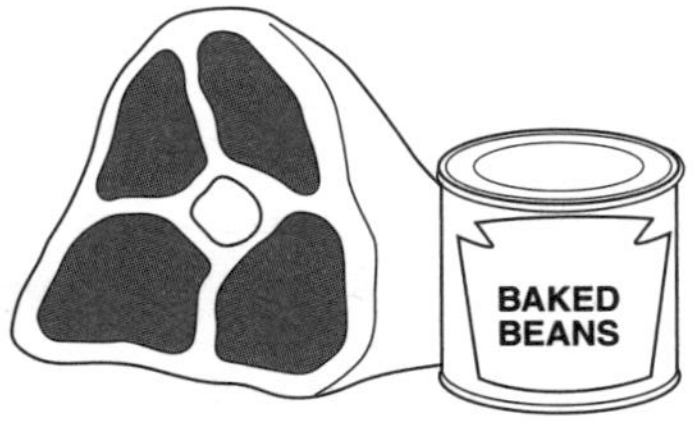

Protein quality can be assessed quantitatively as

$$\textbf{Apparent digestibility of protein} = \frac{\text{N (nitrogen) intake} - \text{N in faeces}}{\text{N intake}}$$

and

$$\textbf{Biological value of growth and maintenance} = \frac{\text{N intake} - \text{N in faeces} - \text{N in urine}}{\text{N intake} - \text{N in faeces}}$$

Digested in stomach, duodenum and ileum and absorbed as **amino acids.**

Minerals have a range of **specific** roles (direct structural components, e.g. Ca^{2+} in bones; constituents of macromolecules, e.g. PO_4^{3-} in DNA; part of pumping systems, e.g. Na^+ in glucose uptake; enzyme cofactors, e.g. Fe^{3+} in catalase; electron transfer, e.g. Cu^{2+} in cytochromes) and **collectively** help to maintain solute concentrations essential for control of water movement. They are usually ingested with other foods – dairy products and meats are particularly important sources.

Water is required as a solvent, a transport medium, a substrate in hydrolytic reactions and for lubrication. A human requires 2-3 dm^2 of water daily, most commonly from drinks and liquid foods.

Obesity: possible health problems associated with being overweight.

Men who are 20% overweight have a 25% increase in mortality – mainly from diabetes, coronary heart disease and cerebrovascular disorders.

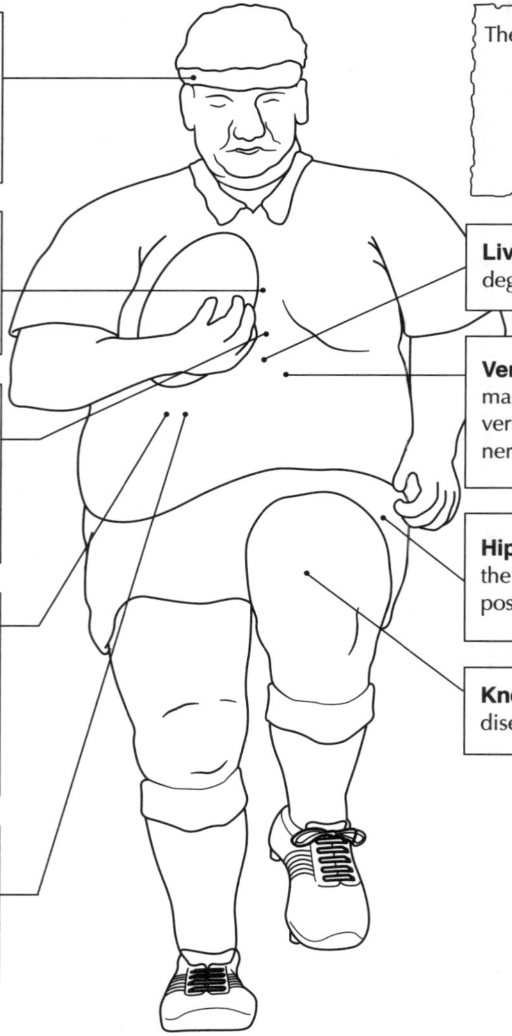

Arteries/arterioles: arteriosclerosis and hypertension → increased risk of stroke or brain haemorrhage. Greater danger of thrombosis.

Heart: cardiac arrest is more likely (coronary artery occlusion and high blood pressure).

Lungs: infections are more likely since greater body mass impairs movement of the diaphragm and therefore limits tidal flow in and out of lungs.

Pancreas: incidence of pancreatic cancer increases. Endocrine function impaired – obesity causes increased onset of diabetes and associated complications.

Gall bladder: Gallstones are more frequent – these are largely cholesterol stones.

There's also increased risk of:
- **hiatus hernia**
- **breast cancer**
- **endometrial cancer**
- **prostate enlargement**
- **menstrual abnormalities.**

Liver: increased fatty degeneration of liver tissue.

Vertebral column: greater body mass causes compression wear to vertebrae – scoliasis and sciatic nerve trap may result.

Hip joints: fracture of the head of the femur – particularly in obese, post-menopausal females.

Knee joints: degenerative disease more likely.

The Energy Balance : Intake in the diet and Demand for metabolism

intake > demand

Thus **body mass increases:**
- may follow giving up smoking
- may reflect lack of exercise (1hour cycling or jogging to 'burn up' 100 g of chocolate)
- may have genetic basis.

intake < demand

Thus **body mass decreases:**
- may be caused by eating disorders
- may result from excessive exercise (and promotion of metabolic rate)
- may result from high thermogenesis in brown fat.

Body Mass Index: Are You Obese?

Tables of desirable weights, with allowance for age, sex and height, may be consulted ~ 20% greater than ideal weight is considered **morbid obesity**.

$$\textbf{body mass index} = \frac{\text{mass in kg}}{(\text{height in metres})^2}$$

	Ideal	Morbid obesity
Male	20–25	30+
Female	18.5–23.5	28.5+

Treatment

- **Dietary control**

Long term use of a balanced diet is likely to be successful: 'crash' diets are to be avoided.

- **Drug therapy**

– may depress appetite with **serotonin** mimics
– may promote metabolism with amphetamines. Long-term use is addictive.

- **Surgery**

Jaw wiring limits food intake.

Gastric stapling increase feeling of satiation.

Bacterial growth may be stopped by antibiotics:

These compounds may affect several sites in the bacterium

The original definition of an **antibiotic** was 'a compound produced by one microorganism which is capable of killing or inhibiting another'. Nowadays many compounds with antibiotic properties are synthesised in the organic laboratory. The ideal antibiotic is **specific** i.e. it inhibits some bacterial activity without having any effect on the host organism (i.e. the infected human being).

Bacteria may develop resistance to antibiotics

Bacteria become resistant in two ways.

Mutation (frequently about 1 in 10^7): the antibiotic does not **induce** resistance but it does **select resistant strains.**

Bacteria breed rapidly and populations are enormous. Within these populations rare **mutations** may produce cells which are **antibiotic-resistant.** Over-use of antibiotics may destroy 'normal' bacterial cells but allow 'resistant' cells to survive. These may multiply to form a resistant population.

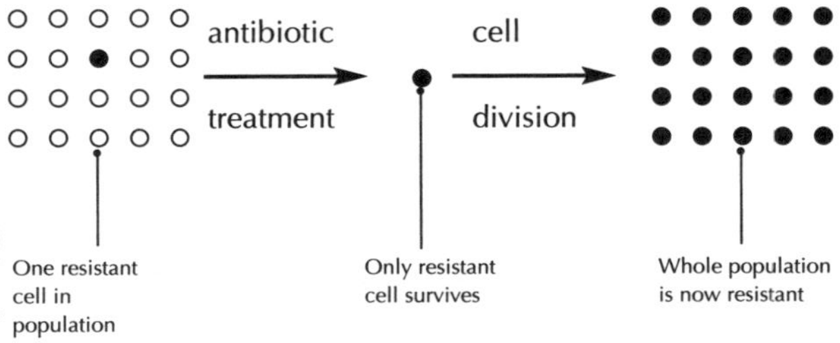

This is a form of **artificial selection,** and the antibiotic resistance may be passed to subsequent generations by **vertical gene transmission.**

Acquired resistance – the collection of resistant genes from another bacterium (even another species). This may be by **conjugation** (transfer of genes including the 'resistant' R-plasmid) by a primitive form of sex, **transduction** in which the genes are transferred by a virus or by **transformation** in which 'naked' genes are transferred without a vector.

This is the passage of antibiotic resistance by **horizontal gene transmission.**

The resistance may be due to:

inactivation – for example, the enzyme , ß-lactamase which breaks open the lactam ring, destroying the activity of the antibiotic;

molecular change – for example, an amino acid replacement in one of the ribosomal proteins confers resistance to streptomycin;

impermeability – for example, tetracycline may be excluded from cells by a membrane pump which actively pumps the antibiotic out of the susceptible cells.

Protein biosynthesis: although the process is essentially similar in bacteria and in host cells, there are differences in ribosomal structure such that there are specific inhibitors of bacterial protein synthesis which affect the intiation and translocation/elongation steps.

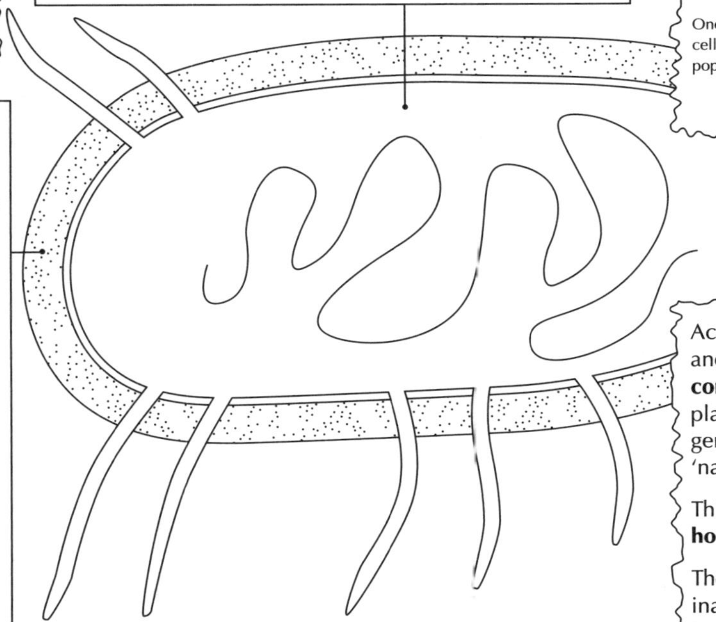

Cell wall biosynthesis: the bacterial cell wall contains peptidoglycan molecules (up to 50% in Gram-positive species, up to 20% in Gram-negative species) – these molecules are long chains of alternating N-acetylglucosamine (NAG) and N-acetylmuramic acid (NAM), cross-linked by tetrapeptide chains. The first effective antibiotics, the **penicillins** or **ß-lactams,** inhibit the assembly of these peptide links causing the bacterial cell wall to weaken. If water enters by osmosis the cytoplasm expands and the cell may burst – this is osmotic lysis.

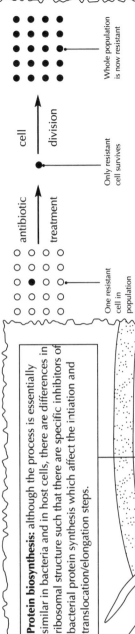

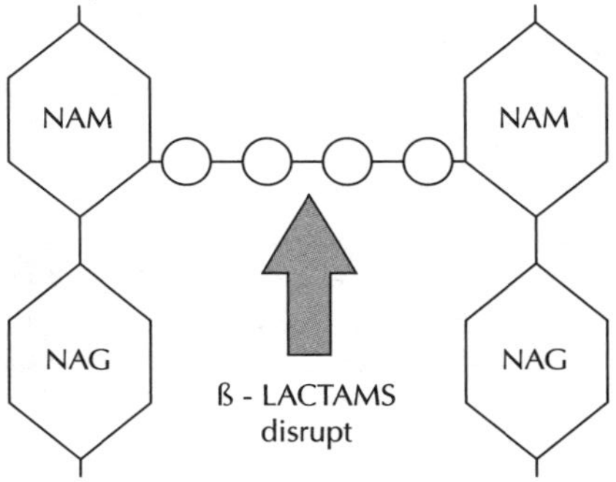

Because antibiotics of this type work during the synthesis of the cell wall they are effective against bacteria only when they are growing.

Because no mammalian cells have a peptidoglycan cell wall these antibiotics are selectively toxic against bacteria.

Dangerous examples of resistance include:

strains of MTB resistant to rifampicin cause difficulty in treating tuberculosis;

MRSA (Methicillin Resistant Staphylococcus Aureus) is a common bacterium resistant to may widely-used antibiotics.

Human immunodeficiency virus (HIV) is a retrovirus

The most common methods of transmission of HIV are:

Unprotected sexual intercourse with an infected partner. Anal intercourse is particularly dangerous. *Use a condom!*

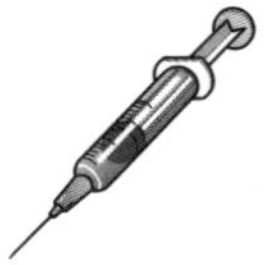

Sharing needles, usually for intravenous drug abuse, with an infected person. *Do not share needles!*

Infection from blood products, and vertical transmission from an infected mother to her foetus, are now almost eliminated as risk factors.

virus crosses protective mucous membrane

↓

virus enters 'host' bloodstream

Treatment and prevention

- inhibitors of viral enzymes (AZT inhibits reverse transcriptase)
- no alcohol, no smoking and 'healthy' diet for HIV+ individuals
- safe sexual practices, or abstinence
- vaccines are not yet very successful.

The virus

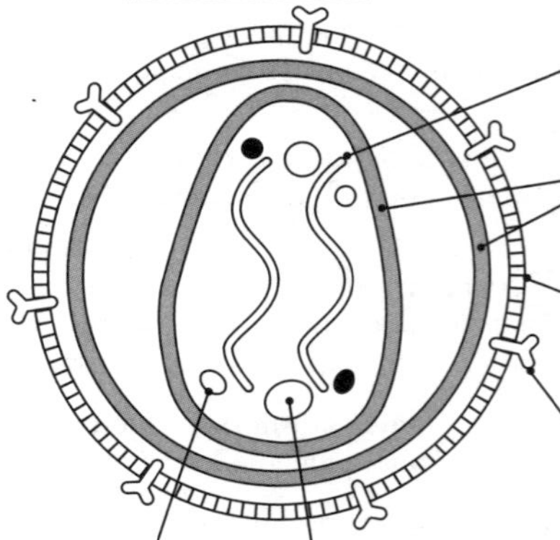

- Two strands of RNA: viral genes.
- Two protein coats: protect the RNA.
- Lipid envelope fuses with host cell membrane.
- Glycoprotein spikes: allow the virus to bind to protein receptors on susceptible host cells.
- Integrase: enzyme to 'cut and paste' viral DNA into host DNA.
- Reverse transcriptase: produces viral DNA for integration into host genetic material.

Life cycle: HIV replicates in host cells

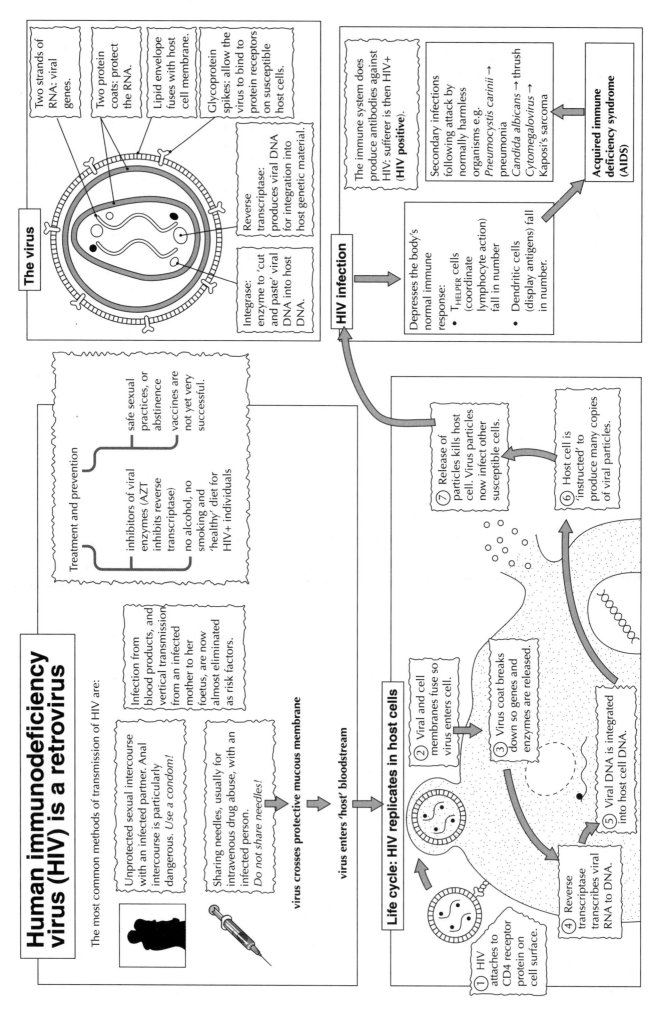

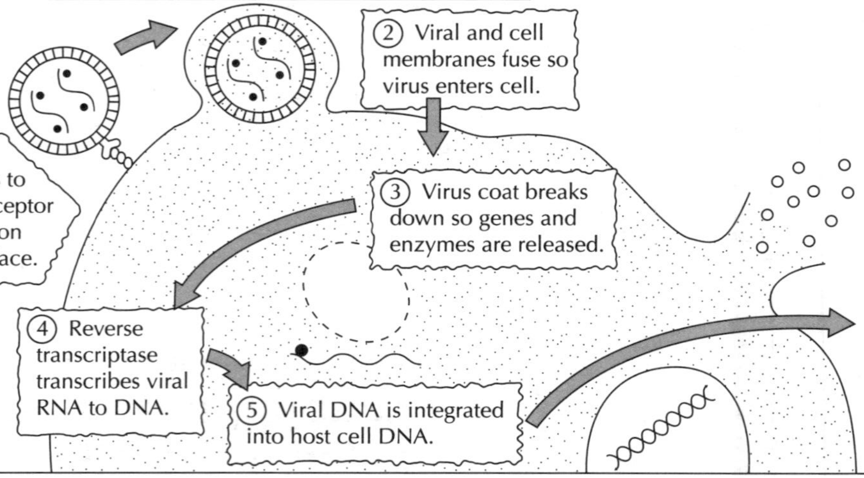

1. HIV attaches to CD4 receptor protein on cell surface.
2. Viral and cell membranes fuse so virus enters cell.
3. Virus coat breaks down so genes and enzymes are released.
4. Reverse transcriptase transcribes viral RNA to DNA.
5. Viral DNA is integrated into host cell DNA.
6. Host cell is 'instructed' to produce many copies of viral particles.
7. Release of particles kills host cell. Virus particles now infect other susceptible cells.

HIV infection

Depresses the body's normal immune response:

- T_{HELPER} cells (coordinate lymphocyte action) fall in number
- Dendritic cells (display antigens) fall in number.

The immune system does produce antibodies against HIV: sufferer is then HIV+ (**HIV positive**).

Secondary infections following attack by normally harmless organisms e.g.
Pneumocystis carinii → pneumonia
Candida albicans → thrush
Cytomegalovirus → Kaposi's sarcoma

Acquired immune deficiency syndrome (AIDS)

Malaria

has killed more humans than has any other disease.

Control involves an understanding of the parasite, *Plasmodium*, and the vector, *Anopheles*.

Mosquitoes breed in bodies of still water, such as lakes and ponds. Rafts of eggs are laid and hatch into larvae, then pupae, which hang just below the water's surface to obtain oxygen through a short air tube.

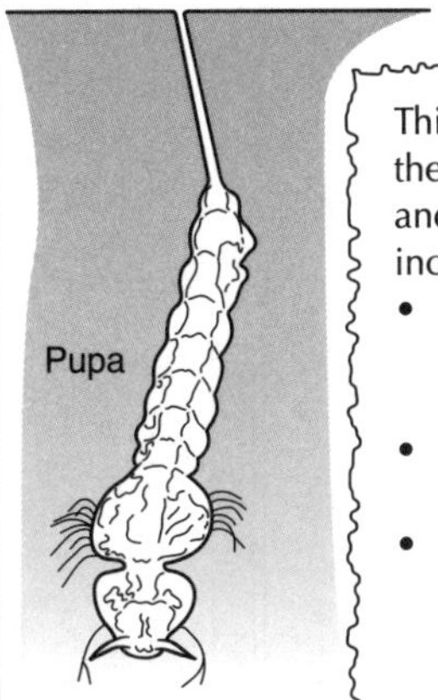

This is a weak point in the vector's life cycle, and control methods include:

- spraying oil on the surface, to block the breathing tubes;
- draining marshes and swamps;
- introducing small fish, called mosquito fish, which feed on larvae and pupae.

The *Anopheles* mosquito is the **vector** for the pathogen, *Plasmodium*. *Plasmodium* multiplies in the stomach wall of the mosquito, then migrates to the salivary glands.

Adult mosquitoes can be controlled by insecticides, such as DDT, sprayed onto their resting places. This may select **pesticide-resistant** strains of mosquito.

Infected mosquito passes on parasites to uninfected human. *Plasmodium* passes down piercing mouthparts from salivary glands.

Control methods include:

- sleeping under mosquito nets to prevent biting;
- spraying of insect-repellant chemicals onto the skin;
- wearing long-sleeved clothes during the evening (most likely time to be bitten).

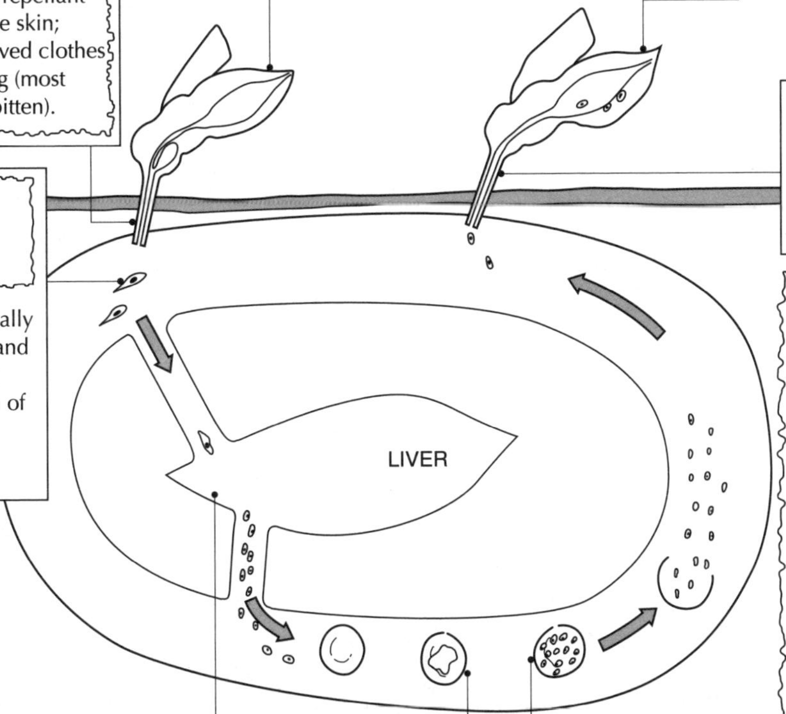

An uninfected mosquito picks up parasites from the blood of an infected person.

Plasmodium is the protoctist which causes malaria.

It can reproduce sexually inside the mosquito, and the resulting variation makes the production of successful vaccines very difficult.

Vaccines have limited success

- The parasite is 'hidden' from the immune system in liver and red blood cells.
- Antigens on the surface of the parasite 'evolve' so that antibodies cannot recognise them.

Parasites multiply rapidly in the liver, and are then released into the blood.

Drugs such as Paludrin and Larium control the disease by limiting entry to the liver and reducing multiplication of the parasite.

Parasite released from the liver invade red blood cells where they feed on haemoglobin and divide by **multiple fission**. The red cells filled with parasites burst and release the parasites into the blood.

Effects on the red cells cause many of the symptoms of malaria:

- tiredness, since fewer red cells means less oxygen is transported;
- fever, as the red cells burst.

People with sickle cell anaemia are protected from malaria. The parasite cannot feed so well inside 'sickle' cells.

Tonic water contains **quinine**, an effective anti-malarial drug. Early colonists of malarial areas could justify drinking gin and tonic!

Growth of a bacterial population

Koch's postulates: rules for proving that a particular microbe is responsible for a particular disease.
- It must be shown that the microbe in question is always present in individuals with the disease.
- The microbe must be isolated from the diseased host and grown in pure culture containing only that one species of microbe.
- Microbes obtained from the pure culture, when injected into a healthy susceptible host, must produce the disease in that host.
- Microbes must be isolated from the experimentally infected host, grown in pure culture, and compared with the microbes in the original culture.

Bacterial disease may be:
- **Infective**: the growing population directly damages tissues, sometimes by competition for oxygen and nutrients.
- **Toxic**: the bacterial cells release poisons (**toxins**) which irritate or damage cells of the host organism.

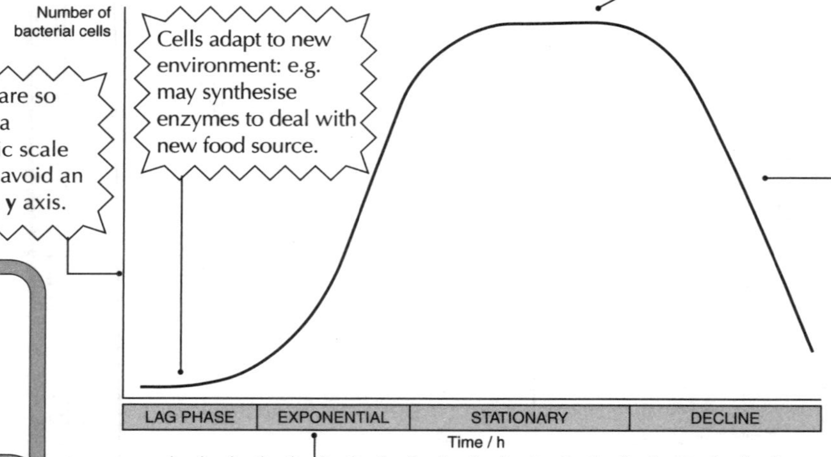

Numbers are so great that a logarithmic scale is used to avoid an enormous **y** axis.

Cells adapt to new environment: e.g. may synthesise enzymes to deal with new food source.

Population stable as cell deaths balanced by continued cell division. It is often during this phase that **enterotoxins** (gut poisons) are produced and released.

Death rate > rate of division:
- all nutrients used
- toxins produced kill cells directly
- oxygen concentration reduced for aerobic species.

During this phase the population doubles with each unit of time – 1 cell → 2 → 4 → 8 → 16 → 32 → 64 → 128
doubling time may be as little as 20 min – at this rate one bacterial cell → 8000 kg in 24 h!
Food preservation techniques hold population early in this phase by imposing **environmental resistance** i.e. by **limiting availability of factors for growth**.

Complete cycle may be no more than a few hours.

Water:
- required for hydrolysis reactions and for cell expansion

▲ control bacterial cell division by limiting access to water e.g. by dehydration or vacuum packing of foods.

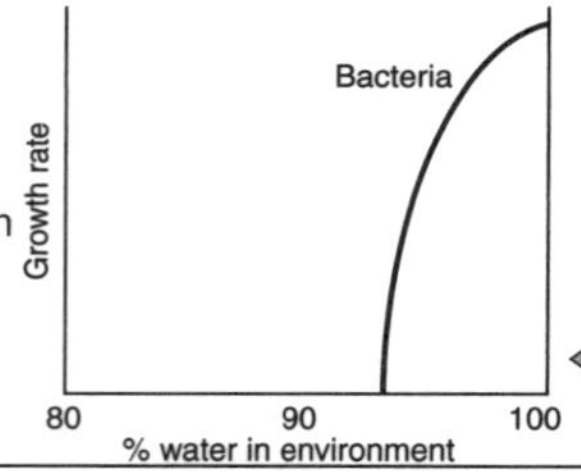

pH:
- bacteria can tolerate a wide range of pH – enzymes are quite resistant.

▲ low pH (acid environment) limits bacterial growth, e.g. HCl in stomach, 'pickled' foods.

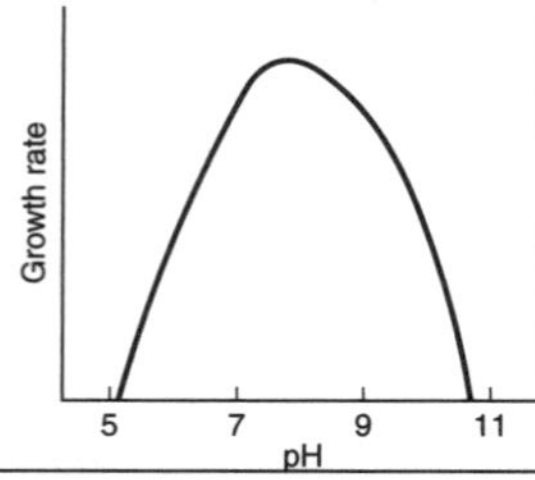

Temperature:
- affects enzyme activity, fluidity of membranes, solubility of molecules, energy release by respiration

▲ reduce temperature by refrigeration/freezing (food preservation); increase temperature during fever.

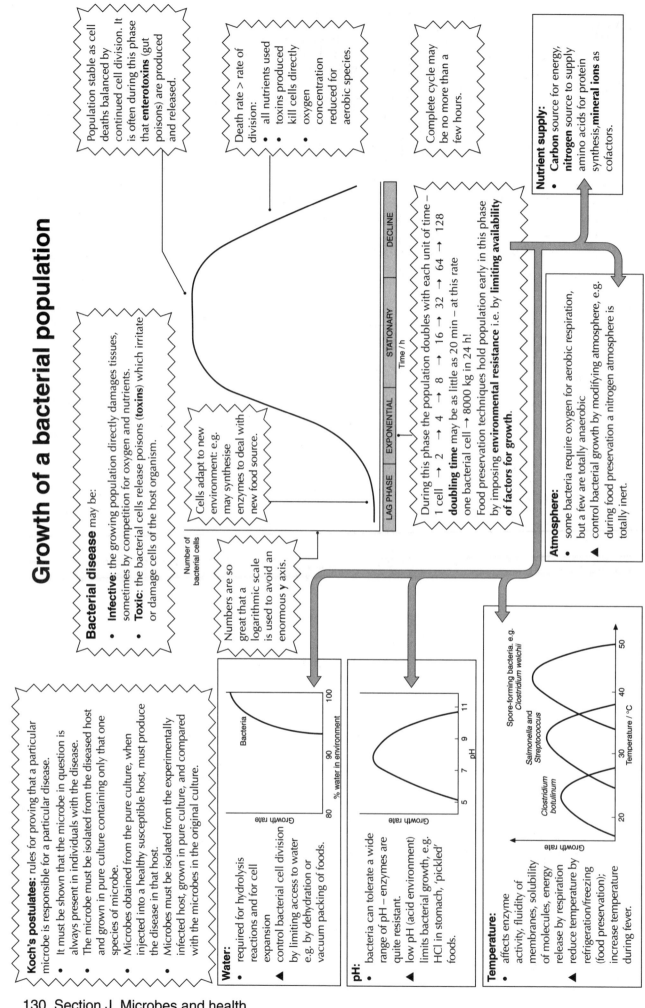

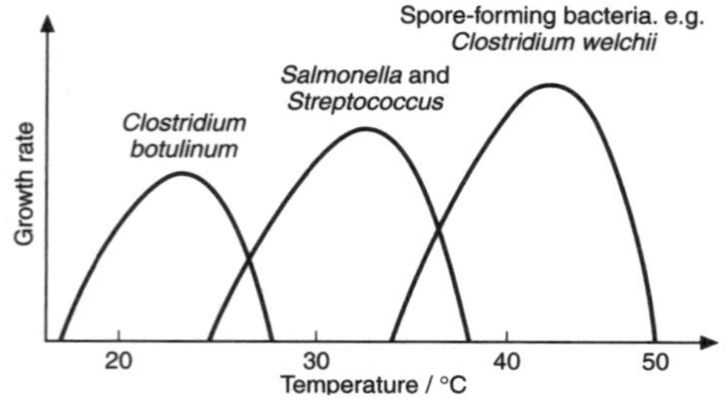

Atmosphere:
- some bacteria require oxygen for aerobic respiration, but a few are totally anaerobic

▲ control bacterial growth by modifying atmosphere, e.g. during food preservation a nitrogen atmosphere is totally inert.

Nutrient supply:
- **Carbon** source for energy, **nitrogen** source to supply amino acids for protein synthesis,**mineral ions** as cofactors.

Microbes such as bacteria can be grown in a **growth medium** containing **nutrients**, **buffers** to provide an appropriate pH and with **aeration** to provide **mixing** and, where necessary, **oxygen**. The growth medium and the containers are **sterilised** to avoid infection by competitive or dangerous microbes.

The medium may be **liquid** or mixed with **agar** to form a **gel** that can be spread inside **petri dishes**.

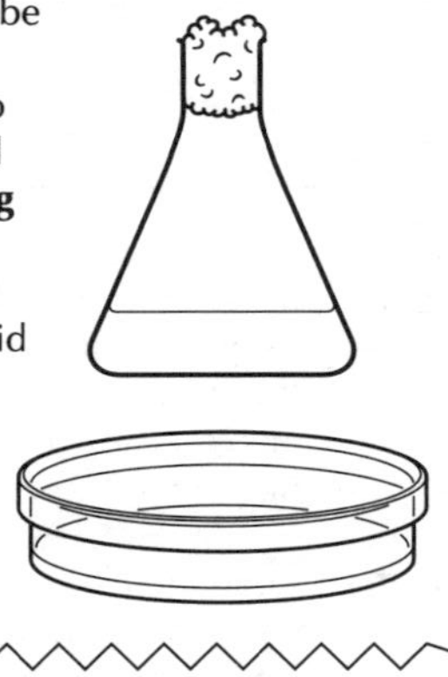

Any labelling should be on the bottom of the plate, just in case the lids become mixed up!

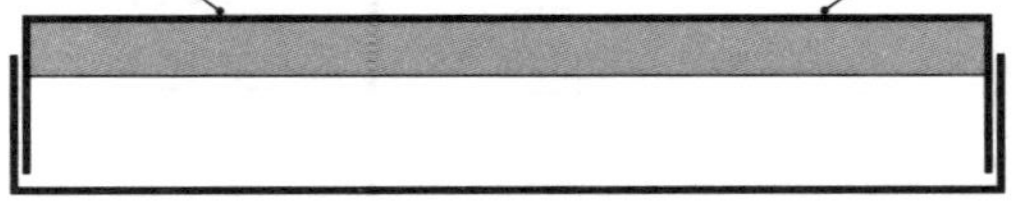

The dish is
- sealed with tape
- placed upside down (to prevent any condensation falling back onto the agar).
- in an incubating oven – 25°C allows growth of many microbes but not those which colonise the human body.

Aseptic techniques

- minimise danger to humans
- minimise contamination of cultures.

Microbes can be collected from any source, e.g.

- food, such as milk
- water samples
- skin
- pure cultures, obtained from microbiological suppliers, and transferred to a fresh, sterile agar plate.

Transfer of microbes: uses an **inoculating loop**, made of wire held in a steel handle.

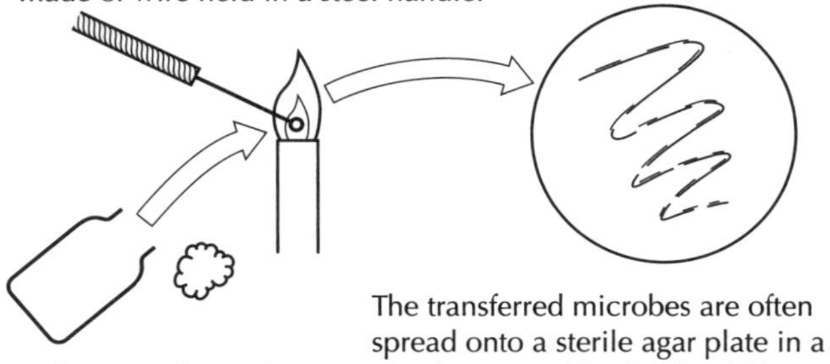

Loop, and top of culture vessel, are held in Bunsen flame to sterilise them.

The transferred microbes are often spread onto a sterile agar plate in a **streak pattern**. The lid should be removed at the last moment to reduce the risk of contamination by airborne microbes.

Counting cells in a population of microbes

Total count: records all cells, whether they are alive (so capable of reproduction) or dead.

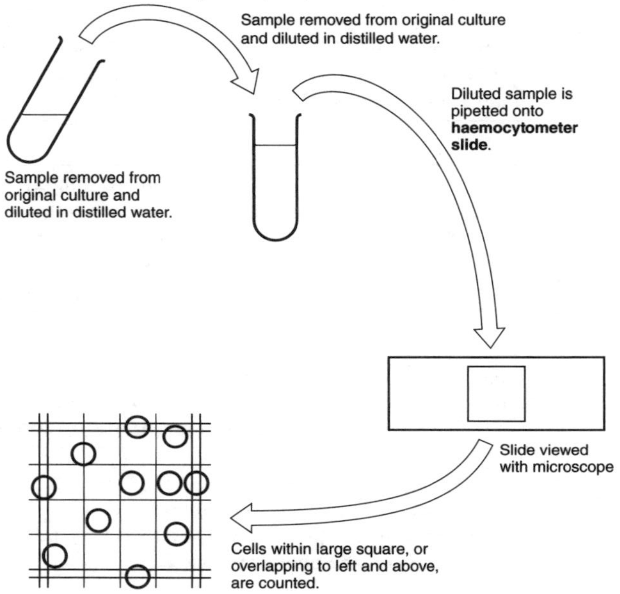

$$\text{Original population density} = \frac{\text{Mean number in square}}{\text{Volume of square}} \times \text{Dilution factor}$$

Viable (live) count: records only cells that are **alive and capable of reproduction**.
A **serial dilution** is used to provide individual cells – these divide to produce colonies on agar plates.

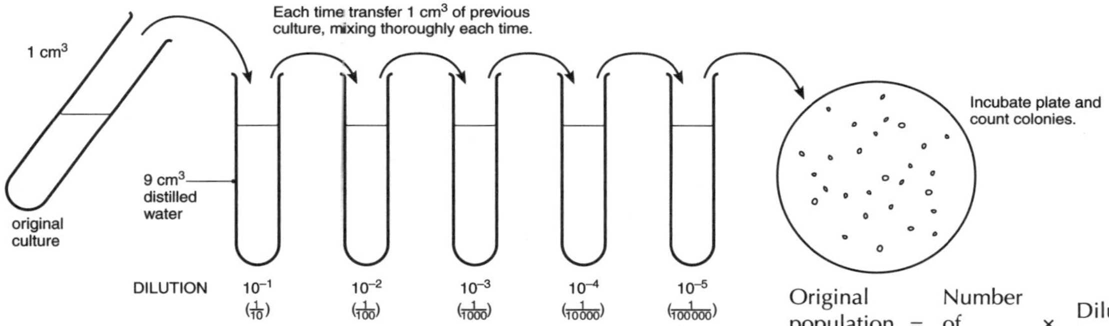

$$\text{Original population density} = \text{Number of colonies} \times \text{Dilution factor}$$

Bioreactors/fermenters exploit microbes for commercial reasons.

Why use micro-organisms?
- carry out reactions at moderate temperatures
  → **great savings on fuel**
- are very efficient
  → **less waste and purer products**
- can be 'genetically engineered' to produce **compounds needed for humans**.

Input for microbial culture: the organisms which will carry out the fermentation process are cultured separately until they are growing well.

Nutrient input: the micro-organisms require:
- an **energy source** - usually carbohydrate
- **growth materials** - amino acids, or ammonium salts which can be converted to amino acids, are required for protein synthesis.

Constant temperature water jacket: the temperature is controlled so that it is high enough to promote enzyme activity but not so high that enzymes and other proteins in the microbes are denatured.
- This initially involves **heating** to initiate the fermentation.
- Fermentation releases heat so that later stages may require **cooling**.

Paddle stirrers: continuously mix the contents of the bioreactor
- ensure micro-organisms are always in contact with nutrients
- ensures an even temperature throughout the fermentation mixture
- for aerobic (oxygen-requiring) fermentations the mixing may be carried out by an **airstream**.

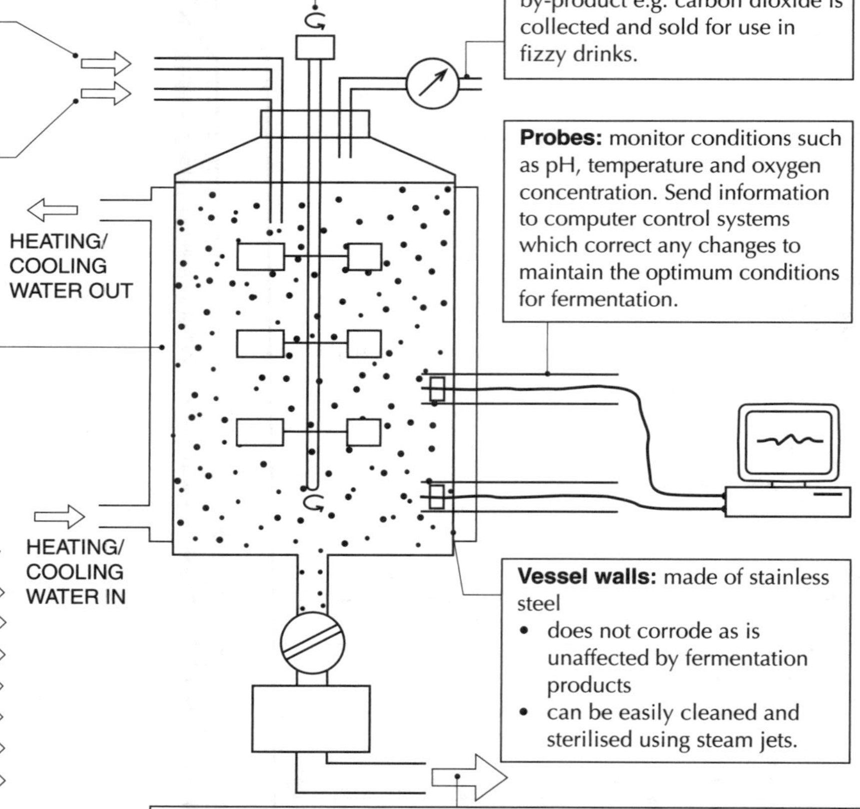

Gas outlet: gas may be evolved during fermentation. This must be released to avoid pressure damage, and may be a valuable by-product e.g. carbon dioxide is collected and sold for use in fizzy drinks.

Probes: monitor conditions such as pH, temperature and oxygen concentration. Send information to computer control systems which correct any changes to maintain the optimum conditions for fermentation.

Vessel walls: made of stainless steel
- does not corrode as is unaffected by fermentation products
- can be easily cleaned and sterilised using steam jets.

'Batch' or 'continuous' culture

'Batch'	'Continuous'
fixed period	can continuously add nutrients and oxygen – microbial culture can be kept in the exponential phase
vessels are versatile, so can be switched quickly to other products	smaller (so less expensive to heat, sterilise) vessels can be used as the process is more productive
any contamination → limited losses (only one batch)	may be practical problems e.g. foaming → blocking of inlets and outlets
vessels must be cleaned and sterilised between batches	any contamination may ruin enormous quantities of product.

'Batch' is more expensive but less risky – choose for low volume, high selling price products.

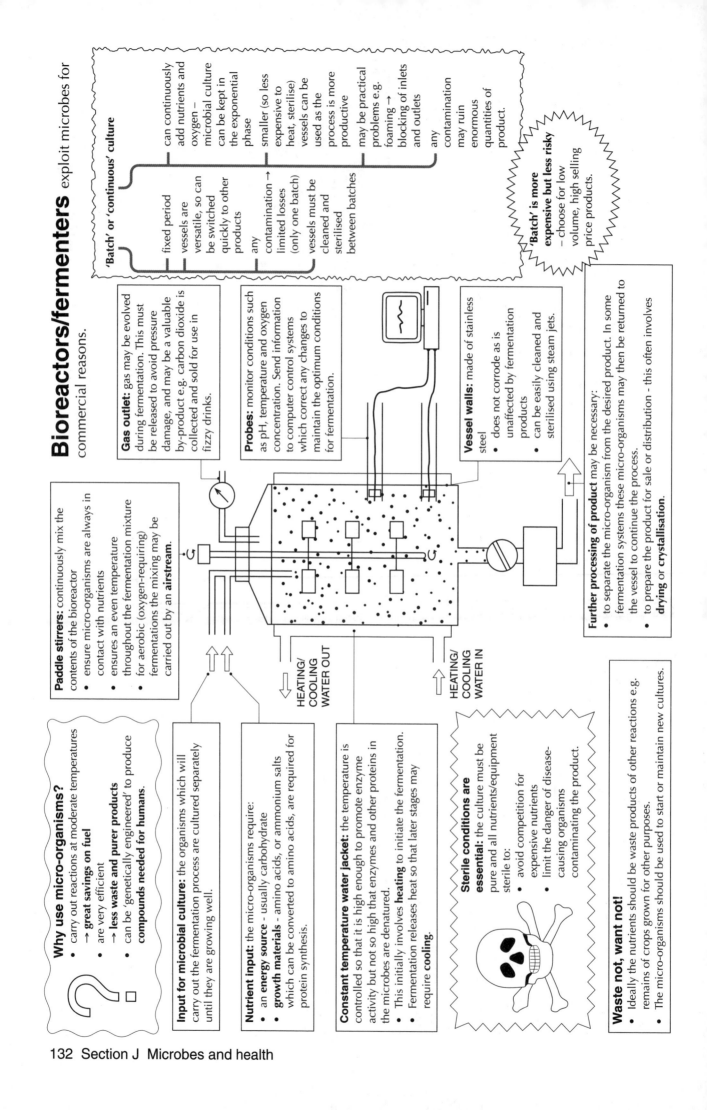

Sterile conditions are essential: the culture must be pure and all nutrients/equipment sterile to:
- avoid competition for expensive nutrients
- limit the danger of disease-causing organisms contaminating the product.

Further processing of product may be necessary:
- to separate the micro-organism from the desired product. In some fermentation systems these micro-organisms may then be returned to the vessel to continue the process.
- to prepare the product for sale or distribution - this often involves **drying** or **crystallisation**.

Waste not, want not!
- Ideally the nutrients should be waste products of other reactions e.g. remains of crops grown for other purposes.
- The micro-organisms should be used to start or maintain new cultures.

Section K Populations

Ecology is the 'study of living organisms in relation to their environment'.

A more recent definition is 'the scientific study of the interactions that determine the distribution and abundance of organisms'.

Synecology is the study of groups of organisms associated to form a functional unit of the environment.

Two useful terms are:

Community (biotic community): all of the populations occupying a given, defined physical area, e.g. all the organisms within a rock pool.

Ecosystem: the biotic community together with the physical (non-living or abiotic) environment, e.g. a rock pool.

Autecology is the study of **single organisms** or **populations of single species** and their relationship to their environment, e.g. Common limpet (*Patella vulgaris*) – an animal of the rocky shore.

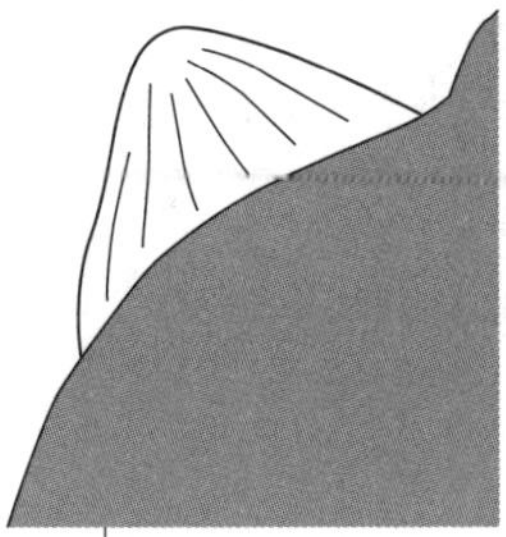

What does it **feed on**?

How does it **avoid drying out**?

How does it **minimise damage from wave action**?

What are its **predators**?

How does it **reproduce**?

How do its young **disperse**?

What are its **competitors**?

The niche of an organism
Answering some of these questions about a single species can describe its **ecological niche** – the part the organism plays in its ecosystem, e.g. the limpet is a rocky shore herbivore.

The **niche** describes:

- where an organism lives;
- how the organism feeds.

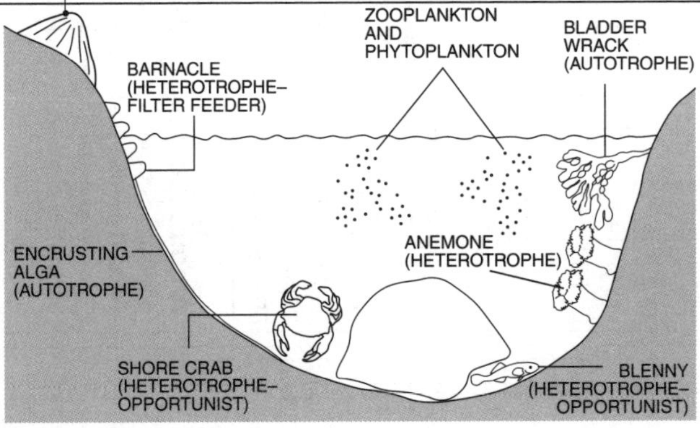

N.B. The rock pool is not self-sustaining and relies on the twice-daily tidal cycle for maintenance of optimum abiotic (e.g. temperature, salinity) and biotic (e.g. nutrients, new colonisers) factors.

A rock pool ecosystem is maintained by a series of **interactions.**

Abiotic factors (including temperature, volume of water, pH, salinity, dissolved oxygen substrate).

interacting with

The biotic community

Autotrophes – the **producers –** which require an input of light and inorganic nutrients. Include algae and phytoplankton.

interacting with

Heterotrophes – the **consumers** – which include herbivores, carnivores, omnivores and decomposers.

Estimating populations

To study the dynamics of a population, or how the distribution of the members of a population is influenced by a biotic or an abiotic factor, it is necessary to estimate the population size. In other words, it will be necessary to count the number of individuals in a population. Such counting is usually carried out by taking **samples** (in which the organisms are in the same proportion as in the whole population) because:

1. counting the whole population would be extremely laborious and time-consuming
2. counting the whole population might cause unacceptable levels of damage to the habitat, or to the population being studied.

The samples must be representative. They should be:

1. of the same size (e.g. a 0.25 m^2 area of grassland)
2. randomly selected – for example, samples may be taken at predetermined points on an imaginary grid laid over the sampling area. The coordinates of the points may be selected using random numbers generated by a calculator
3. non-overlapping.

Quadrat sampling

Quadrats are sampling units of a known area. They are most often square, and are usually constructed of wood or metal. The quadrat can be used in simple form, or it may have wire subdivisions to produce a number of sampling points.

Reliable sampling with quadrats requires answers to three questions:

1. What size of quadrat should be used?
2. How many quadrats should be used?
3. Where should the quadrats be positioned?

What size quadrat?

If individuals within a population are truly randomly dispersed, then any quadrat size should be equally efficient in the estimation of that population. However, environmental factors are rarely evenly distributed so that the living organisms dependent on them tend to occur in clumps.

Small quadrats are more efficient in estimating populations (more can be taken, and they can cover a wider range of habitat than larger ones) but there are practical considerations to be taken into account (a small quadrat might not include a dominant tree in a woodland). Optimum quadrat size is determined by counting the number of different species present in quadrats of increasing size.

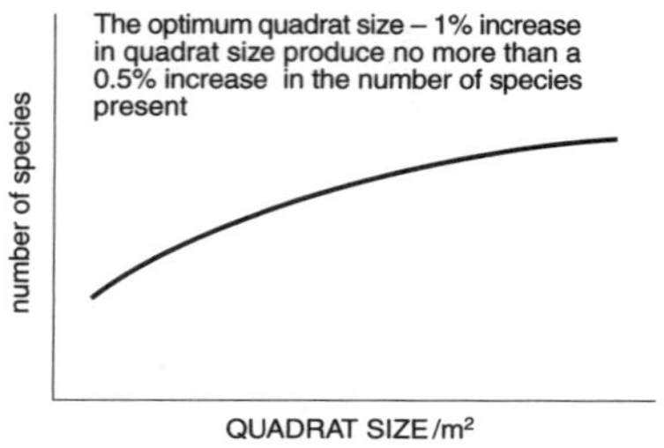

How many quadrats?

Too few might be unrepresentative, and too many might be tedious and time-consuming. To determine the optimum number, a series of quadrats of the optimum size is placed randomly across the sampling area – the cumulative number of species is recorded after each increase in quadrat number.

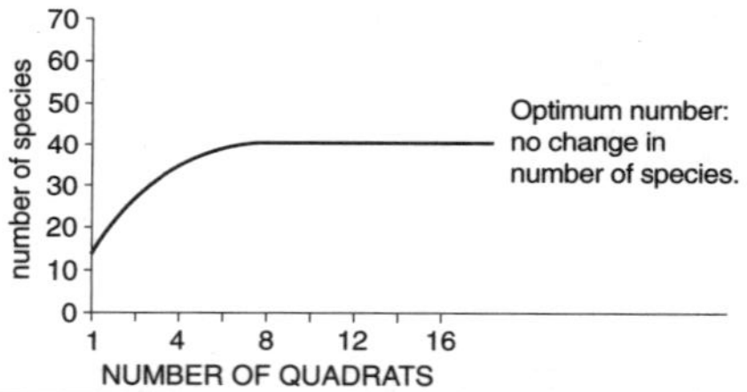

Random positioning of quadrats

The position can be chosen using random co-ordinates, as described earlier.

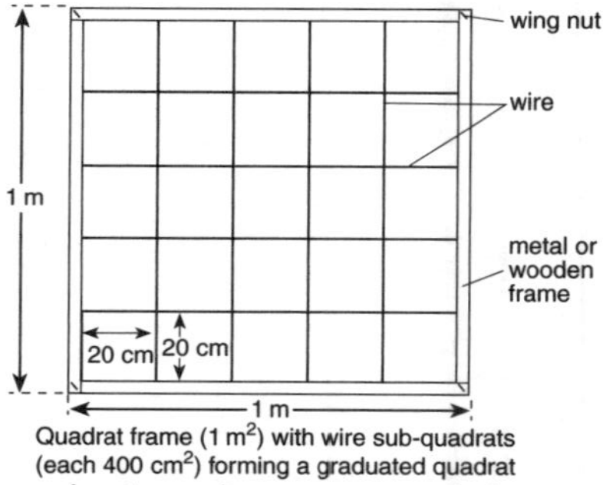

Quadrat frame (1 m^2) with wire sub-quadrats (each 400 cm^2) forming a graduated quadrat

A quadrat is used most commonly for estimating the size of plant populations, but may also be valuable for the study of populations of sessile or slow-moving animals (e.g. limpets).

The occurrence of a species within a quadrat can be expressed in several ways:

- **Density:** count the number of individuals in a fixed number of quadrats, and express results as average number of individuals per unit area
- **Percentage cover:** proportion of the ground area covered by the above ground parts of the species
- **Percentage frequency:** percentage of the quadrats in which the species occurs.

Transects are used to describe the distribution of species in a straight line across a habitat. Transects are particularly useful for describing **zonation** of species, for example around field or pond margins, or across a marsh. A simple **line transect** records all of the species which actually touch the rope or tape stretched across the habitat, a **belt transect** records all of those species present between two lines (perhaps 0.5 m^2 apart), and an **interrupted belt transect** records all of those species present in a number of quadrats placed at fixed points along a line stretched across the habitat.

A point quadrat is used for sampling plant populations in short grassland. It consists of pointed needles pushed through a horizontal wooden or metal frame, usually in groups of ten. Each plant touched by the point of a needle is recorded.

A home-made quadrat frame. Two of the ten needles have been lowered.

Sampling motile species

Quadrats and line transects are ideal methods for estimating populations of plants or sedentary animals. Motile animals, however, must be captured before their populations can be estimated. Once more, a representative sample of the population will be counted and the total population estimated from the sample. One important technique is the **mark-recapture** (also known as mark-release-recapture) **method.** This method involves:

1. capturing the organism
2. marking in some way which causes no harm (e.g. beetles can be marked with a drop of waterproof paint on their wing cases, and mice may have a small mark clipped into their fur)
3. releasing the organism to rejoin its population
4. a second sample group from this population is captured and counted at a later date
5. the population size is estimated using the Lincoln Index:

$$\text{Population size} = \frac{n_1 \times n_2}{n_m}$$

where n_1 is the number of individuals marked and released ('1' because it was the first sample), n_2 is the number of individuals caught the second time round ('2' because it was the second sample) and n_m is the number of marked individuals in the second sample ('m' standing for marked).

This method depends on a number of assumptions – failure of any of these to hold up can lead to poor estimates being made. These assumptions are:

- the marked organisms mix randomly back into the normal population (allow sufficient time for this to occur, bearing in mind the mobility of the species);
- the marked animals are no different to the unmarked ones – they are no more prone to predation, for example;
- changes in population size due to births, deaths, immigration and emigration are negligible;
- the mark does not wear off or grow out during the sampling period.

Tullgren Funnel: used to collect small organisms from the air spaces of the soil or from leaf litter. The lamp is a source of heat and dehydration – organisms move to escape from it and fall through the sieve (the mesh is fine enough to retain the soil or litter). The animals slip down the smooth-sided funnel and are immobilised in the alcohol. They may then be removed for identification.

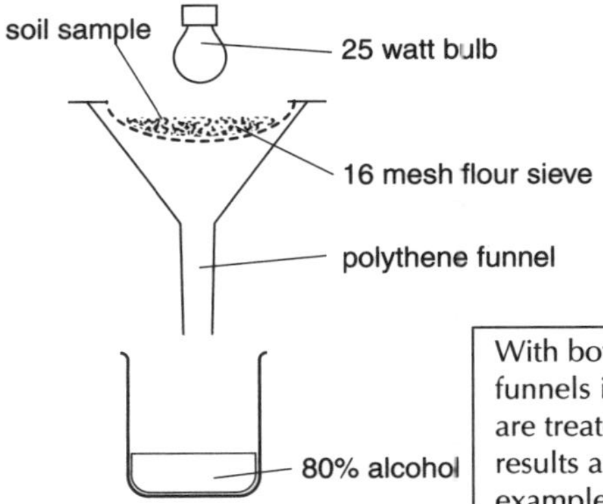

Baermann Funnel: works on a similar principle to the Tullgren funnel, but extracts organisms living in the soil water. The heat source drives the animals out of the muslin bag and into the surrounding water. Examples of water can be released at intervals, and the organisms in the sample collected and identified.

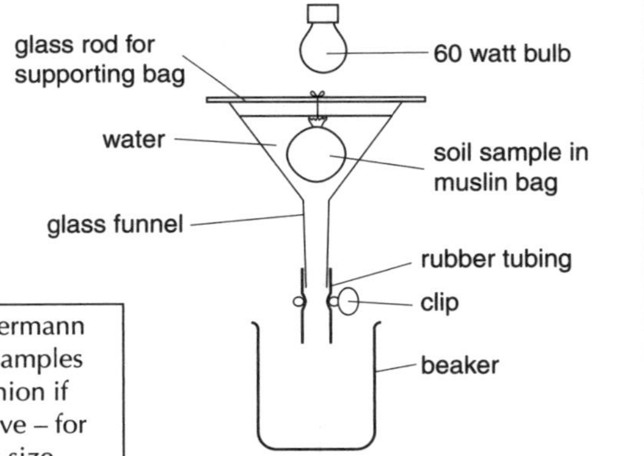

With both Tullgren and Baermann funnels it is essential that samples are treated in identical fashion if results are to be comparative – for example, use fixed sample size, length of exposure to heat source and wattage of lamp.

Pitfall Traps: used to sample arthropods moving over the soil surface.

The roof prevents rainfall from flooding the trap, and also limits access to certain predators. The activities of trapped predators can be prevented by adding a small quantity of methanol to the trap. Bait of meat or ripe fruit can be placed in the trap.

Pitfall traps are often set up on a grid system to investigate the movements of ground animals more systematically.

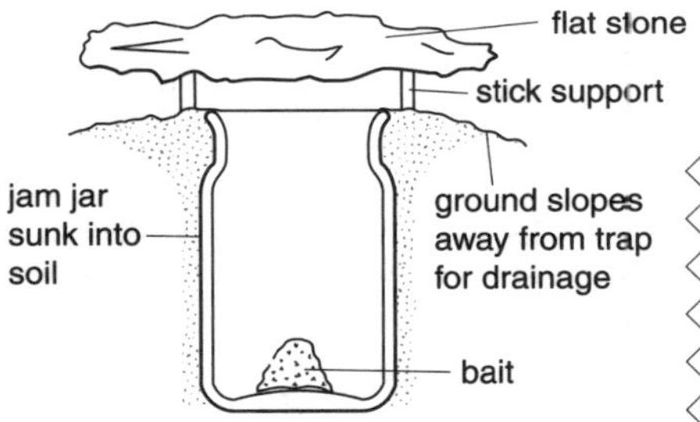

Pooter: used to collect specimens of insects and other arthropods which have been extracted from trees or bushes by beating the vegetation over a sheet or tray. Collection in the pooter does not harm the organism, and it can then be returned to its natural habitat.

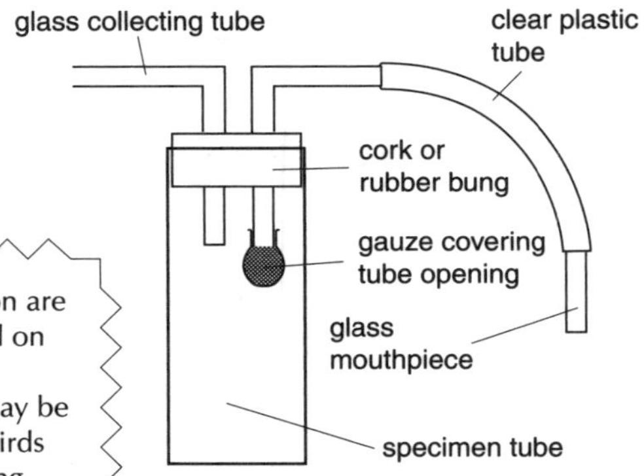

Other methods of collection are numerous. Many are based on some form of netting – for example, large mist nets may be used to collect migrating birds for identification and ringing, and sweep nets may be used to capture aerial or aquatic arthropods.

Factors affecting population growth

The **carrying capacity** (**K**) of the environment is the maximum number of a species which can be accommodated within a defined space or habitat : it is determined by the availability of *nutrients, shelter* and *breeding sites*. If the carrying capacity is increased (K → K*) then the size of the population tends to increase to take advantage.

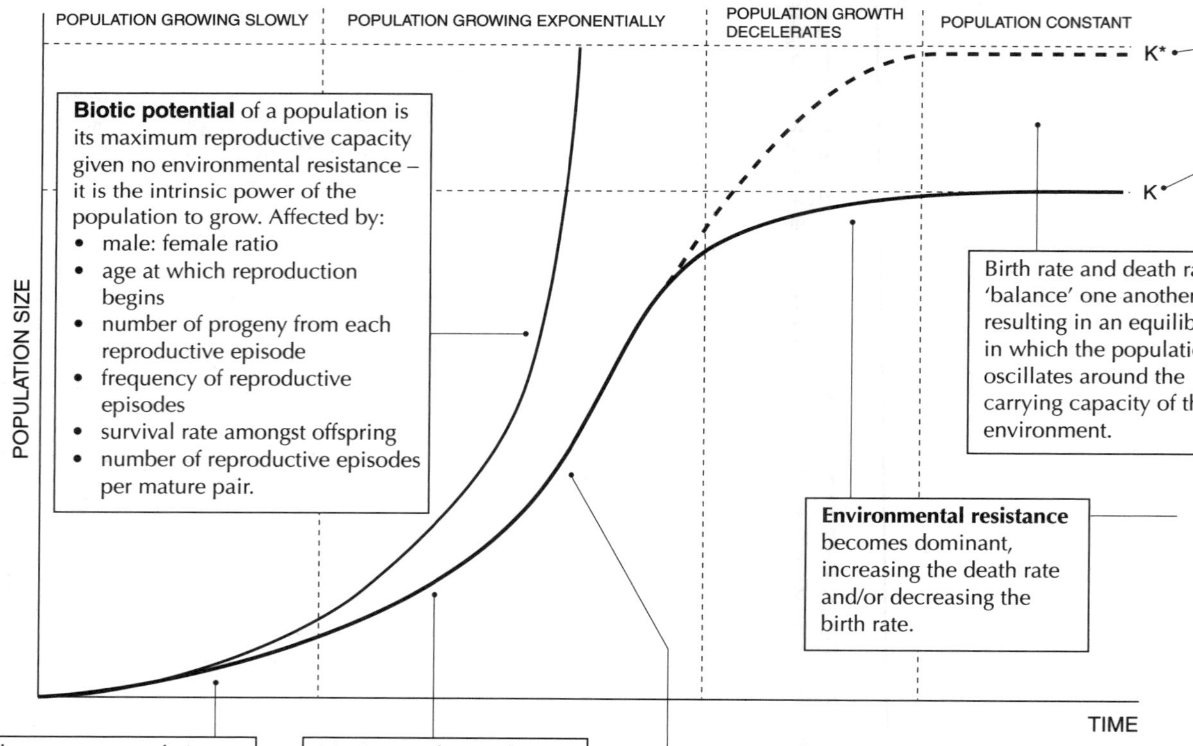

Biotic potential of a population is its maximum reproductive capacity given no environmental resistance – it is the intrinsic power of the population to grow. Affected by:

- male: female ratio
- age at which reproduction begins
- number of progeny from each reproductive episode
- frequency of reproductive episodes
- survival rate amongst offspring
- number of reproductive episodes per mature pair.

Birth rate and death rate 'balance' one another resulting in an equilibrium in which the population oscillates around the carrying capacity of the environment.

Environmental resistance becomes dominant, increasing the death rate and/or decreasing the birth rate.

Slow start to population growth as the number of mature, reproducing individuals is low and they may be widely dispersed.

Maximum observed growth rate – compromise between biotic potential and initial environmental resistance.

Population density, the number of individuals per unit area, is determined by:

- carrying capacity
- biotic potential
- time.

Doubling time is the period required for population size to increase by 100%.

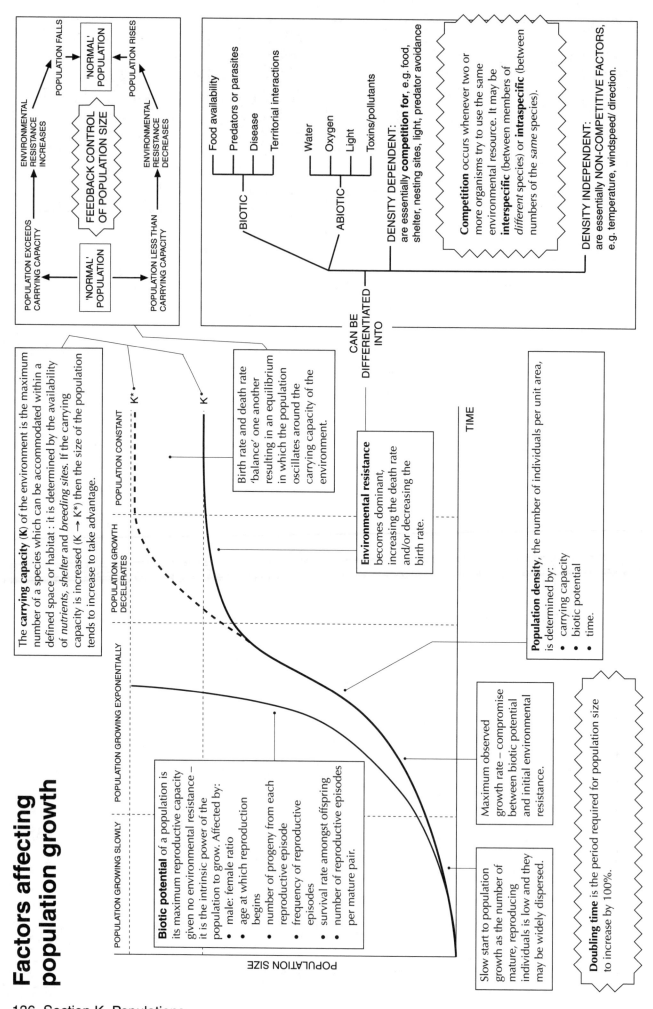

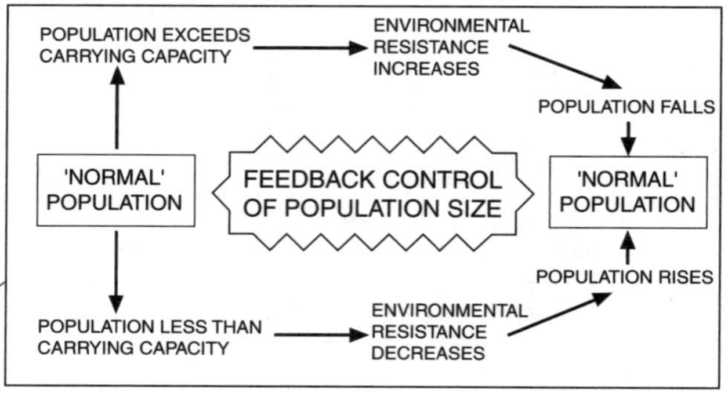

CAN BE DIFFERENTIATED INTO

- BIOTIC
  - Food availability
  - Predators or parasites
  - Disease
  - Territorial interactions
- ABIOTIC
  - Water
  - Oxygen
  - Light
  - Toxins/pollutants
- DENSITY DEPENDENT: are essentially **competition for**, e.g. food, shelter, nesting sites, light, predator avoidance
- DENSITY INDEPENDENT: are essentially NON-COMPETITIVE FACTORS, e.g. temperature, windspeed/ direction.

Competition occurs whenever two or more organisms try to use the same environmental resource. It may be **interspecific** (between members of *different* species) or **intraspecific** (between numbers of the *same* species).

Human population growth: I

Human success
Measured as:
- worldwide distribution
- large number of individuals
- dominance over other species largely due to behavioural skills which:
  - allow solution of complex problems
  - allows control of/modification of environment leading to changes in carrying capacity of the environment.

Three significant phases in human population growth

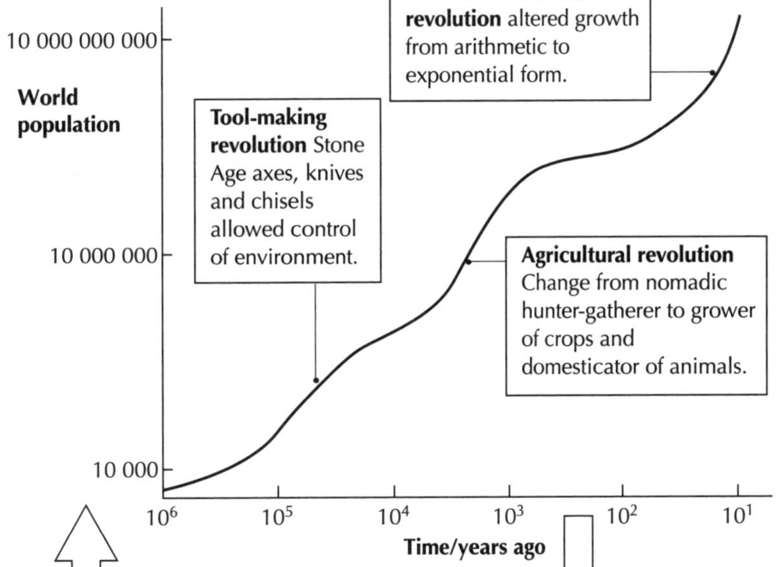

Tool-making revolution Stone Age axes, knives and chisels allowed control of environment.

Agricultural revolution Change from nomadic hunter-gatherer to grower of crops and domesticator of animals.

Scientific-industrial revolution altered growth from arithmetic to exponential form.

Factors affecting growth of the U.K. population:
The demographic transition model

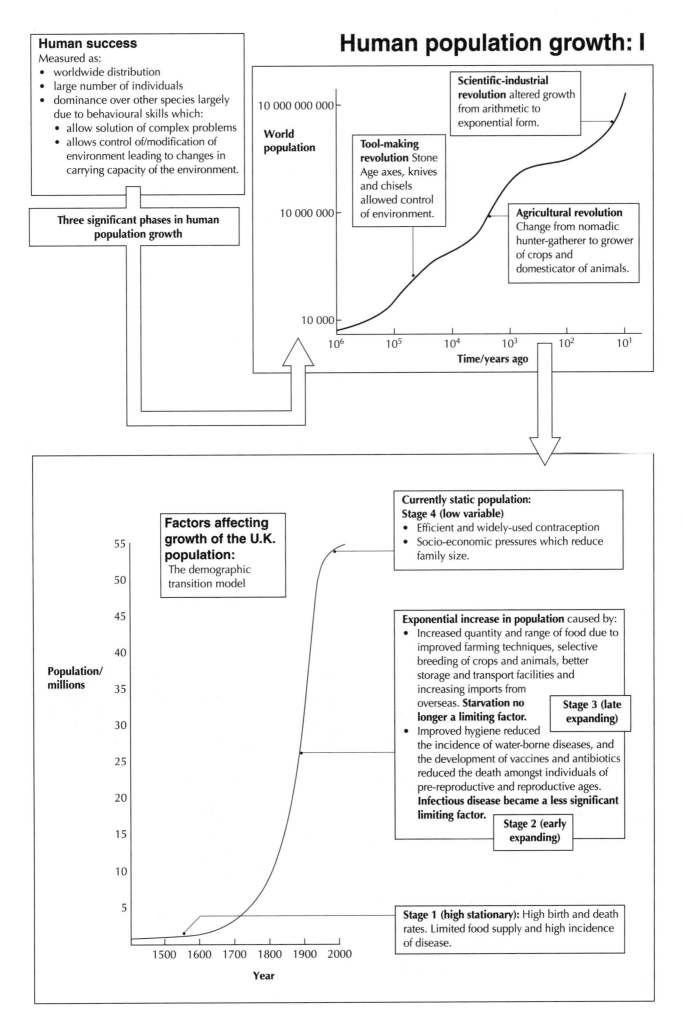

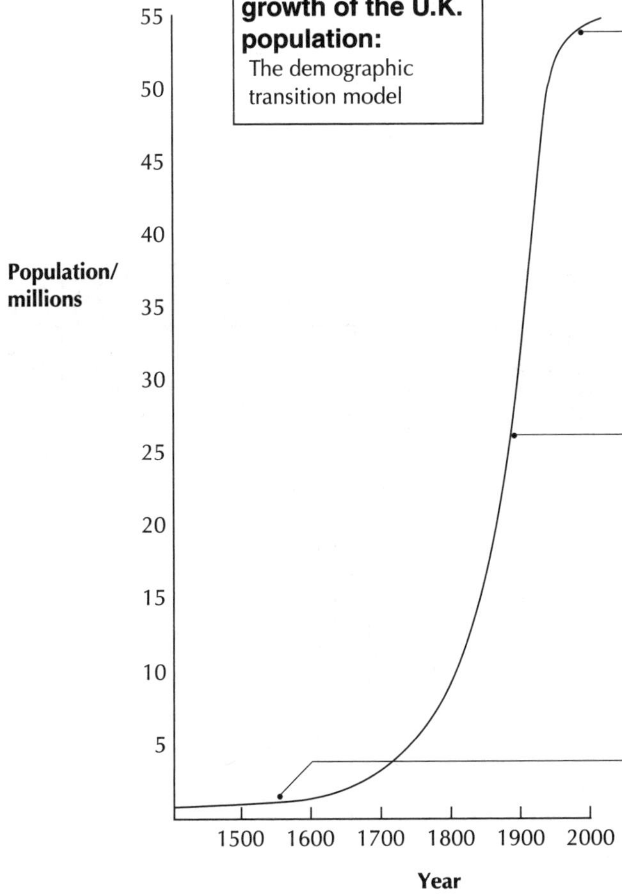

Stage 1 (high stationary): High birth and death rates. Limited food supply and high incidence of disease.

Stage 2 (early expanding)

Stage 3 (late expanding)

Exponential increase in population caused by:
- Increased quantity and range of food due to improved farming techniques, selective breeding of crops and animals, better storage and transport facilities and increasing imports from overseas. **Starvation no longer a limiting factor.**
- Improved hygiene reduced the incidence of water-borne diseases, and the development of vaccines and antibiotics reduced the death amongst individuals of pre-reproductive and reproductive ages. **Infectious disease became a less significant limiting factor.**

Currently static population:
Stage 4 (low variable)
- Efficient and widely-used contraception
- Socio-economic pressures which reduce family size.

Human population growth: II

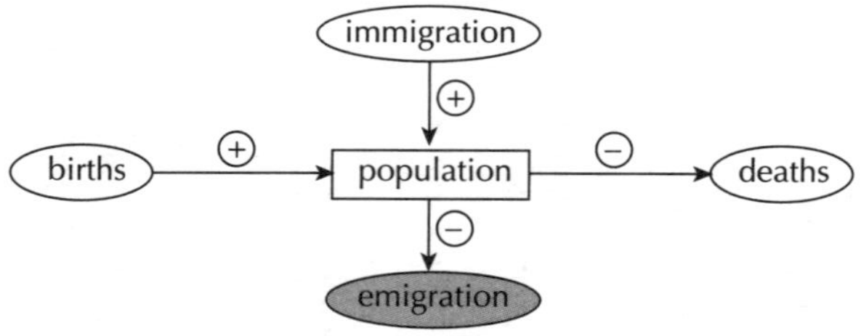

Factors affecting population size

Dealing with population data

Population growth rate = $\frac{\text{population change in given period of time}}{\text{population at start of period}} \times 100\%$

Rate of natural increase (RNI) = $\frac{\text{(number of births − number of deaths) in one year}}{\text{total population for that year}} \times 100\%$

Doubling time = $\frac{\text{number of years for current population to double}}{\text{(assuming a constant RNI)}}$

RNI may be misleading! This figure takes no account of rates of immigration or emigration: **MEDNs (more economically developed nations)** have population growth rate **increased** by economic immigration. **LEDNS (less economically developed nations)** have population growth rate **decreased** by economic emigration.

Age (population) pyramids show the age structure of a population.

Affected by:
- food availability
- hygiene
- medical provision
- shelter
- working conditions

Can be compared with respect to:
- width of base (proportion of young people in population)
- height (number of individuals surviving to old age)
- angle of sides (death rate affects slope).

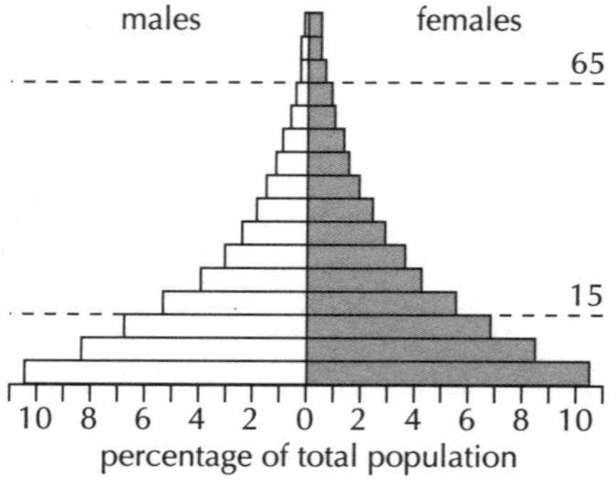

LEDN (e.g. Kenya)
- wide base
- low height
- sloping sides

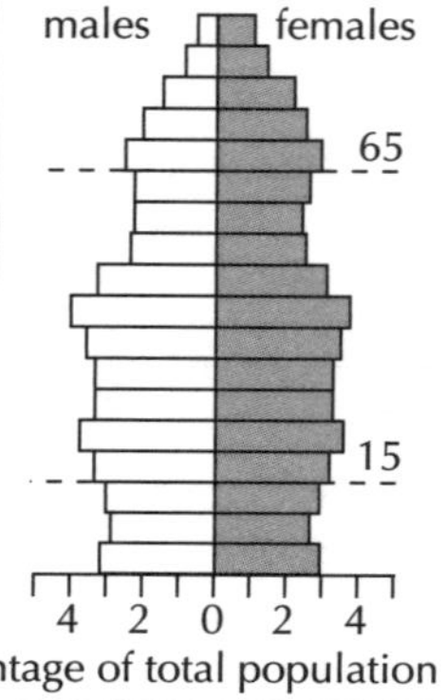

MEDN (e.g. UK)
- narrow base
- considerable height
- vertical sides

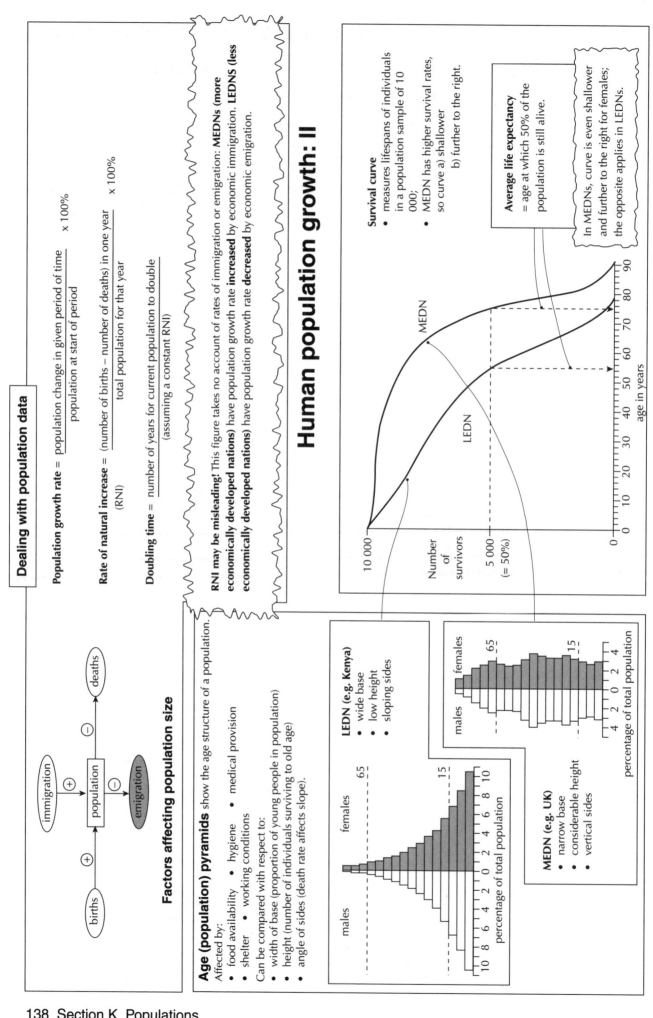

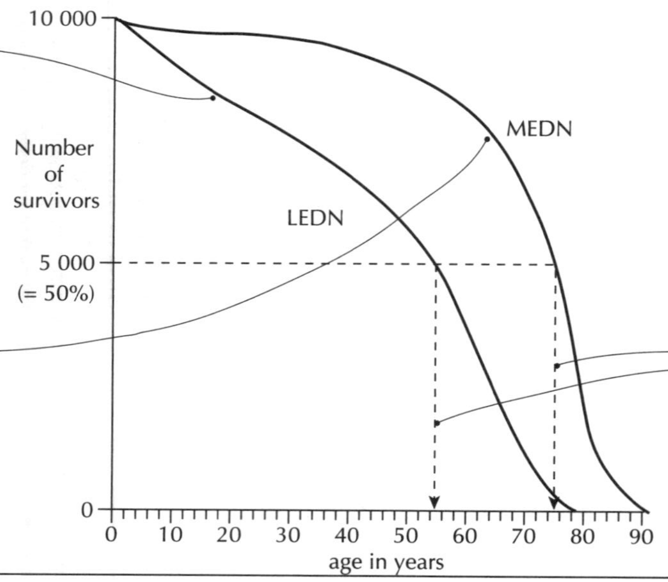

Survival curve
- measures lifespans of individuals in a population sample of 10 000;
- MEDN has higher survival rates, so curve a) shallower
  b) further to the right.

Average life expectancy
= age at which 50% of the population is still alive.

In MEDNs, curve is even shallower and further to the right for females; the opposite applies in LEDNs.

Section L Energy and ecosystems

ATP: the energy currency of the cell

This part of the molecule acts like a 'handle' – it has a shape which can be recognised by highly specific enzymes.

This part of the molecule contains anhydride bonds (O-P) which can be hydrolysed in reactions which are **exergonic** (energy-yielding) and can be coupled to **endergonic** (energy-demanding) reactions.

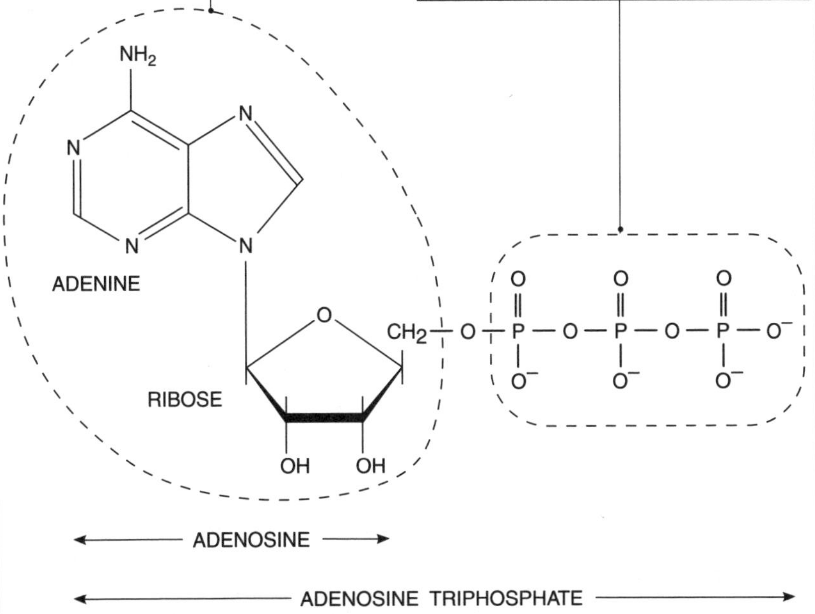

- ATP is **A**denosine **T**ri **P**hosphate.
- It is produced by adding phosphate (P) to ADP.
- When it is hydrolysed, energy is released to drive biological processes.

Substrate level phosphorylation: A phosphate group is transferred from a phosphorylated compound to ADP.

e.g.

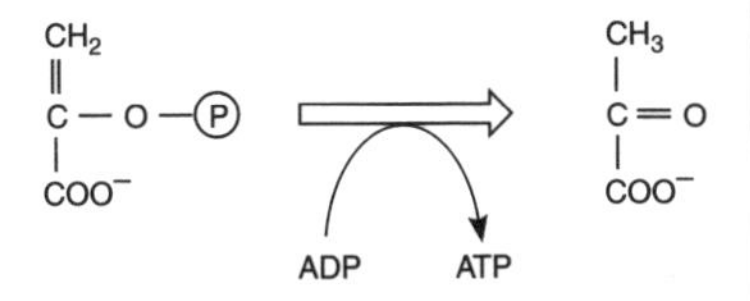

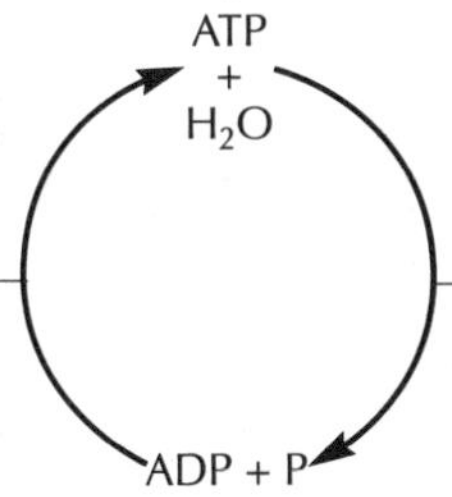

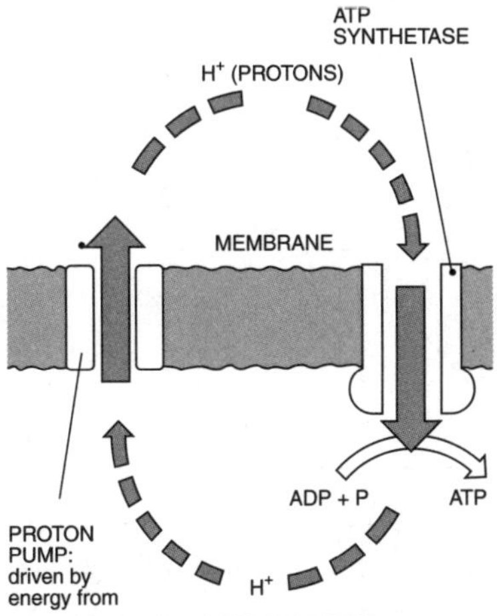

PROTON PUMP: driven by energy from RESPIRATION OF FOODS (OXIDATIVE PHOSPHORYLATION) or from ABSORPTION OF LIGHT (PHOTOPHOSPHORYLATION).

Chemiosmosis: A proton gradient across an impermeable membrane is dissipated and the energy released is used to drive the phosphorylation of ADP.

Muscle contraction: ATP hydrolysis changes the position of the myosin 'head' relative to actin.

Urea synthesis: ATP hydrolysis drives the ornithine cycle which removes toxic ammonia.

$2NH_3 + CO_2 + 3ATP + 3H_2O \rightarrow \text{urea} + AMP + 2ATP$

Protein synthesis: ATP is used to 'load' amino acids onto transfer RNA.

Active transport systems are driven by the phosphorylation of membrane-bound proteins.

Calvin cycle (dark stage of photosynthesis): ATP hydrolysis drives the cyclic reduction of CO_2 to triose phosphate.

$3CO_2 + 6ATP \rightarrow \text{triose phosphate}$

Nitrogen fixation involves the ATP-driven reduction of molecular nitrogen.

$N_2 + 8[H] + 12ATP \rightarrow 2NH_4^+ + 12\ ADP + 12P$

Bioluminescence: ATP hydrolysis drives the oxidation of luciferin which releases some energy as visible light – useful for fireflies!

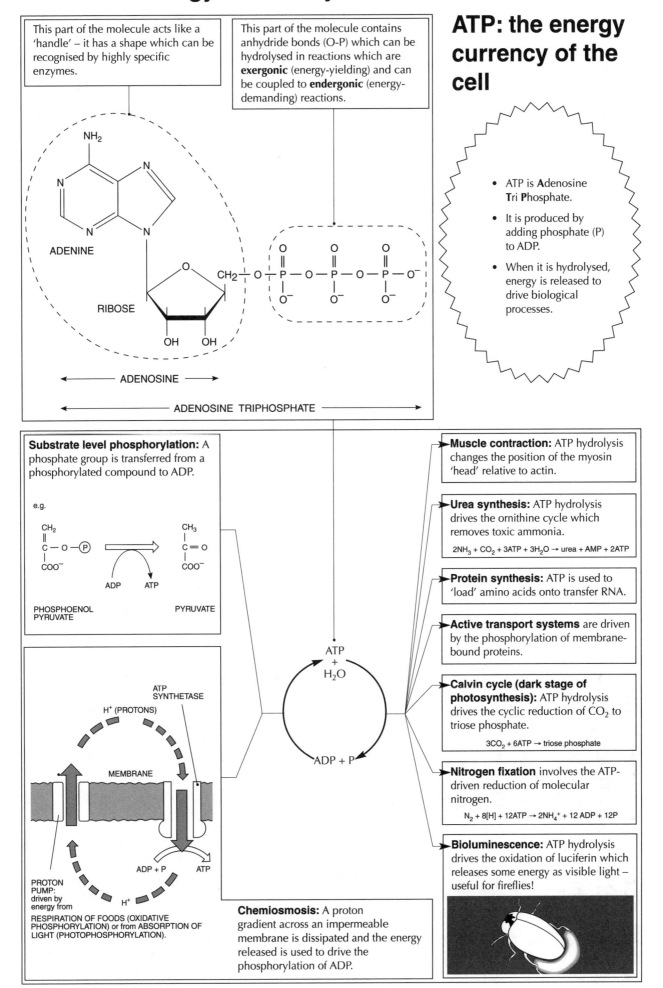

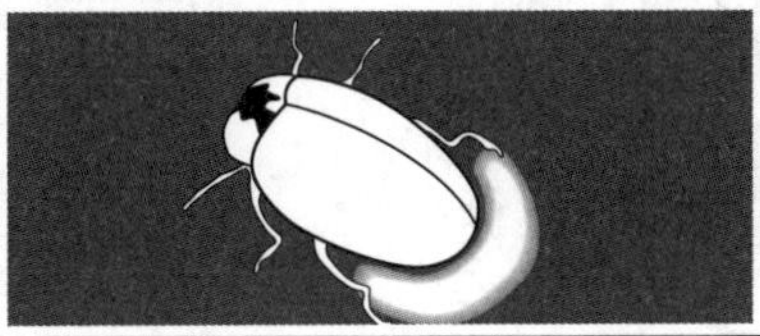

Light-dependent reaction: non-cyclic photophosphorylation

Light energy excites electrons, resulting in the splitting of water and the synthesis of ATP and $NADPH_2$.

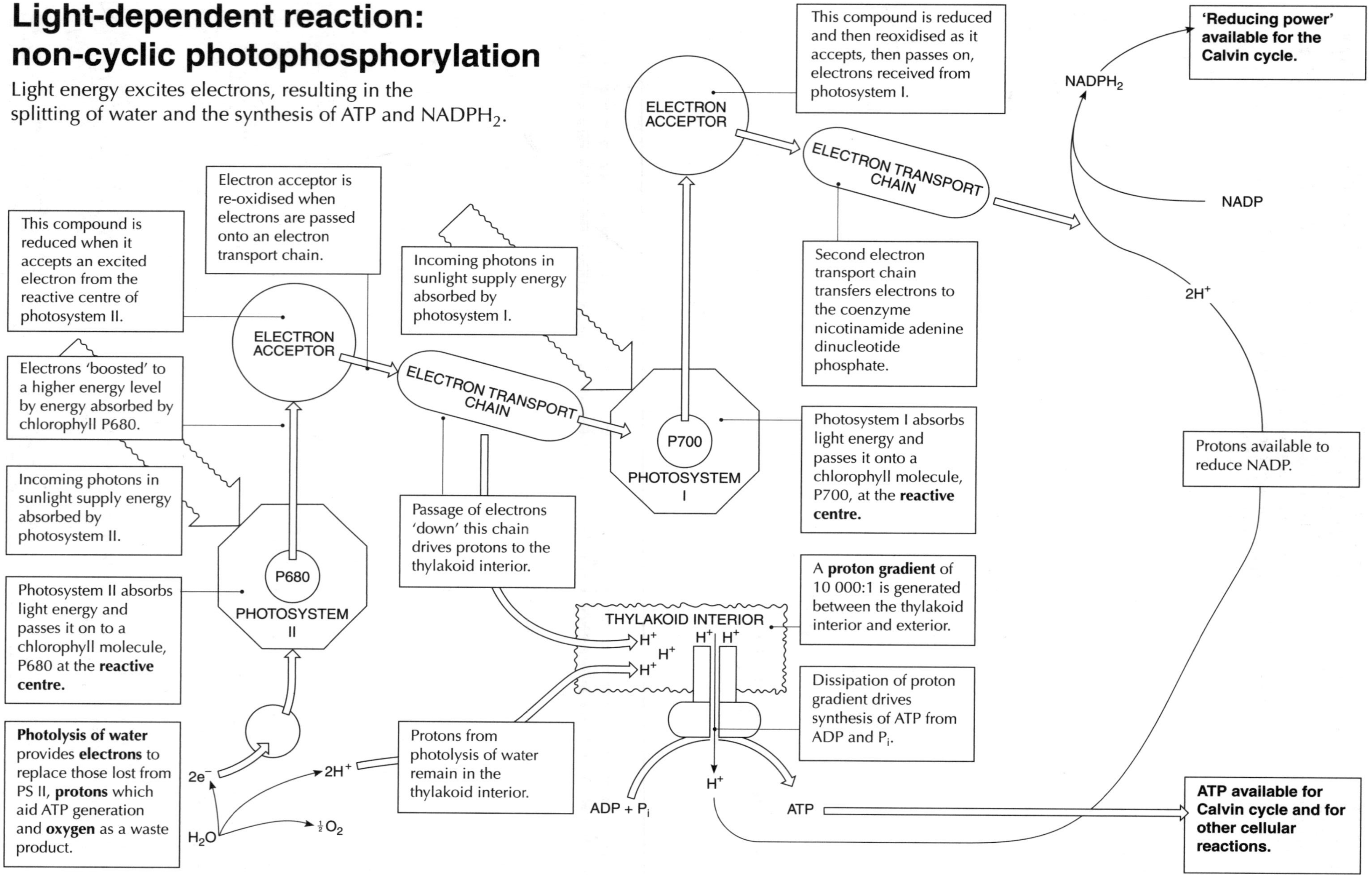

'Reducing power' available for the Calvin cycle.

This compound is reduced and then reoxidised as it accepts, then passes on, electrons received from photosystem I.

Second electron transport chain transfers electrons to the coenzyme nicotinamide adenine dinucleotide phosphate.

Photosystem I absorbs light energy and passes it onto a chlorophyll molecule, P700, at the reactive centre.

A proton gradient of 10 000:1 is generated between the thylakoid interior and exterior.

Dissipation of proton gradient drives synthesis of ATP from ADP and P_i.

ATP available for Calvin cycle and for other cellular reactions.

Incoming photons in sunlight supply energy absorbed by photosystem I.

Electron acceptor is re-oxidised when electrons are passed onto an electron transport chain.

Passage of electrons 'down' this chain drives protons to the thylakoid interior.

Protons from photolysis of water remain in the thylakoid interior.

This compound is reduced when it accepts an excited electron from the reactive centre of photosystem II.

Electrons 'boosted' to a higher energy level by energy absorbed by chlorophyll P680.

Incoming photons in sunlight supply energy absorbed by photosystem II.

Photosystem II absorbs light energy and passes it on to a chlorophyll molecule, P680 at the reactive centre.

Photolysis of water provides electrons to replace those lost from PS II, protons which aid ATP generation and oxygen as a waste product.

ELECTRON TRANSPORT CHAIN

ELECTRON ACCEPTOR

P700

PHOTOSYSTEM I

THYLAKOID INTERIOR

NADP

$2H^+$

Protons available to reduce NADP.

$NADPH_2$

ELECTRON TRANSPORT CHAIN

ELECTRON ACCEPTOR

P680

PHOTOSYSTEM II

H^+ H^+

H^+

H^+ H^+

H^+

ADP + P_i

ATP

$2e^-$

$2H^+$

$\frac{1}{2}O_2$

H_2O

Light-independent reaction: the Calvin cycle

Carbon dioxide fixation: this is the key step in the 'dark' stage – inorganic CO_2 is 'fixed' into an organic compound by covalently bonding to the 5 carbon acceptor molecule, ribulose 1,5-bisphosphate. The enzyme which catalyses the reaction, *RuBP carboxylase,* is the most abundant of all enzymes – there are probably about 40 × 10^{12} g of this protein on Earth (about 0.2% of all the Earth's protein!).

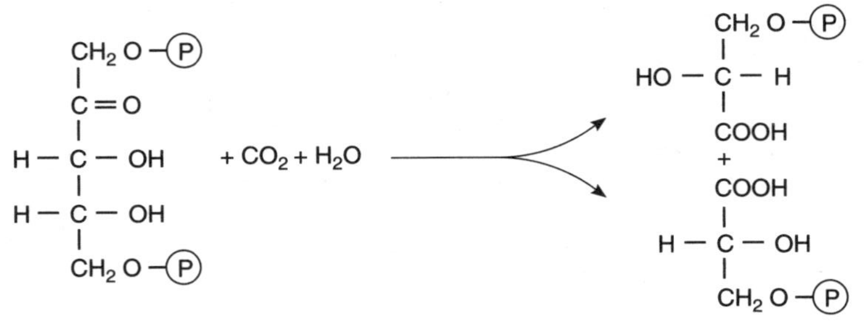

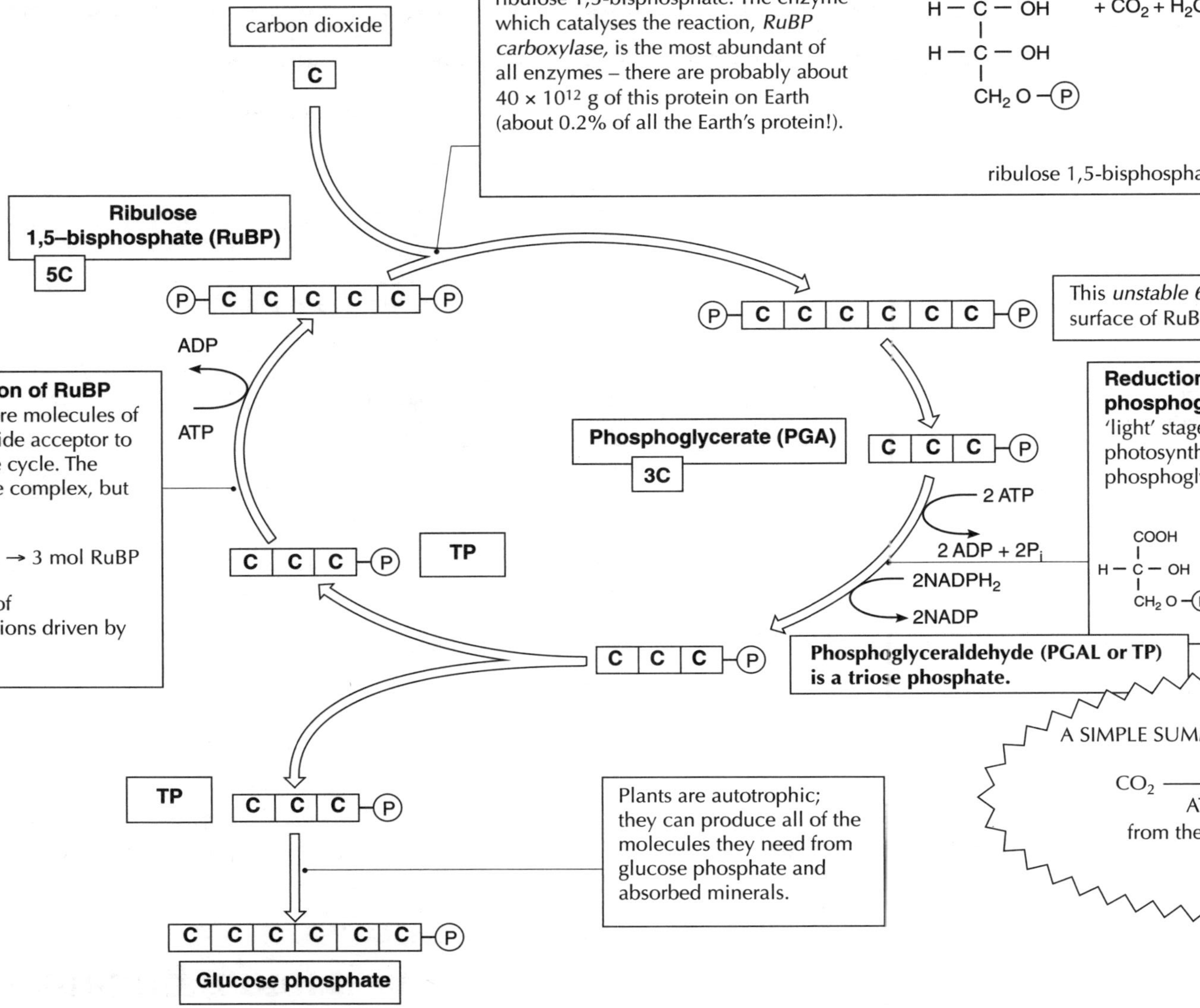

This *unstable 6 carbon intermediate* only exists on the surface of RuBP carboxylase – it has never been isolated.

Reduction of phosphoglycerate to phosphoglyceraldehyde: uses ATP and $NADPH_2$ from the 'light' stage to produce the first carbohydrate product of photosynthesis, the 3-carbon compound, phosphoglyceraldehyde (PGAL), a triose phosphate (TP).

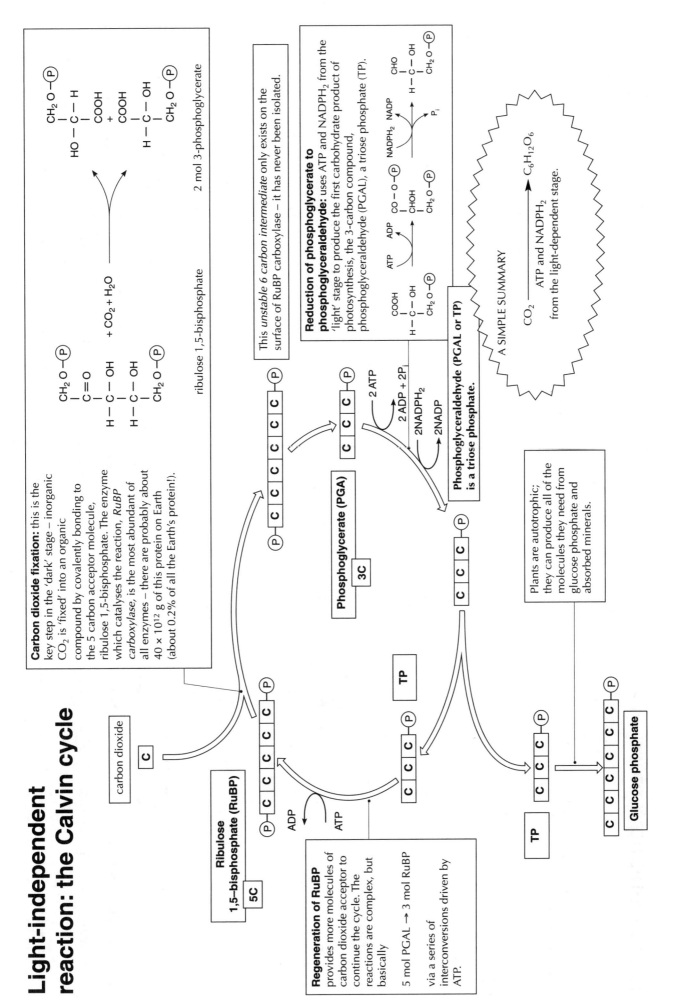

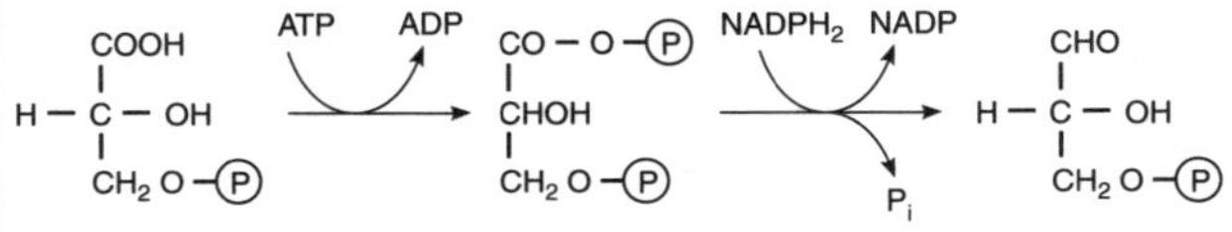

Phosphoglyceraldehyde (PGAL or TP) is a triose phosphate.

Regeneration of RuBP provides more molecules of carbon dioxide acceptor to continue the cycle. The reactions are complex, but basically

5 mol PGAL → 3 mol RuBP

via a series of interconversions driven by ATP.

Plants are autotrophic; they can produce all of the molecules they need from glucose phosphate and absorbed minerals.

A SIMPLE SUMMARY

$$CO_2 \xrightarrow[\text{from the light-dependent stage.}]{\text{ATP and NADPH}_2} C_6H_{12}O_6$$

Law of limiting factors

Blackman stated: 'when a process is affected by more than one factor its rate is limited by the factor which is nearest its minimum value: it is that **limiting factor** which directly affects a process if its magnitude is changed.'

Photosynthesis is a multi-stage process – for example the Calvin cycle is dependent on the supply of ATP and reducing power from the light reactions – and the principle of limiting factors can be applied.

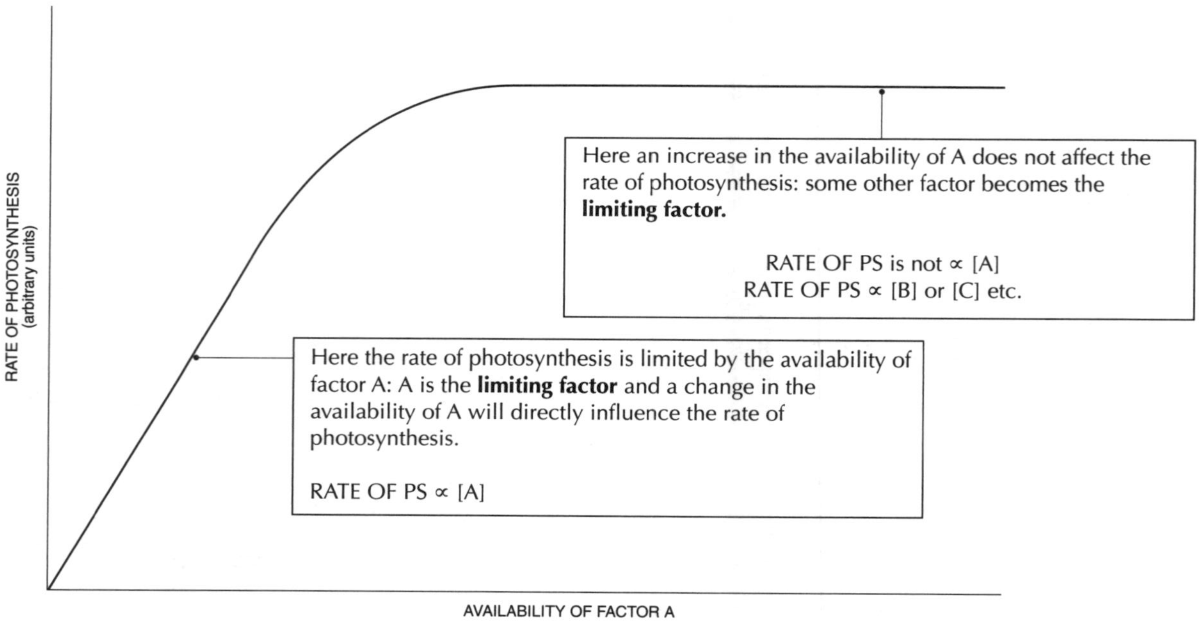
RATE OF PHOTOSYNTHESIS (arbitrary units)
AVAILABILITY OF FACTOR A
Here an increase in the availability of A does not affect the rate of photosynthesis: some other factor becomes the limiting factor.
RATE OF PS is not ∝ [A]
RATE OF PS ∝ [B] or [C] etc.
Here the rate of photosynthesis is limited by the availability of factor A: A is the limiting factor and a change in the availability of A will directly influence the rate of photosynthesis.
RATE OF PS ∝ [A]

The rate of a multi-stage process may be subject to different limiting factors at different times. Photosynthesis may be limited by **temperature** during the early part of a summer's day, by **light intensity** during cloudy or overcast conditions or by **carbon dioxide concentration** at other times. The principal limiting factor in Britain during the summer is **carbon dioxide concentration:** the atmospheric [CO_2] is typically only 0.04%. Increased CO_2 emissions from combustion of fossil fuels may stimulate photosynthesis.

The **mechanism of photosynthesis** is made clearer by studies of limiting factors – the fact that **light** is a limiting factor indicates a **light-dependent stage,** the effect of **temperature** suggests that there are **enzyme-catalysed** reactions, the **interaction of [CO_2]** and **temperature** suggests an enzyme catalysed **fixation of carbon dioxide.** The existence of more than one limiting factor suggests that **photosynthesis is a multi-stage process.**

The **limiting factors** which affect **photosynthesis** are:

Light intensity: light energy is necessary to generate ATP and $NADPH_2$ during the light-dependent stages of photosynthesis.

Carbon dioxide concentration: CO_2 is 'fixed' by reaction with ribulose bisphosphate in the initial reaction of the Calvin cycle.

Temperature: the enzymes catalysing the reactions of the Calvin cycle and some of the light-dependent stages are affected by temperature.

Water availability and **chlorophyll concentration** are not normally limiting factors in photosynthesis.

The study of limiting factors has **commercial and horticultural applications.** Since [CO_2] is a limiting factor, crop production in greenhouses is readily stimulated by raising local carbon dioxide concentrations (from gas cylinders or by burning fossil fuels). It is also clear to horticulturalists that expensive increases in energy consumption for lighting and heating are not economically justified if neither of these is the limiting factor applying under any particular set of conditions.

Managing ecosystems: horticulture

can be profitable if limiting factors can be overcome.

Carbon dioxide concentration: is a major limiting factor in photosynthesis. In a greenhouse the plants may photosynthesise very quickly and CO_2 it rapidly used up. The CO_2 concentration is usually raised to about 0.1% of the atmosphere (about three times that in air). This gives a significant (about 50%) increase in crop yield. The extra CO_2 can be provided by burning paraffin (which also raises the temperature) or, more accurately, by releasing it from a cylinder.

Pollinating insects: bees may be introduced to ensure high rates of fertilisation and therefore fruit formation.

Humidity: this affects the rate of transpiration, and therefore the rate at which the plants must be supplied with water. Thus a high humidity reduces the need for additional water and generally favours growth (NB **too** high humidity favours the growth of fungal pests). Humidity is reduced by opening ventilators and raised using an automatic mist spray.

Illumination: it is important to control:

- **intensity** – this will influence the rate of food production by photosynthesis. Greater intensity → more photosynthesis until some other limiting factors intervenes.
- **quality** – photosynthesis is most efficient at red and blue wavelengths so that 'white' light contains some wavelengths ('green') which are not useful.
- **duration** – if fruit is the desired product the plant must flower. Flowering is controlled by daylength (the duration of light in a 24 hour period).

Sunlight provides some illumination but artificial lighting systems are more controllable (but more expensive).

Computer control: this is widely used in large commercial greenhouses. Sensors provide information about air temperature, CO2 concentration, water available to roots, humidity and mineral concentration, and the control centre ensures that any changes are corrected.

e.g. temperature sensor → 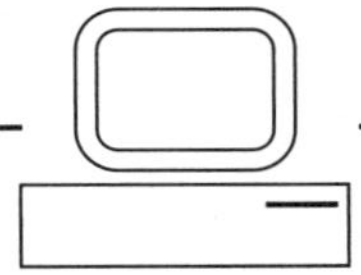 → heater

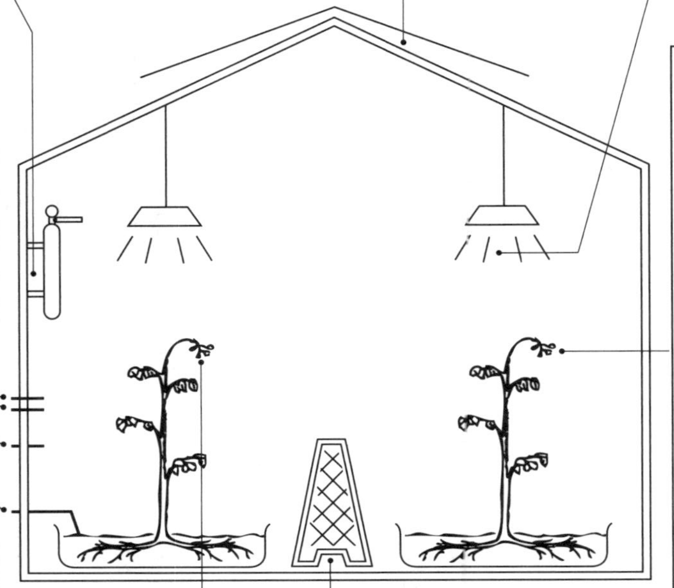

Pest control: is particularly important since a pest species could spread very rapidly through a greenhouse containing only a single crop.

- **Fungi** can be a problem since high humidity/high temperature encourages spore germination and growth of fungus. Control with **fungicide spray** or by **reduction of humidity**.
- **Weed plants** could compete with the crop species for light, water and mineral nutrients. They are removed by hand or with **selective herbicide**.
- **Animal pests** are herbivorous (e.g. caterpillars) or sap-sucking (e.g. aphids) insects. These may be controlled using **insecticides** but
  - these can be inefficient, especially against waxy-coated pupae
  - they may need to be reapplied, sop can be expensive
  - they may leave residues so that crop will need washing

  or **biological control** which
  - uses a natural predator of the pest
  - usually only requires a single application
  - does not leave residues.

  Important examples of biological control are
  ladybirds which eat aphids
  wasp larvae which live as parasites on **whitefly larvae**.

High-yielding strains of crop: selective breeding and/or genetic engineering can develop crop strains which

- have a high yield;
- produce fruit of a desirable colour/texture;
- produce fruit at the same time;
- may have genetic pest resistance.

Temperature: affects plant growth because of its affect on the enzymes of photosynthesis. NB high temperatures may also speed up the life cycle of pests.
Sunlight provides some heat (shading may be necessary in summer) but more control is available using thermostatically-regulated heaters.

Measurement of photosynthesis:

Audus' photosynthometer.

Principle: bubbles of evolved gas are collected over a known period of time.

$$\textbf{Rate of photosynthesis} = \frac{\text{collected volume / mm}^3}{\text{time / min}}$$

= mm^3 O_2 evolved / min at a known temperature, t°C.

Syringe: the plunger should be pushed well in at the start of the experiment. The syringe can be used to draw the collected bubbles of evolved gas along the tube so that the length can be measured against the scale, and then towards the syringe where it will not interfere with further collections and measurements of gas.

Problems:
1. Are **controlled variables** at their optimum?
2. Are all plant samples comparable – same size, age, activity?
3. Is apparatus reliable – clean, leakproof?
4. Are collected bubbles all oxygen, all produced by photosynthesis?

Thermometer to check temperature of water bath. Temperature should be recorded and rate of photosynthesis defined *at a temperature t.*

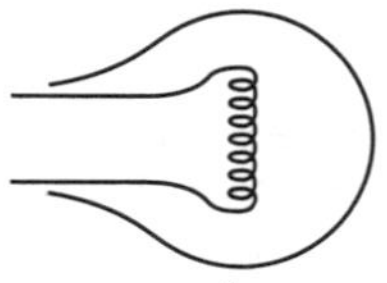

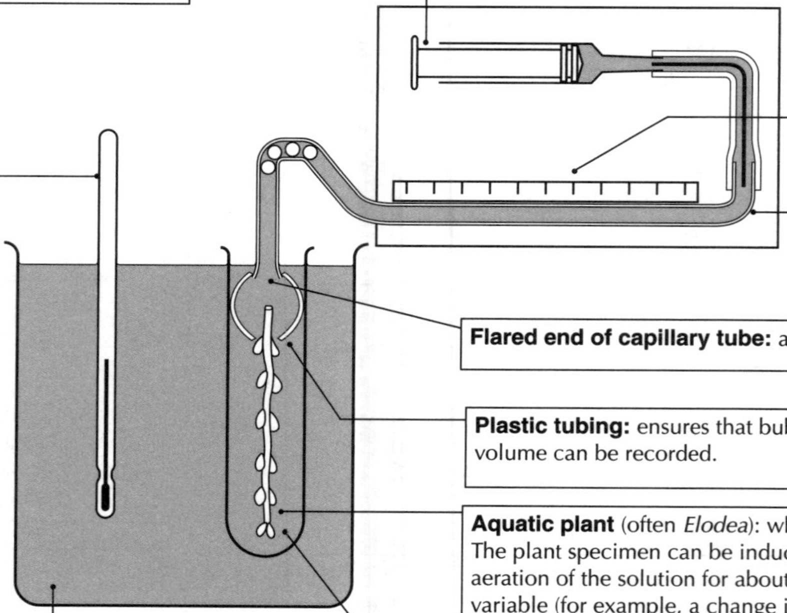

Graduated scale: allows length (l) of collected bubble to be measured. If diameter (d) of the capillary tube is known the volume (v) of gas evolved can be calculated from:

$$v = \frac{\pi d^2}{4} l$$

Capillary tube: the narrow diameter means that a small volume of evolved oxygen will register as a long bubble in the tube – more accurate measurement possible.

Flared end of capillary tube: aids attachment of plastic tubing and consequent capture of bubbles.

Plastic tubing: ensures that bubbles released from the cut end of the plant are trapped so that their volume can be recorded.

Aquatic plant (often *Elodea*): when photosynthesising, releases oxygen from a cut end of the stem. The plant specimen can be induced to photosynthesise actively by prior illumination, and gentle aeration of the solution for about an hour before the experiment. Any adjustment of the manipulated variable (for example, a change in light intensity), should be followed by a short period (ca. 5 min) to allow the plant to re-adjust (it allows the system to equilibrate).

Light source: for most applications the wavelength should be fixed (e.g. a white light source) but the apparatus can be used to investigate the effect of wavelength on rate of photosynthesis, in which case the wavelength can be varied by the insertion of appropriate filters between the lamp and the photosynthesising plant. If **light intensity** is to be the manipulated variable then the light may be moved to a known distance from the plant (in which case light intensity $\propto 1/d^2$) or a rheostat may be incorporated into the lamp circuit.

Water bath: may serve as a heat filter if an incandescent light source is used (so that **temperature** remains a controlled variable) or as a thermostatically controlled water bath if temperature is to be the manipulated variable.

Hydrogen carbonate solution is a source of carbon dioxide of known concentration. The concentration may be varied if **concentration of carbon dioxide** is to be the manipulated variable, or should be fixed at about 5 × optimum [CO_2] if it is to be a controlled variable but not a limiting factor. The [HCO_3^-] should not be great enough to markedly alter pH.

Cellular respiration

(glycolysis) occurs in a series of localised stages.

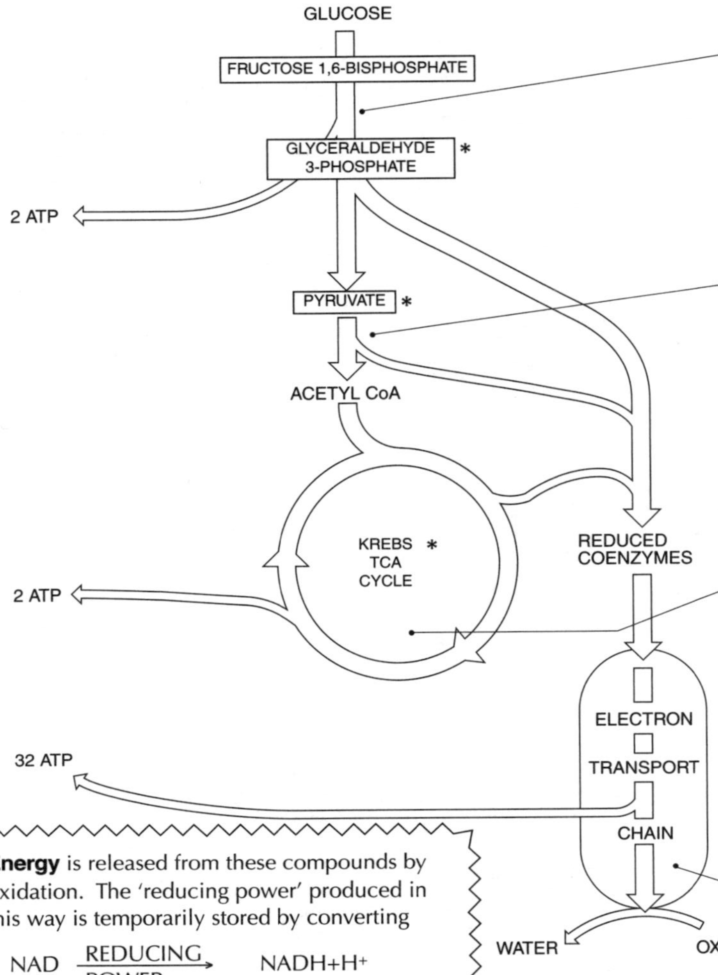

First stage: glycolysis is a series of about ten steps by which glucose is degraded ('lysed') to two molecules of pyruvate. Two molecules of ATP and two molecules of reduced coenzyme are generated. This stage can occur anaerobically.

Second stage: 'activation' of pyruvate to acetyl CoA 'drives' the pyruvate molecule towards the TCA cycle. Two molecules of reduced coenzyme are generated.

Third stage: the Krebs (tricarboxylic acid) cycle is a series of dehydrogenations, decarboxylations and isomerisations. One molecule of ATP and four molecules of reduced coenzyme are generated for **each turn of the cycle.** (The cycle is 'turned' twice for each glucose molecule.) Aerobic.

Fourth stage: oxidative phosphorylation in which a proton gradient is generated and its electrochemical potential is used to drive the synthesis of 32 molecules of ATP. Aerobic.

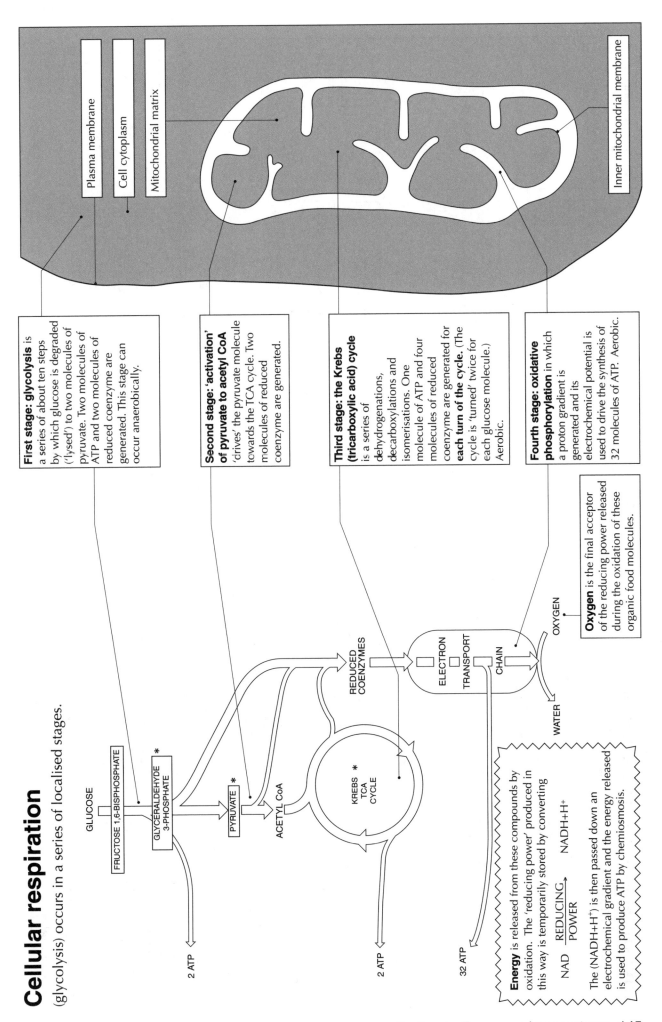

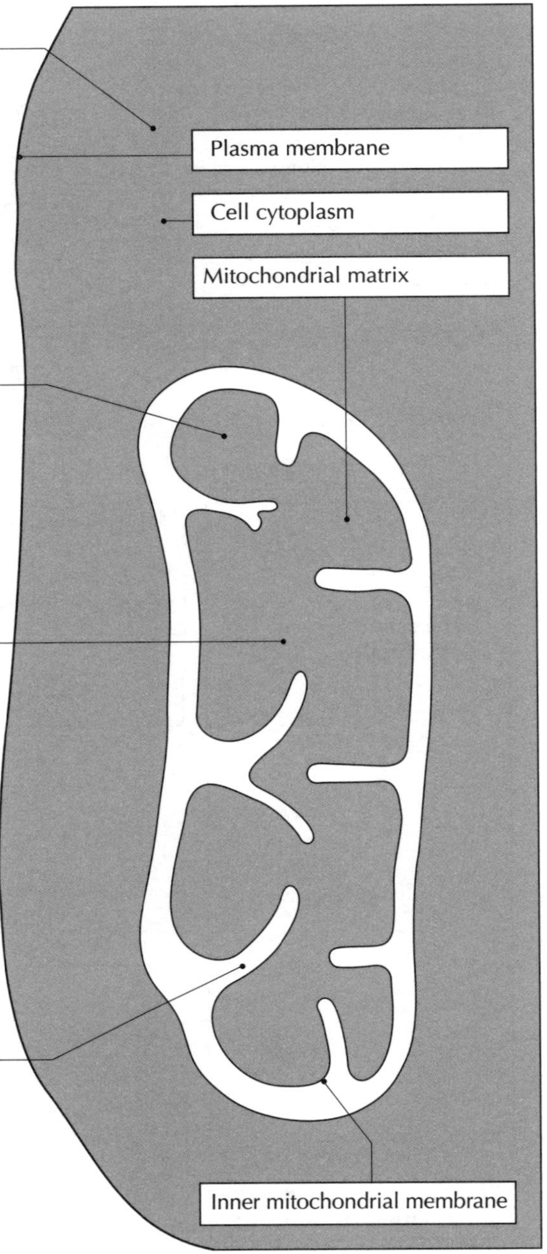

Energy is released from these compounds by oxidation. The 'reducing power' produced in this way is temporarily stored by converting

$$NAD \xrightarrow[\text{POWER}]{\text{REDUCING}} NADH+H^+$$

The ($NADH+H^+$) is then passed down an electrochemical gradient and the energy released is used to produce ATP by chemiosmosis.

Oxygen is the final acceptor of the reducing power released during the oxidation of these organic food molecules.

Glycolysis generates ATP, reduced electron carriers and pyruvate. It occurs in the cytoplasm.

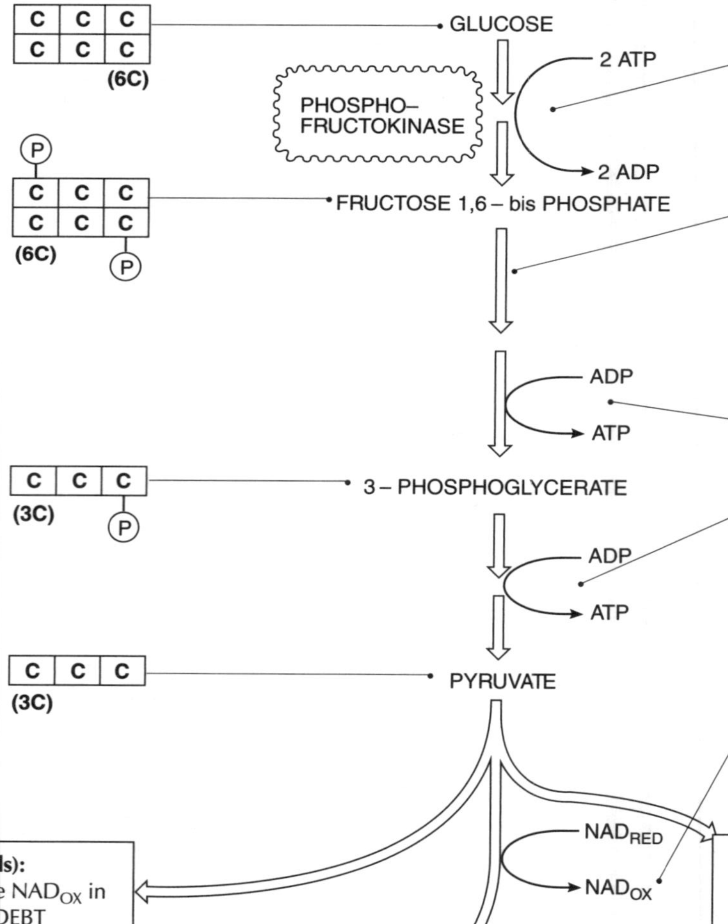

These steps:

- activate the glucose molecule but consume two molecules of ATP in doing so;
- commit the glucose molecule to the glycolytic pathway (phosphofructokinase is the first enzyme specific to glycolysis).

- This is the **glycolytic** step in which the 6-carbon unit is split into two 3-carbon sugars: **triose phosphate**
- An oxidation occurs as the 3-C sugar is formed: NAD_{OX} is reduced to NAD_{RED}.

ATP is produced by **substrate level phosphorylation.**

Glycolysis depends on a supply of NAD_{OX}:

- This is a **limiting factor** for the whole process
- Supplied by regeneration from NAD_{RED}, either

aerobically
or
anaerobically

Lactate (in muscle cells):
The need to regenerate NAD_{OX} in this way → OXYGEN DEBT

Ethanol + CO_2 (in yeasts): this is **alcoholic fermentation,** the basis of **brewing** and **baking.**

Acetyl CoA + CO_2
Only under **aerobic conditions.** Molecular oxygen then reoxidises NAD_{RED} via the **electron transport system.**

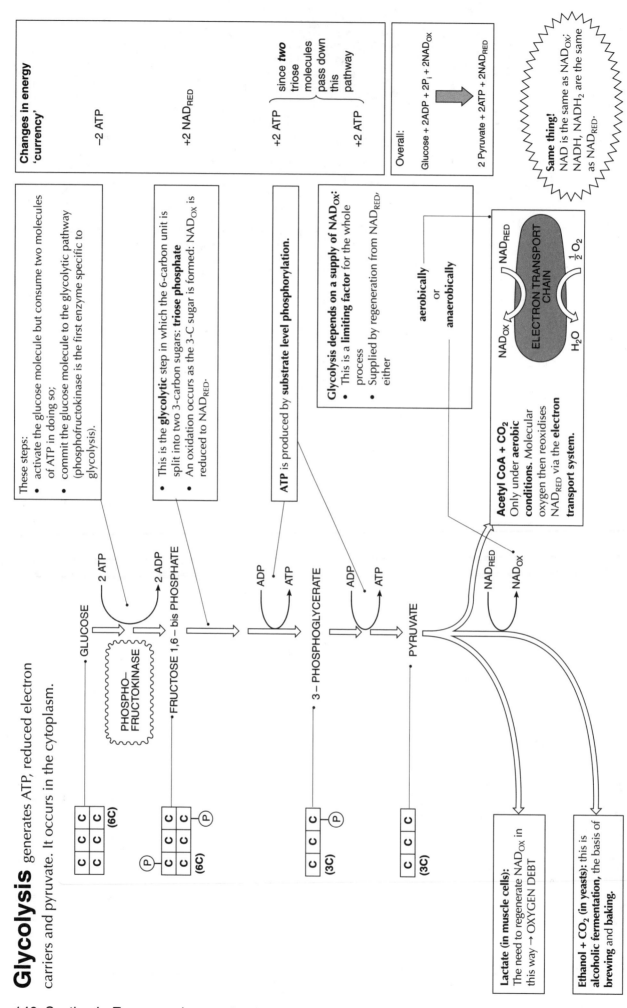

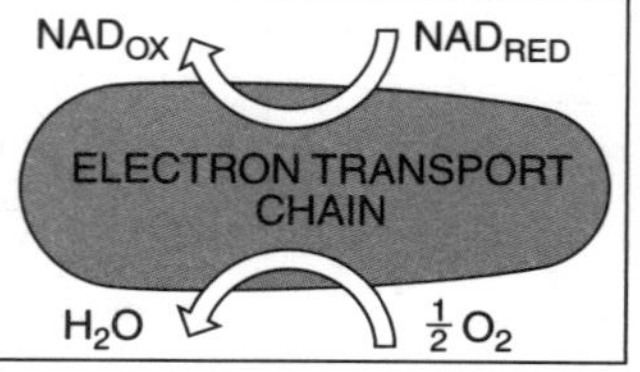

Changes in energy 'currency'

−2 ATP

+2 NAD_{RED}

+2 ATP
+2 ATP
since ***two*** triose molecules pass down this pathway

Overall:

Glucose + 2ADP + $2P_i$ + $2NAD_{OX}$
↓
2 Pyruvate + 2ATP + $2NAD_{RED}$

Same thing!
NAD is the same as NAD_{OX}; NADH, $NADH_2$ are the same as NAD_{RED}.

The Krebs Cycle

- takes place in the matrix of the mitochondrion;
- breaks down pyruvate by oxidation and decarboxylation;
- produces carbon dioxide and ATP.

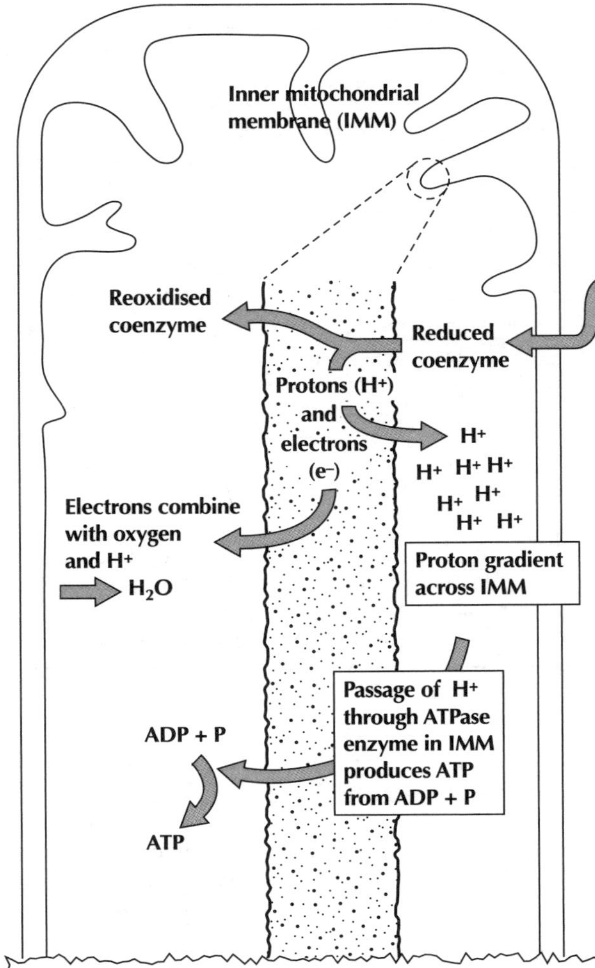

Reduced coenzymes

- 4 molecules of NAD_{RED} and 1 molecule of FAD_{RED} from each molecule of pyruvate which is **oxidatively decarboxylated** in the Krebs Cycle
- pass on protons and electrons to the **electron transfer chain** in the inner mitochondrial membrane.

The link reaction

- 'forces' pyruvate towards the Krebs Cycle
- Pyruvate + coenzyme A → acetyl coenzyme A + carbon dioxide.

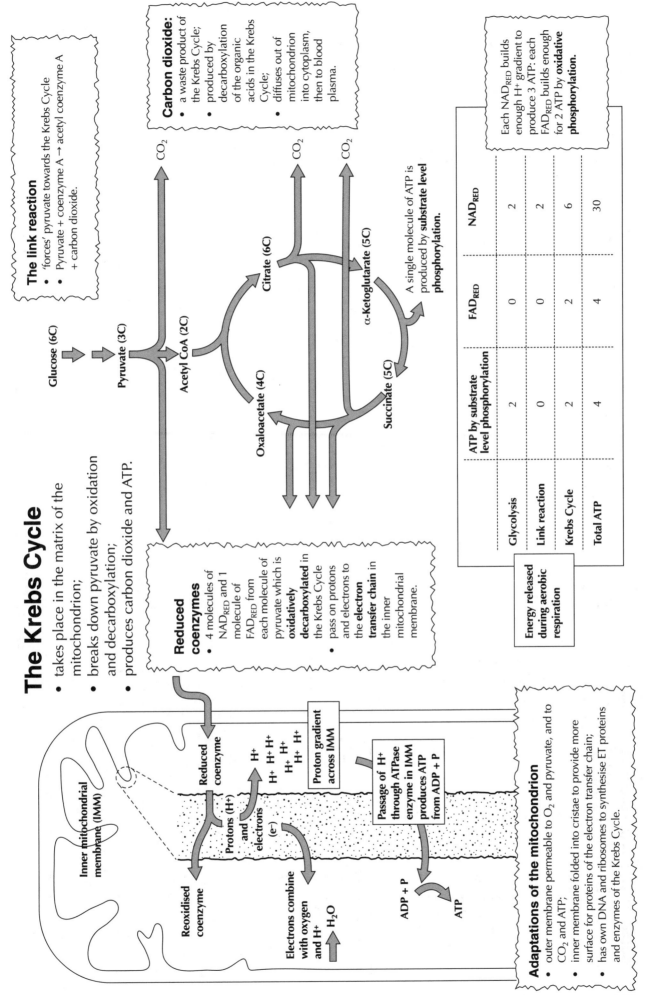

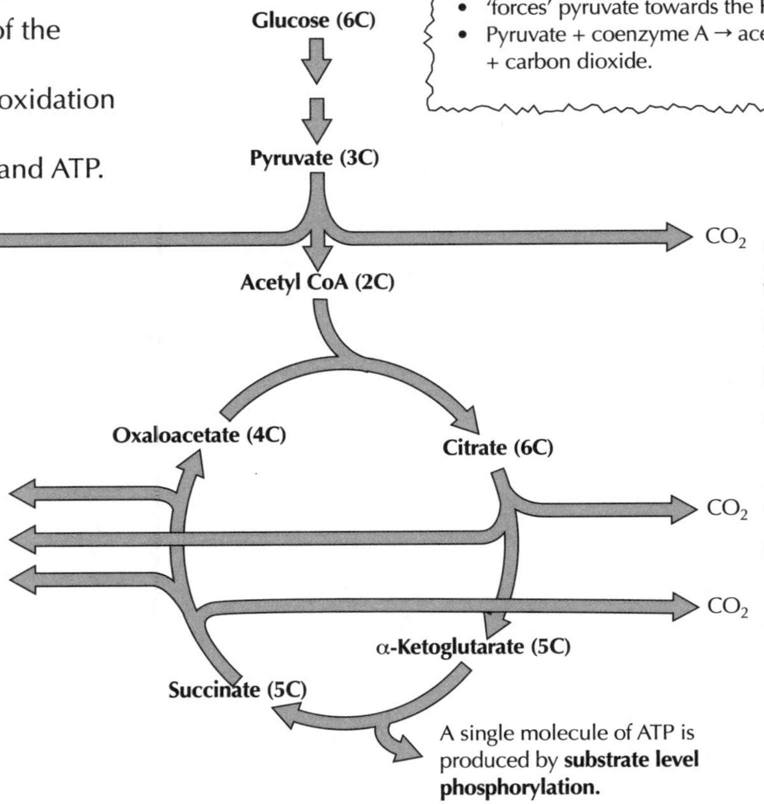

Carbon dioxide:

- a waste product of the Krebs Cycle;
- produced by decarboxylation of the organic acids in the Krebs Cycle;
- diffuses out of mitochondrion into cytoplasm, then to blood plasma.

Energy released during aerobic respiration

	ATP by substrate level phosphorylation	FAD_{RED}	NAD_{RED}
Glycolysis	2	0	2
Link reaction	0	0	2
Krebs Cycle	2	2	6
Total ATP	4	4	30

Each NAD_{RED} builds enough H^+ gradient to produce 3 ATP: each FAD_{RED} builds enough for 2 ATP by **oxidative phosphorylation.**

Adaptations of the mitochondrion

- outer membrane permeable to O_2 and pyruvate, and to CO_2 and ATP;
- inner membrane folded into cristae to provide more surface for proteins of the electron transfer chain;
- has own DNA and ribosomes to synthesise ET proteins and enzymes of the Krebs Cycle.

Structure and function of the mitochondrion

Mitochondrial matrix (M compartment) contains the enzymes of the TCA cycle and is thus the site of oxidation of pyruvate to carbon dioxide and the reduced coenzymes NAD_{RED} and FAD_{RED}.

Ribosomes of the mitochondrion are more typical of prokaryote than eukaryote cells. This, together with the presence of circular DNA and an outer membrane, offer support for the **symbiotic invasion hypothesis** for the origin of mitochondria.

Mitochondria are 'cigar-shaped':
This gives a large surface area for transport of O_2, pyruvate, CO_2 and ATP and short diffusion distances.

Cristae are folding of the IMM and increase the surface area available for the redox and phosphorylation reactions leading to ATP generation.
The number and form of the cristae may vary from tissue to tissue.

Outer mitochondrial membrane (OMM) is permeable to pyruvate, oxygen, ATP and many other molecules.

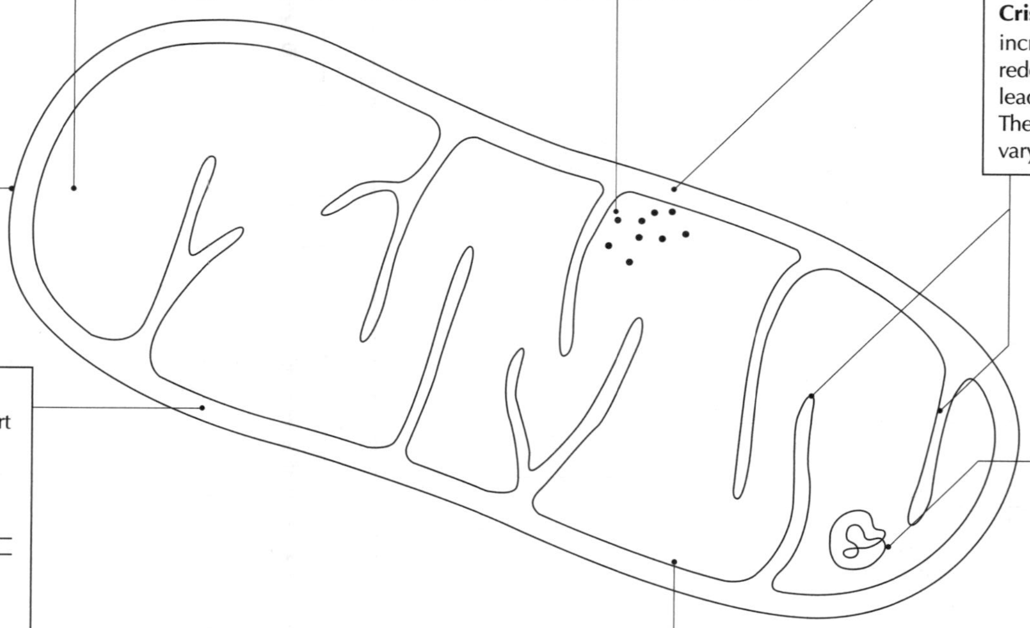

Mitochondrial DNA carries the genes for the synthesis of many, but not all, mitochondrial proteins. The DNA is loosely coiled and is very prone to mutation by oxidation – this leads to a reduction in mitochondrial efficiency and may be a contributory factor in cell ageing.

It's my Mum's!
All of the mitochondrial DNA is maternal: at fertilisation only nuclear DNA enters from the sperm! Ancestral history can be traced much more reliably from this 'undiluted' maternal DNA!

Intermembrane space (O cavity) has a high proton concentration generated by the electron transport system of the IMM, and maintained by the proton impermeability of the IMM.

A proton gradient of $10^3 – 10^4 : 1$ can exist between the O cavity and the matrix

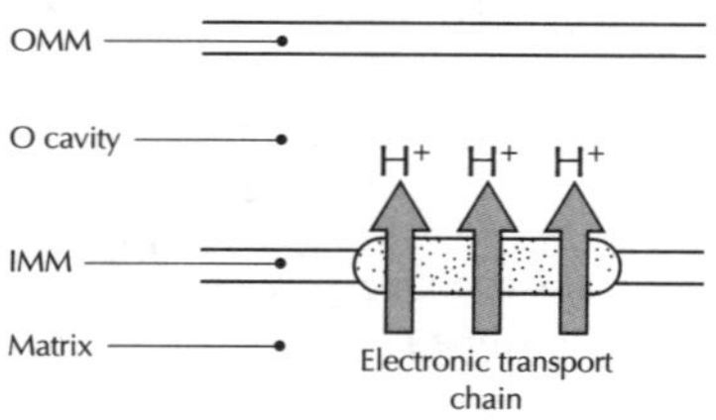

Innermitochondrial membrane (IMM) has visible 'stalked particles' with the head pointing into the matrix.

The IMM contains the electron transport system (which generates a proton gradient) and the ATPase complex (which uses the proton gradient to generate ATP from ADP).
Thus the IMM is the site of **oxidative phosphorylation.**

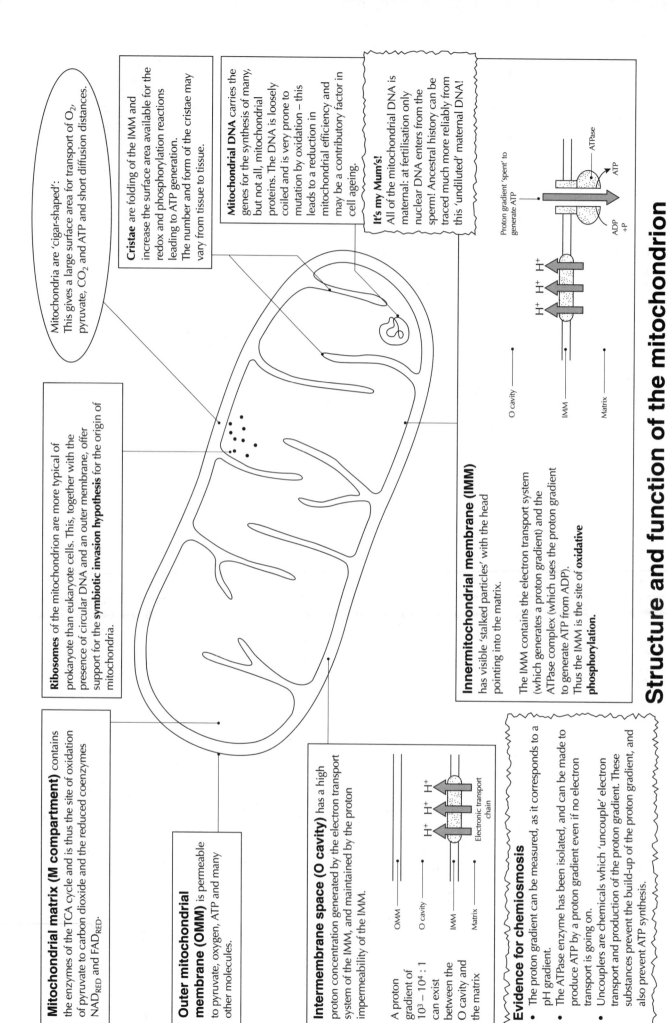

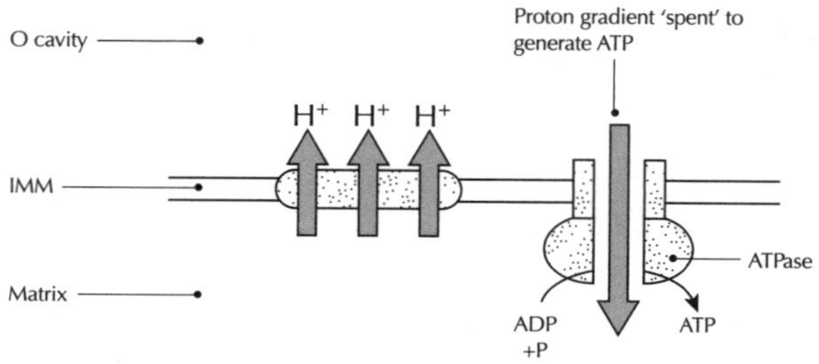

Evidence for chemiosmosis
- The proton gradient can be measured, as it corresponds to a pH gradient.
- The ATPase enzyme has been isolated, and can be made to produce ATP by a proton gradient even if no electron transport is going on.
- Uncouplers are chemicals which 'uncouple' electron transport and production of the proton gradient. These substances prevent the build-up of the proton gradient, and also prevent ATP synthesis.

Proton gradient is generated by the action of proton pumps and maintained by the impermeability of the inner mitochondrial membrane to protons.

Proton pumps are proteins of the inner mitochondrial membrane.

Protein flow

Proton channel or F_0 unit: allows proton flux across the IMM.

Inner mitochondrial membrane (IMM) is impermeable to protons.

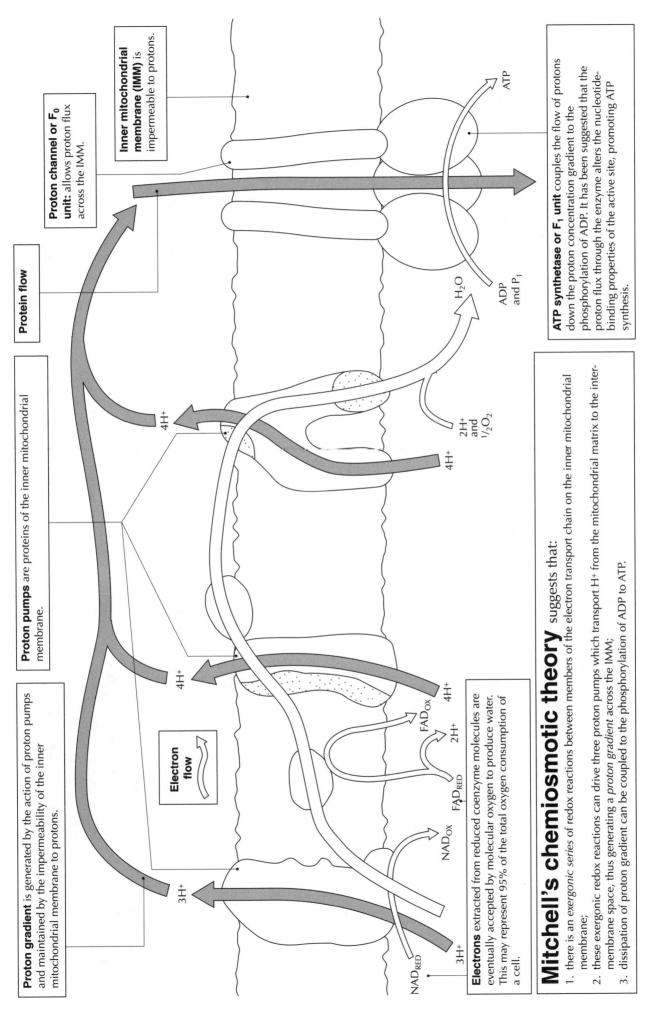

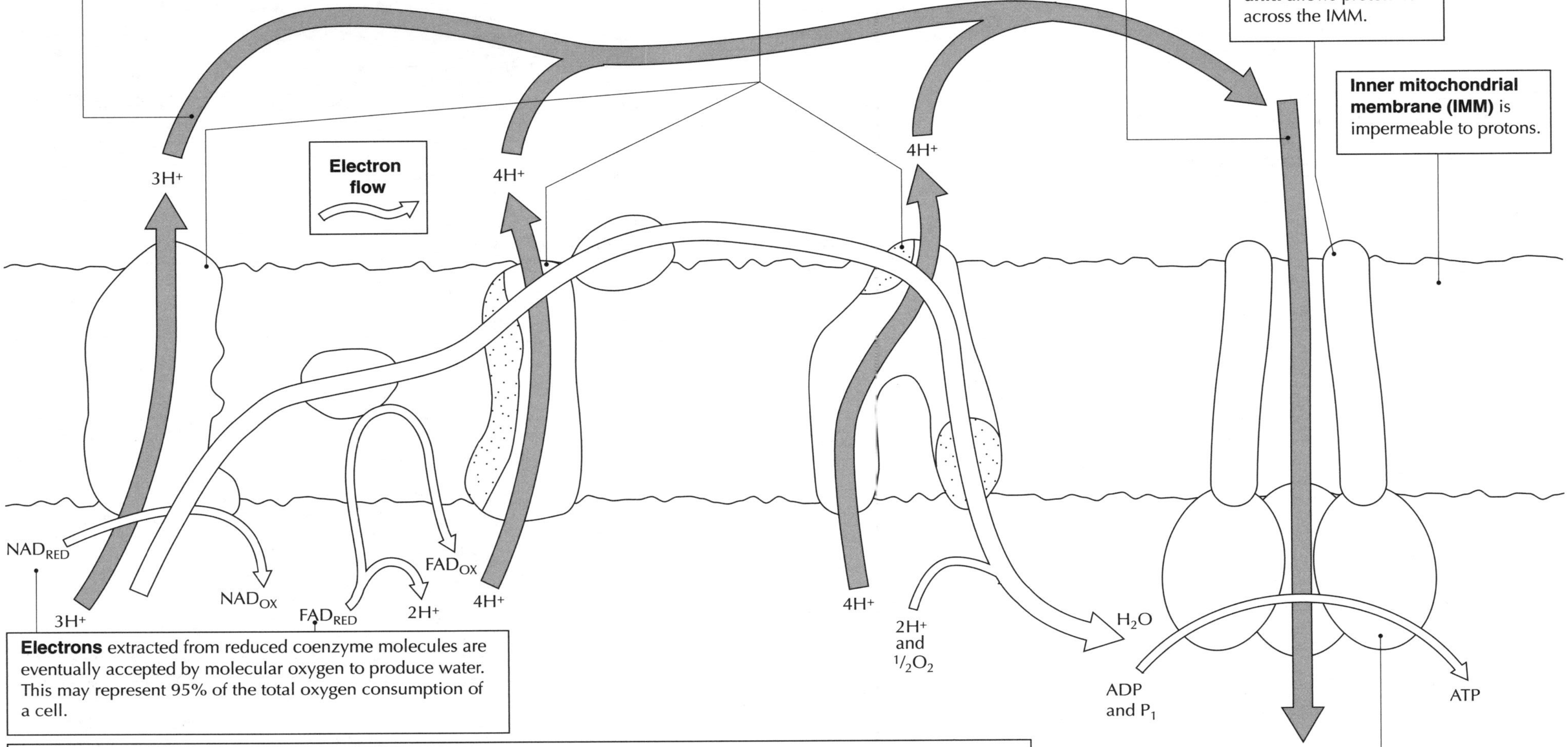

Electrons extracted from reduced coenzyme molecules are eventually accepted by molecular oxygen to produce water. This may represent 95% of the total oxygen consumption of a cell.

ATP synthetase or F_1 unit couples the flow of protons down the proton concentration gradient to the phosphorylation of ADP. It has been suggested that the proton flux through the enzyme alters the nucleotide-binding properties of the active site, promoting ATP synthesis.

Mitchell's chemiosmotic theory suggests that:

1. there is an *exergonic series* of redox reactions between members of the electron transport chain on the inner mitochondrial membrane;
2. these exergonic redox reactions can drive three proton pumps which transport H^+ from the mitochondrial matrix to the inter-membrane space, thus generating a *proton gradient* across the IMM;
3. dissipation of proton gradient can be coupled to the phosphorylation of ADP to ATP.

Energy flow through an ecosystem: I

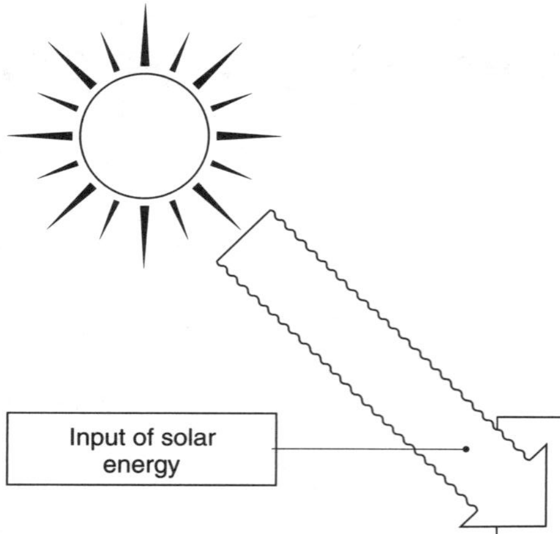

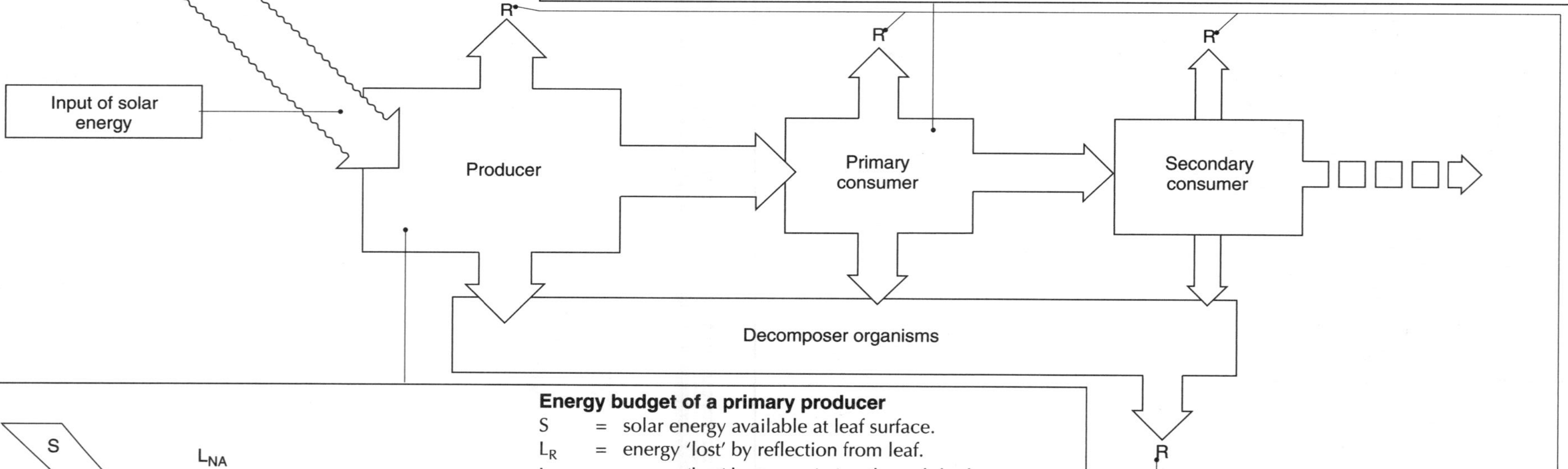

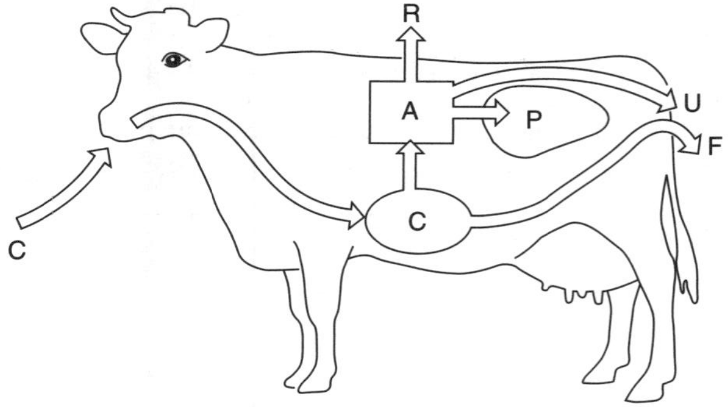

Energy budget of a primary consumer

C = energy consumed. N.B. This does *not* equal NPP of producer since feeding is inefficient.
A = energy assimilated.
R = energy 'consumed' by respiration and eventually 'lost' as heat.
U = energy 'lost' in urine.
F = energy 'lost' in faeces.
P = production of biomass, i.e. energy available for secondary consumer.

Thus
$C = P + R + U + F$

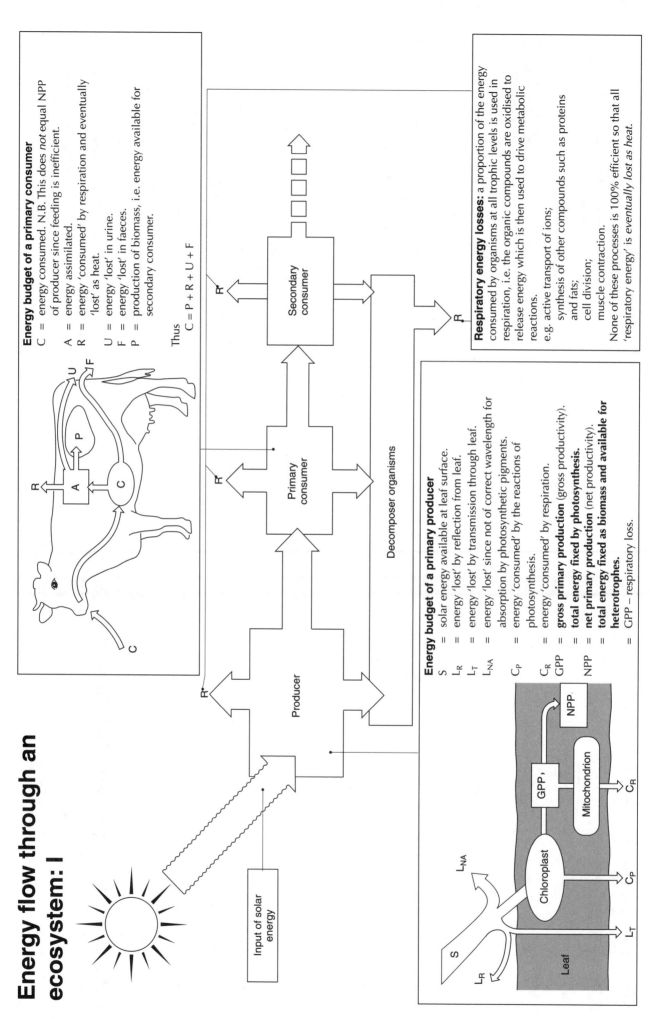

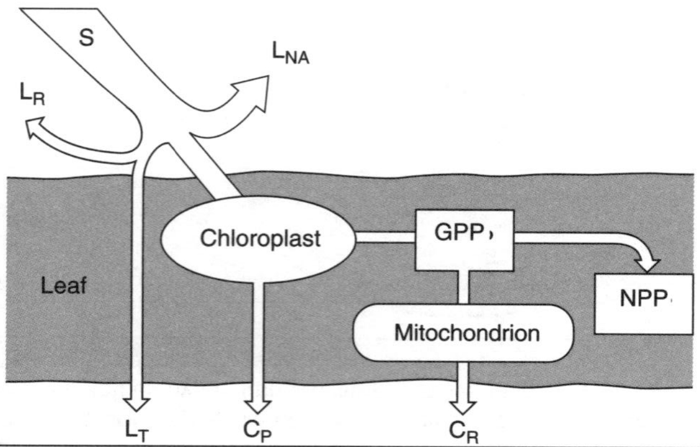

Energy budget of a primary producer

S = solar energy available at leaf surface.
L_R = energy 'lost' by reflection from leaf.
L_T = energy 'lost' by transmission through leaf.
L_{NA} = energy 'lost' since not of correct wavelength for absorption by photosynthetic pigments.
C_P = energy 'consumed' by the reactions of photosynthesis.
C_R = energy 'consumed' by respiration.
GPP = **gross primary production** (gross productivity).
= **total energy fixed by photosynthesis.**
NPP = **net primary production** (net productivity).
= **total energy fixed as biomass and available for heterotrophes.**
= GPP – respiratory loss.

Respiratory energy losses: a proportion of the energy consumed by organisms at all trophic levels is used in respiration, i.e. the organic compounds are oxidised to release energy which is then used to drive metabolic reactions.

e.g. active transport of ions;
synthesis of other compounds such as proteins and fats;
cell division;
muscle contraction.

None of these processes is 100% efficient so that all 'respiratory energy' is *eventually lost as heat.*

Energy flow through an ecosystem: II

A closed ecosystem is rare: migratory animals may deposit faeces, fruits and seeds may enter or leave during dispersal, and leaves may blow in from surrounding trees.

The **gross primary production (GPP), gross productivity,** – the total energy fixed by photosynthesis – represents only about 0.5–1% of the light energy available to the leaf.

At the equator, the **solar flux** (sunlight which reaches the Earth's upper atmosphere) is almost constant. Most of this incoming sunlight energy is reflected by the atmosphere, heats the atmosphere and Earth's surface or causes the evaporation of water. Less than 0.1% actually falls on leaves and is available for photosynthesis.

Energy transfer from producer to primary consumer is typically in the order of 5–10% of NPP.
This is because:

- Much of plant biomass (NPP) is indigestible to herbivores – there are no animal enzymes to digest lignin and cellulose.
- Much of the plant biomass may not be consumed by any individual herbivore species – roots may be inaccessible or trampled grass may be considered uneatable.

Energy transfer from primary consumer (herbivore) to secondary consumer (carnivore) is typically 10–20% of herbivore biomass. This is more efficient than producer → consumer because:

- animal tissue is more digestible than plant tissue;
- animal tissue has a higher energy value;
- carnivores may be extremely specialised for prey consumption;

but is still considerably less than 100% because:

- some animal tissue – bone, hooves and hide for example – is not readily digestible;
- feeding is not 100% efficient – much digestible material (e.g. food fragments and blood) may be lost to the environment.

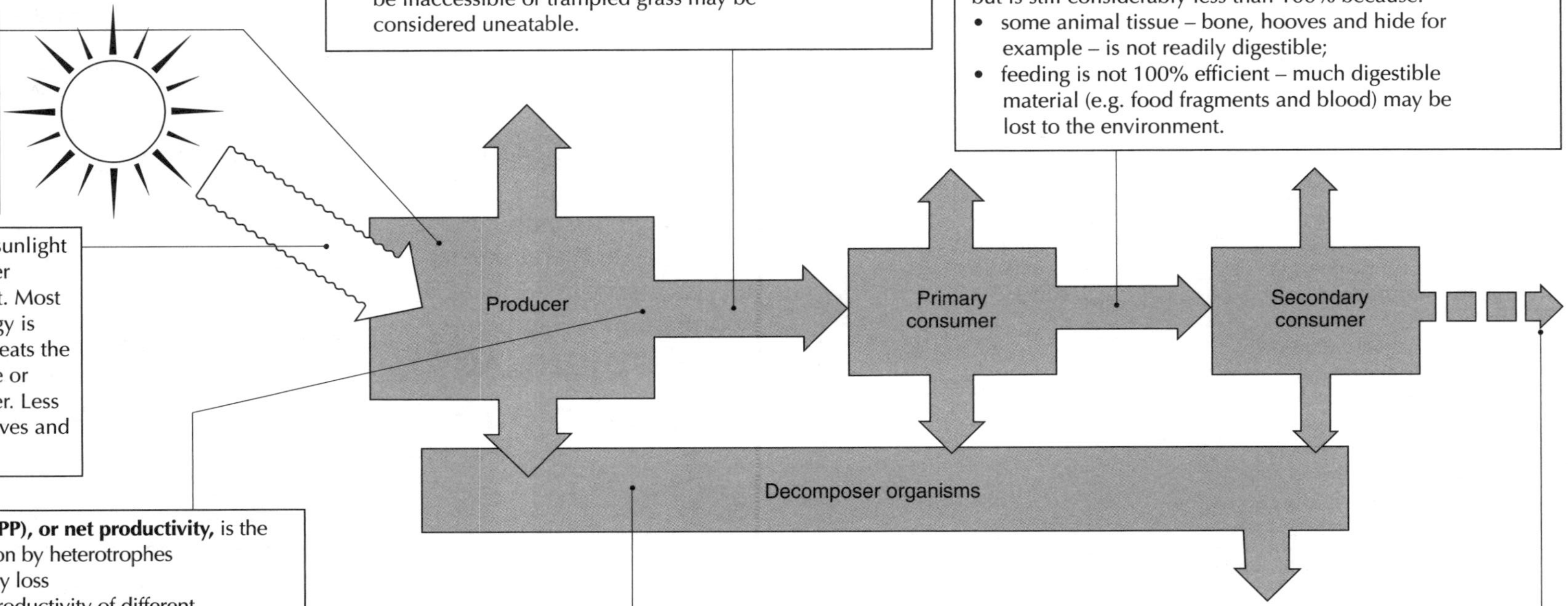

The **net primary production (NPP), or net productivity,** is the energy available for consumption by heterotrophes
= gross productivity – respiratory loss
NPP can be used to compare productivity of different ecosystems.

Modern farming practices increase productivity.

Productivity may be expressed as **units of energy** (e.g. kJ m^{-2} yr^{-1}) or **units of mass** (e.g. kg m^{-2} yr^{-1}).

The **decomposers** are fungi and bacteria which obtain energy and raw materials from animal and plant remains. In some situations 80% or more of the productivity at any trophic level may go through a decomposer pathway (e.g. forest floors of tropical forests). In some ecosystems – peat bogs, for example – the cold, wet, acidic conditions inhibit decomposition to such an extent that only about 10% of the material entering the decomposer food chain is broken down. The remainder accumulates as peat.

The limit to the number of trophic levels is determined by:

- the total producer biomass;
- the efficiency of energy transfer between trophic levels (only 10%).

In practice, the energy losses limit the number of levels to 3 or 4, very rarely 5 or 6. The longest food chains can only be supported by an enormous producer biomass, e.g. a 6 level chain will only have about 10% x 10% x 10% x 10% of NPP available to the top carnivores. The enormous volume of the oceans can provide sufficient biomass to support the longest food chains.

Ecological pyramids provide information about trophic levels in ecosystems. Which pyramid is chosen for examination depends on the type and quantity of data collected.

Pyramid of numbers: a diagrammatic representation of the numbers of different organisms at each trophic level in an ecosystem **at any one time.**

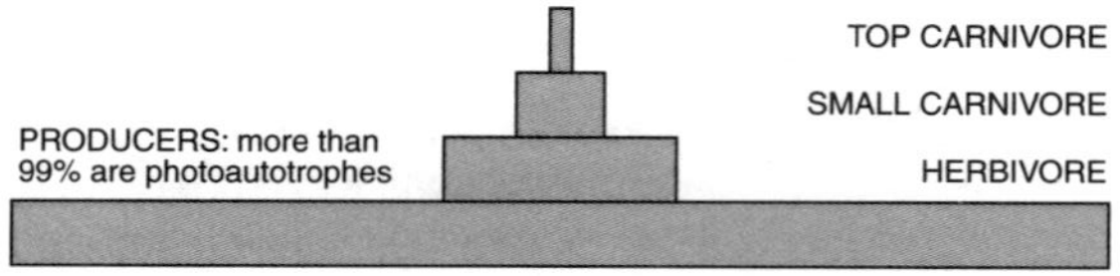

N.B.
- The number of organisms at any trophic level is represented by the length (or the area) of a rectangle.
- Generally, as the pyramid is ascended, the **number** of organisms **decreases,** but the **size** of each individual **increases.**

Problems
- The range of numbers may be enormous – 500 000 grass plants may only support a single top carnivore – so that drawing the pyramid may be very difficult.
- Pyramids may be **inverted,** particularly if the **producer is very large** (e.g. an oak tree) or **parasites feed on the consumers** (e.g. bird lice on owl).

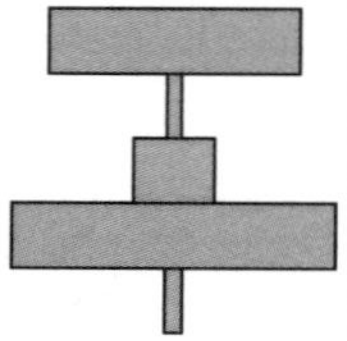

SO WHY NOT USE

Pyramid of biomass ...
... which represents the **biomass** (number of individuals x mass of each individual) at each trophic level **at any one time.**
This *should* eliminate the inversion and scale problems encountered when constructing a pyramid of numbers.

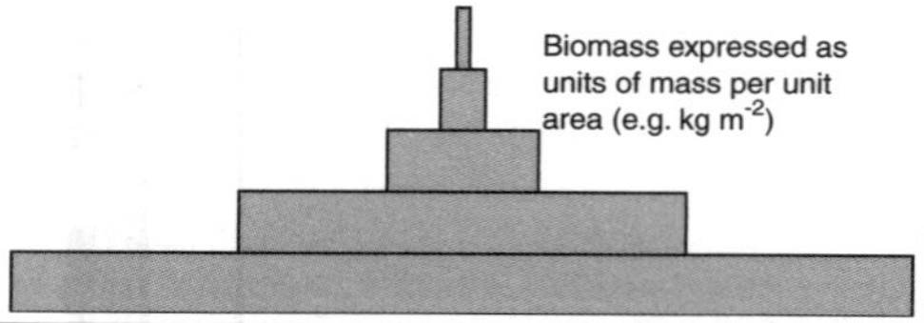

EVEN SO, THERE ARE STILL PROBLEMS!

An inverted pyramid of biomass can occur if the producer level includes organisms **with a high turnover rate** (rapid reproduction) so that they have a high productivity **over a period of time.**

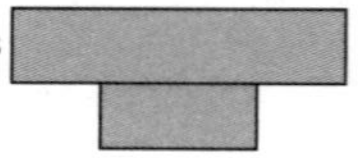

A pyramid of biomass refers to a sampling situation **at any one time** and therefore deals with **standing crop** rather than with **productivity** – the capacity of any trophic level to produce biomass over a period of time, e.g. zooplankton in a marine ecosystem.

SO IT MIGHT BE BETTER TO CONSIDER

Pyramid of energy, which represents the flow of energy through each trophic level of an ecosystem **during a fixed time period** (usually one year, to account for seasonal effects). The energy values are expressed in kJ m^{-2} yr^{-1}.

Advantages
- It compares **productivity** because a time factor is incorporated.
- Biomass may not be equivalent to energy value, e.g. 1 g of fat has many more kJ than 1 g of cellulose or lignin.
- The solar input of energy may be included as an extra rectangle at the base.
- No inverted pyramids are obtained.

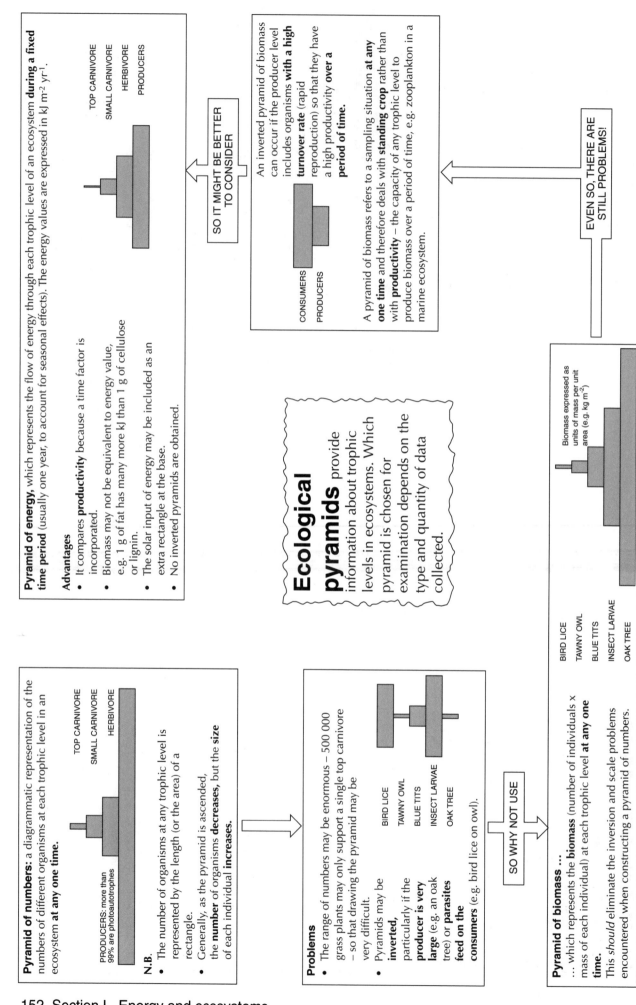

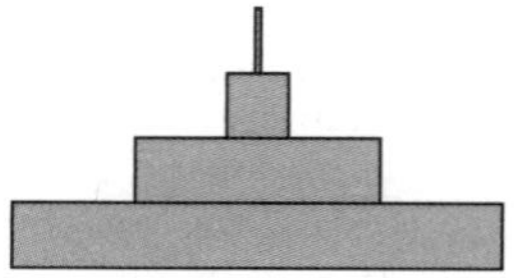

Biological pest control reduces the population of one species to levels at which it is no longer a pest by the use of one of the pest species' natural predators.

Ideal relationship between pest and its control agent

The **pest species** becomes the **prey** of the control agent: it is the **target** in the system of biological control.

Pest population falls due to **predation** by control agent.

Population size above which the pest is **economically harmful:** often determined by the expected yield and potential value of the crop.

Population of control agent falls because of a food shortage caused by reduction in prey (pest) numbers.

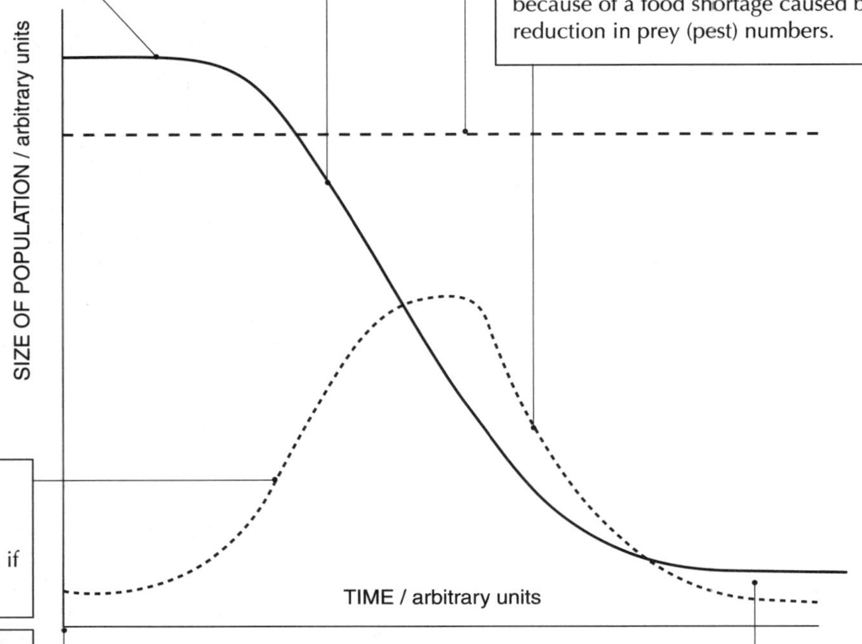

Control agent selected according to the criteria outlined above is the **predator** on the prey **pest** species. Population rises as the agent breeds, if conditions are appropriate.

Introduction of control agent: the size of the introduced population must be great enough to ensure a rise in numbers which is rapid enough to ensure control of the pest population within an economic time, e.g. before a crop plant has been extensively damaged.

A **dynamic equilibrium** is set up in which a moderate residual population of the control agent is able to permanently restrict the population of the pest. N.B. the pest species **must not be entirely eliminated** or the control agent will die out and a further introduction will be necessary to prevent re-establishment of economically damaging pest populations.

Typical biological control programme

1. Identify the pest and **trace its origins,** i.e. where did it come from?
2. Investigate original site of pest and **identify natural enemies** of the pest.
3. Test the potential **control agent** under careful quarantine to ensure that it:
   a. is **specific** (does not prey on other species)
   b. will not change its prey species and **become a pest itself**
   c. has a life cycle which will allow it to **develop a population large enough to act as an economic control.**
4. **Mass culture** of the control agent.
5. Development of the most **effective distribution/release method** for the control agent.

Principal techniques in biological control

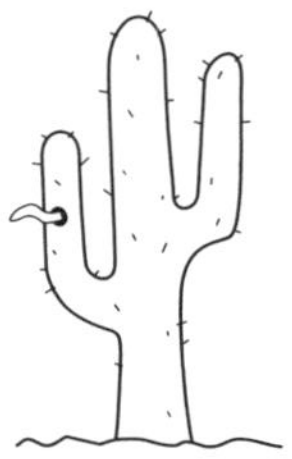

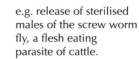

1. Use a **herbivore** to control a **weed** species,
   e.g. *Cactoblastis* larvae on prickly pear.
2. Use a **carnivore** to control a **herbivorous pest,**
   e.g. hoverfly larvae on aphids.
3. Use a **parasite** to control its **host,**
   e.g. *Encharsia,* a parasitic wasp, on the greenhouse whitefly, *Trialeurodes vaporariorum.*
4. Disrupt the **breeding cycle** of a pest **if it mates once only in its life,**

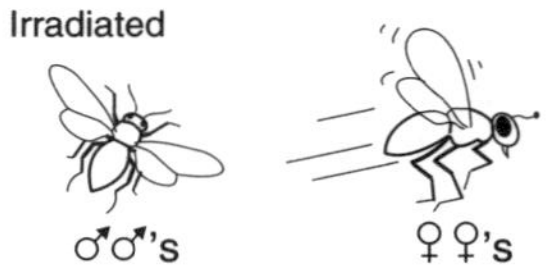

e.g. release of sterilised males of the screw worm fly, a flesh eating parasite of cattle.

5. Control of **pest behaviour,** e.g. sex attractant pheromones are used to attract apple codling moths into lethal traps.

Chemical pest control may involve the use of:

herbicides – for control of weeds;
insecticides – for control of insects;
fungicides – for control of fungi;
molluscicides – for control of slugs and snails.

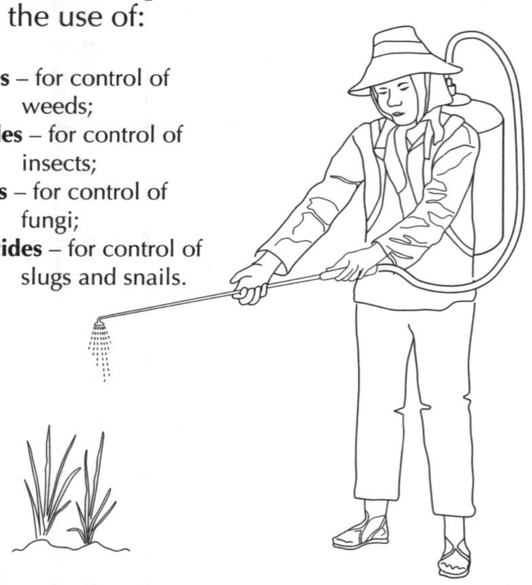

Herbicides may be:

1. **Pre-emergent,** i.e. applied **before** emergence of crop.
   a. Contact herbicides, e.g. **Paraquat,** which kill all above-ground parts of all plants.
   b. Residual herbicides, e.g. **Linuron,** which bind to soil particles and kill weed seedlings as they emerge.

Pre-emergent herbicides can be **non-selective** and are ideal for clearing ground prior to cultivation.

2. **Post-emergent** is applied to both crop and weed, and therefore must be **selective.** Many, such as 2,4-D, are growth regulators.

Systemic herbicides, such as **glyphosate,** are absorbed by weeds and translocated to the meristems where they typically act by inhibition of cell division.

Problems with insecticides: these arise since the principal idea behind chemical control is to **kill as many of the pests as possible** – the effects on harmless or beneficial organisms were not studied or were ignored.

1. **Direct killing:** accidental misuse of toxic chemicals may cause death in humans or in domestic animals.
2. **Non-specificity:** non-target species, particularly natural predators of the pest species, may be killed by some wide-spectrum insecticides, e.g. large doses of **dieldrin** killed many birds as well as the Japanese beetle pest which was the intended target organism.
3. **Pest resistance:** genetic variation means that each pest population contains a *few* resistant individuals. The pesticide eliminates the non-resistant forms and thus a resistant population is selected for and may quickly develop (since many pests reproduce rapidly).
4. **Pest replacement:** most crops are susceptible to attack by more than one species – a **pest complex** and the use of a pesticide to eliminate one species may simply allow another species to assume major pest proportions (since a pesticide may be more deadly to one species than another).
5. **Pest resurgence:** non-specific pesticides may kill natural predators as well as pests – a small residual pest population may now multiply without check, creating a worse problem than initially was present.
6. **Bioaccumulation of toxins:** pesticides or their products may be toxic.
   - They may seriously affect micro-organisms and thus alter decomposition in soils;
   - They may pass along food chains, becoming more concentrated in organisms further up the chain.

e.g. DDT used as an insecticide accumulates in the fatty tissues of carnivorous animals, inhibiting cytochrome oxidase and limiting reproductive success (especially thin eggshells in birds of prey).

Phytoplankton 1 → Mayfly larvae 4 → Trout 50 → Osprey 800

Relative DDT concentration along an aquatic food chain.

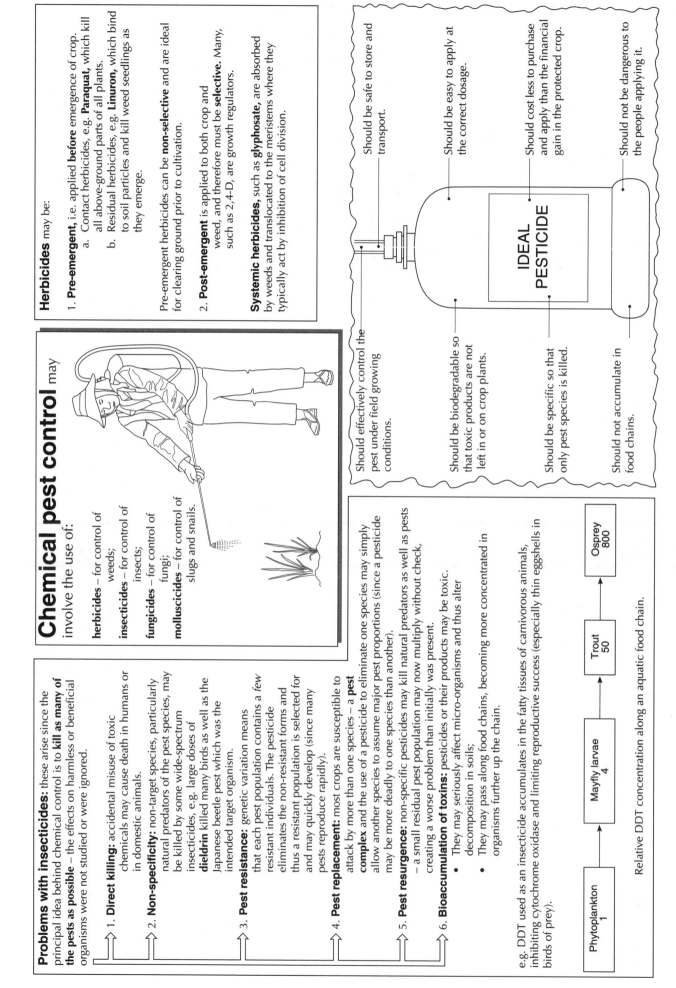

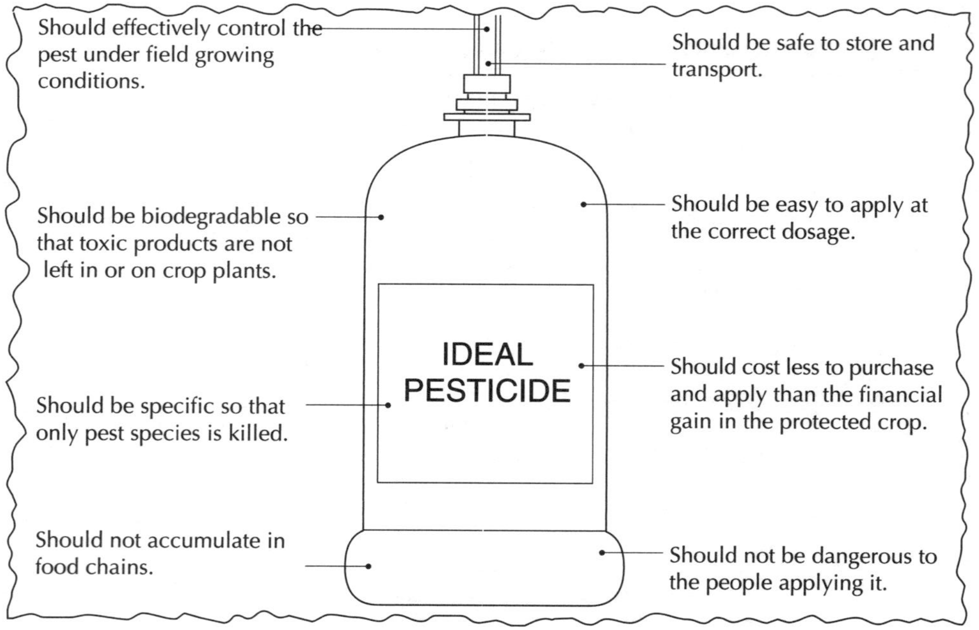
Should effectively control the pest under field growing conditions.
Should be safe to store and transport.
Should be biodegradable so that toxic products are not left in or on crop plants.
Should be easy to apply at the correct dosage.
IDEAL PESTICIDE
Should cost less to purchase and apply than the financial gain in the protected crop.
Should be specific so that only pest species is killed.
Should not accumulate in food chains.
Should not be dangerous to the people applying it.

Pesticides and pest control

Pesticides are important in crop protection

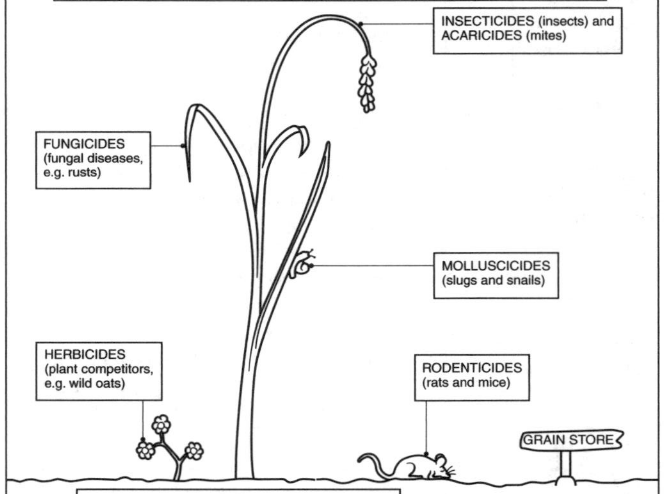

… and for wildlife!

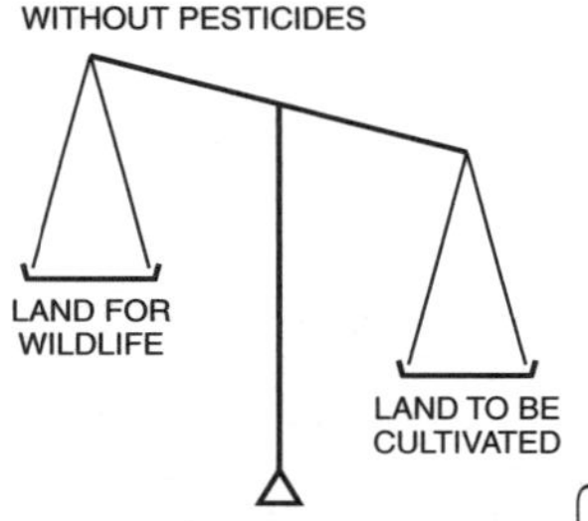

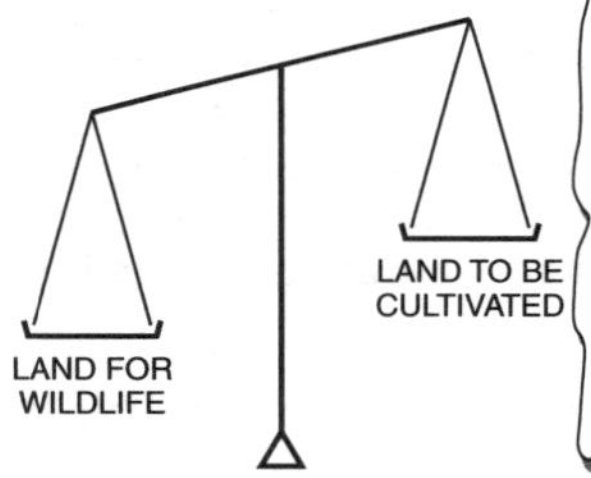

- There is an increasing demand for food.
- Pesticides increase yield.
- Demand can be satisfied by smaller land area.

MORE LAND FOR WILDLIFE!

As with all conservation issues, there must be a **compromise**.

Replanting of forests is good for wildlife – but uses herbicides and insecticides to allow trees to become established!

Pest: Any organism that competes with humans, or with domestic animals and plants.

Techniques of integrated pest management (IPM)

- **Encouraging predators**
  In the garden, flowers growing near vegetables can encourage hover flies, some of which are predators of aphids. However, very little work has been done to determine what the best flowers are and how effective the predation is. We do not know whether increasing hover fly or other predator numbers will result in better control of aphids or simply result in increased predation of the predators. We do not know where the equilibrium position lies. Work is under way at a number of research institutes to try to answer these questions.

- **Resistant varieties**
  Plant breeders are always trying to breed resistance to various plant diseases into plants. Canker *(Phoma lingam)* of oilseed rape has reduced greatly in importance over the last 10 years due to new resistant varieties. Resistance to insects also occurs in the wild but has been much more difficult to use. A potato variety bred a few years ago was resistant to aphids but after years of development it had to be scrapped when it was found to contain toxic levels (to humans) of solanine (an alkaloid). Genetic engineering can now allow the gene for the insecticidal toxin produced by *Bacillus thuringiensis* to be transferred to and expressed in cotton plants. Whether such developments are good or bad needs to be evaluated carefully.

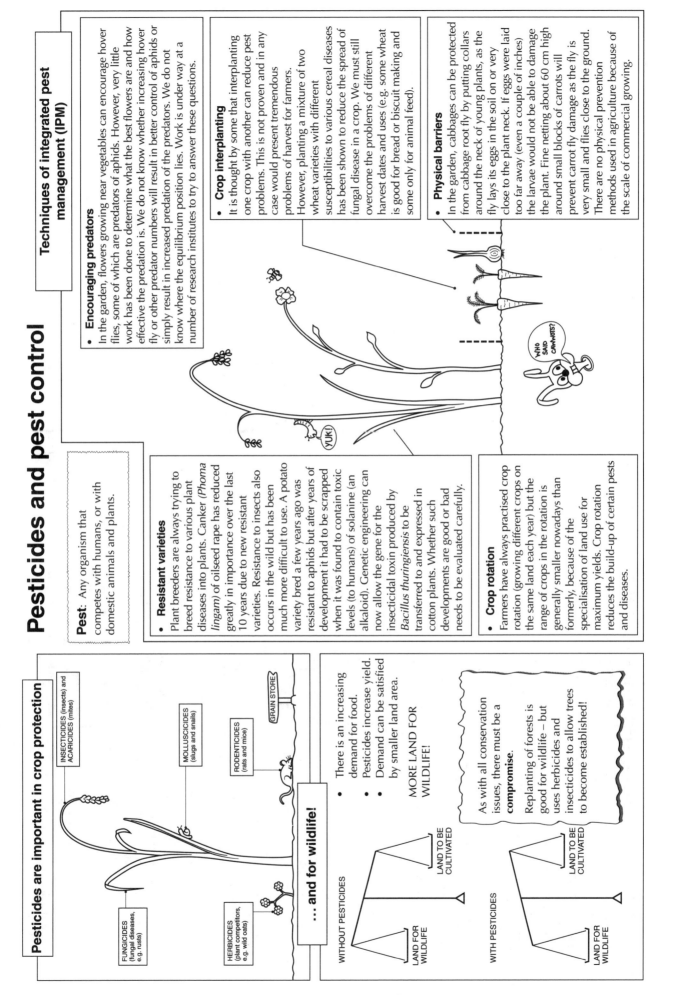

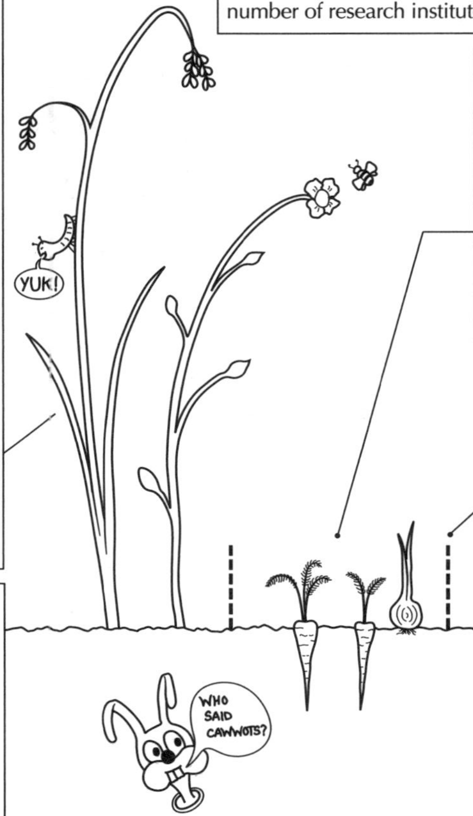

- **Crop interplanting**
  It is thought by some that interplanting one crop with another can reduce pest problems. This is not proven and in any case would present tremendous problems of harvest for farmers. However, planting a mixture of two wheat varieties with different susceptibilities to various cereal diseases has been shown to reduce the spread of fungal disease in a crop. We must still overcome the problems of different harvest dates and uses (e.g. some wheat is good for bread or biscuit making and some only for animal feed).

- **Crop rotation**
  Farmers have always practised crop rotation (growing different crops on the same land each year) but the range of crops in the rotation is generally smaller nowadays than formerly, because of the specialisation of land use for maximum yields. Crop rotation reduces the build-up of certain pests and diseases.

- **Physical barriers**
  In the garden, cabbages can be protected from cabbage root fly by putting collars around the neck of young plants, as the fly lays its eggs in the soil on or very close to the plant neck. If eggs were laid too far away (even a couple of inches) the larvae would not be able to damage the plant. Fine netting about 60 cm high around small blocks of carrots will prevent carrot fly damage as the fly is very small and flies close to the ground. There are no physical prevention methods used in agriculture because of the scale of commercial growing.

Managing ecosystems: animal husbandry

Intensive farming aims to minimise cost/maximise productivity by reducing environmental resistance.

Temperature control: essential so that costly heat energy is not wasted.
If **too high,** animals are uncomfortable and will not feed.
If **too low,** food intake is 'wasted' on heat production to maintain body temperature.

Controlled light regime/photoperiod: may influence growth rate since can permit longer feeding period; may control reproductive cycle so, for example, milk production may be stabilised through the year.

Shelter:
Prevent entry of **predators**
Eliminate/control **competitors** for food
Protect against **climatic extremes.**

Food input:
Control content
- high protein for growth
- minimal fat to suit customer demand for lean meat
- include growth hormone to increase growth rate
- add copper ions which reduce energy consumption for heat production.

Often use dried milk power or single cell protein, with mineral and vitamin supplements.

Strain of animal: selective breeding for animals with high **conversion ratio** i.e. most efficient transfer of food intake to body mass.
Minimise movement: less energy consumption and thus more efficient 'conversion'.
NB can upset **social interactions,** and stress can lead to poor growth/unpleasant 'cortisone' taste.

Hygienic conditions: most animals are free of gut parasites and debilitating bacteria – healthy animals grow more quickly and meat is more saleable.
Slurry (faeces/urine) can be dried and recycled for use as fertiliser.

Growth curve: may indicate optimum time for marketing animals

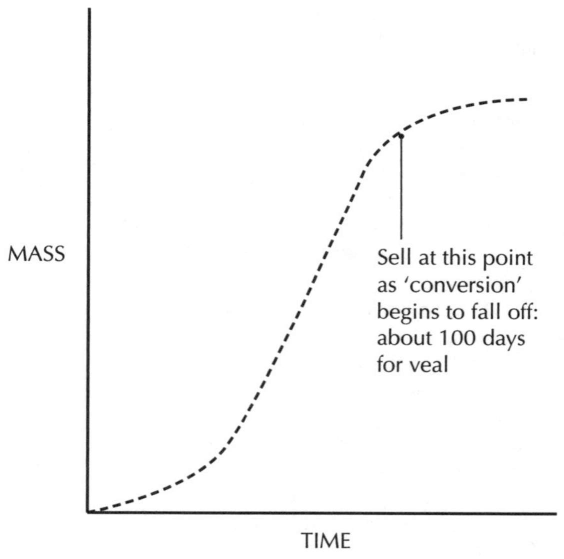
MASS
TIME
Sell at this point as 'conversion' begins to fall off: about 100 days for veal

Veterinary care:
Antibiotics to reduce bacterial infections
Vaccination to minimise viral infections
Hormone/vitamin supplements can be administered more accurately than in the diet
Artificial insemination techniques can reduce costs (no need to keep bulls in dairy farms).

Nitrates are significant pollutants of water.

Effects on human health

1. In the stomach

   NO_3^- ⟶ NITROSAMINES

   Nitrosamines are highly carcinogenic, and some studies have linked high $[NO_3^-]$ in water supplies with increased incidence of stomach and oesophageal cancer.
2. **Blue baby syndrome** (in children younger than 3 months)

   NO_3^- —(bacteria in gut or water supply)⟶ NITRITE (NO_2^-)

   Haemoglobin in babys red blood cells

   methaemoglobin (has Fe^{II} oxidised to Fe^{III}) which reduces oxygen-carrying capacity of baby's blood → **'blue' appearance.**

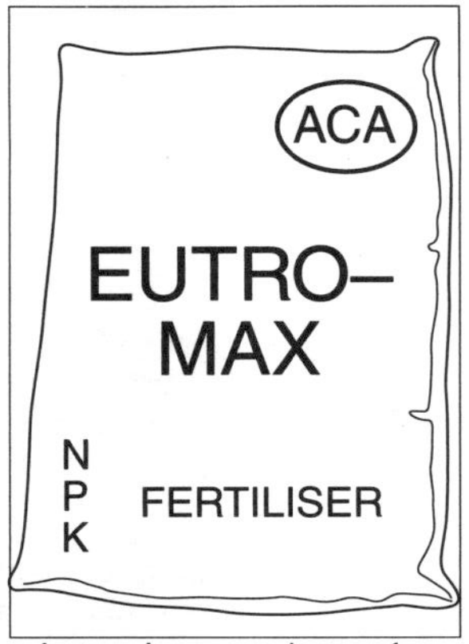

The EU has set a limit of 11.3 p.p.m. total nitrogen in drinking water – this is exceeded in some parts of East Anglia and Cleveland.

There's also a problem with pesticides!

Bioaccumulation of toxins: pesticides or their products may be toxic:

- they may seriously affect micro-organisms and thus alter decomposition in soils;
- they may pass along food chains, becoming more concentrated in organisms further up the chain.

e.g. DDT used as an insecticide accumulates in the fatty tissues of carnivorous animals, inhibiting cytochrome oxidase and limiting reproductive success (especially thin eggshells in birds of prey).

Phytoplankton 1 → Mayfly larvae 4 → Trout 50 → Osprey 800

Relative DDT concentration along an aquatic food chain.

Don't mix up eutrophication (fertilisers) and bioaccumulation (pesticides)!

Eutrophication – nutrient enrichment of ponds, lakes and rivers – is responsible for **biological oxygen demand (BOD)**

BOD is the mass of oxygen consumed by micro-organisms in a sample of water determined by measuring oxygen concentration with an oxygen electrode **before and after** a period of microbial respiration: indicates the oxygen **not available** to more advanced organisms.

Input of **raw sewage**

Leaching of **inorganic fertilisers** from farmland

Increased concentration of **nitrate** and **phosphate** in bodies of water

Algae and green protists use nutrients to multiply rapidly = **algal bloom.**

Reduction of light for bottom-growing plants

DIE

DIE

Large quantities of **organic material.**

Aerobic decomposers (mainly bacteria) multiply and **consume oxygen.**

Aerobic organisms (fish and invertebrates) die of **lack of oxygen.**

POSITIVE FEEDBACK

* N.B. The leaching of **phosphates** into ponds and rivers is at least as important as nitrates in causing eutrophication.

The carbon cycle depends upon both biochemical and physical processes.

Atmospheric carbon dioxide represents the most accessible source of carbon: although only about 0.03% of the atmosphere is CO_2 this represents about 700 x 10^{12} kg (= 700 000 million tonnes).

Respiration (most commonly glucose + oxygen → carbon dioxide + water): this is the most important process for returning CO_2 to the atmosphere.

Carbon fixation involves the reduction of carbon dioxide to large organic molecules.
Simply:
$CO_2 + 4H \rightarrow CH_2O + H_2O$
It is carried out by **photosynthesis** (99%) and by **chemosynthesis** (1%).
The amount of carbon fixed into organic molecules is enormous: more than 50% is in soil organisms and more than 30% in long-lived trees.

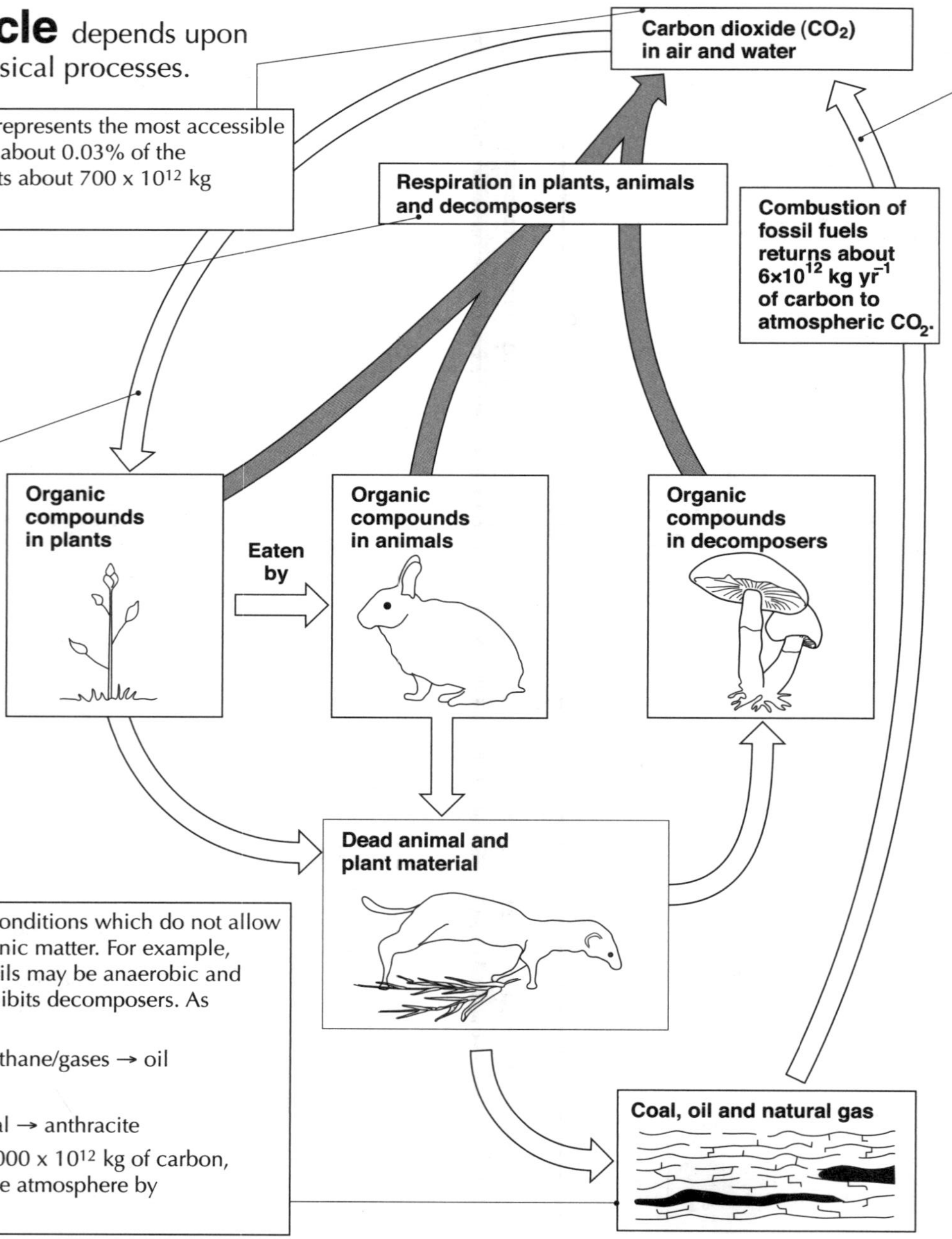

Carbon dioxide (CO_2) in air and water
Respiration in plants, animals and decomposers
Combustion of fossil fuels returns about 6x10^{12} kg yr^{-1} of carbon to atmospheric CO_2.
Organic compounds in plants
Eaten by
Organic compounds in animals
Organic compounds in decomposers
Dead animal and plant material
Coal, oil and natural gas

This is the most rapidly-changing factor which is affecting the concentration of CO_2 in the atmosphere.

Fossil fuels are formed under conditions which do not allow oxidation/decomposition of organic matter. For example, deep oceans and waterlogged soils may be anaerobic and increasing acidity (fall in pH) inhibits decomposers. As pressure increases

plant debris → peat ↗ methane/gases → oil
↘ coal → anthracite

Fossil fuels may 'lock up' 3 000 000 x 10^{12} kg of carbon, which can only be returned to the atmosphere by combustion.

Saprobiotic organisms are **decomposers,** and obtain their nutrients from **dead** bodies or organic wastes of animals and plants – they are **primary consumers** in **detritus food chains**.

e.g. *Rhizopus* (Pin mold)

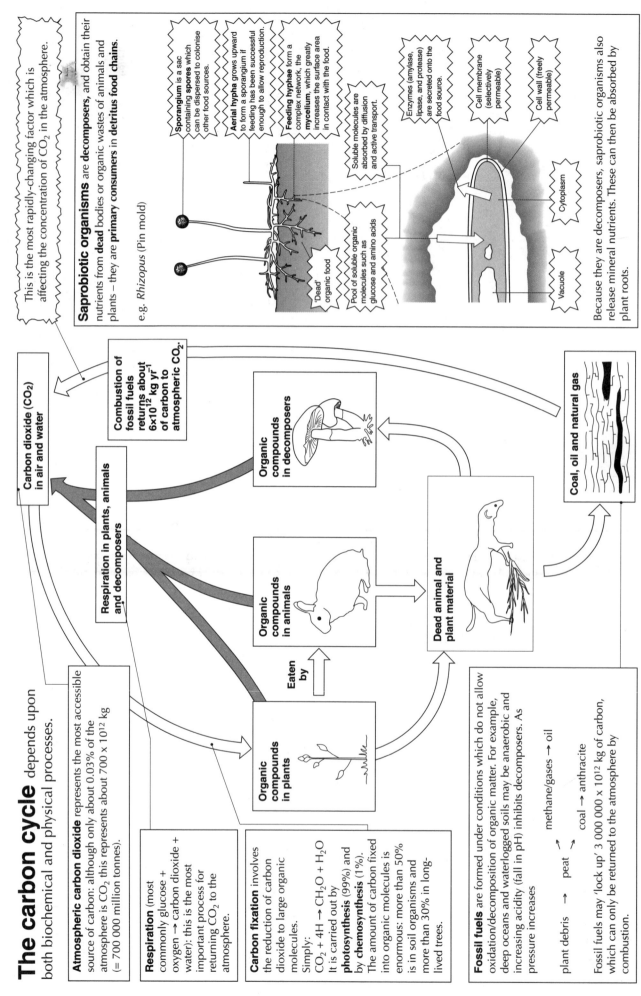

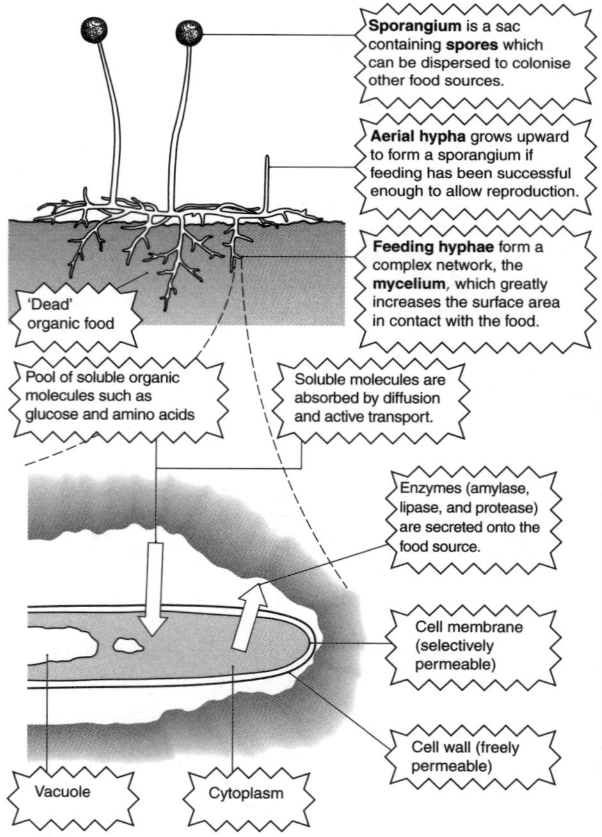

Sporangium is a sac containing spores which can be dispersed to colonise other food sources.
Aerial hypha grows upward to form a sporangium if feeding has been successful enough to allow reproduction.
Feeding hyphae form a complex network, the mycelium, which greatly increases the surface area in contact with the food.
'Dead' organic food
Pool of soluble organic molecules such as glucose and amino acids
Soluble molecules are absorbed by diffusion and active transport.
Enzymes (amylase, lipase, and protease) are secreted onto the food source.
Cell membrane (selectively permeable)
Cell wall (freely permeable)
Vacuole
Cytoplasm

Because they are decomposers, saprobiotic organisms also release mineral nutrients. These can then be absorbed by plant roots.

The nitrogen cycle depends on micro-organisms.

Nitrate reduction: occurs in two stages, each catalysed by a different enzyme and occurring in different parts of the cell:

$NO_3^- \xrightarrow[\text{Nitrate reductase in cytoplasm}]{NADH_2 \rightarrow NAD} NO_2^- \xrightarrow[\text{Nitrite reductase in plastids}]{\text{ferrodoxin-}H_2 \rightarrow \text{ferrodoxin}} NH_4^+$

Denitrification: represents a loss of 'available' nitrogen from the ecosystem, and may occur under **anaerobic conditions** such as might occur in a waterlogged soil.
Examples:

$2NO_3^- \longrightarrow N_2 + 3O_2$ (this O_2 required to respire carbohydrate)

Lightning may fix atmospheric nitrogen. High-energy discharge may combine nitrogen and oxygen, and the products may dissolve in rainwater and fall to the ground as weakly acidic solutions.

$N_2 + 2O_2 \longrightarrow 2NO_2 \xrightarrow{H_2O} 2HNO_3$

$N_2 + O_2 \longrightarrow 2NO \xrightarrow{H_2O} 2HNO_2$

Nitrogen fixation: this process reduces nitrogen from the rhizosphere (soil atmosphere) to ammonium ions. Basic process:
There are some free-living nitrogen-fixing bacteria, but the vast majority of nitrogen fixation is carried out by mutualistic bacteria.

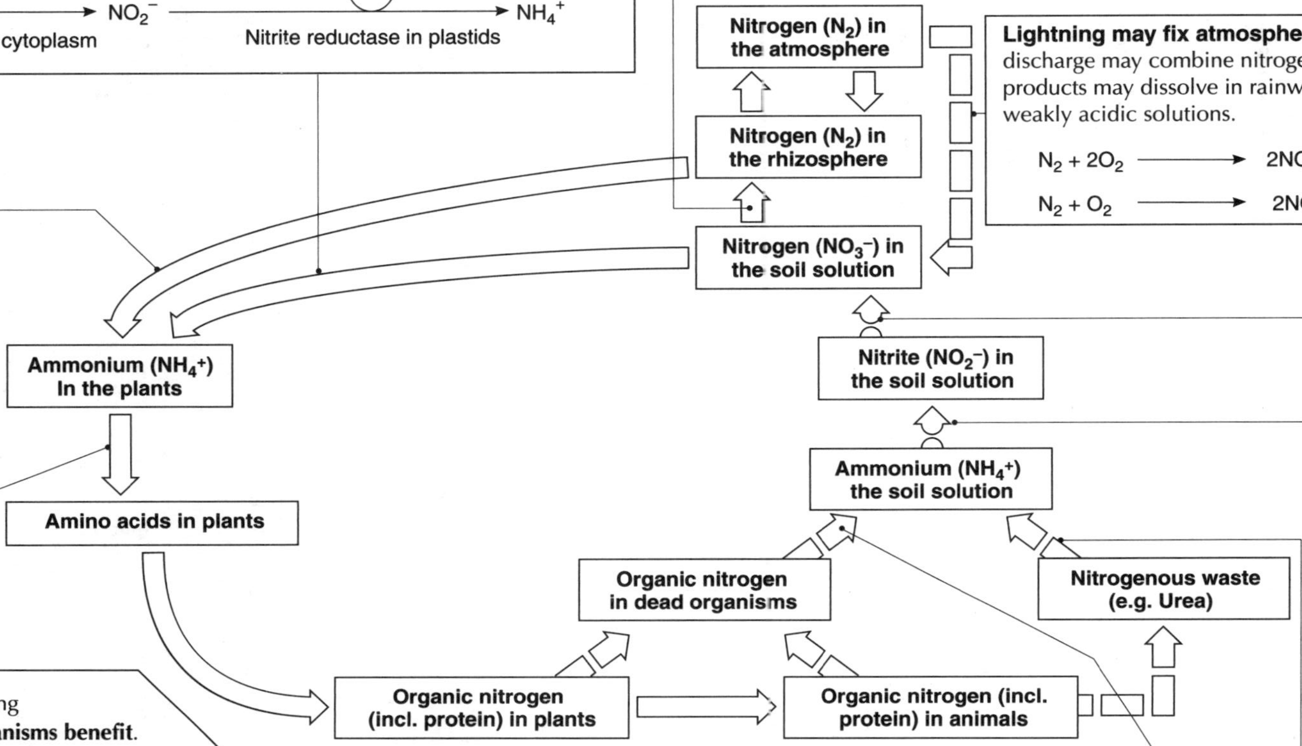

Nitrification: a sequence of **oxidations** which convert ammonium compounds to nitrate and release energy for the chemoautotrophic bacteria which carry them out.

e.g.

$NH_4^+ \xrightarrow{Nitrosomonas} NO_2^- + \text{energy}$

$NO_2^- \xrightarrow{Nitrobacter} NO_3^- + \text{energy}$

These reactions offer nitrogen as *nitrate*, the form in which it is most easily absorbed by plant roots.

Putrefaction/decomposition:
- carried out by saprobiotic bacteria and fungi
- releases NH_4^+ from protein, amino acids and urea.

Mutualism describes a feeding relationship **in which both organisms benefit**.

e.g. the nitrogen-fixing bacterium *Rhizobium* gains carbohydrates from phloem in the root of plants of the pea family.
In return, the plant receives excess amino acids. These plants can do well in nitrogen-deficient soils as a result of this relationship.

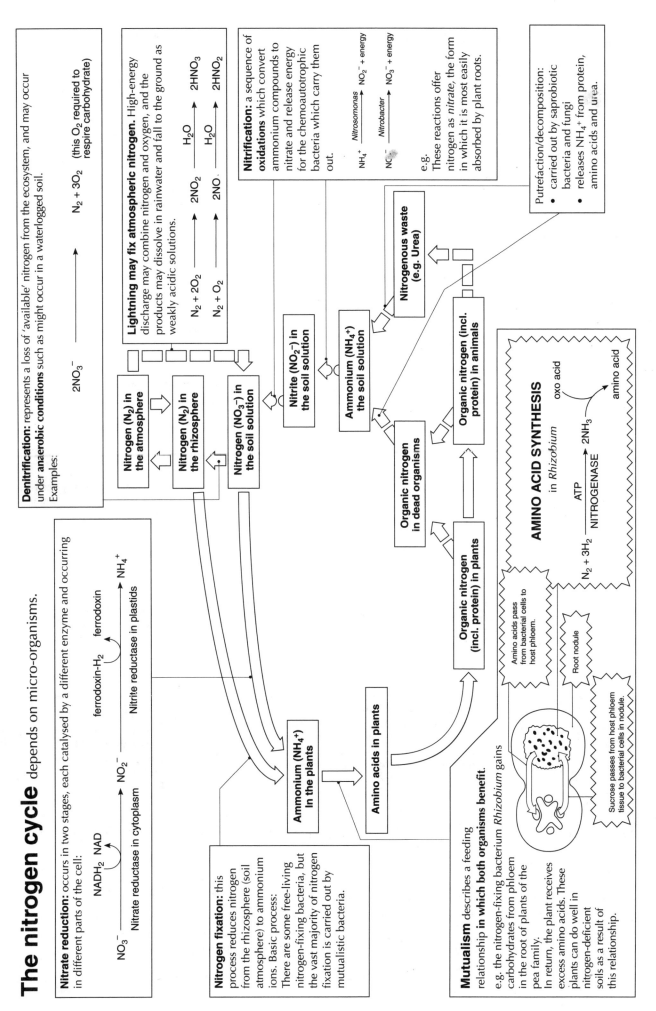

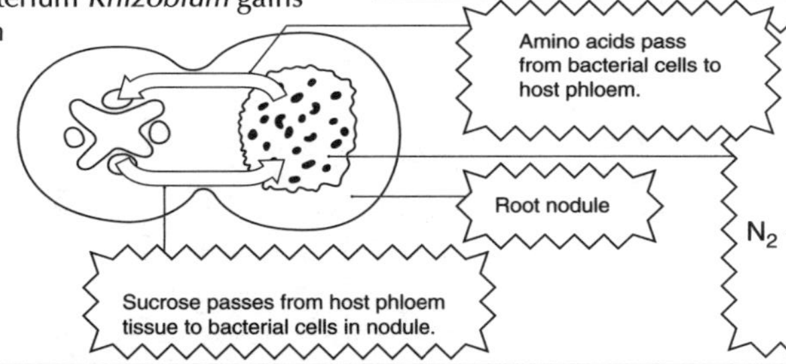

AMINO ACID SYNTHESIS
in *Rhizobium*

$N_2 + 3H_2 \xrightarrow[\text{NITROGENASE}]{ATP} 2NH_3$

$2NH_3$ + oxo acid → amino acid

Ecological succession proceeds via **several stages** to a **climax community** and is characterised by:

1. an increase in **species diversity** and in **complexity of feeding relationships;**
2. a progressive increase in **biomass;**
3. completion when **energy input (community photosynthesis) = energy loss (community respiration).**

A community (all the species present in a given locality at any given time), is the group of interacting populations (all the members of a species in a place at a given time) which represents the biotic component of an ecosystem, and is seldom static. The relative abundance of different species may change, new species may enter the community and others may leave. There are reasons for these changes.

Catastrophes: may be natural (e.g. flooding, volcanic eruption) or caused by people (e.g. oil spill, deforestation).

Seasons: changes in temperature, rainfall, light intensity and windspeed, for example, may alter the suitability of a habitat for particular species.

Succession: long-term directional change in the composition of a community **brought about by the actions of the organisms themselves.**

CAN BE EITHER

Primary succession occurs when the community develops on bare, uncolonised ground **which has never had any vegetation growing on it ,** e.g. mud in river deltas, lava flows, sand dunes, artificial ponds and newly erupted volcanic islands.

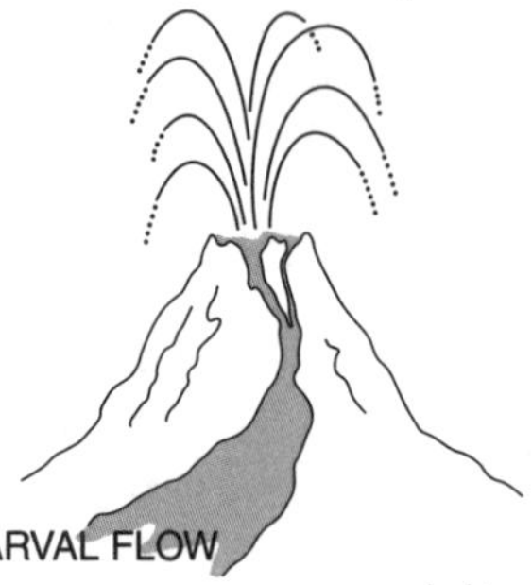

OR

Secondary succession occurs on ground **which had previously been colonised** but is now available because the community has been destroyed, typically by fire, flood or as a result of human agricultural or industrial activities. Such ground will not be 'virgin' but will include remnants of soil, organic debris, seeds and even resistant animals and plants which have survived the changes, e.g. fire debris may be rich in minerals, particuarly phosphate.

... BUT ALWAYS PROCEEDS VIA A SERIES OF STAGES

BARE GROUND ⟹ PIONEER COMMUNITY (COLONISERS) ⟹ SECONDARY COMMUNITY ⟹ CLIMAX COMMUNITY

Migration: the arrival of seeds and spores. If conditions are suitable immigrant species may become established.

These species are simple plants, e.g. lichens and algae with minimal environmental demands. May show symbiotic relationships to aid their establishment. The community is **open,** i.e. space for further colonisers.

The number of species has risen – further stabilising soil and adding nutrients. May still be an input of new species so that there will be both **intra-** and **inter-specific competition.** Pioneer species are often poor competitors and will be replaced by higher, more demanding plants such as grasses, shrubs and, eventually, trees.

The end point of succession: the community is now in equilibrium with the environment and is stable. Composition is often determined by one dominant species e.g. **oak woodland.**

During succession each species modifies the environment, making it **more** suitable for new species and **less** suitable for those already there.

Each of these stages is called a **seral stage** and the complete succession is called a **sere.**

Hydrosere: succession in an aquatic environment.
Xerosere: succession on dry land.
Halosere: succession in a salty environment.
Lithosere: succession on a rocky surface.

A xerosere illustrates that animal succession follows plant succession.

First colonisers of bare rock are mosses and lichens which can adhere directly to rock surfaces - they slowly (may be thousands of years) allow the accumalation of soil in crevices.

Climax community is dominated by oak which shades out the birch. Once woodland is established the shady but relatively humid environment may encourage the establishment of more tolerant species such as lime, holly, avens and bluebells.

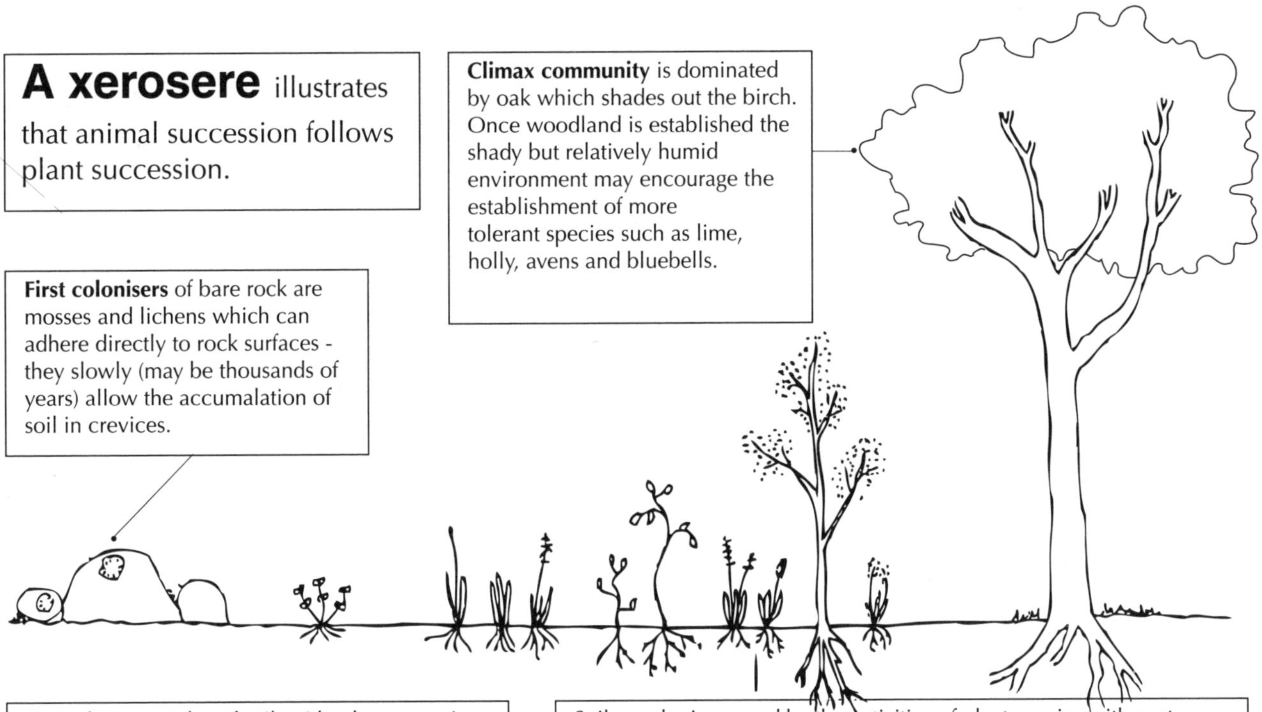

Once finer particles of soil, with a better nutrient and water balance are available, undemanding dicotyledonous species followed by the more demanding grass species begin to form **open grassland community.**

Soil may be improved by the activities of plant species with root nodules. Grassland is invaded by scrub (especially once roosting birds are available to disperse seeds and deposit nutrient rich droppings) – common mid-successional species are hawthorn, birch and eventually oak. Each of these species requires high light intensities to become established.

Unspecialised **herbivore**s, **omnivores** and **detritivores** such as nematodes, earthworms, woodlice and snails join the spiders and mites. Field mice and voles are often the first vertebrates.

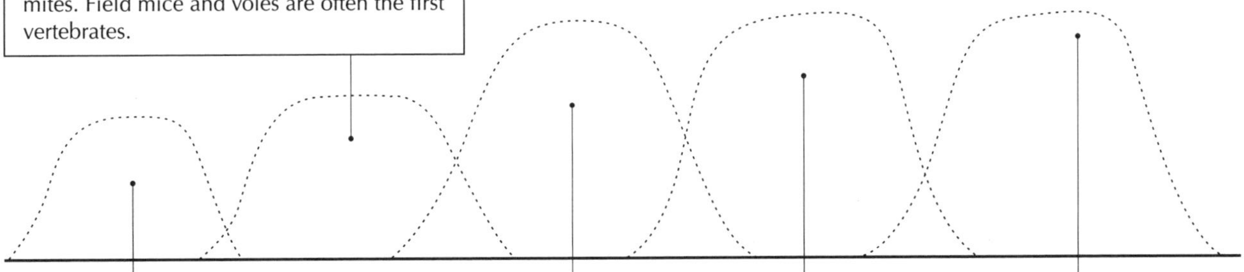

First animal colinisers are usually **carnivores** such as spiders which scavenge on insects which have been blown into the rocky environment. Animal remains help the formation of soil cover.

Typical **bird** of grassland is the skylark - can obtain both **food** and **nesting sites** in the grassland. Large vertebrates tend to be visitors foraging for food. Rabbits may become established if burrowing sites available.

Typical scrubland birds are warblers which nest low down but feed on high trees. Extensive mouse and vole populations encourage visiting predators such as hen harrier and short-eared owl.

Mature woodland has a complex community of many resident animal species: one good indicator species is the chaffinch which both nests and feeds in the tall trees. Predators such as fox, sparrowhawk and tawny owl may now be resident.

Please note that...

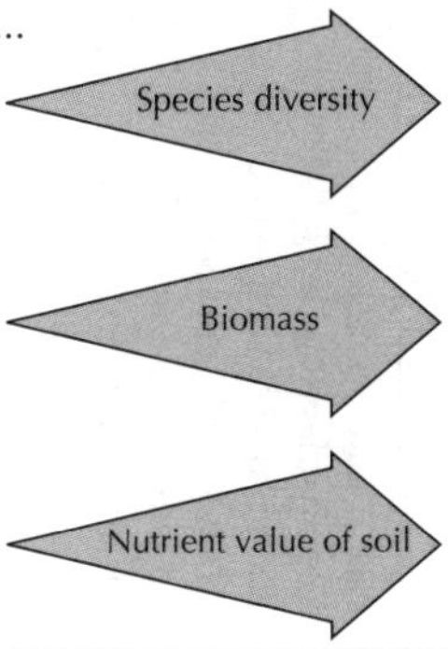

Climatic climax community
mature and self-sustaining community dominated by one vegetation type which is determined by climatic conditions in the region where the succession takes place.

Man may limit succession

- **Agriculture** maintains a sub climax commonly at the equivalent of a grassland stage (e.g. cereal crops).
- **Mowing lawn or cutting a hedge** prevents development of a climax community.
- **Management of a nature reserve** often prevents development of a climax community e.g. maintaining open grassland or forest glades for butterflies.

Plant growth substances

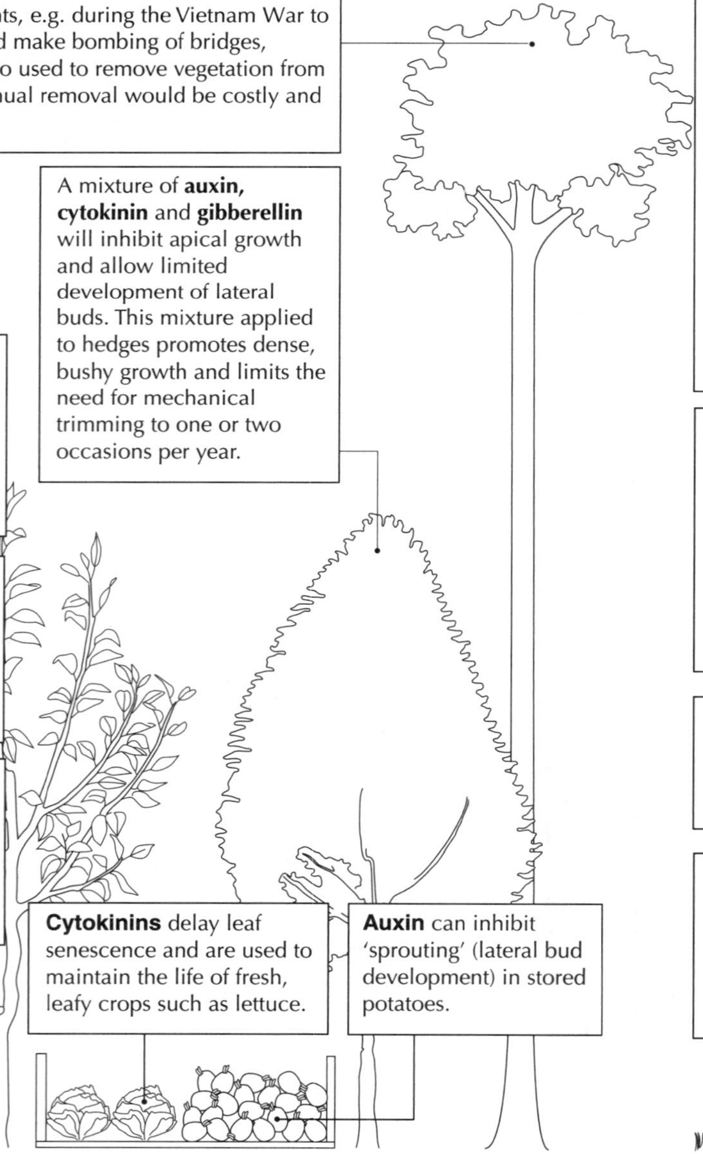

Auxins are used as defoliants, e.g. during the Vietnam War to clear areas of vegetation and make bombing of bridges, roads, and troops easier. Also used to remove vegetation from overhead power lines – manual removal would be costly and dangerous.

A mixture of **auxin, cytokinin** and **gibberellin** will inhibit apical growth and allow limited development of lateral buds. This mixture applied to hedges promotes dense, bushy growth and limits the need for mechanical trimming to one or two occasions per year.

Gibberellic acid may mimic red light: control of flowering time (promote long-day species, inhibit short-day species) means flowers can be available 'out of season'.

Ethene sprayed onto day-neutral species such as pineapple can synchronise flowering/fruiting so that crop picking can be more efficient.

Auxin can act as a selective lawn weed killer since broad leaved 'weed' species are killed by auxin concentrations which do not affect monocotyledons.

Cytokinins delay leaf senescence and are used to maintain the life of fresh, leafy crops such as lettuce.

Auxin can inhibit 'sprouting' (lateral bud development) in stored potatoes.

Natural plant growth hormones

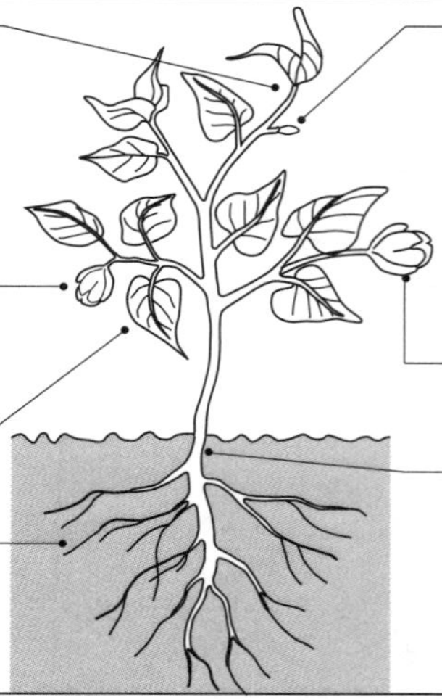

Growth of stem: cell enlargement is promoted by **auxin** and **gibberellin.** Redistribution of **auxin** causes phototropism.

Seed dormancy is maintained by **abscisic acid** but is broken by **gibberellic acid.**

Leaf fall is promoted by **abscisic acid.**

Root growth is **promoted** by **auxin** at **low** concentration but **inhibited** at **high auxin** concentration.

Lateral bud development is inhibited by **auxin** but promoted by **cytokinin (antagonism).**

Stomatal closure under stress may be promoted by **abscisic acid.**

Flowering may be triggered by **florigen.**

Root growth of adventitious roots is promoted by **auxin.**

N.B. Many commercial applications of these growth phenomena rely on **plant growth regulators,** which are synthetic derivatives of the natural compounds, but are usually more effective in lower concentrations because they are degraded less rapidly by the plant.

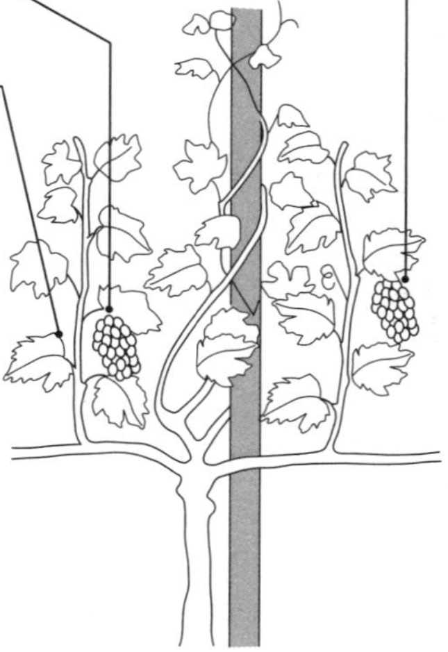

Cytokinin can promote fruit growth, and synergistically with **auxin** and **gibberellin** can promote parthenocarpy. This is useful if seed fails to 'set' due to poor pollinating conditions, and 'seedless' (parthenocarpic) fruits are popular with consumers.

Gibberellins increase fruit side in grapes if open since some ovules abort – 'crowding' is reduced allowing more nutrients to reach remaining fruits and limiting fungal infections.

Auxin can prevent premature fruit drop (windfall losses) since it is antagonistic to **abscisic acid.**

Auxin can act as a selective lawn weed killer since broad leaved 'weed' species are killed by auxin concentrations which do not affect monocotyledons.

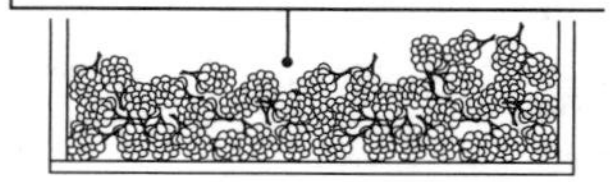

Ethene is used to accelerate ripening – ideal for grapes which can be picked earlier and thus have a longer drying period for forming raisins. Ripening can be delayed by keeping fruits in an oxygen-free atmosphere: ethene can then induce ripening as required.

Plant growth responses

are controlled by IAA.

Darwin: the light stimulus is perceived at the tip of the stem.

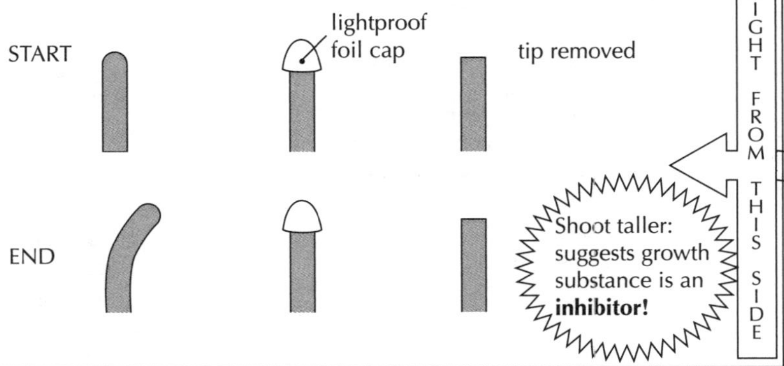
START
END
lightproof foil cap
tip removed
LIGHT FROM THIS SIDE
Shoot taller: suggests growth substance is an inhibitor!

Boyd-Jensen: a growth-promoting substance is produced in the tip and diffuses back to cells below where the growth response takes place.

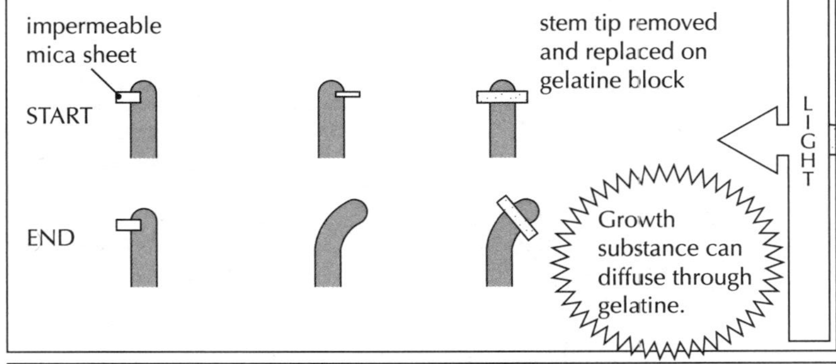
impermeable mica sheet
START
END
stem tip removed and replaced on gelatine block
LIGHT
Growth substance can diffuse through gelatine.

Went: the growth substance is distributed unevenly – more is found on the shaded side, which grows more quickly.

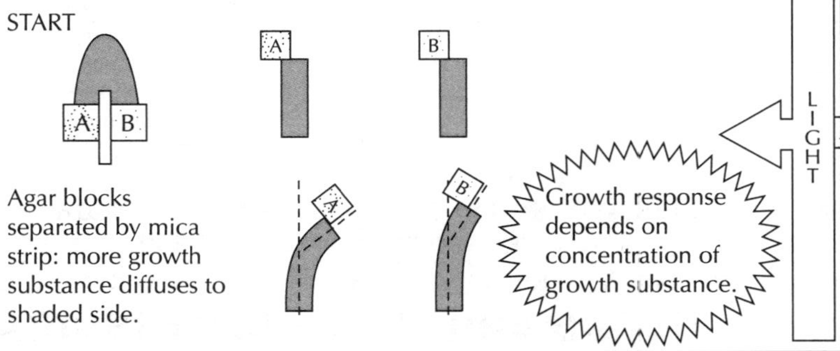
START
A
B
Agar blocks separated by mica strip: more growth substance diffuses to shaded side.
LIGHT
Growth response depends on concentration of growth substance.

Shoots are positively phototropic

They show growth **movements towards light.**

- **Mechanism:** the plant growth substance called **auxin** is inhibited by light. It therefore accumulates on the shaded side of shoots **causing faster growth on the shaded side** and thus **'bending' towards light**.
- **Benefit:** leafy shoots tilt towards light to improve rate of photosynthesis and therefore increase rate of growth.

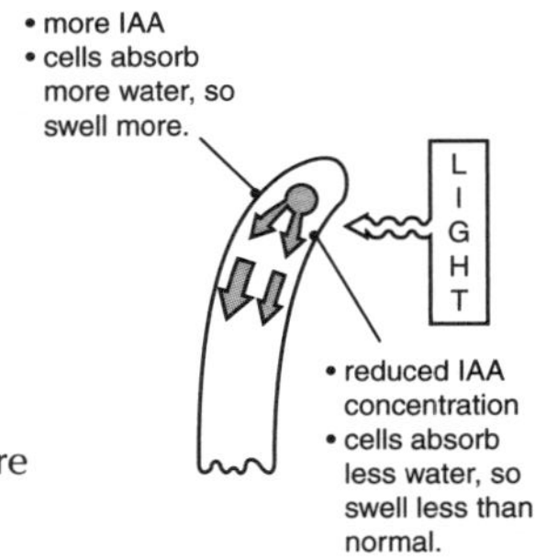
• more IAA
• cells absorb more water, so swell more.
LIGHT
• reduced IAA concentration
• cells absorb less water, so swell less than normal.

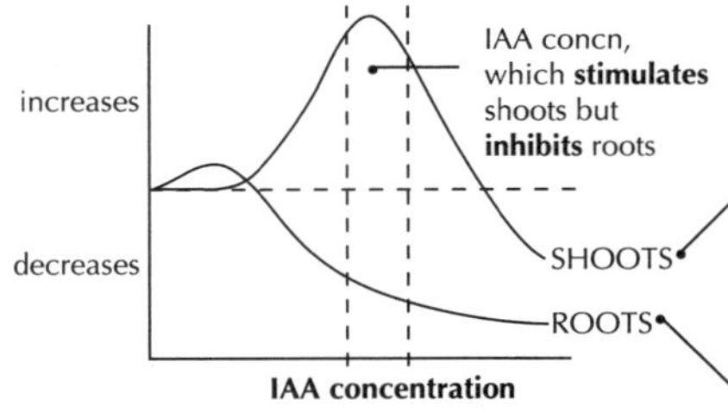
increases
decreases
IAA concn, which **stimulates** shoots but **inhibits** roots
SHOOTS
ROOTS
IAA concentration

Roots are positively geotropic (and negatively phototropic)

They show growth **movements towards gravity**.

- **Mechanism:** gravity causes large starch grains called **statoliths** to sink towards cells on lower side of root. This causes **auxin** to accumulate on lower side of root. Unlike shoot growth, root growth is **inhibited** by this concentration of auxin so that **lower side grows more slowly** and root **'bends' towards gravity**.
- **Benefit:** root grows into soil to anchor plant firmly and improve uptake of minerals and water.

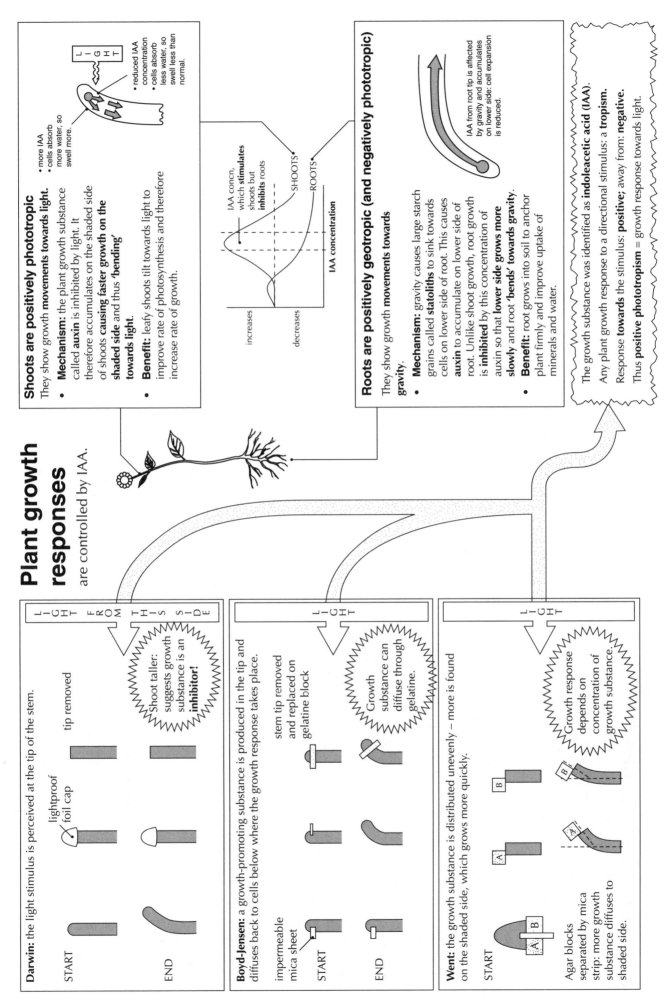

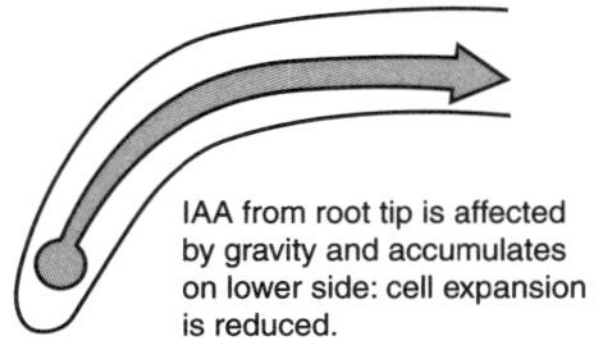
IAA from root tip is affected by gravity and accumulates on lower side: cell expansion is reduced.

The growth substance was identified as **indoleacetic acid (IAA)**.

Any plant growth response to a directional stimulus: a **tropism.**

Response **towards** the stimulus: **positive;** away from: **negative.**

Thus **positive phototropism** = growth response towards light.

Spinal cord and reflex action

Reflex actions are rapid responses to stimuli which allow the body to avoid damage. They all involve the sequence:

RECEPTOR → SENSORY NEURONE → CNS → MOTOR NEURONE → EFFECTOR

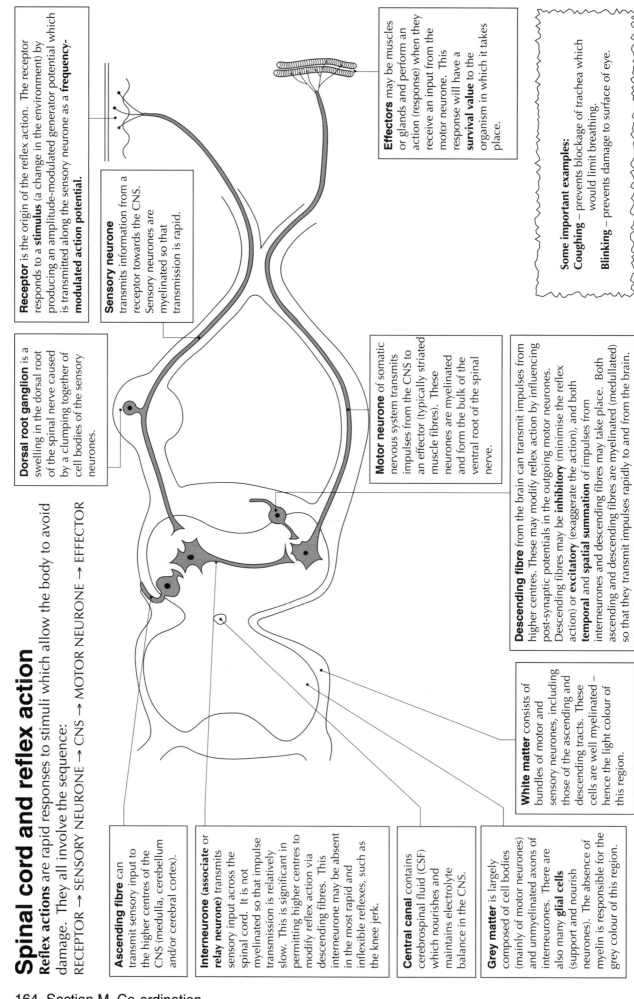

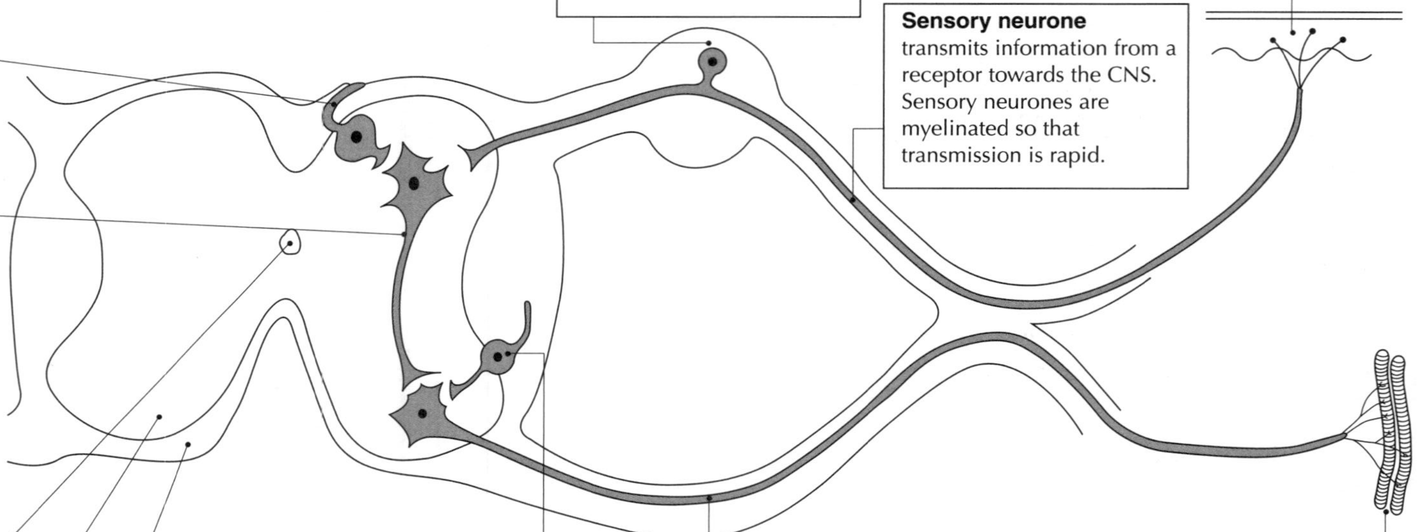

Ascending fibre can transmit sensory input to the higher centres of the CNS (medulla, cerebellum and/or cerebral cortex).

Interneurone (associate or relay neurone) transmits sensory input across the spinal cord. It is not myelinated so that impulse transmission is relatively slow. This is significant in permitting higher centres to modify reflex action via descending fibres. This interneurone may be absent in the most rapid and inflexible reflexes, such as the knee jerk.

Central canal contains cerebrospinal fluid (CSF) which nourishes and maintains electrolyte balance in the CNS.

Grey matter is largely composed of cell bodies (mainly of motor neurones) and unmyelinated axons of interneurones. There are also many **glial cells** (support and nourish neurones). The absence of myelin is responsible for the grey colour of this region.

White matter consists of bundles of motor and sensory neurones, including those of the ascending and descending tracts. These cells are well myelinated – hence the light colour of this region.

Dorsal root ganglion is a swelling in the dorsal root of the spinal nerve caused by a clumping together of cell bodies of the sensory neurones.

Receptor is the origin of the reflex action. The receptor responds to a **stimulus** (a change in the environment) by producing an amplitude-modulated generator potential which is transmitted along the sensory neurone as a **frequency-modulated action potential.**

Sensory neurone transmits information from a receptor towards the CNS. Sensory neurones are myelinated so that transmission is rapid.

Motor neurone of somatic nervous system transmits impulses from the CNS to an effector (typically striated muscle fibres). These neurones are myelinated and form the bulk of the ventral root of the spinal nerve.

Descending fibre from the brain can transmit impulses from higher centres. These may modify reflex action by influencing post-synaptic potentials in the outgoing motor neurones. Descending fibres may be **inhibitory** (minimise the reflex action) or **excitatory** (exaggerate the action), and both **temporal** and **spatial summation** of impulses from interneurones and descending fibres may take place. Both ascending and descending fibres are myelinated (medullated) so that they transmit impulses rapidly to and from the brain.

Effectors may be muscles or glands and perform an action (response) when they receive an input from the motor neurone. This response will have a **survival value** to the organism in which it takes place.

Some important examples:
Coughing – prevents blockage of trachea which would limit breathing.
Blinking – prevents damage to surface of eye.

Sensory cells, including the Pacinian Corpuscle, convert stimuli to electrical impulses.

The Pacinian Corpuscle is

- located in the **dermis** of the skin
- **multilamellar** (made of many layers)
- **encapsulated** (walled off from surrounding tissues).

As a result it is easily studied because

- it is large
- it can easily be separated from other tissues.

Using the Pacinian Corpuscle it has been possible to measure …

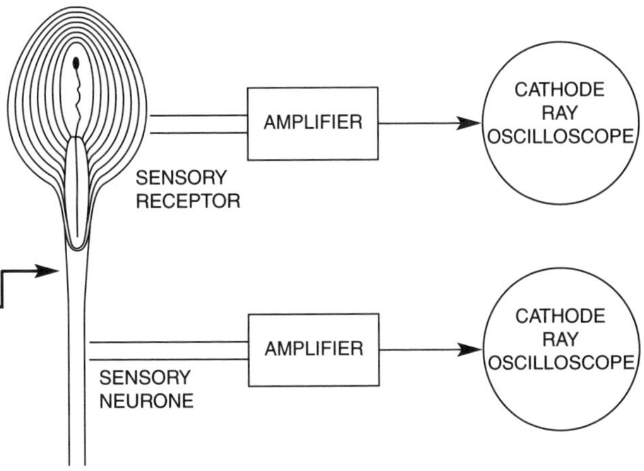

… effect of a stimulus on a sensory receptor

… effect of stimulus intensity on polarisation of sensory neurone membrane

How a sensory receptor works

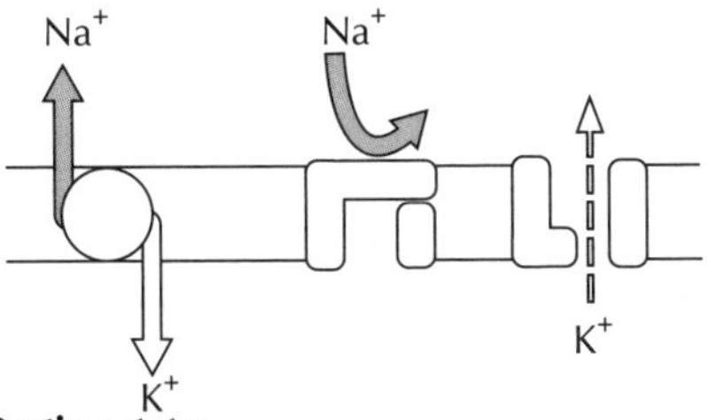

Membrane is polarised due to Na^+/K^+ pump, outward leakage of K^+, and impermeability to Na^+.

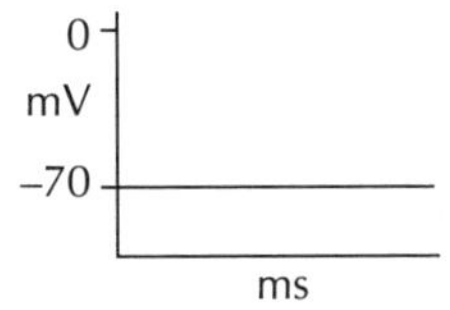

Resting state:

Mild stimulus: mild depolarisation due to some Na^+ movement inward through stretch-mediated sodium channels. There is a small **generator potential.**

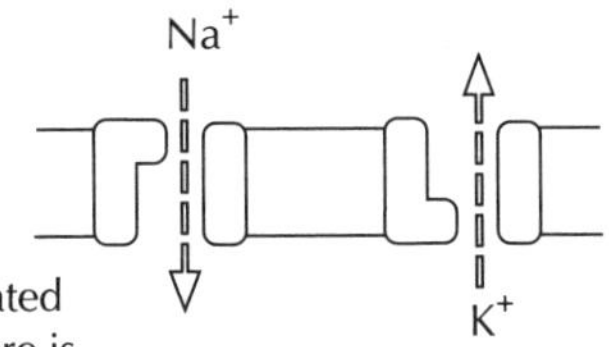

Stimulus is **sub-threshold** so no action potentials are recorded.

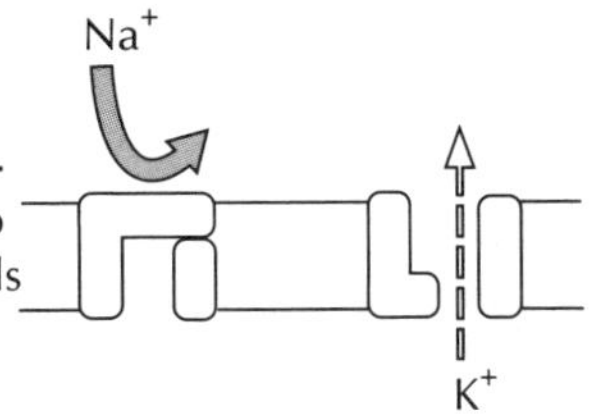

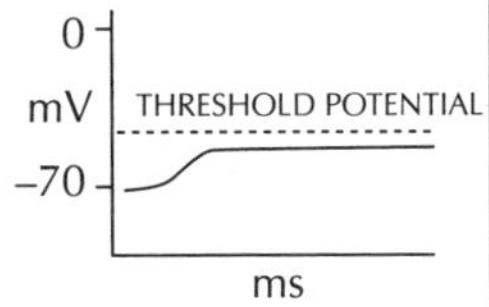

Larger stimulus: opening of pressure-sensitive Na^+ gates produces large **generator potential**.

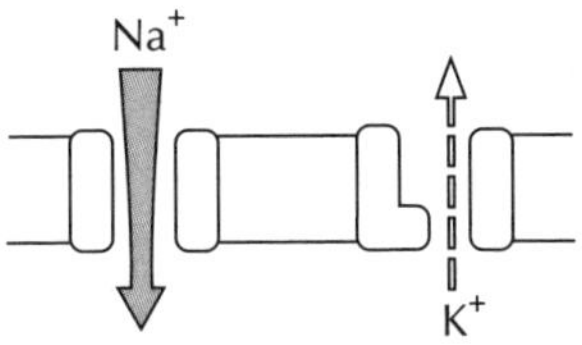

Generator potentials open ion-gated Na^+ channels. $Na^+ \rightarrow$ action potentials recorded.

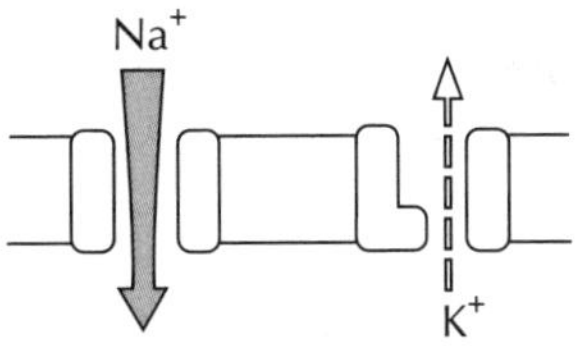

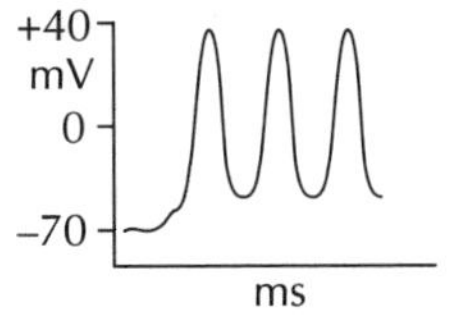

RECEPTOR CELL MEMBRANE

SENSORY NEURONE MEMBRANE

Summary:

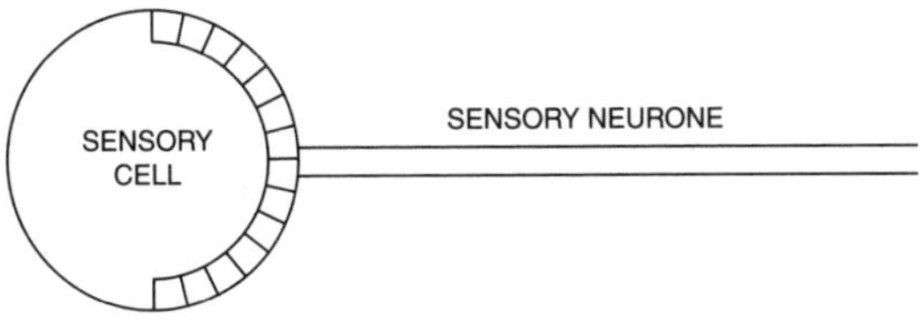

Increased intensity of stimulus ⟶ larger depolarisation, i.e. the **generator (receptor) potential is amplitude modulated**.

Inreased intensity of stimulus ⟶ more frequent but equal-sized action **potentials**, i.e. **action potential is frequency modulated**.

A sensation is an awareness of an external or internal stimulus. **Variety of sensation** results from:

- more frequent action potentials;
- more abundant connections in sensory areas of brain;
- stimulation of more sensory neurones.

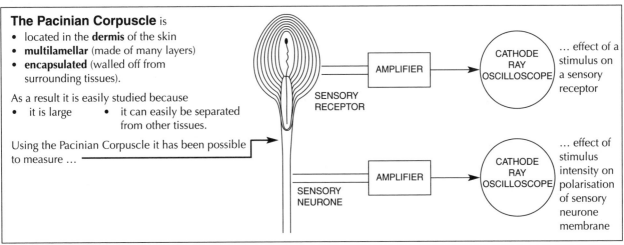

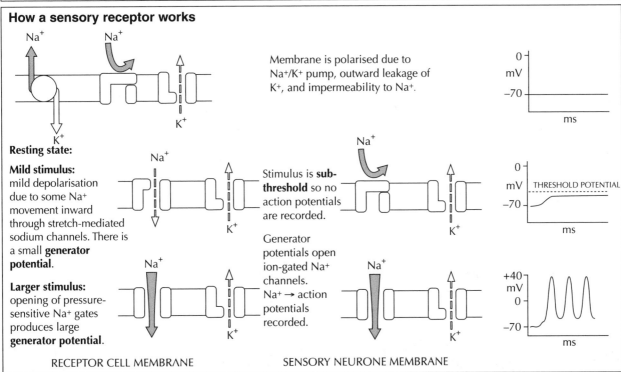

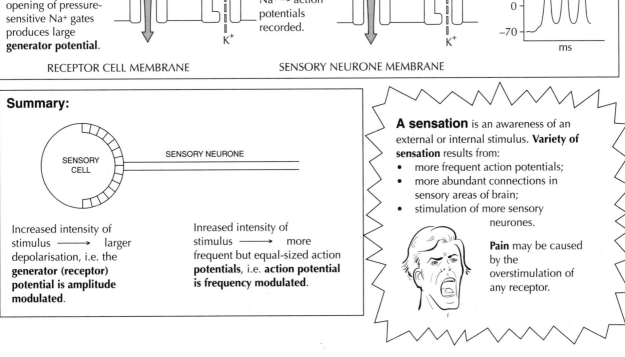

Pain may be caused by the overstimulation of any receptor.

Photoreceptors

in the retina produce an **amplitude modulated generator potential.**

Structure and function in rod cells

Outer segment contains numerous discs or lamellae which are packed with the photosensitive compound **rhodopsin**. Light falling on rhodopsin induces a conformational change in the protein part of the molecule which triggers a cascade which eventually hyperpolarises the rod cell membrane.

Neck Region contains a modified cilium and basal body, and connects the metabolic centre of the inner segment to the photo-sensitive region.

Inner segment contains numerous mitochondria (ATP for active transport of Na^+ ions), ribosomes (for synthesis of opsin) and an extensive endoplasmic reticulum (Ca^{2+} storage and membrane assembly).

Cell Body with nucleus

Synaptic Terminal releases an inhibitory neurotransmitter. There is greatest release of this compound **in the dark,** so stimulation by light **reduces** the release of the neurotransmitter and the rod cell membrane becomes **hyperpolarised**.

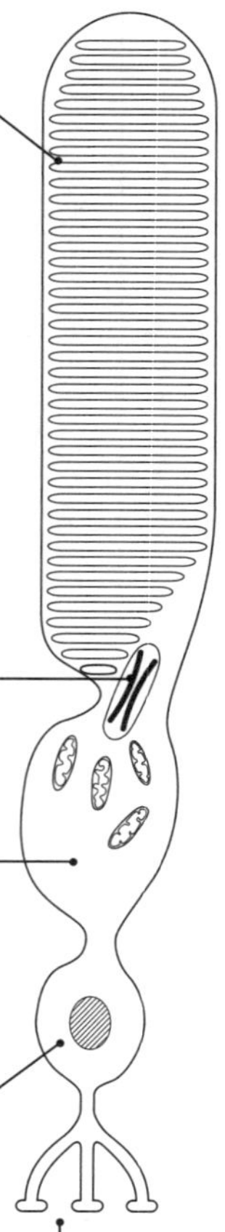

The visual cycle

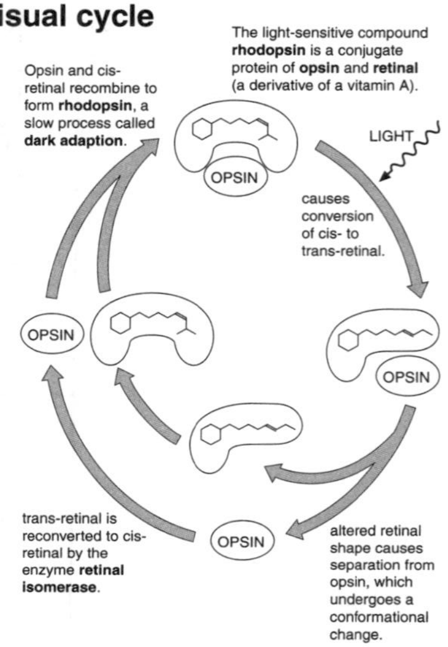

The light-sensitive compound **rhodopsin** is a conjugate protein of **opsin** and **retinal** (a derivative of a vitamin A).
LIGHT
causes conversion of cis- to trans-retinal.
altered retinal shape causes separation from opsin, which undergoes a conformational change.
trans-retinal is reconverted to cis-retinal by the enzyme **retinal isomerase**.
Opsin and cis-retinal recombine to form **rhodopsin**, a slow process called **dark adaption**.
OPSIN

Trichromatic theory of colour vision suggests that there are three variants of iodopsin, each of which is sensitive to a different range of wavelengths, corresponding to the three primary colours **red, blue** and **green**.

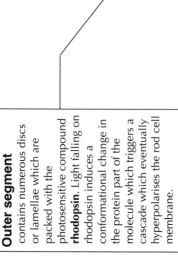

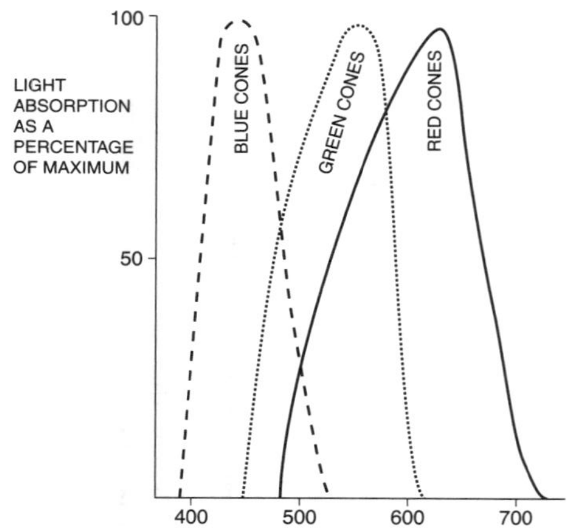

LIGHT ABSORPTION AS A PERCENTAGE OF MAXIMUM
100
50
BLUE CONES
GREEN CONES
RED CONES
400
500
600
700

Each type of iodopsin is probably located in different cones, and different colours are perceived in the brain from the sensory input from combinations of the three 'primary cones'.

Cones and colour vision

The visual cycle in cones is very similar to that in rods, except that the protein component of the photosensitive compound (**iodopsin** or **photopsin**) is different. In particular the dissociation of the pigment which leads to the generator potential requires a higher energy input – thus cones are not very sensitive in dim light.

PRIMARY CONES STIMULATED			COLOUR PERCEIVED IN BRAIN
BLUE (440 nm)	GREEN (550 nm)	RED (600 nm)	
+	+	+	WHITE
+			BLUE
	+		GREEN
		+	RED
+	+		CYAN
	+	+	ORANGE/YELLOW
+		+	MAGENTA

The trichromatic theory is supported by studies using recombinant DNA techniques which demonstrated separate genes for blue-sensitive, green-sensitive and red-sensitive opsin.

Photoreceptor cells in the retina

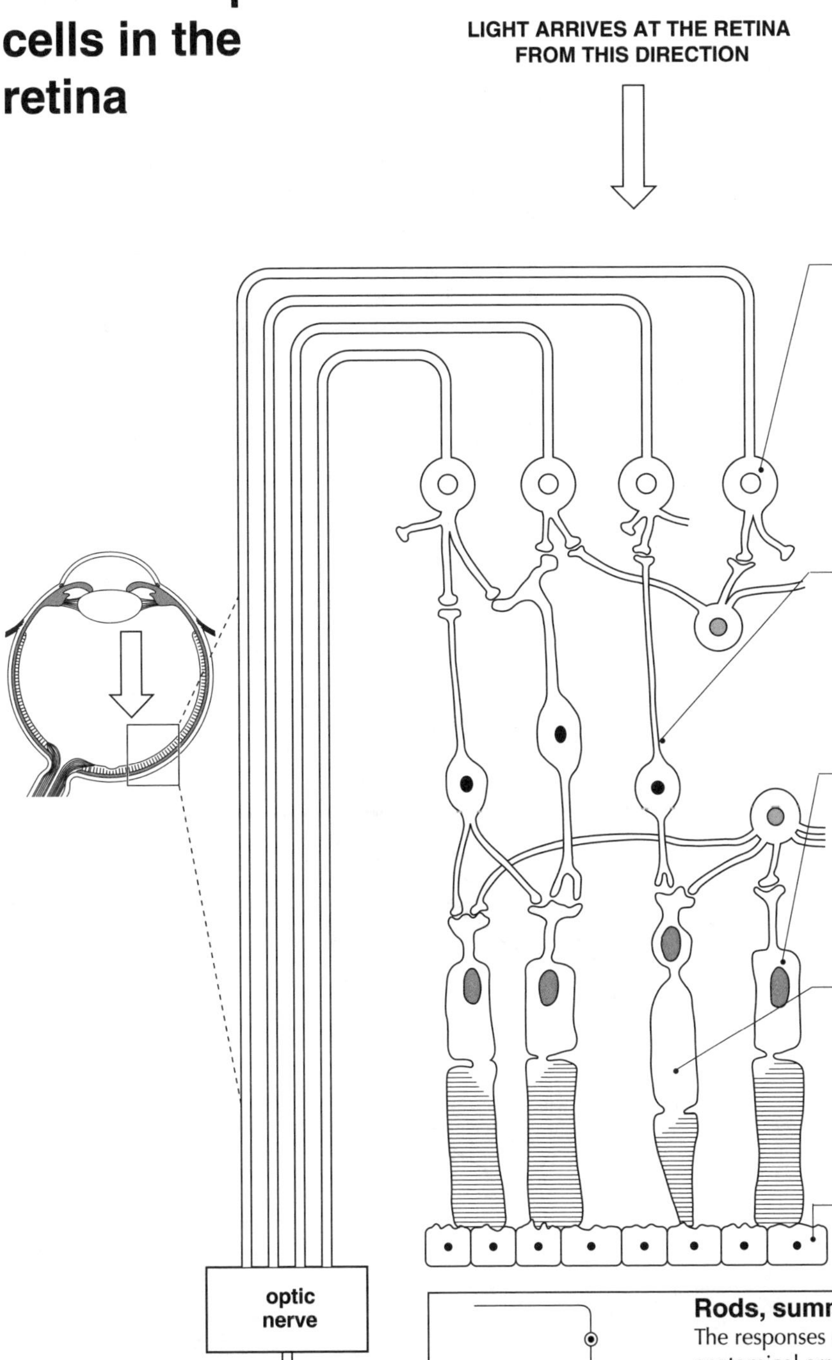
LIGHT ARRIVES AT THE RETINA FROM THIS DIRECTION
optic nerve

Ganglion cell: transmits action potentials (with frequency depending on generator potential of photo-sensitive cells) along neurones of optic nerve, i.e. this information is **frequency modulated.**

Bipolar cell: forms a link between photoreceptors and ganglion cells, converting **hyperpolarisation** in receptors to **depolarisation** in the sensory neurones.

Rod cell: contains photosensitive pigment, RHODOPSIN, which generates a signal at low light intensities but does not distinguish between different wavelengths. The brain only detects 'black and white'.

Rods and cones are photoreceptors which become hyperpolarised to produce an **amplitude modulated generator potential.**

Cone cell: contains pigment called **iodopsin** which is only sensitive to high levels of illumination but exists in different forms, so these cells can send signals which allow the brain to detect 'colours'.

Pigmented epithelium of retina

No image is 'seen' until action potentials arrive at the visual association centre: we 'see' **with our brain.**

Visual association cortex

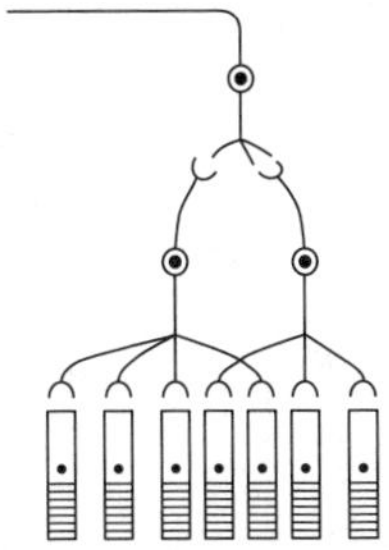

Rods, summation and sensitivity

The responses of many rods may be 'summed' by the anatomical arrangement of the bipolar cells, which may synapse with several rods but only a single ganglion cell. This **synaptic convergence** permits great **visual sensitivity,** and rods are thus of great value for **night vision.** Since rods are more abundant away from the fovea, objects are often seen more clearly at night by not looking directly at them.

Cones and visual acuity

Each cone is connected, via a bipolar cell, to a single ganglion cell. Since cones are packed closely together, especially at the fovea, these cells are able to discriminate between light stimuli which arrive in close proximity – the retina is able to resolve two light sources falling on cones separated only by a single other cone.
The **absence of synaptic convergence** offers **acuity** but **poor sensitivity.**

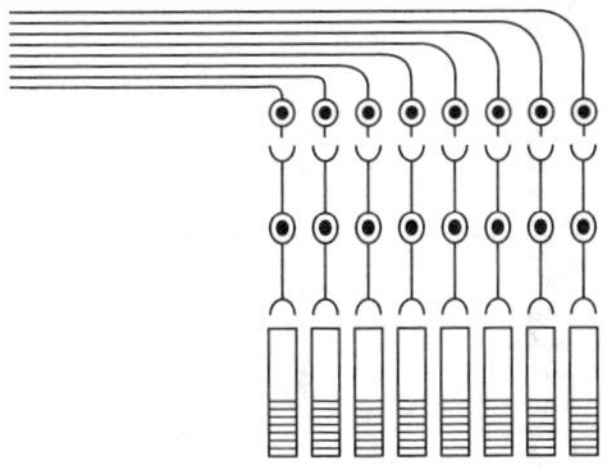

Control of the heartbeat

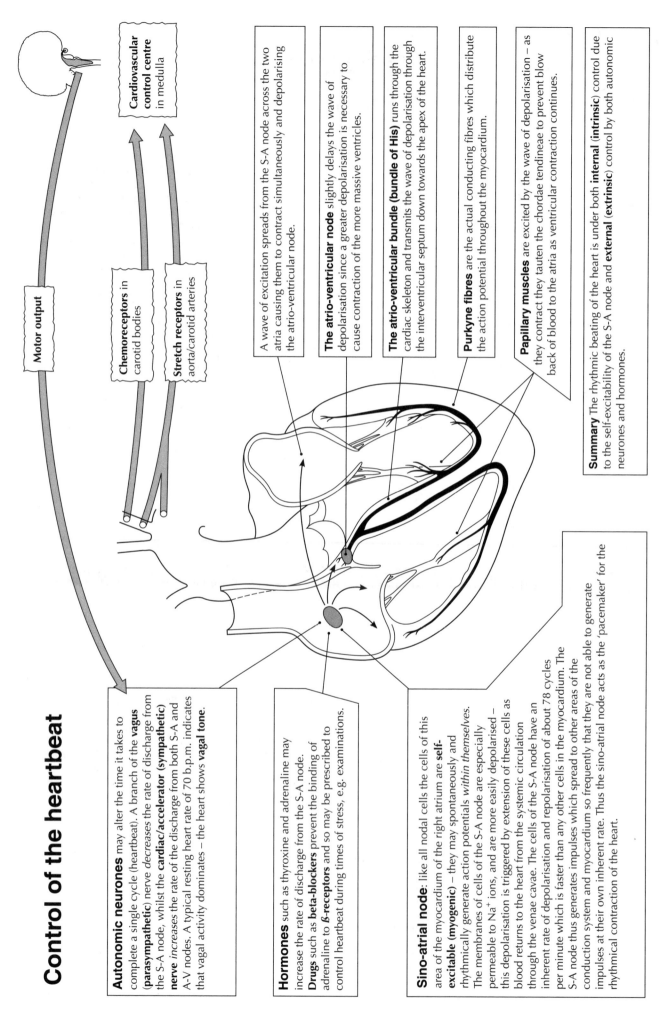

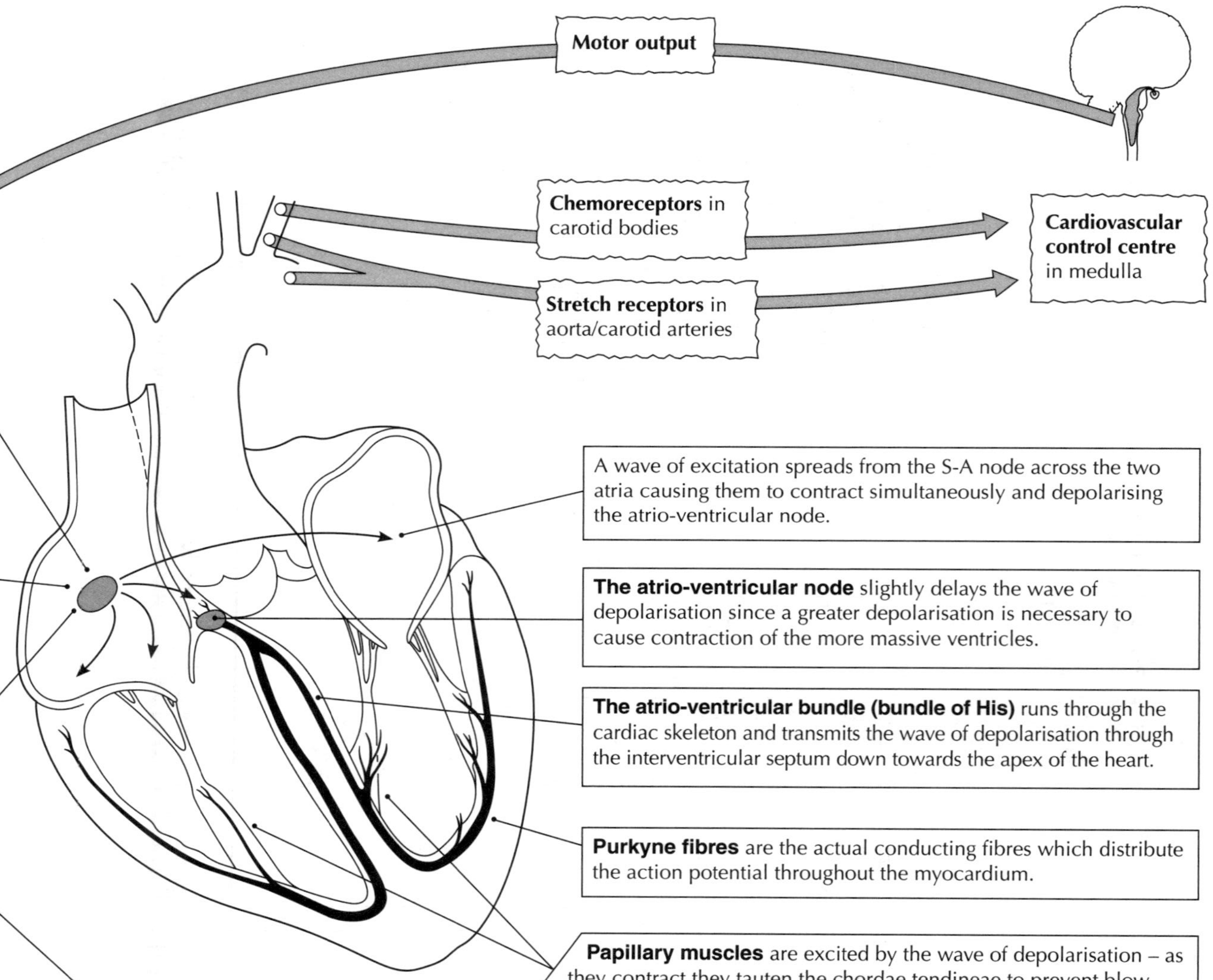

Autonomic neurones may alter the time it takes to complete a single cycle (heartbeat). A branch of the **vagus** (**parasympathetic**) nerve *decreases* the rate of discharge from the S-A node, whilst the **cardiac/accelerator (sympathetic) nerve** *increases* the rate of the discharge from both S-A and A-V nodes. A typical resting heart rate of 70 b.p.m. indicates that vagal activity dominates – the heart shows **vagal tone**.

Hormones such as thyroxine and adrenaline may increase the rate of discharge from the S-A node.
Drugs such as **beta-blockers** prevent the binding of adrenaline to **ß-receptors** and so may be prescribed to control heartbeat during times of stress, e.g. examinations.

Sino-atrial node: like all nodal cells the cells of this area of the myocardium of the right atrium are **self-excitable (myogenic)** – they may spontaneously and rhythmically generate action potentials *within themselves*. The membranes of cells of the S-A node are especially permeable to Na^+ ions, and are more easily depolarised – this depolarisation is triggered by extension of these cells as blood returns to the heart from the systemic circulation through the venae cavae. The cells of the S-A node have an inherent rate of depolarisation and repolarisation of about 78 cycles per minute which is faster than any other cells in the myocardium. The S-A node thus generates impulses which spread to other areas of the conduction system and myocardium so frequently that they are not able to generate impulses at their own inherent rate. Thus the sino-atrial node acts as the 'pacemaker' for the rhythmical contraction of the heart.

A wave of excitation spreads from the S-A node across the two atria causing them to contract simultaneously and depolarising the atrio-ventricular node.

The atrio-ventricular node slightly delays the wave of depolarisation since a greater depolarisation is necessary to cause contraction of the more massive ventricles.

The atrio-ventricular bundle (bundle of His) runs through the cardiac skeleton and transmits the wave of depolarisation through the interventricular septum down towards the apex of the heart.

Purkyne fibres are the actual conducting fibres which distribute the action potential throughout the myocardium.

Papillary muscles are excited by the wave of depolarisation – as they contract they tauten the chordae tendineae to prevent blow back of blood to the atria as ventricular contraction continues.

Summary The rhythmic beating of the heart is under both **internal** (**intrinsic**) control due to the self-excitability of the S-A node and **external** (**extrinsic**) control by both autonomic neurones and hormones.

Behaviour may be innate (instinctive) or learned

MOMMA!

Taxes and kineses are innate behaviours

A **taxis** (pl. = **taxes**) is an **orientation movement** towards or away from a **directional stimulus**.

- May be **negative** (**away from** the stimulus) or **positive** (**towards** the stimulus)
- e.g. *Euglena* shows positive phototaxis (towards light)

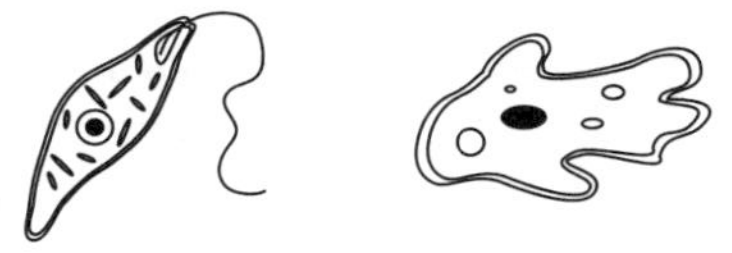

Amoeba shows negative chemotaxis (away from a chemical stimulus, e.g. increased [CO_2]).

A **kinesis** (pl. = **kineses**) is a random movement related to **intensity of stimulus** but **not** to its direction.

- **orthokinesis** – changes in **speed of movement**
- **klinokinesis** – changes in **rate of turning**

e.g. Planaria (flatworms) detect food (meat or rotting leaves).

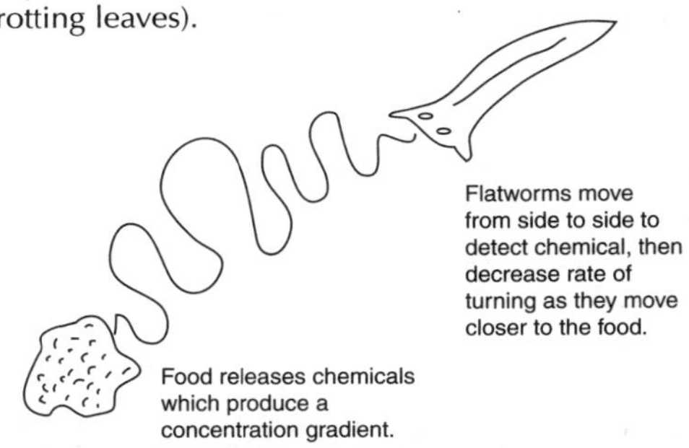

Flatworms move from side to side to detect chemical, then decrease rate of turning as they move closer to the food.
Food releases chemicals which produce a concentration gradient.

Innate and learned behaviour

Innate	Learned
inherited, i.e. due to genes	not inherited (although **capability** to learn may have some genetic basis).
inflexible, and not changed by environment	flexible ('plastic'), i.e quickly adapts to new circumstances/environments
similar in all members of the same species	may markedly differ between members of the same species.

Follows a classical **fixed action plan** (**FAP**)

STIMULUS or RELEASER

⇩

INNATE BEHAVIOURAL PROGRAMME ⇨ FAP: a predictable behavioural response

Migration

Migration is a long-distance directional movement related to season/food availability:

- is **innate** in small, short-lived species, e.g. warblers;
- is a combination of **learned** and **innate** in longer-lived species, i.e. geese and swans;

HABITAT 1: food in winter ⇄ HABITAT 2: food in summer

⇧

CHANGING PHOTOPERIOD

- related to **position of Sun** in **daytime migrants**, and **position of stars** in **night migrants;**
- Earth's **magnetic field** may also be important;
- **biological clock** allows compensation for changes in position of Sun and stars.

Imprinting

- is a form of **learned** behaviour that occurs during a short, genetically determined **critical period** in the life of an animal;
- normally involves young animal learning characteristics of parents/members of the same species, or adult animals recognising their own offspring;
- can be harmful if an unusual stimulus is presented during the critical period, e.g. ducklings may imprint on a human present during the hatching period.

Conditioning

- occurs when an animal learns to associate an unnatural stimulus with a particular response, e.g. **conditioned reflex**: dog salivates when it hears a bell if the bell is always sounded when food is presented.

Operant conditioning: animal is rewarded for an action (**operation**) that it does naturally from time to time, e.g. rat pressing a lever which releases food.

CHOCCO DROPS

The human nervous system

can be subdivided into central and peripheral components.

Central nervous system (CNS)

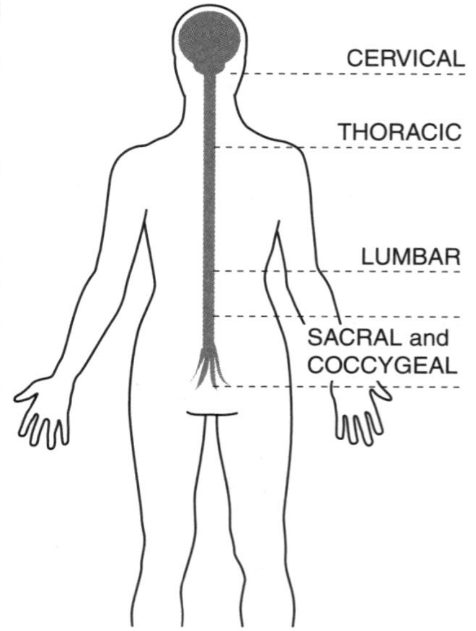

Spinal Cord:
- relay centre for reflex activities
- connection between brain and the peripheral nervous system
- extends from base of brain to lumbar region, where it branches into many descending nerves
- enclosed within the neural arches of the vertebrae which make up the vertebral column, or 'backbone'
- connection to the peripheral nervous system is via 33 pairs of spinal nerves which exist between adjacent vertebrae
- each nerve innervates a group of muscles (**myotome**) and an area of skin (**dermatome**).

Brain: the primary **integration** and **control organ** of the body
- receives information (**sensory input**) from sense organs in the head and from other body organs via ascending fibres of the spinal cord;
- communicates via **cranial nerves,** the **descending fibres** of the spinal cord, and through regulation of endocrine secretions via the **pituitary gland;**
- utilises many neurotransmitters, including dopamine, serotonin, histamine, GABA and endorphins.

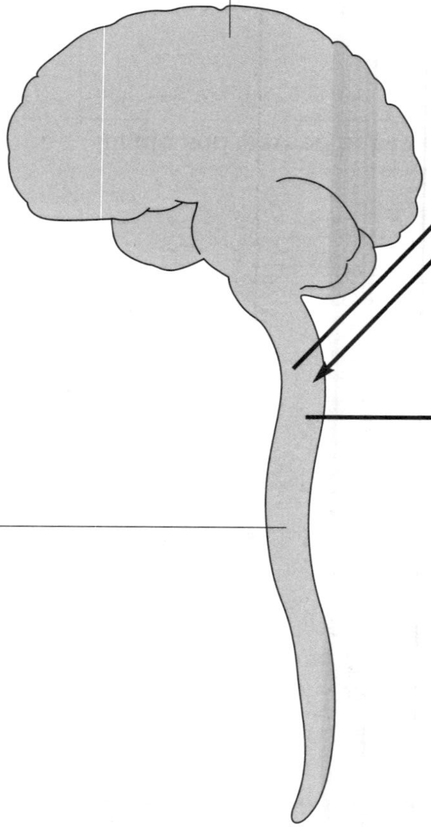

Peripheral nervous system (PNS)

Somatic nervous system:
- has both **sensory** and **motor** parts;
- regulates **voluntary activity** e.g. actions of limbs, neck, lips, tongue and eyelids and inhibition of bladder emptying at socially unacceptable times!
- principal neurotransmitter is **acetylcholine.**

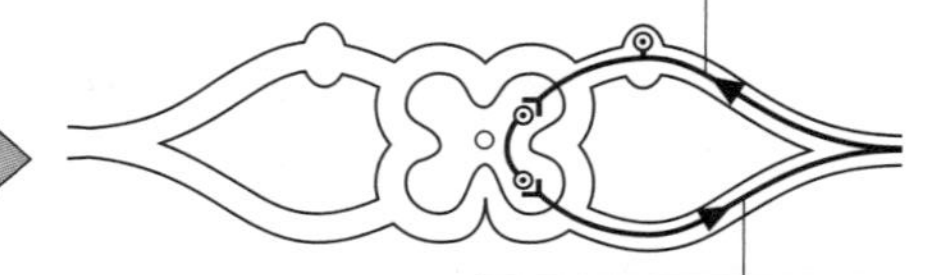

Autonomic nervous system:
- has only **motor** parts;
- regulates **subconscious physiological activities** e.g. heart rate, distribution of blood, salivation;
- uses both **acetylcholine** and **nor-adrenaline** as neurotransmitters;
- has two divisions, the **sympathetic** and the **parasympathetic.**

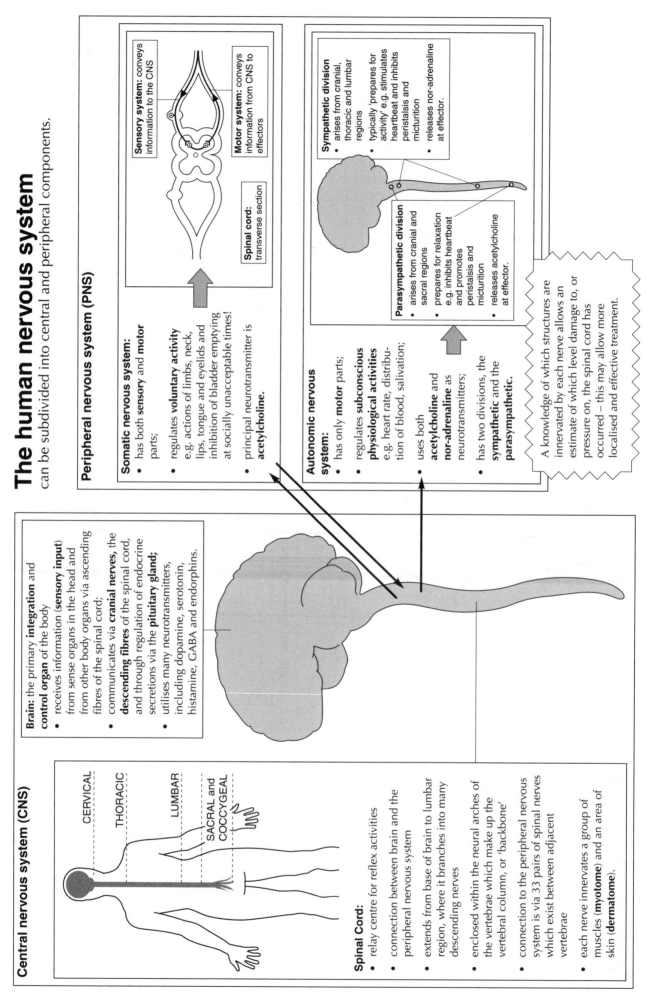

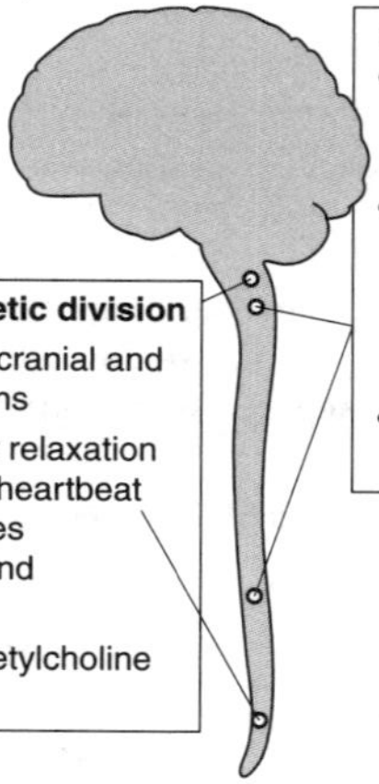

Parasympathetic division
- arises from cranial and sacral regions
- prepares for relaxation e.g. inhibits heartbeat and promotes peristalsis and micturition
- releases acetylcholine at effector.

Sympathetic division
- arises from cranial, thoracic and lumbar regions
- typically 'prepares for activity' e.g. stimulates heartbeat and inhibits peristalsis and micturition
- releases nor-adrenaline at effector.

A knowledge of which structures are innervated by each nerve allows an estimate of which level damage to, or pressure on, the spinal cord has occurred – this may allow more localised and effective treatment.

The brain is an integrator — it is able to accept sensory information from a number of sources, compare it with previous experience (learning) and make sure that the appropriate actions are initiated.

Meninges: membranes which line the skull and cover the brain. They help to protect and nourish the brain tissues, but may be infected either by a virus or a bacterium to cause the potentially fatal condition **meningitis**.

Cerebral cortex: has motor areas to control voluntary movement, sensory areas which interpret sensations and association areas to link the activity of motor and sensory regions. The centre of intelligence, memory, language and consciousness.

Forebrain: here many emotions are localised, and damage to this area may cause aggression, apathy, extreme sexual behaviour and other emotional disturbances.

Hypothalamus: contains centres which control thirst, hunger and thermoregulation.

Pituitary gland: is a link between the central nervous system and the endocrine system. Secretes a number of hormones, including follicle stimulating hormone which regulates development of female gametes, and anti-diuretic hormone which controls water retention by the kidney.

Medulla: the link between the spinal cord and the brain, and relays information between these two structure. Has a number of reflex centres which control:

- vital reflexes which regulate heartbeat, breathing and blood vessel diameter;
- non-vital reflexes which co-ordinate swallowing, salivation, coughing and sneezing.

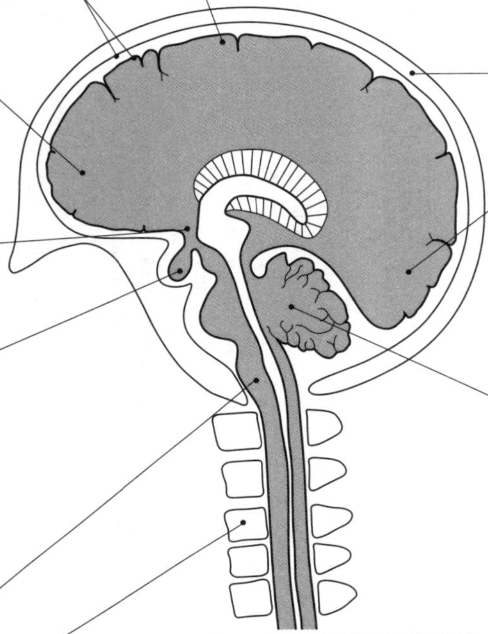

Skull: the cranium is a bony 'box' which encloses and protects the brain.

Visual centre: this area of the cerebral cortex

- interprets impulses along the optic nerve i.e. responsible for vision;
- has the connector neurones for both accommodation and the pupil reflex.

Cerebellum: co-ordinates movement using sensory information from position receptors in various parts of the body.
Helps to maintain posture using sensory information from the inner ear.
Can control learned sequences of activity involved in dancing, athletic pursuits and in the playing of musical instruments.

Spine (vertebral column): composed of 33 separate vertebra which surround and protect the spinal cord; between each pair of vertebrae two **spinal nerves** carry sensory information into the spinal cord and motor information out of it. Dislocation of the vertebrae may compress the spinal nerves, causing great pain, or even crush the spinal cord, leading to paralysis.

Don't forget

- In addition to its being a centre of learning the brain is the relay centre for **cranial reflexes** such as blinking.
- **Conditioned reflexes** involve setting up nerve pathways in the brain which link a substitute stimulus to a certain response e.g. hearing a lunch bell may set off the production of saliva.
- The evolutionary advantages of mammals may be partly due to a constant body temperature allowing the brain to 'work' for longer periods so that learning is more successful.

Drugs and brain function

Alcohol has a progressive effect on the brain.

- 1–2 units release inhibitions by affecting emotional centres in the forebrain.
- 5–6 units affect co-ordinated movements by influencing motor areas of the cerebral cortex.
- 7–8 units cause stupor and insensitivity to pain as sensory areas, including the visual centre, are impaired.
- More than 10 units may be fatal as the vital centres of the medulla and hypothalamus are severely inhibited.

Heroin can cause euphoria and insensitivity to pain as pain and emotion centres are inhibited.

L.S.D. may cause hallucinations by altering the balance of brain neurotransmitters and causing loss of the medulla's ability to 'filter' information reaching the cerebral cortex.

Neurones: the dendrites (antennae), axon (cable), synaptic buttons (contacts) are serviced and maintained by the cell body.

Dendrites are extensions of the cell body containing all typical cell body organelles. They provide a large surface area to receive information which they then pass on towards the cell body. The plasma has a high density of **chemically gated ion channels**, important in impulse transmission.

Motor neurone

Cell body contains a well-developed nucleus and nucleolus and many organelles such as lysosomes and mitochondria. There is **no mitotic apparatus** (centriole/spindle) in neurones more than six months old, which means that damaged neurones can never be replaced (although they may regenerate – see below).

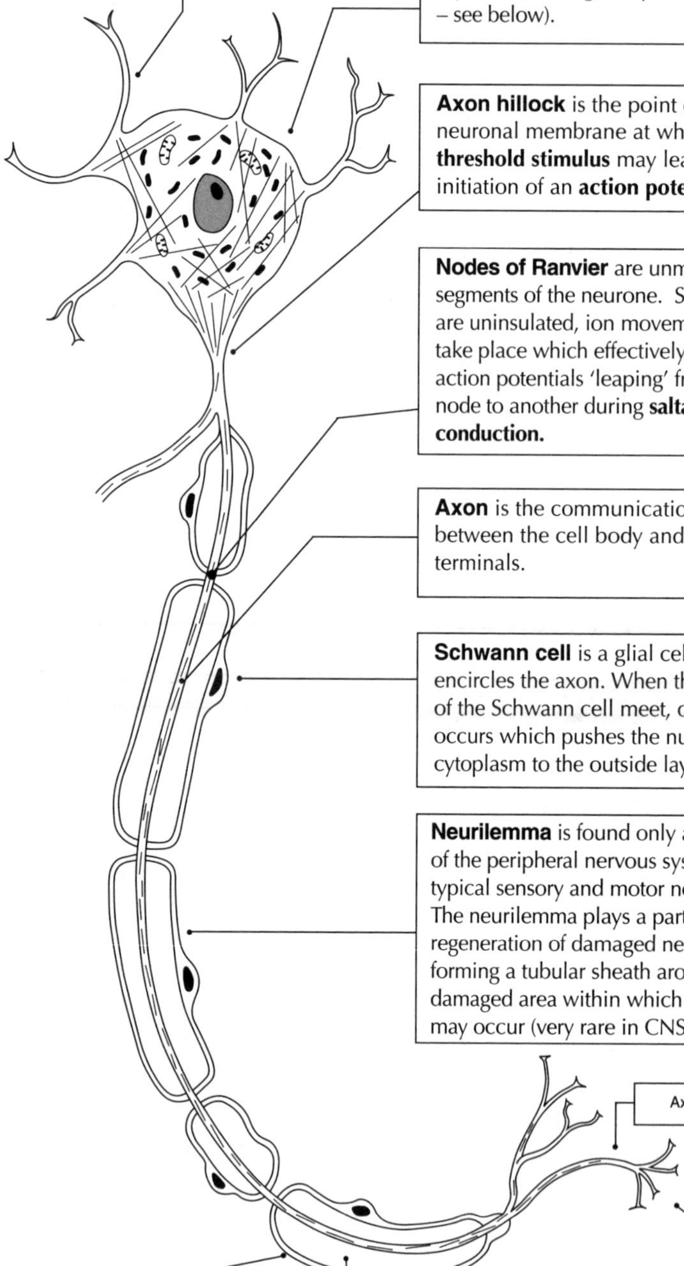

Axon hillock is the point on the neuronal membrane at which a **threshold stimulus** may lead to the initiation of an **action potential.**

Nodes of Ranvier are unmyelinated segments of the neurone. Since these are uninsulated, ion movements may take place which effectively lead to action potentials 'leaping' from one node to another during **saltatory conduction.**

Axon is the communication route between the cell body and the axon terminals.

Schwann cell is a glial cell which encircles the axon. When the two 'ends' of the Schwann cell meet, overlapping occurs which pushes the nucleus and cytoplasm to the outside layer.

Neurilemma is found only around axons of the peripheral nervous system, i.e. typical sensory and motor neurones. The neurilemma plays a part in the regeneration of damaged nerves by forming a tubular sheath around the damaged area within which regeneration may occur (very rare in CNS).

DIRECTION OF IMPULSE TRANSMISSION

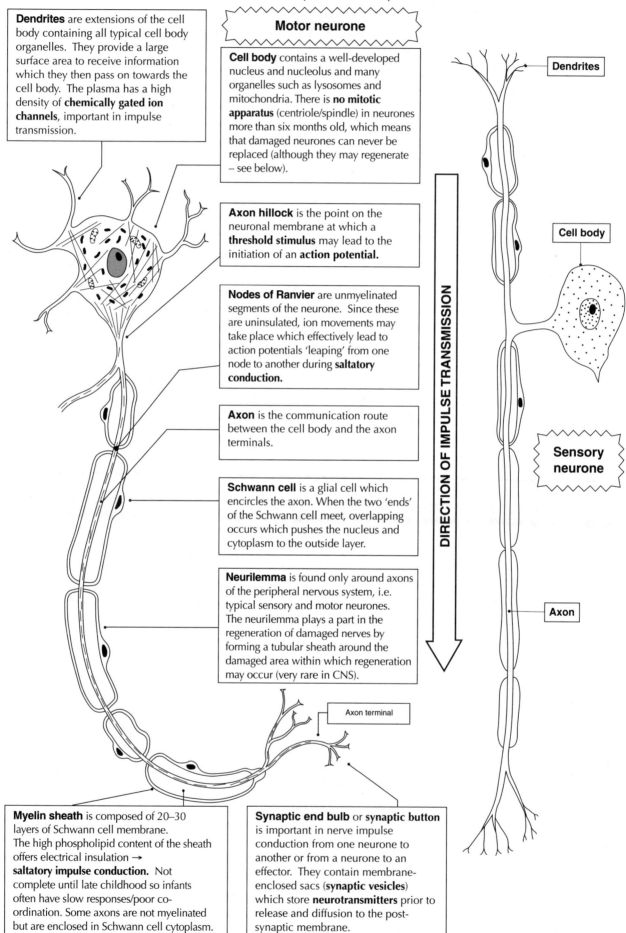

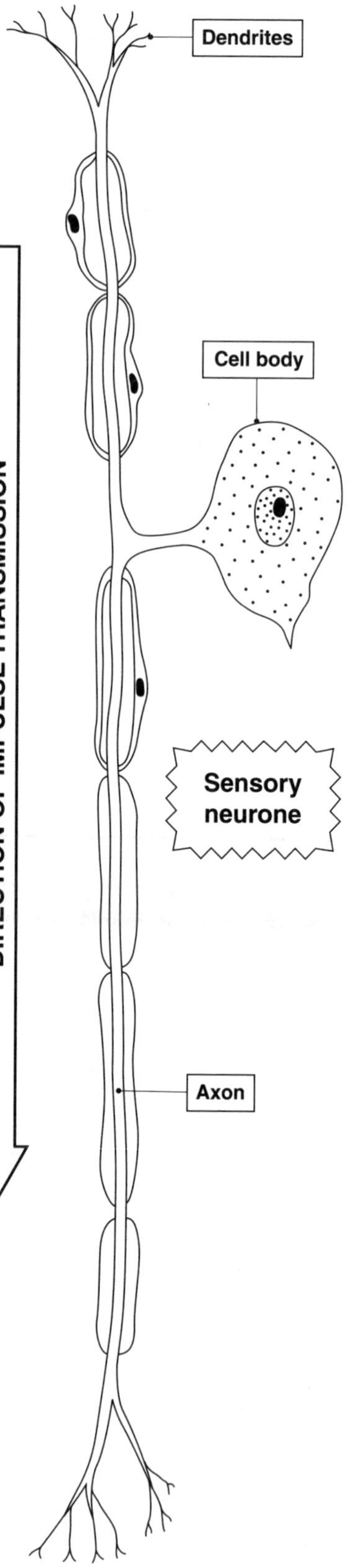

Sensory neurone

Myelin sheath is composed of 20–30 layers of Schwann cell membrane. The high phospholipid content of the sheath offers electrical insulation → **saltatory impulse conduction.** Not complete until late childhood so infants often have slow responses/poor co-ordination. Some axons are not myelinated but are enclosed in Schwann cell cytoplasm.

Synaptic end bulb or **synaptic button** is important in nerve impulse conduction from one neurone to another or from a neurone to an effector. They contain membrane-enclosed sacs (**synaptic vesicles**) which store **neurotransmitters** prior to release and diffusion to the post-synaptic membrane.

Synapse structure and function

Transmission of an action potential across a chemical synapse involves a uni-directional release of molecules of neurotransmitter from pre-synaptic to post-synaptic membranes.

Drugs and poisons may interfere with synaptic transmission by:

1. **mimicry of neurotransmitter,** e.g. **nicotine** mimics both acetylcholine and noradrenaline;
2. **reduced degradation,** e.g. **cocaine** inhibits re-uptake of noradrenaline;
3. **blocking receptors,** e.g. ***β*-blockers** may help to control rapid heartbeat by blocking receptors on the heart muscle. These β-receptors are normally sensitive to adrenaline;
4. **reduced release of neurotransmitter,** e.g. **alcohol** alters sleeping patterns by reducing release of serotonin.

(1) **Increase in local Ca^{2+} concentration:** depolarisation of membrane at synaptic button affects 'calcium channels' so that Ca^{2+} ions flow quickly into synaptic button from tissue fluid.

Mitochondria are abundant in the synaptic button: release energy for refilling of synaptic vesicles and possibly for pumping of Ca^{2+} to re-establish Ca^{2+} concentration gradient across neurone membrane.

(2) **Synaptic vesicles** containing molecules of neuro-transmitter move towards the presynaptic membrane.

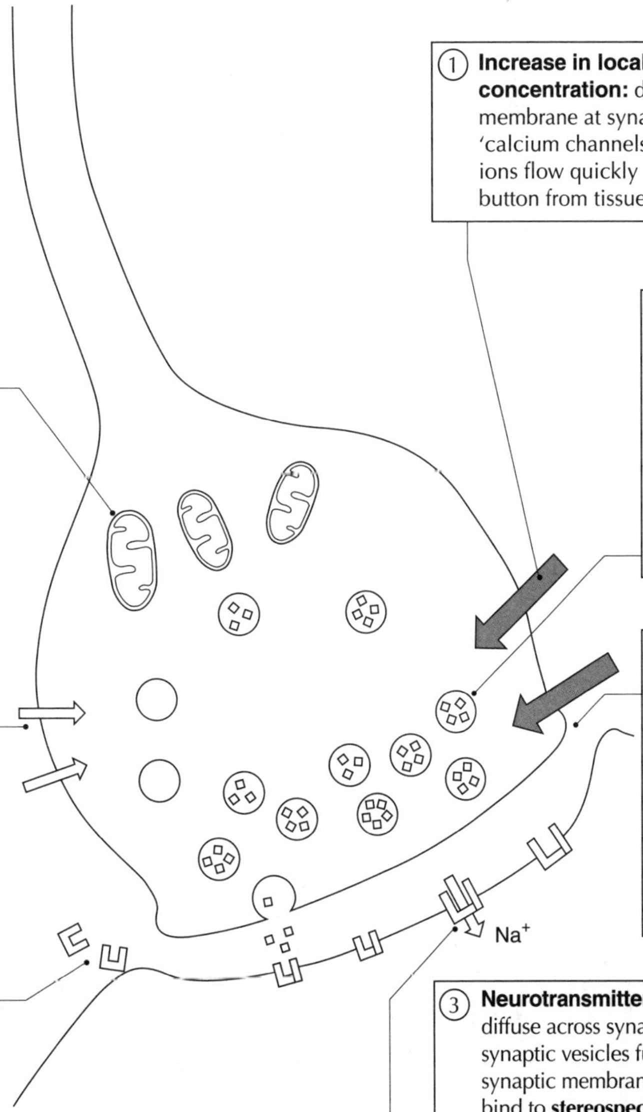

(6) **Reabsorption of neurotransmitter or products of degradation.** Molecules are resynthesised and reincorporated into synaptic vesicles. **Catecholamines** are often reabsorbed without degradation.

Synaptic cleft represents a barrier to the direct passage of the wave of depolarisation from pre-synaptic to post-synaptic membranes.

(5) **Enzymes degrade neurotransmitter.** These degradative enzymes, which are released from adjacent glial cells or are located on the post-synaptic membrane, remove neurotransmitter molecules so that their effect on the chemically gated ion channels is only short-lived. They include:

- **monoamine oxidase** (degrades **catecholamines)**
- **acetylcholine esterase** (degrades **acetylcholine).**

(3) **Neurotransmitter molecules** diffuse across synaptic gap when synaptic vesicles fuse with pre-synaptic membrane. Molecules bind to **stereospecific receptors** in the post-synaptic membrane. **Catecholamines** such as adrenaline are released from **adrenergic nerve endings, acetylcholine** from **cholinergic nerve endings,** and **GABA** (*gamma*-aminobutyric acid) and **serotonin** at synapses in the brain.

(4) **Chemically gated ion channels** on post-synaptic membrane – allow influx of Na^+ and efflux of K^+ → depolarisation of post-synaptic membrane. Ion channels are 'opened' when triggered by binding of neurotransmitter.

Excitatory post-synaptic potentials result if the neurotransmitter binding to the receptors on the post-synaptic membrane **opens** chemically gated ion channels, making **depolarisation more likely.**

Inhibitory post-synaptic potentials result if the neurotransmitter binding to the receptors on the post-synaptic membrane **keeps** chemically gated ion channels **closed,** promoting **hyperpolarisation** and making **depolarisation less likely.**

Resting potential and action potential:

The **resting potential** of -70mV is largely the result of a potassium ion (K+) equilibrium.

SODIUM CHANNELS are pore proteins which permit VERY LIMITED DIFFUSION of Na^+ from the extracellular fluid to the cytoplasm of the nerve cell.

ATP-DEPENDENT Na^+-K^+ PUMP transports ions across the plasamembrane of the nerve cell. Three Na^+ ions are moved from the inside to the outside of the cell for every two K^+ which are moved in the opposite direction, thus establishing ION CONCENTRATION GRADIENTS across the membrane.

POTASSIUM CHANNELS are pore proteins which permit K^+ ions to diffuse from the nerve cell cytoplasm to the extracellular fluid along a concentration gradient. These channels ensure that membrane permeability to K^+ is about 100x greater than that to Na^+ ions. K^+ movement continues until concentration gradient is 'balanced' by electrochemical gradient.

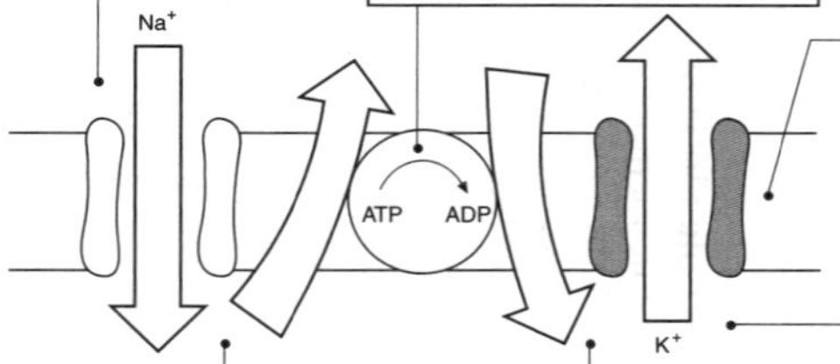

THE CYTOPLASM OF THE NERVE CELL contains many negative ions, mostly protein anions. The nerve cell membrane is almost impermeable to these ions.

NET Na^+ ION MOVEMENTS generate a lesser INSIDE - OUTSIDE K^+ ION GRADIENT

NET Na^+ ION MOVEMENTS generate a high OUTSIDE - INSIDE Na^+ ION GRADIENT

As a result of
a. Na^+/K^+ pump
b. K^+ diffusion outwards
c. negative ions in the nerve cell cytoplasm

the plasamembrane of the nerve cell becomes POLARISED: the inside is about -70mV with respect to the outside. This represents the RESTING POTENTIAL of the nerve cell.

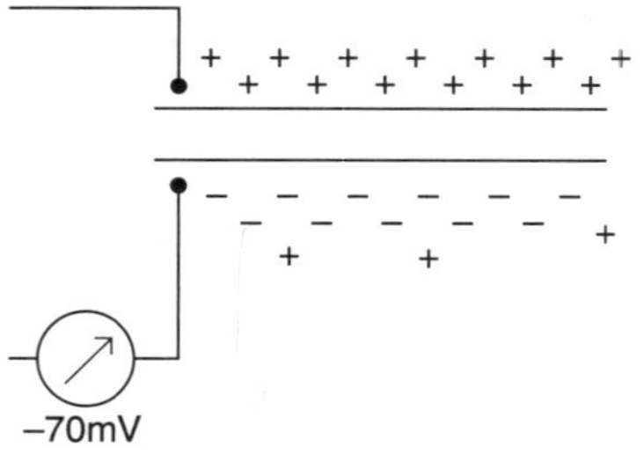

Return to resting potential: although the action potential involves Na^+ and K^+ movements, the changes in absolute ion concentrations are very small (probably no more than 1 in 10^7). Many action potentials could be transmitted before concentration gradients are significantly changed – the sodium–potassium pump can quickly restore resting ion concentrations to 'pre-impulse' levels.

An **action potential** is the depolarisation–repolarisation cycle at the neurone membrane following the application of a threshold stimulus. The depolarisation is about 110 mV resulting from **an inward flow of Na+ ions.** Since the depolarisation–repolarisation depends upon ion concentration gradients and upon time of ion channel opening, both of which are effectively fixed, **all action potentials are of the same size.** Thus a nerve cell obeys the **all-or-nothing principle:** if a stimulus is strong enough to generate an action potential, the impulse is conducted along the entire neurone **at a constant and maximum strength** for the existing conditions.

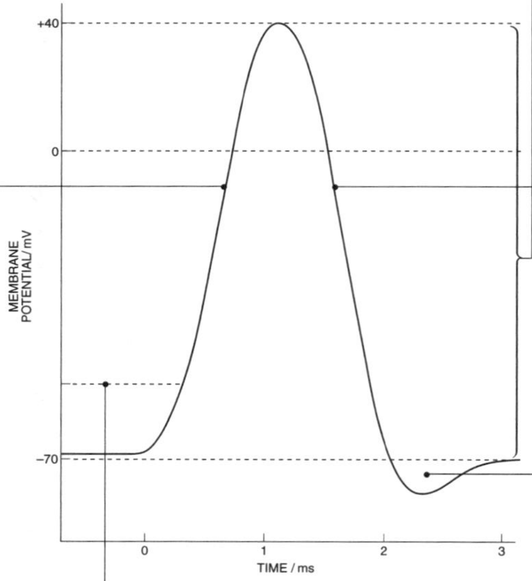

Threshold value: any stimulus strong enough to initiate an impulse is called a **threshold** or **liminal stimulus.** The stimulus begins the depolarisation of the neuronal membrane – once a sufficient number of voltage-gated sodium channels is opened, positive feedback will ensure a complete depolarisation. Any stimulus weaker than a threshold stimulus is called a **sub-threshold** or **subliminal stimulus.** Such a stimulus is incapable of initiating an action potential, but a series of such stimuli **may** exert a cumulative effect which may be sufficient to initiate an impulse. This is the phenomenon of **summation of impulses.**

Depolarisation: the voltage-gated sodium channels open so that Na^+ ions can move **into the axon.**

a. down a $\mathbf{Na^+}$ **concentration gradient**

b. down an **outside–inside electrical gradient**

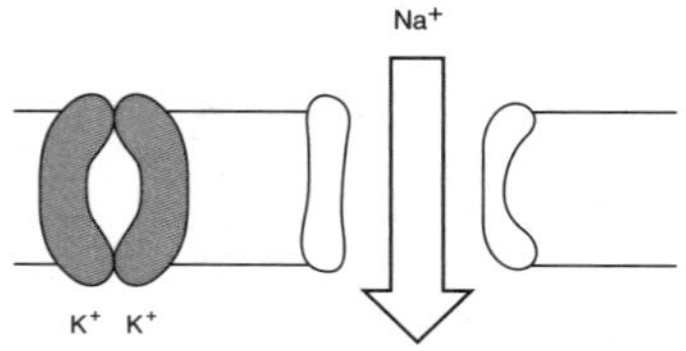

The inward movement of Na^+ ions during depolarisation is an example of a **positive feedback system.** As Na^+ ions continue to move inward depolarisation increases, which opens more sodium channels so more Na^+ ions enter causing more depolarisation and so on.

Repolarisation: sodium channels are closed but potassium channels are open so that K^+ ions are able to move **out of the axon.**

a. down a K^+ **concentration gradient**

b. down an **electrochemical gradient**

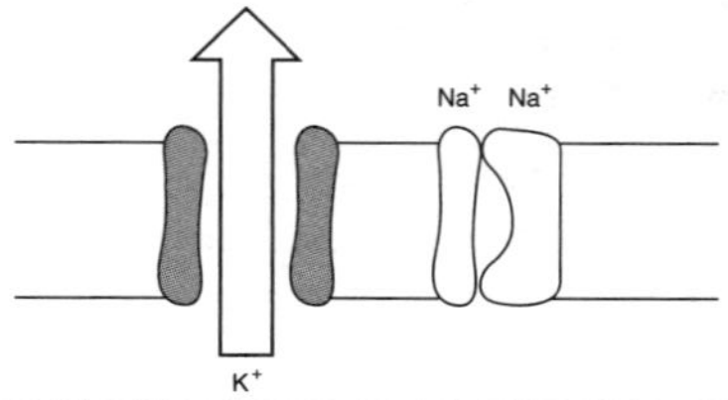

Hyperpolarisation and refractory period: Potassium channels close and short term conformational changes in the pore proteins of the sodium channels mean that these voltage-gated sodium channels are **inactivated.** As a result the neuronal membrane becomes **refractory** – unable to respond to a stimulus which would normally trigger an action potential. This keeps each action potential **discrete** (separate from the next one).

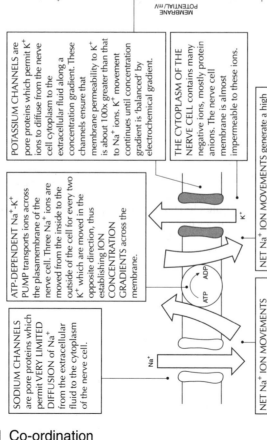

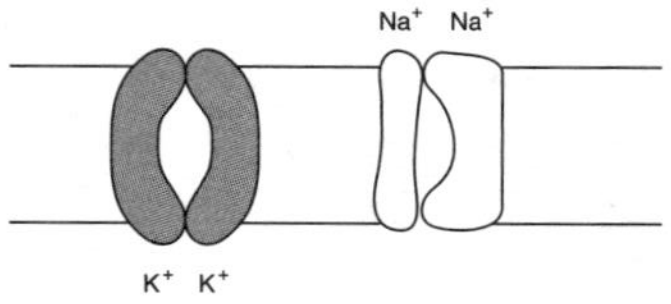

Propagation of action potentials

In this sector the voltage-gated sodium channels are **open** so that Na^+ ions can move into the axon and the neuronal membrane becomes **depolarised**. Positive feedback and K^+ ion outflow cause an action potential which begins to depolarise sodium channels if they are in the active state.

This sector is in its **resting state** (at **resting potential**) with the voltage-gated sodium channels **closed but active**.

In this sector the voltage-gated sodium channels are **closed but active**: the neuronal membrane is no longer in its **refractory period**.

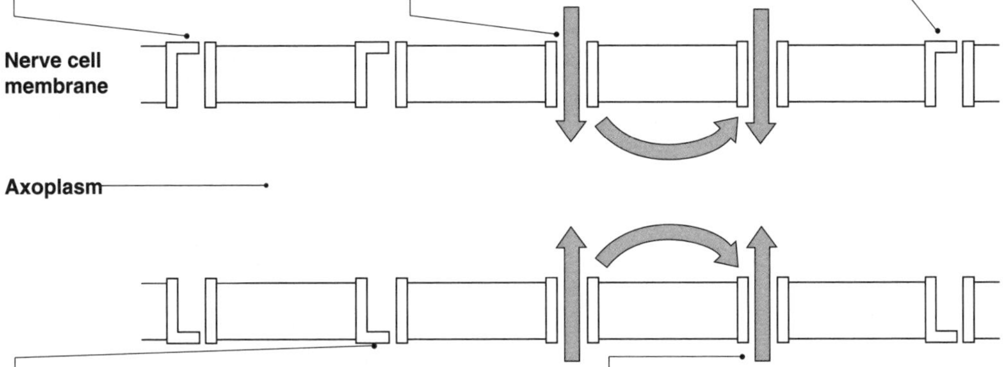

In **continuous conduction** an impulse is transmitted as a **wave of depolarisation** in which action potentials can be detected in immediately adjacent sectors of **unmyelinated neurones.**

In this sector the voltage-gated sodium channels are **closed and inactive**. The neuronal membrane is within its **refractory period** – it is unable to respond to depolarisation in adjacent sectors. This ensures that the waves of depolarisation – the **impulse** – can only travel in **one direction**.

Local depolarisations begin to open voltage-gated sodium channels which are in the **closed but active** state. This can initiate an action potential in this sector.

Direction of propagation of action potential →

Myelin sheath contains **myelin**, a phospholipid which does not conduct an electric current. There are no transmembrane ion movements across myelinated membranes.

Node of Ranvier is a region of unmyelinated nerve cell membrane. Voltage gated ion channels may allow Na^+ and K^+ movement across membrane so that an **action potential** can be generated.

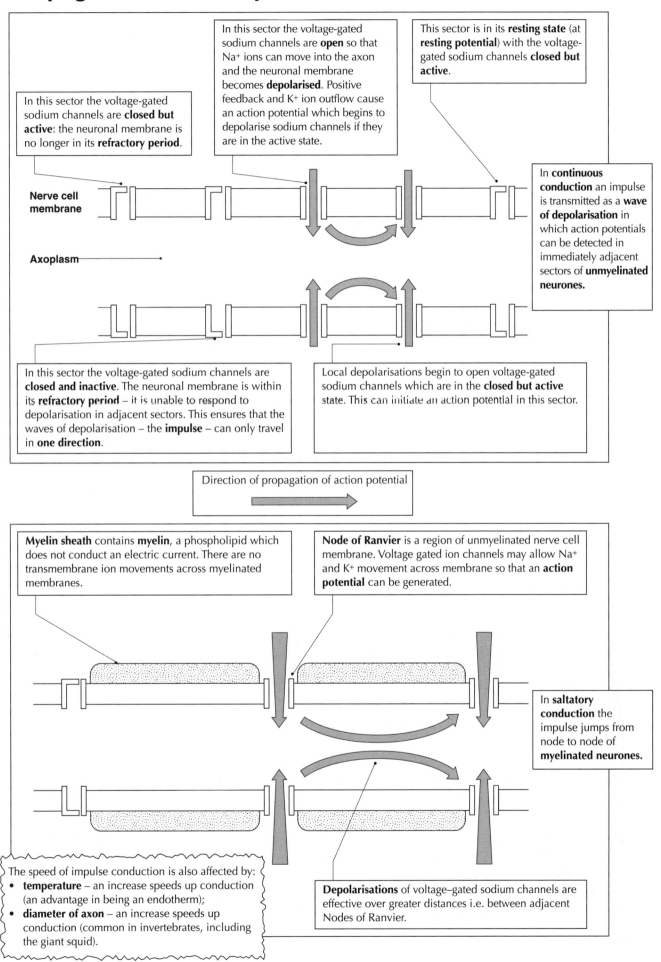

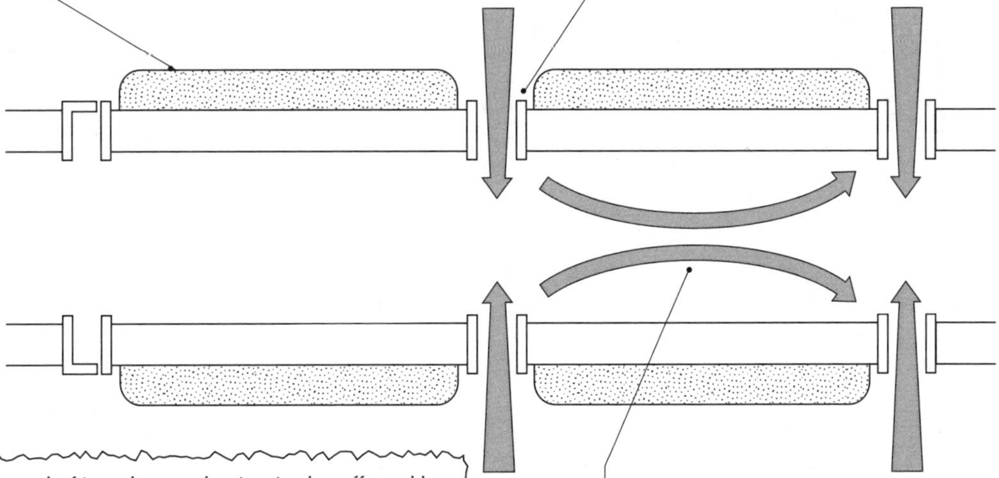

In **saltatory conduction** the impulse jumps from node to node of **myelinated neurones.**

The speed of impulse conduction is also affected by:

- **temperature** – an increase speeds up conduction (an advantage in being an endotherm);
- **diameter of axon** – an increase speeds up conduction (common in invertebrates, including the giant squid).

Depolarisations of voltage–gated sodium channels are effective over greater distances i.e. between adjacent Nodes of Ranvier.

Functions of synapses

Unidirectionality
The synapses acts like a 'valve': action potentials can only cross in one direction (the 'correct' one for the functioning of the body). This is because neurotransmitter molecules can only diffuse one way – there are no vesicles containing these chemicals on the post-synaptic side of the cleft.

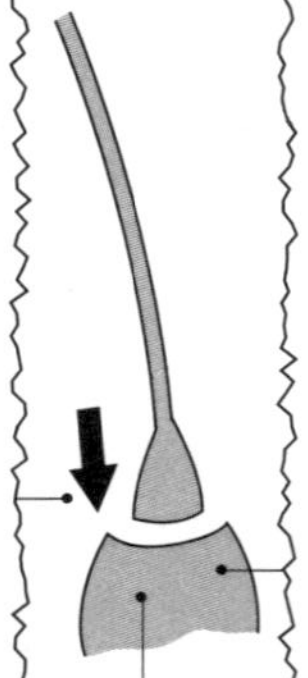

Inhibition
- NT diffuses across cleft
- binding to receptor reduces leakage of Na^+
- post-synaptic membrane becomes **HYPERPOLARISED**

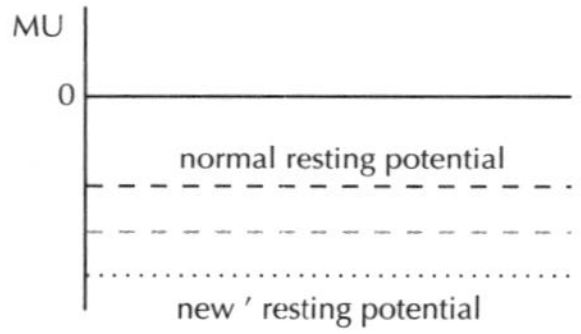

- post-synaptic membrane has further to depolarise: now more difficult for action potential to carry on.

Summation
The frequency of action potentials arriving at the synapse may be very low: not enough NT is released to set off a potential in the post-synaptic membrane.
The effects of separate action potentials can be added together; this is **summation**.

Summation can be **temporal** or **spatial**

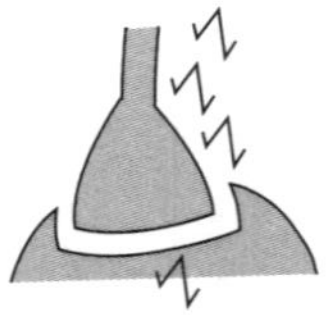

- Several action potentials arrive close after one another.
- NT released is added together.
- Potential carries on in post-synaptic neurone.

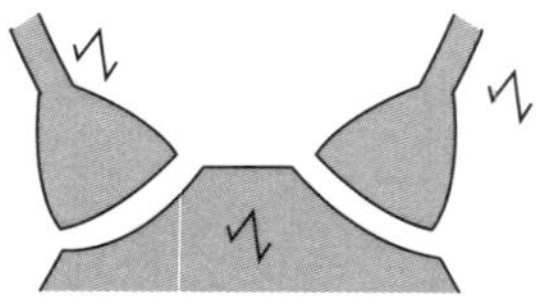

- Several action potentials arrive simultaneously from different axon terminals.
- NT released is added together.
- Potential carries on in post-synaptic neurone.

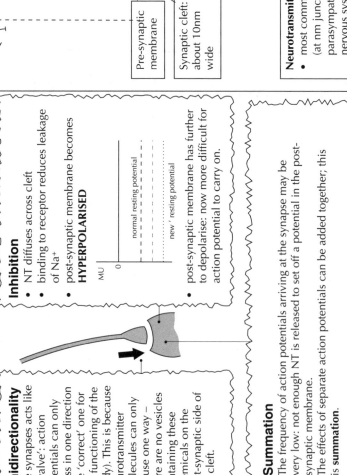

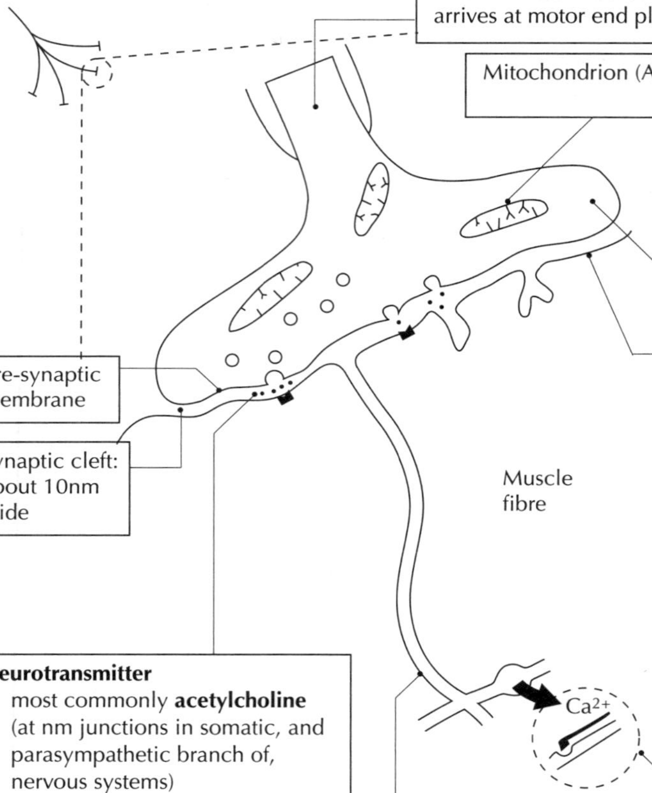

Axon of motor neurone: action potential arrives at motor end plate.

Mitochondrion (ATP for
- synthesis/packaging of neurotransmitter)
- pumping of Ca^{2+} to set up Ca^{2+} concentration gradients.

Motor end plate: large surface area in contact with muscle.

Post-synaptic membrane
- is **sarcolemma** – membrane around the muscle fibre
- has receptors which bind specifically to neurotransmitter molecules diffusing across the cleft.

Myasthenia gravis is a condition in which a reduced number of receptors means muscle does not contract effectively.

Pre-synaptic membrane

Synaptic cleft: about 10nm wide

Release of Ca^{2+} from **sarcoplasmic reticulum** is the signal for 'nodding' of myosin heads and muscle contraction.

Neurotransmitter
- most commonly **acetylcholine** (at nm junctions in somatic, and parasympathetic branch of, nervous systems)
- nor-**adrenaline** (sympathetic N.S.).

t-tubule (Transverse tubule) allows wave of depolarisation to penetrate seep into muscle.

In **muscular dystrophy**, a Ca^{2+}-pumping protein is missing or not functioning well: muscle contraction is inefficient and death usually follows.

Useful drugs at n.m. junctions
- **Curare** competes with A.Ch. at somatic muscles: relaxes muscles during surgery
- **Atropine** competes with A.Ch. at parasymp. nm junctions: allows pupil to dilate during eye examinations.

The neuromuscular junction
is a synapse between a motor neurone and a muscle fibre.

Movement of the forelimb illustrates the action of **muscle groups.** Skeletal muscles produce movement by exerting forces of contraction on tendons, which in turn pull on bones.

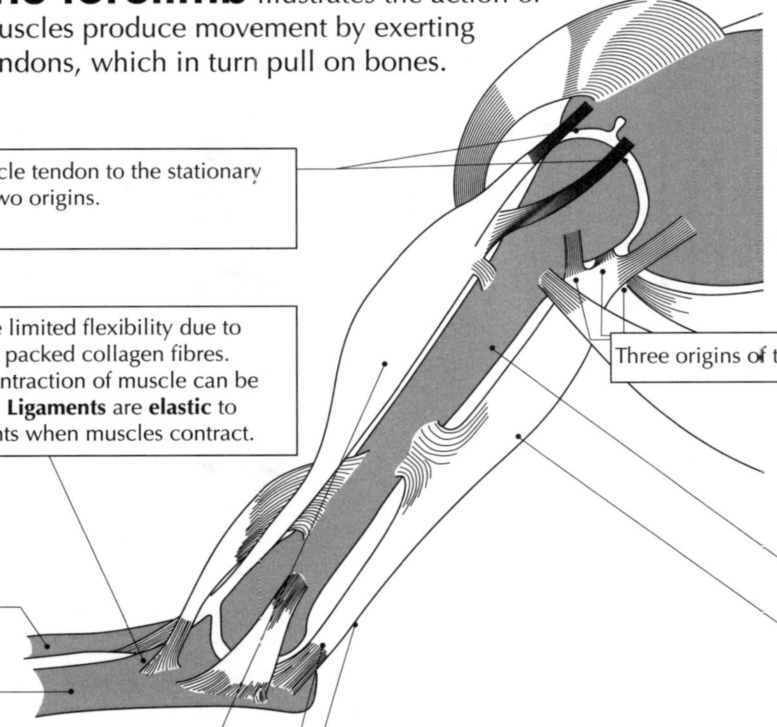

Origin: the attachment of a muscle tendon to the stationary bone. Note that the *bi*ceps has two origins.

Tendon: inelastic but with some limited flexibility due to parallel arrangements of densely packed collagen fibres. Inelasticity is essential so that contraction of muscle can be transmitted to the moving bone. **Ligaments** are **elastic** to allow movement of bones at joints when muscles contract.

Three origins of the *tri*ceps.

Radius

Ulna

Biceps brachii (commonly 'biceps') is the principal **flexor** of the elbow joint. During flexation of the elbow the biceps is the **prime mover** or **agonist** and the triceps brachii is the **antagonist.**
The biceps also supinates the palm and forearm (turns them upward or forward).

Insertion: the attachment of a muscle tendon to the movable bone. This tendon runs across the joint.

The fleshy part of a muscle (the **belly** or **gaster**) does not generally cover the moving part – instead the inserting tendon extends across the joint which permits the movement.

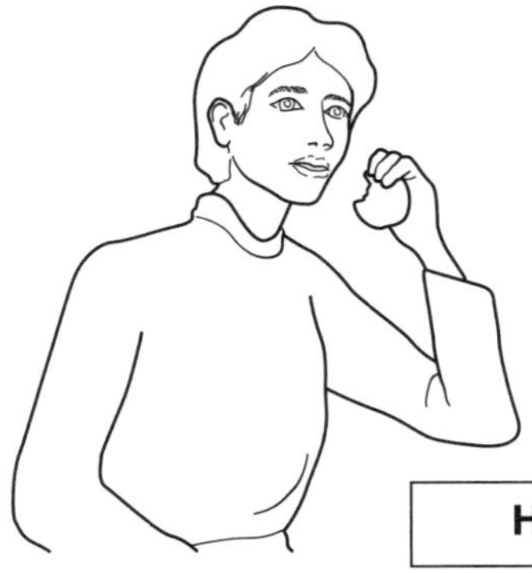

Length of forelimb in humans permits accurate use of arms in feeding.

Human forelimb

Separate ulna and radius allow considerable rotation of forearm. This allows wide range of movement at wrist – vital for efficient use of tools.

Humerus

Triceps is the main **extensor** of the elbow joint. When the biceps contracts and the elbow flexes, the triceps is the **antagonist** of the biceps.

Contraction of biceps causes flexion

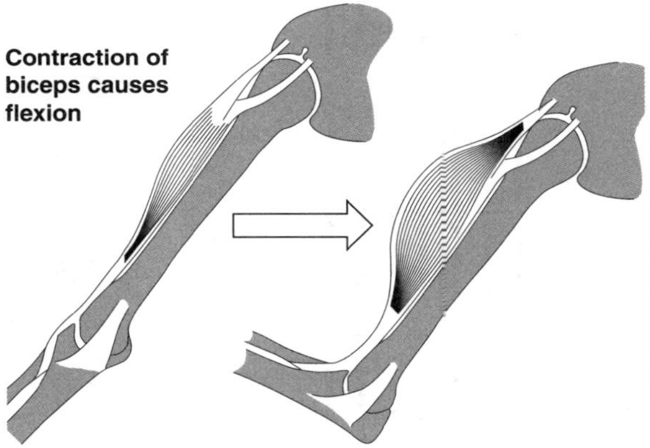

Contraction of triceps causes extension

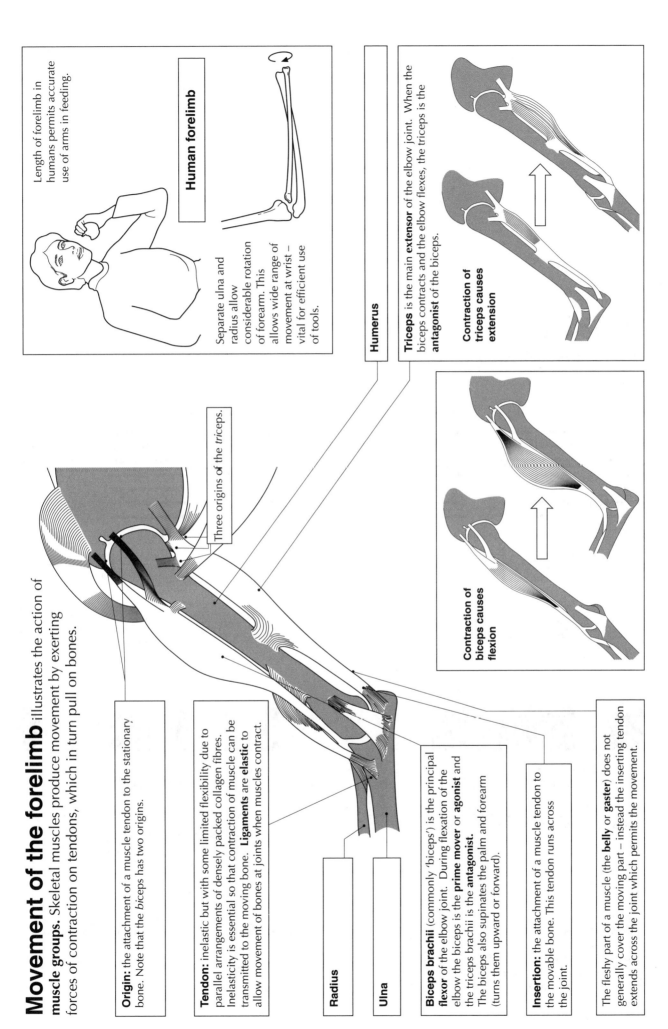

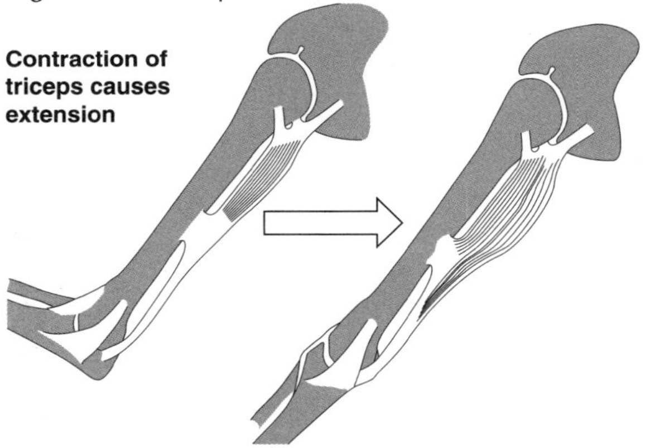

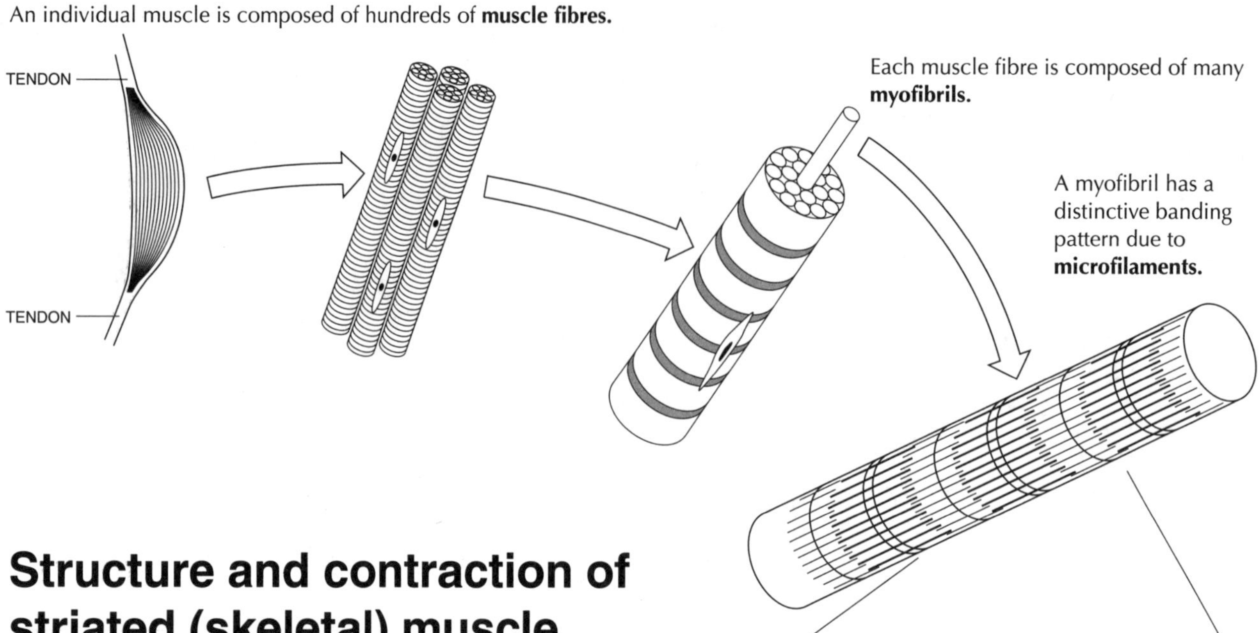
An individual muscle is composed of hundreds of **muscle fibres.**
TENDON
TENDON
Each muscle fibre is composed of many **myofibrils.**
A myofibril has a distinctive banding pattern due to **microfilaments.**

Structure and contraction of striated (skeletal) muscle

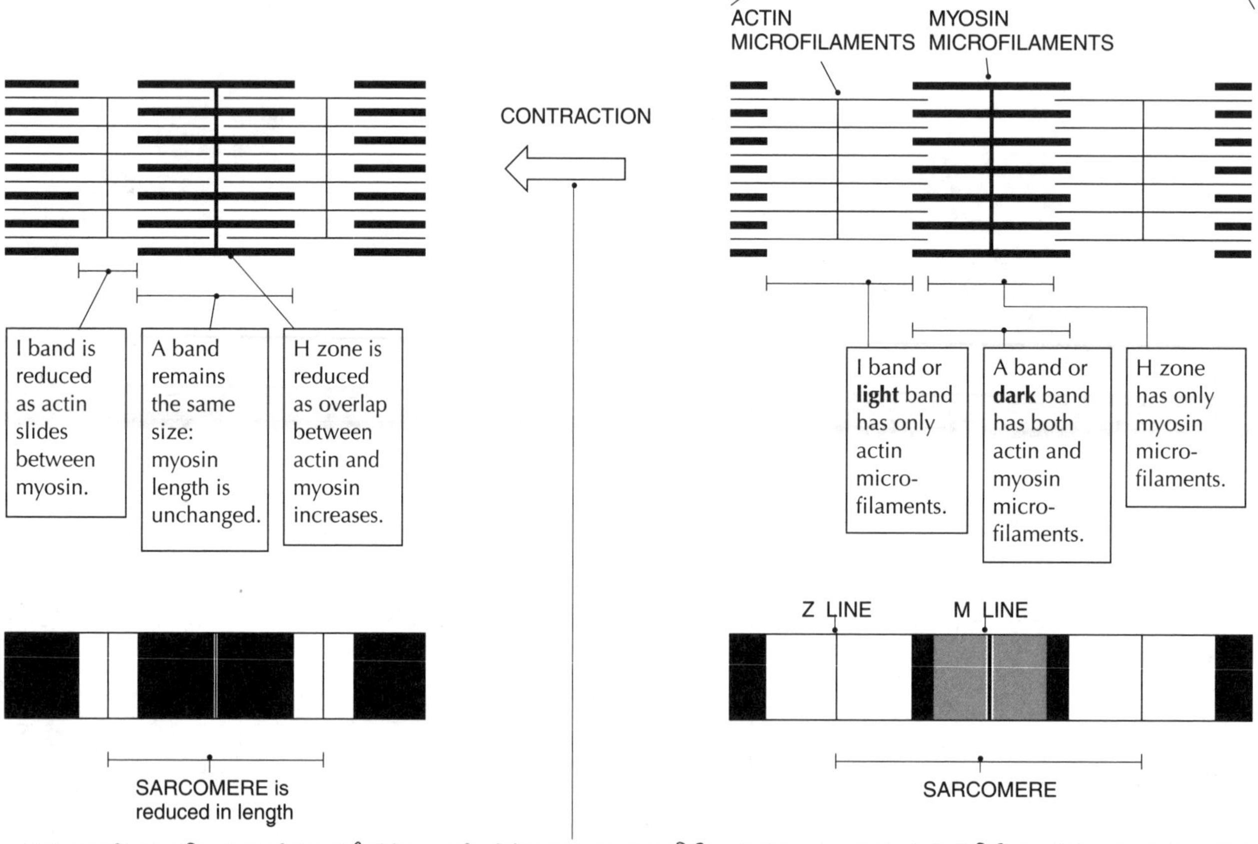
ACTIN MICROFILAMENTS
MYOSIN MICROFILAMENTS
CONTRACTION
I band is reduced as actin slides between myosin.
A band remains the same size: myosin length is unchanged.
H zone is reduced as overlap between actin and myosin increases.
I band or **light** band has only actin micro-filaments.
A band or **dark** band has both actin and myosin micro-filaments.
H zone has only myosin micro-filaments.
Z LINE
M LINE
SARCOMERE is reduced in length
SARCOMERE

Muscle contraction requires Ca^{2+} ions and ATP

Remove the regulatory protein **tropomyosin** which otherwise blocks the binding site on the actin molecule to which the 'nodding head' of myosin can attach during the sliding of the actin and myosin filaments between one another.

Alters the shape of the myosin 'head' so that actin and myosin filaments can slide relative to one another.

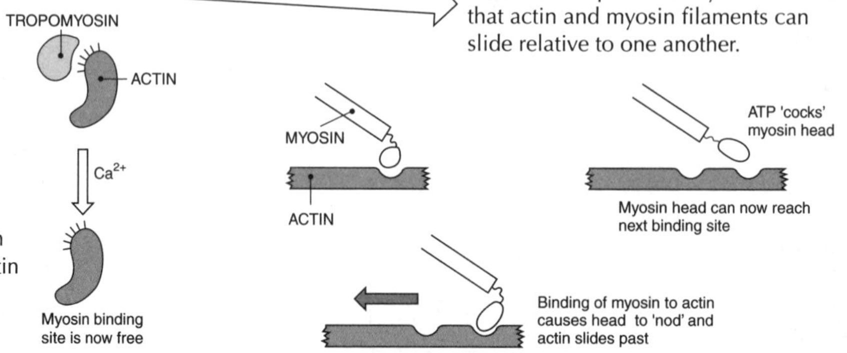
TROPOMYOSIN
ACTIN
Ca2+
Myosin binding site is now free
MYOSIN
ACTIN
ATP 'cocks' myosin head
Myosin head can now reach next binding site
Binding of myosin to actin causes head to 'nod' and actin slides past

Energy for exercise – the source of energy for muscle contraction.

The immediate source of energy for the contraction of muscle (via the formation of actino-myosin cross bridges) is adenosine triphosphate (ATP). A problem is encountered when stimulation is continuous, since there is only a low concentration of ATP in a muscle cell – enough for about 7–8 'twitches' only. ATP is replaced by the transfer of a phosphate group from **creatine phosphate** to ADP.

Many athletes consume creatine to improve their capacity for work.

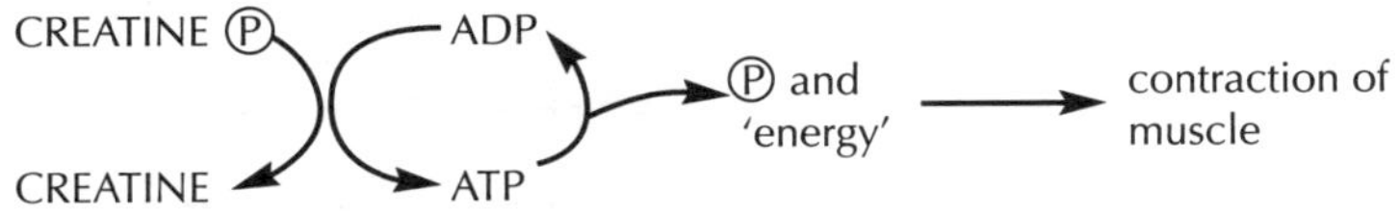

CREATINE Ⓟ
CREATINE
ADP
ATP
Ⓟ and 'energy'
contraction of muscle

Fast or slow muscle?

Type	Fast/white: sprint (little myoglobin)	Slow/red: marathon (much myoglobin)
Location	Close to body surface	Deep inside the limbs
Main use	Immediate, **fast** contraction before circulatory system has adjusted to new levels of activity: **Used for locomotion**	**Slow**, sustained contraction without exceeding oxygen delivery by circulatory system: **Important for maintenance of posture**
Characteristics	Fatigue quickly Little myoglobin, few mitochondria Glycogen main respiratory substrate Quickly become anaerobic, and build up oxygen debt	Do not fatigue quickly Much myoglobin, and many mitochondria Respire glucose and fatty acids Only become anaerobic very slowly

Anaerobic respiration and the oxygen debt: The liver removes lactate produced in the muscles.

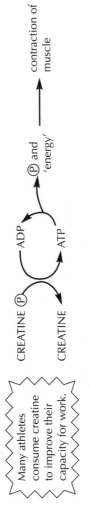

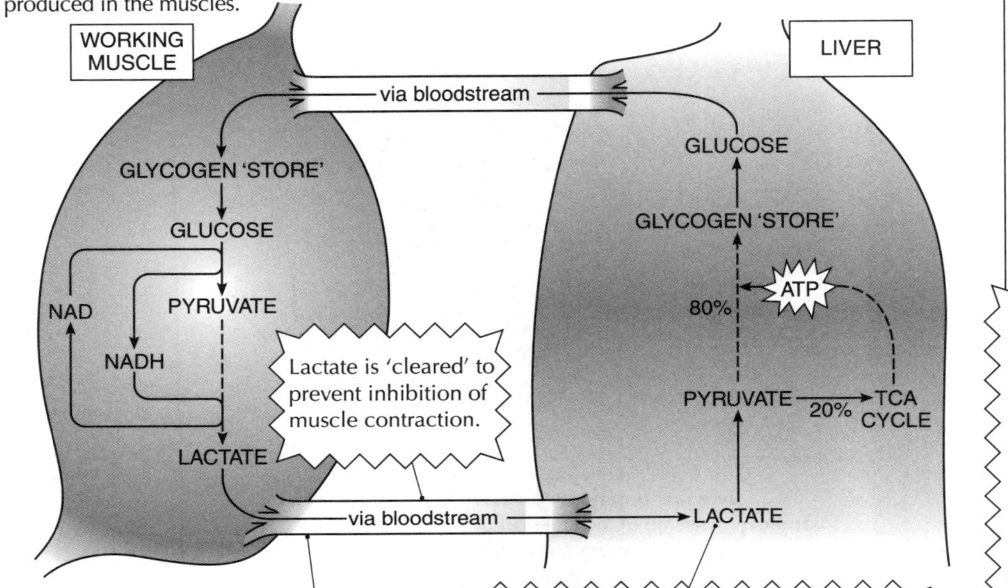

WORKING MUSCLE
LIVER
via bloodstream
GLYCOGEN 'STORE'
GLUCOSE
NAD
PYRUVATE
NADH
LACTATE
Lactate is 'cleared' to prevent inhibition of muscle contraction.
via bloodstream
GLUCOSE
GLYCOGEN 'STORE'
80%
ATP
PYRUVATE
20%
TCA CYCLE
LACTATE

Training improves blood supply to and from muscle and therefore increases the rate at which lactate can be removed.

The additional oxygen required to reoxidise the lactate accumulated during anaerobic exercise is called the **oxygen debt**. Physiologically this is limited to about 20 dm^3, rather more in well-trained athletes.

Fatigue: If a muscle is continuously stimulated over an extended period of time the strength of contraction becomes progressively weaker. This is **fatigue**.

Characterised by:

- ↓ O_2 availability
- ↑ [lactate]: toxic effect on neuromuscular junctions
- ↓ $[K^+]$
- Bohr shift: O_2-Hb curve moves to the right.

The significant factors that contribute to muscular fatigue are:

1. excessive activity, resulting in the accumulation of toxic products;
2. malnutrition, resulting in insufficient supplies of glucose and, therefore, ATP;
3. cardiovascular disturbances that impair the delivery/removal services for the muscles;
4. respiratory disturbances that interfere with the oxygen supply and increase the oxygen debt;
5. alcohol consumption which causes accumulation of high concentrations of lactate.

The DRD4 dopamine receptor is linked to many aspects of Human behaviour.

The **nucleus accumbens** is the region of the forebrain which contains the **dopamine circuit**. This region plays an important role in **pleasure, laughter, reward, addiction** and **fear**.

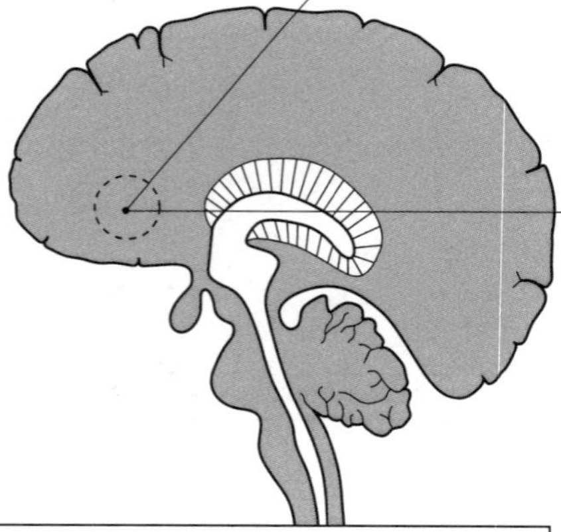

Dopamine is a neurotransmitter of the **catecholamine** type: it is produced from the amino acid **tyrosine**.

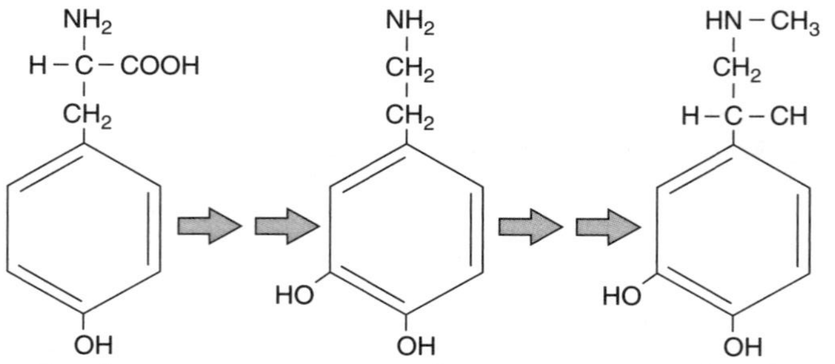

Tyrosine **Dopamine** **Adrenaline**

Dopamine is mainly inhibitory: it is involved in control of voluntary movement, learning, sexual arousal and feelings of pleasure.

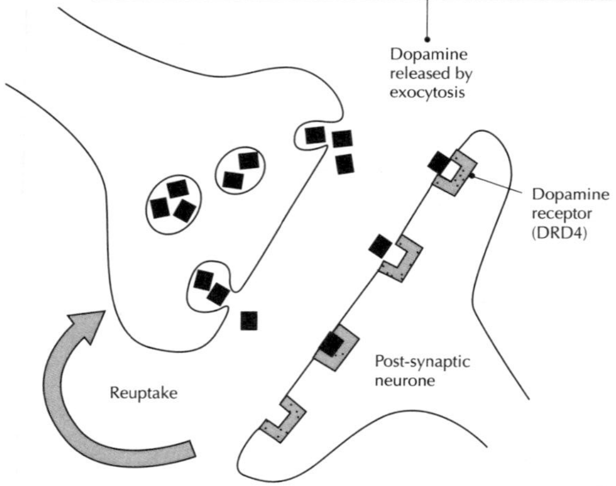

How do we know?
Mice can be genetically-engineered to mutate the DRD4 gene.

- Some have learning disorders (similar to ADHD)
- Many lack ability to become aroused.
- Some develop hypertension (dopamine may help to control salt release in the urine).

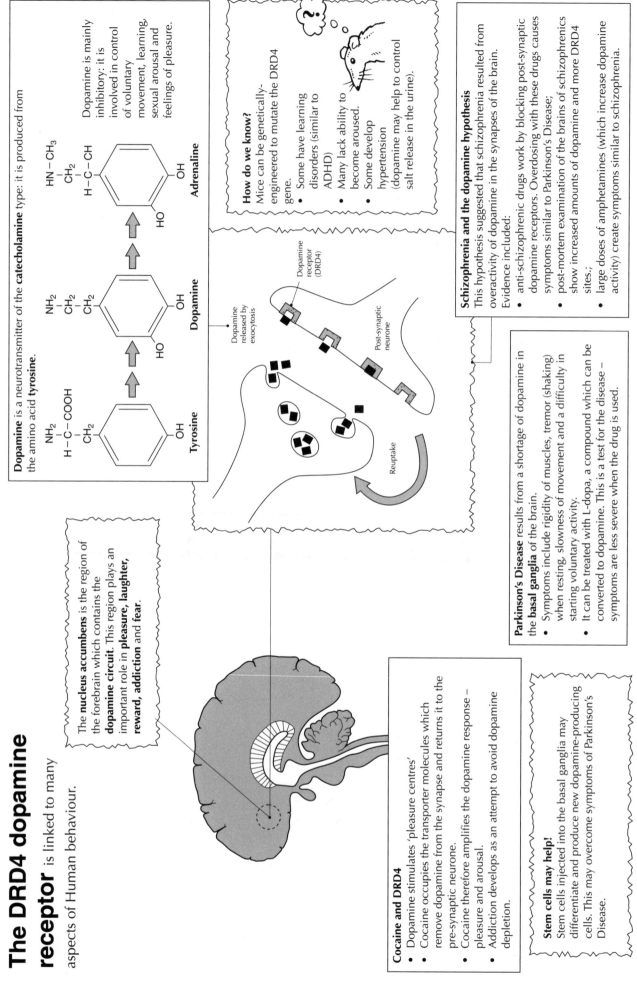

Cocaine and DRD4

- Dopamine stimulates 'pleasure centres'
- Cocaine occupies the transporter molecules which remove dopamine from the synapse and returns it to the pre-synaptic neurone.
- Cocaine therefore amplifies the dopamine response – pleasure and arousal.
- Addiction develops as an attempt to avoid dopamine depletion.

Stem cells may help!
Stem cells injected into the basal ganglia may differentiate and produce new dopamine-producing cells. This may overcome symptoms of Parkinson's Disease.

Parkinson's Disease results from a shortage of dopamine in the **basal ganglia** of the brain.

- Symptoms include rigidity of muscles, tremor (shaking) when resting, slowness of movement and a difficulty in starting voluntary activity.
- It can be treated with L-dopa, a compound which can be converted to dopamine. This is a test for the disease – symptoms are less severe when the drug is used.

Schizophrenia and the dopamine hypothesis
This hypothesis suggested that schizophrenia resulted from overactivity of dopamine in the synapses of the brain. Evidence included:

- anti-schizophrenic drugs work by blocking post-synaptic dopamine receptors. Overdosing with these drugs causes symptoms similar to Parkinson's Disease;
- post-mortem examination of the brains of schizophrenics show increased amounts of dopamine and more DRD4 sites.;
- large doses of amphetamines (which increase dopamine activity) create symptoms similar to schizophrenia.

Learned behaviour: modified by experience

Habituation occurs when an **insignificant stimulus** (i.e. one that is neither **beneficial** nor **harmful**) is repeated, and an animal learns **not** to respond to it,

e.g. sleeping through background traffic noise, but still waking up when an alarm clock rings.

The advantage is that, by ignoring these stimuli, animals spend both time and energy more efficiently.

Classical conditioning occurs when an animal learns to associate a **neutral** stimulus with an important one: the response to the important stimulus is now a **conditioned response**, and is

- shown in the presence of the neutral stimulus (called the **conditioned** stimulus)
- involuntary, and reinforced by repetition
- temporary, and will be **unlearned** if the neutral stimulus completely replaces the important one.

Pavlov: showed dogs would salivate on hearing a bell normally rung when receiving food.

Operant conditioning ('trial and error' learning) occurs when an animal learns to associate a response with a reward of a punishment.

Skinner: used a specially-devised box (a Skinner box) which allowed rats or pigeons to choose an action – through trial and error they learned which action (which button to press) gave them a reward of food.

A mild electric shock could also be used to teach them which button **not** to press.

mmm...
... which one?
RATTY GUMS

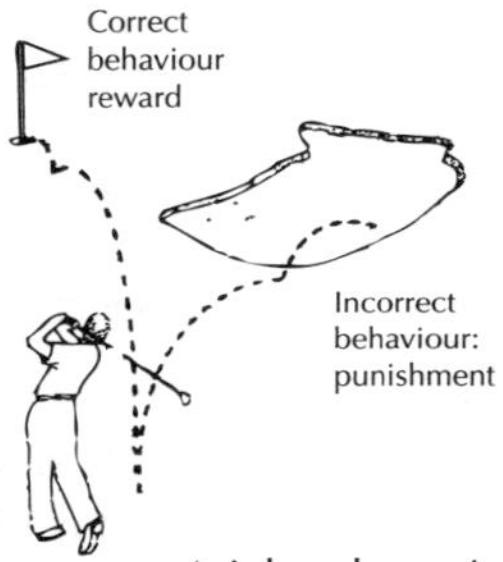
Correct behaviour reward
Incorrect behaviour: punishment

Humans use 'trial and error' learning: a golfer might try a different grip on the club when hitting a golf shot.

Latent learning occurs when an animal acquires knowledge at a certain time, without any positive reinforcement, but does not use it until a later time when the knowledge is needed.

Example: learning to use a screwdriver by watching a parent, without any obvious reward at the time

Example: learning a route when being driven to school.

SHOE SHOP

Insight behaviour (reasoning) is the most complex form of learning:

- is the ability to solve a problem without trial and error – may involve recall of past experiences.

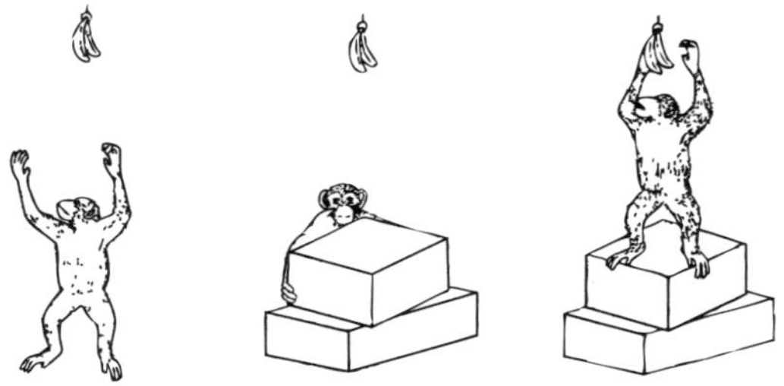

Köhler: showed that chimpanzees could 'work out' (reason) how to reach fruit. Only humans and other primates can do this.

Imprinting involves:

- a combination of innate **and** learned behaviour
- occurs within a limited **critical period** of birth

Lorenz: showed that ducklings would follow a human if the human was present as the duckling hatched. There is **innate** behaviour to follow the mother, and **learned** behaviour about who the mother is.

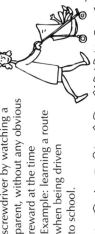

Somebody loves me!
Mama!

Behaviour in primates

What about human social behaviour?
- form social groups which may extend beyond survival e.g. based on income, nationality or cultural interests
- have a highly-developed power of speech, and can use written and drawn symbols for communication
- have an extended range due to their ability to modify their environment – make homes in fixed position
- make complex tools, and have learned to control fire for cooking and heating.

Mothers and infants: form the basic social group for many primates
- mother-infant bonding needed for young to develop into socially-capable adults
- may continue after infancy, with females remaining together and males dispersing to other groups
- combined power of such a female group may mean they are dominant to the alpha-male.

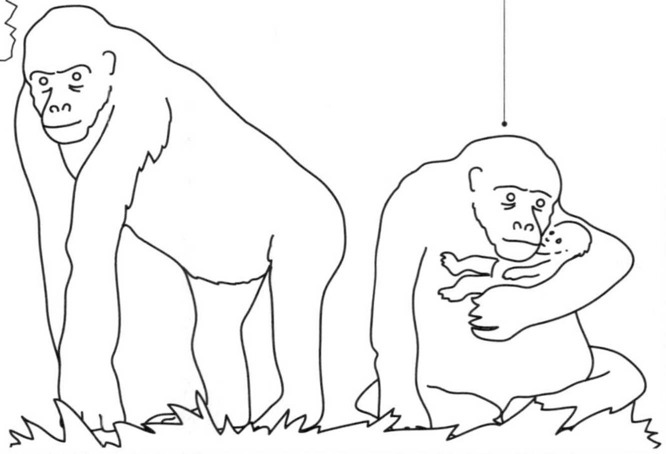

Cultural behaviour: this is learned behaviour that is passed from generation to generation.
e.g.
- Japanese Macaques: learned that sweet potatoes taste better if sand is washed off.
- Chimpanzees: 'termite-fishing' using pointed sticks seems to be learned by all members of a group, and passed on to infants.
- Many examples in Humans, including use of writing and drawing instruments.

Reproduction: all primate females (except for Humans) show a seasonal or cyclical willingness to mate.
- Female usually shows visual changes (e.g. swellings around the anus and/or vulva) so it is clear when she is ready to mate.
- Bonobos (pygmy chimps) may have periods of frequent copulation which may be important in pair bonding.

Social behaviour

Dominance: primates are mainly group-living animals and form **dominance hierarchies.**

- higher-ranked animals displace lower-ranked animals from food, mates and 'space'. They usually have more reproductive success (often through mating more frequently)
- are **not** fixed, and depend on age, sex, aggression and intelligence
- rank is learned through play, fighting and peer support behaviour.

Grooming: grooming of others (allogrooming) is an important support mechanism:
- subordinates groom dominant animals
- males groom females for sexual access
- mothers groom infants to clean their fur.

Communication
- includes scents, vocalisation, postures and gestures
- may be reflexes indicating emotional state e.g. teeth baring/eyelid flashing or may have a more specific purpose e.g. loud territorial calls in howler monkeys
- can be used to reduce aggression e.g. presentation of rear, and offers of copulation in baboons.

There has been some success in teaching sign language to chimpanzees!

Aggression
- may occur **within** the group to keep dominance hierarchies
- may involve communal activity to **protect** the group e.g. baboons cooperate to repel leopards
- some scientists suggest that bipedalism evolved so males could look taller and could beat their chests during displays of aggression.

Chemical signals in communication

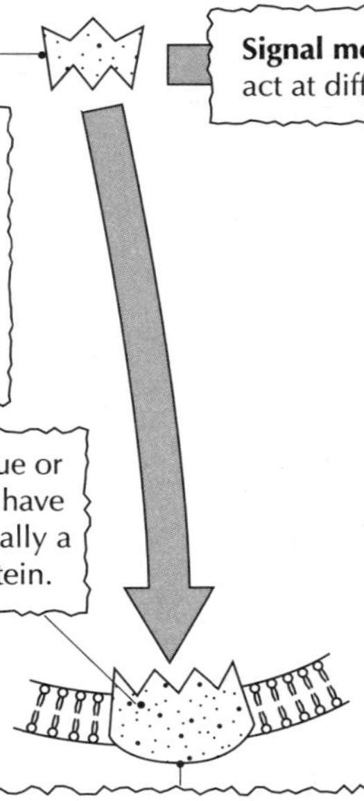

Released chemical acts as a **signal.** Includes fatty acids, steroids, peptides and amines.

Target cell, tissue or organism must have a **receptor,** usually a membrane protein.

The presence of the receptor gives the **specificity** to the response.

Signal molecules may act at different levels.

Within the same cell i.e. a molecule produced in one part of the cell affects an activity somewhere else in the cell, e.g.
- messenger RNA (mRNA) is produced by transcription in the nucleus but is 'read' at the ribosomes
- cyclic AMP – a second messenger – is produced close to membrane-bound adenyl cyclase but changes the activity (by alteration of three dimensional shape) of enzymes in the cytoplasm.

Between cells close to one another – these are **local hormones**

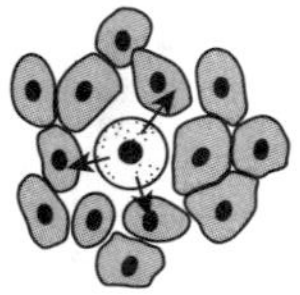

Prostaglandins are a type of fatty acid:
- originally found in the prostate gland (hence their name)
- work by controlling activity of adenyl cyclase, and so adjust levels of cAMP in cells
- involved in control of blood pressure (by affecting contraction of smooth muscle of artery walls), release of acid in the stomach and inflammation.

Medical importance!
Synthetic PGs are used:
- to induce labour by stimulation of uterine muscle;
- to reduce hypertension;
- to control symptoms of asthma;
- to reduce risk of stomach ulcers.

Histamine is made from the amino acid, **histidine:**
- causes vasodilation
- affects capillary permeability during inflammation
- acts as a brain neurotransmitter, where it helps to regulate body temperature and water balance.

Endorphins are polypeptides:
- mimic the effects of opiate drugs such as heroin and morphine
- block pain receptors in the brain
- stimulate dopamine secretion, so increase sensation of pleasure.

Marathon runners have very high levels of endorphins!

Between tissues: these are the typical HORMONES
- include steroids (e.g. oestrogens), amines (e.g. adrenaline) and polypeptides (e.g. insulin)
- control internal environment (homeostasis) e.g. insulin and blood glucose concentration
- control growth rate (e.g. human growth hormone)
- control sexual development and reproductive activity (e.g. oestrogen, testosterone, oxytocin)
- manage the body's response to stress (e.g. adrenaline, cortisone).

Between different individuals of the same species: these are PHEROMONES
- many are fatty acids – they are volatile and disperse quickly into the environment
- can act over long distances (e.g. sex attractants) or very close to producer (e.g. queen substance in a hive)
- may be important in human sexual attraction, although we mask them with perfumes!

Know your place workers
WHIFF!

Nervous and chemical co-ordination compared

	Nervous	Chemical e.g. endocrine
Nature of message	Electrochemical impulses	Chemical compounds (hormones)
Route of transmission	Specific nerve cells	General blood system
Type of effects	Rapid, but usually short-term e.g. blinking	Usually slower, but generally longer-lasting e.g. growth

Note that the hormone **adrenaline** has effects which are rapid but short-term, and therefore partially mimics some effects of the sympathetic nervous system.

Control of body temperature in endotherms

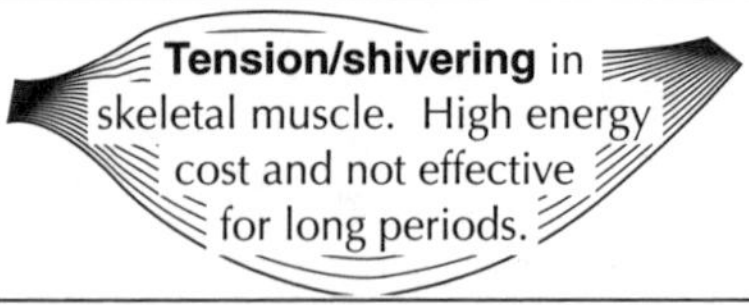

Tension/shivering in skeletal muscle. High energy cost and not effective for long periods.

External temperature high

Pilo-erector muscles relaxed: hair shafts 'flatten' and allow free circulation of air over hairs. Moving air is a good convector of heat.

External temperature low

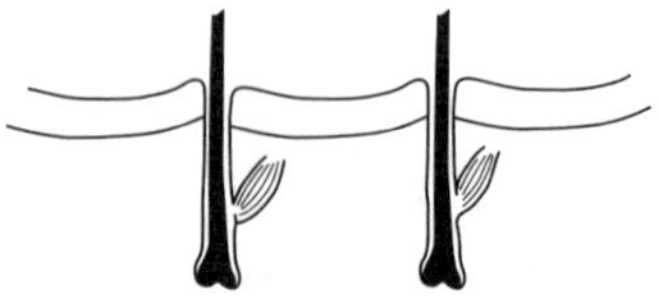

Pilo-erector muscles contracted: hair shafts perpendicular to skin surface. Trapped air is a poor conductor of heat so warm skin is insulated (similar effect by adding layers of clothing).

External temperature high

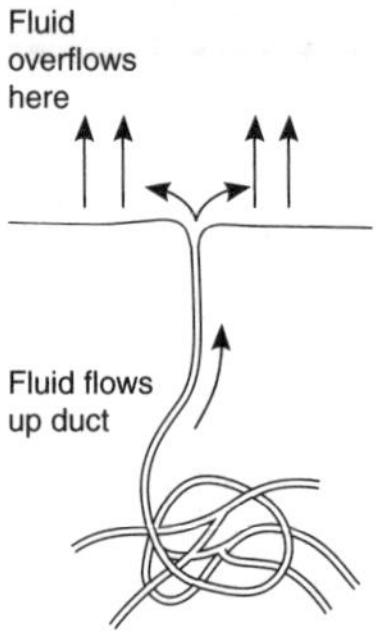

Sweat changed into vapour, taking latent heat of evaporation from the body to do this (about 2.5 kJ for each gram evaporated). Sweat glands extract larger volume of fluid from blood.

External temperature low

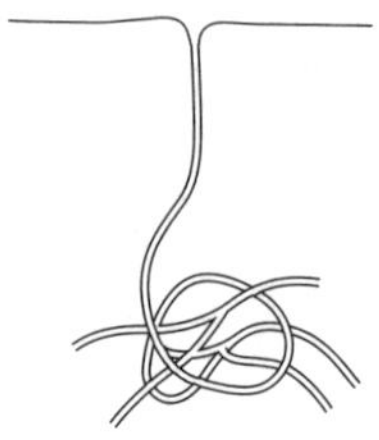

Skin surface comparatively dry – no evaporation and no cooling effect.

Sweat glands extract very little fluid from blood.

Non-shivering heat response triggers a general increase in metabolic rate. This is particularly noticeable in **brown adipose tissue** of newborns and animals that become acclimatised to cold. The **liver** of adults is also affected.

Changes in behaviour

Dressing/undressing; moving in and out of shade; rest/activity cycles may all affect heat loss or production.

Hypothalamus contains the thermoregulatory centre which compares sensory input with a set point and initiates the appropriate motor responses. The set point may be raised by the action of pyrogens during pyrexia (fever).

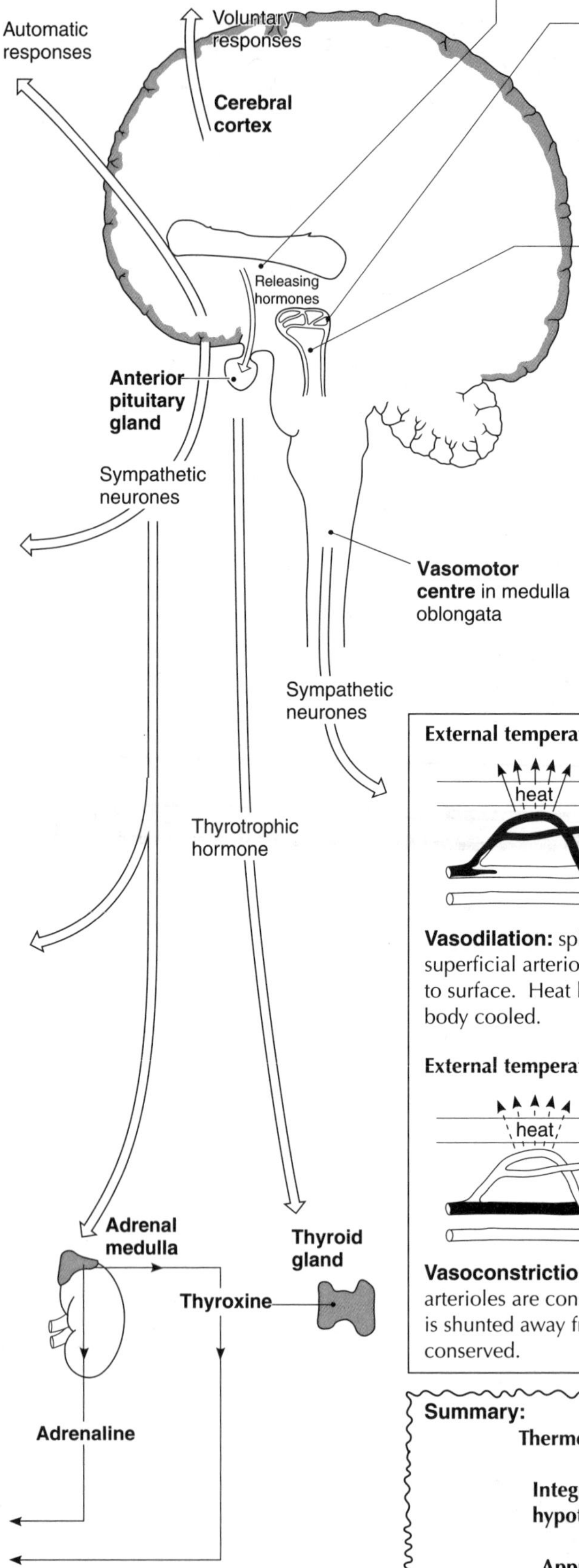

Core temperature affects temperature of circulating blood which is monitored in the thermoregulatory centre.

Skin temperature is detected by skin thermoreceptors which deliver sensory input to hypothalamus via cutaneous nerves.

External temperature high

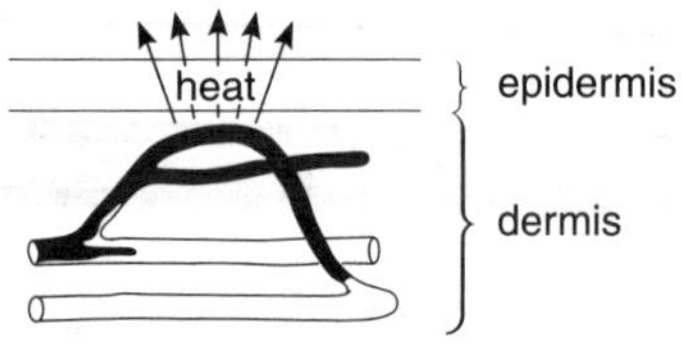

Vasodilation: sphincters/dilation of superficial arterioles allow blood close to surface. Heat lost by radiation – body cooled.

External temperature low

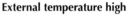

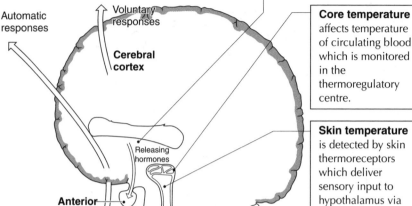

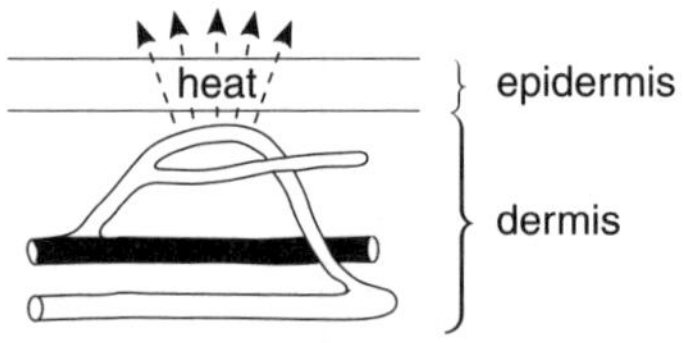

Vasoconstriction: superficial arterioles are constricted so that blood is shunted away from surface – heat is conserved.

Summary:

Thermoreceptors
↓
Integration by hypothalamus
↓
Appropriate responses

Ectotherms attempt to maintain body temperature by **behavioural** rather than **physiological** methods. These methods are less precise and so thermoregulation is more difficult.

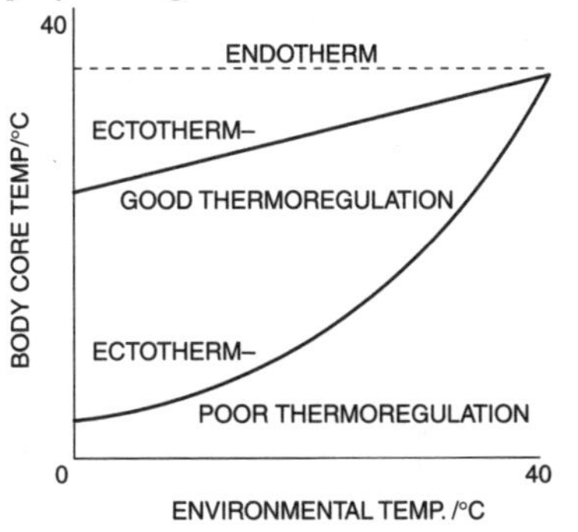

Reorientation of the body with respect to solar **radiation** can vary the surface area exposed to heating. A terrestrial ectotherm may gain heat rapidly by aligning itself at right angles to the Sun's rays but as its body temperature rises it may reduce the exposed surface by reorientating itself parallel to the Sun's rays.

Thermal gaping is used by some larger ectotherms such as alligators and crocodiles. The open mouth allows heat loss by **evaporation** from the moist mucous surfaces. Some tortoises have been observed to use a similar principle by spreading saliva over the neck and front legs which then acts as an evaporative surface.

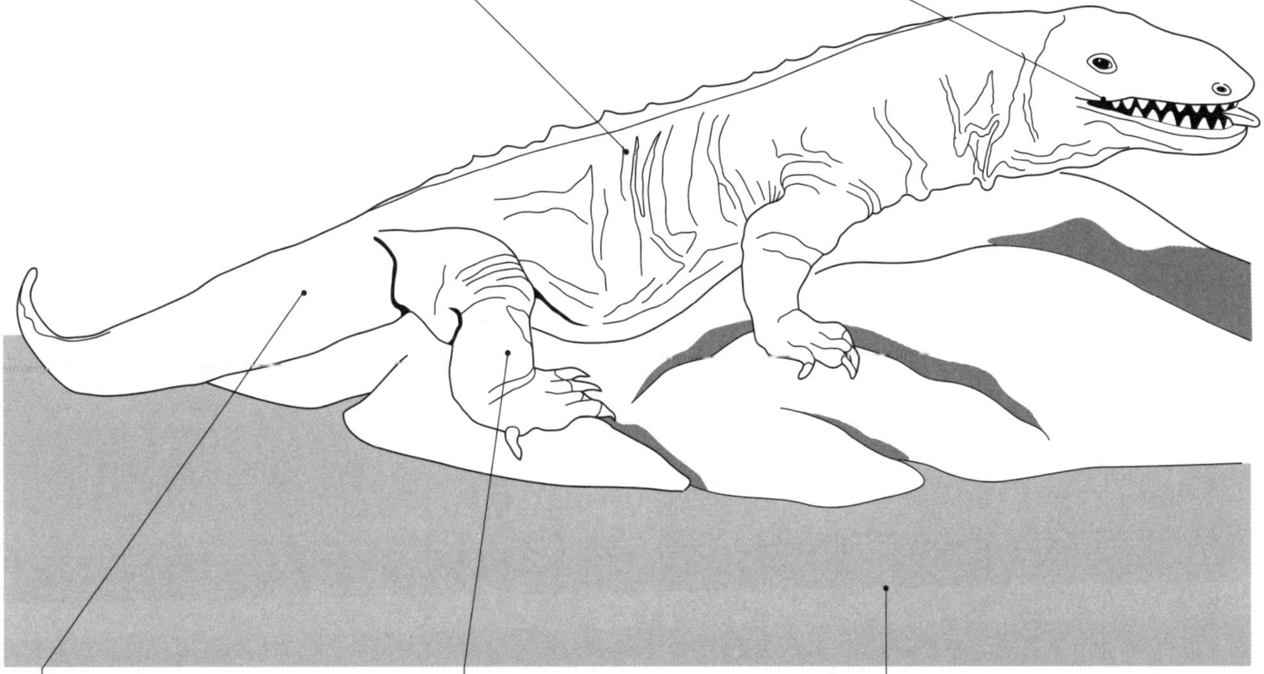

Colour changes of the skin may alter the ability of the body to absorb **radiated** heat energy. A dark-bodied individual will absorb heat more rapidly than a light-bodied one – thus some ectotherms begin the day with a dark body to facilitate 'warming-up' but then lighten the body as the environmental temperature rises.

Body raising is used by ectotherms to minimise heat gains by **conduction** from hot surfaces such as rocks and sand. The whole body may be lifted and the animal may reduce the area of contact to the absolute minimum by balancing on alternate diagonal pairs of feet.

Burrowing is a widely used behavioural device which enables ectotherms to avoid the greater temperature fluctuations on the surface of their habitat. The temperature in even a shallow burrow may only fluctuate by 5°C over a 24 h period whereas the surface temperature may range over 40°C during the same time. Amphibious and semi-aquatic reptiles such as alligators and crocodiles may return to water rather than burrow, since the high heat capacity of water means that its temperature is relatively constant.

The marine iguana and bradycardia

The marine iguana of the Galapagos Islands feeds by browsing on seaweed gathered from the sea around the rocky shores on which it lives. When basking on the rocks it normally maintains a body temperature of 37°C but during the time spent feeding in the sea it is exposed to environmental temperatures of 22–25°C. In order to avoid losing heat rapidly by **conduction and convection** the iguana reduces the flow of blood between its core tissues (at 37°C) and its skin (22°C) by slowing its heart rate (bradycardia).

Heat transfer between the body of an organism and its environment depends on the **magnitude** and **direction** of the **thermal gradient** (i.e. the temperature difference between the organism and its surroundings). Heat may be **lost** or **gained** by
conduction (heat transfer by physical contact)
convection (heat transfer to the air) and
radiation (heat transfer in the form of long-wave, infra-red electromagnetic waves)
but can only be **lost** by
evaporation (heat consumption during the conversion of water to water vapour).

Endocrine control depends upon chemical messengers secreted from cells and binding to specific hormone receptors.

A hormone is a chemical produced in one part of an organism which is transported throughout the organism and produces a specific response in target cells.

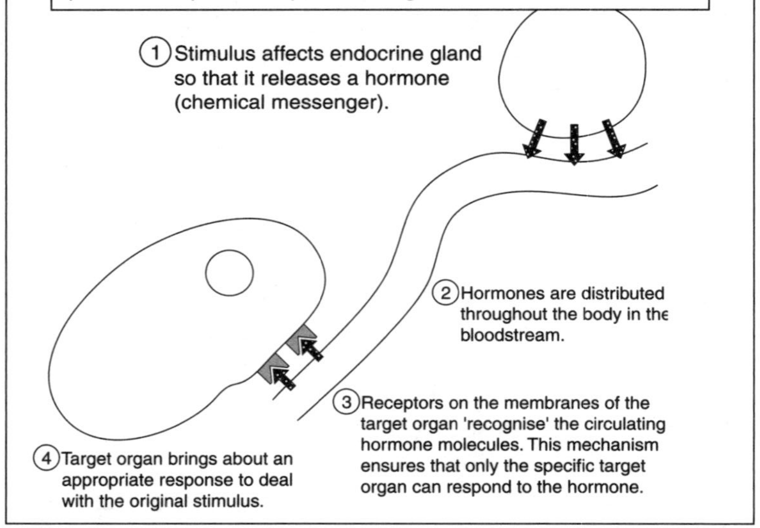

① Stimulus affects endocrine gland so that it releases a hormone (chemical messenger).
② Hormones are distributed throughout the body in the bloodstream.
③ Receptors on the membranes of the target organ 'recognise' the circulating hormone molecules. This mechanism ensures that only the specific target organ can respond to the hormone.
④ Target organ brings about an appropriate response to deal with the original stimulus.

Endocrine or Exocrine

- Endocrine: secretes hormones into the bloodstream via capillaries
- Exocrine: secretes product via a duct

Removal of hormone:

- by excretion in urine (e.g. steroids can be detected in 'cheating' athletes)
- by deamination in the liver
- by enzymic removal of second messenger (caffeine prevents removal of CAMP – cell remains stimulated!) …

… and **hormone production** is closely controlled by **negative feedback.**

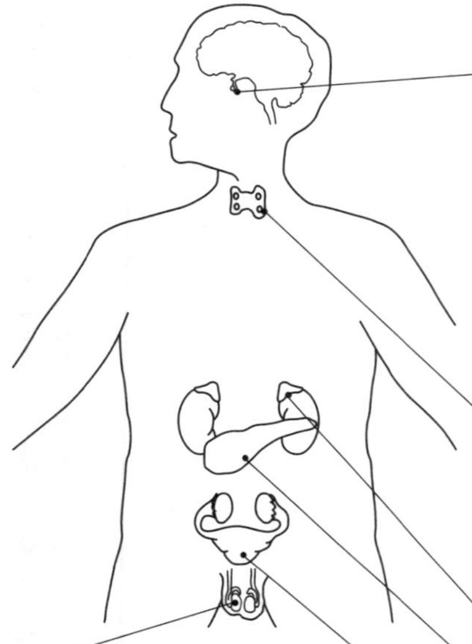

Pituitary gland secretes:

- a number of hormones (trophins) which influence other endocrine organs;
- human growth hormone
- anti diuretic hormone.

Thyroid gland secretes thyroxine which controls metabolic rate and is important in regulation of body temperature.

Testes: testosterone

Ovaries: oestrogen and progesterone

Pancreas: insulin and glucagon

Medulla of adrenal glands secrete **adrenaline** which is called the **fight or flight hormone** because it prepares the body for action:

- more glycogen is converted to glucose
- deeper, more rapid breathing
- faster heartbeat
- diversion of blood from gut to muscles.

The above all help to provide more glucose and more oxygen for the working muscles. There may also be

- dilation of the pupils
- raising of the hair.

Adrenaline mimics the **sympathetic N.S.**

Adrenaline works through a second messenger

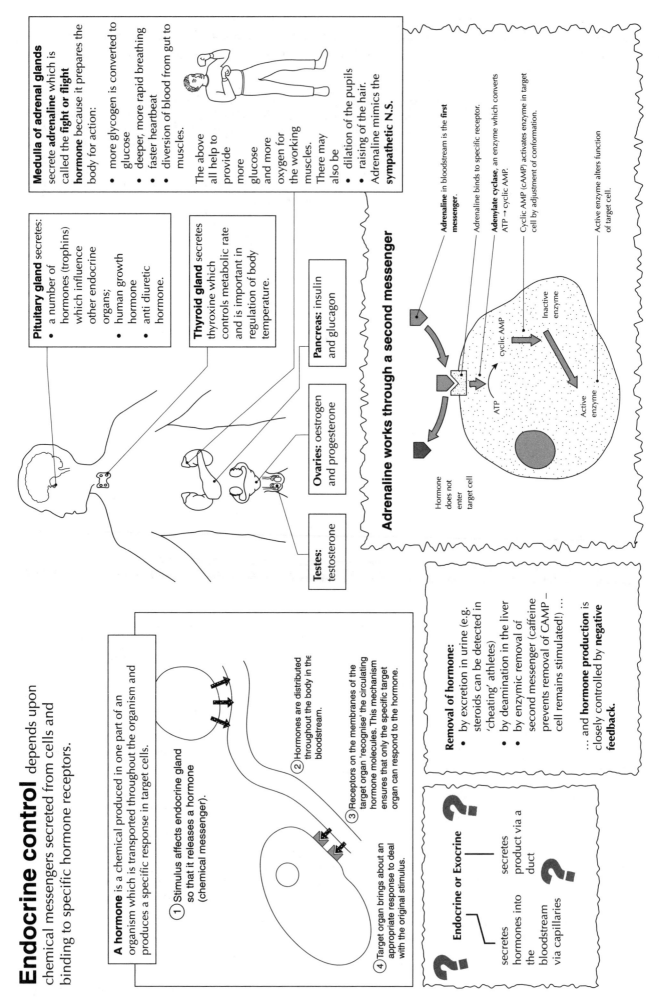

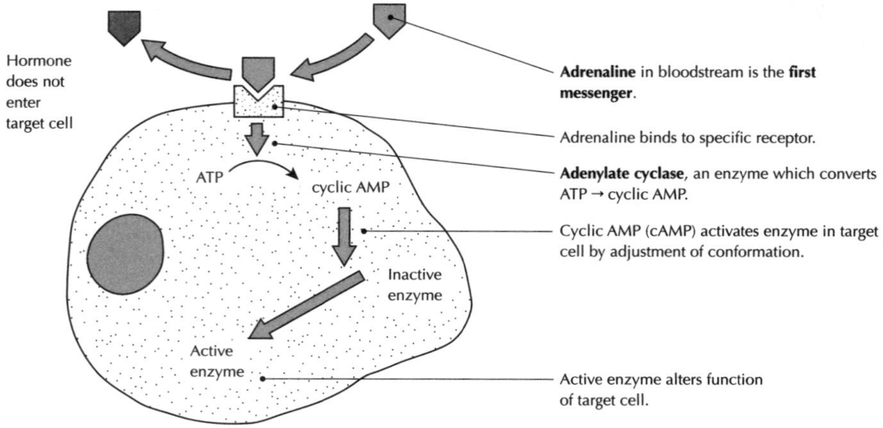

Hormone does not enter target cell
ATP
cyclic AMP
Inactive enzyme
Active enzyme
Adrenaline in bloodstream is the **first messenger**.
Adrenaline binds to specific receptor.
Adenylate cyclase, an enzyme which converts ATP → cyclic AMP.
Cyclic AMP (cAMP) activates enzyme in target cell by adjustment of conformation.
Active enzyme alters function of target cell.

Hormones of the pancreas regulate blood glucose concentration by negative feedback.

Claude Bernard: '… the maintenance of a constant internal environment is a necessity for a free life'.

There are severe consequences of any lengthy diversion of blood glucose concentration from its optimum.

Blood glucose concentration: optimum 1 mg per 1 cm³ blood

Classic homeostatic principle:

stimulus
is
detected
and
regulation system
controls
effectors
which produce
response
to cancel out
stimulus

⇧ Blood [glucose] caused by dietary intake

Reduces initial stimulus

Lowered blood glucose concentration

β CELLS → INSULIN

Increased uptake of glucose and amino acids from blood (particularly into skeletal muscle)
Increased glycogenesis in liver and muscle

Glucose —PHOSPHORYLASE→ Glycogen

Increased lipogenesis, especially in adipose cells
i.e. more glucose → fat
Decreased gluconeogenesis
i.e. less AMINO ACIDS → GLUCOSE
Decreased glycogenolysis
i.e. less GLYCOGEN → GLUCOSE

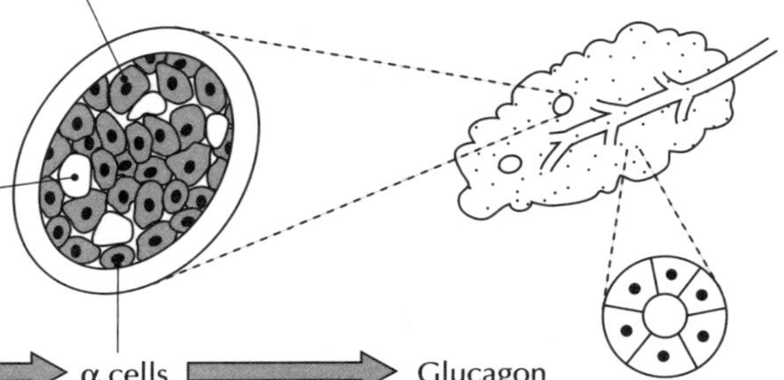

⇩ Blood [glucose] caused by assimilation or loss

α cells → Glucagon

ISLETS OF LANGERHANS IN PANCREAS

Increased release of glucose from the liver into the blood

Increased gluconeogenesis
i.e. more AMINO ACIDS → GLUCOSE
Increased glycogenolysis
i.e. less GLUCOSE → GLYCOGEN
Decreased glycogenesis

Raised blood glucose concentration

Reduces initial stimulus

How does insulin work?

- binds into specific receptor cells on target cells (liver and muscle)
- binding to receptor opens 'gated' glucose channel in membrane.

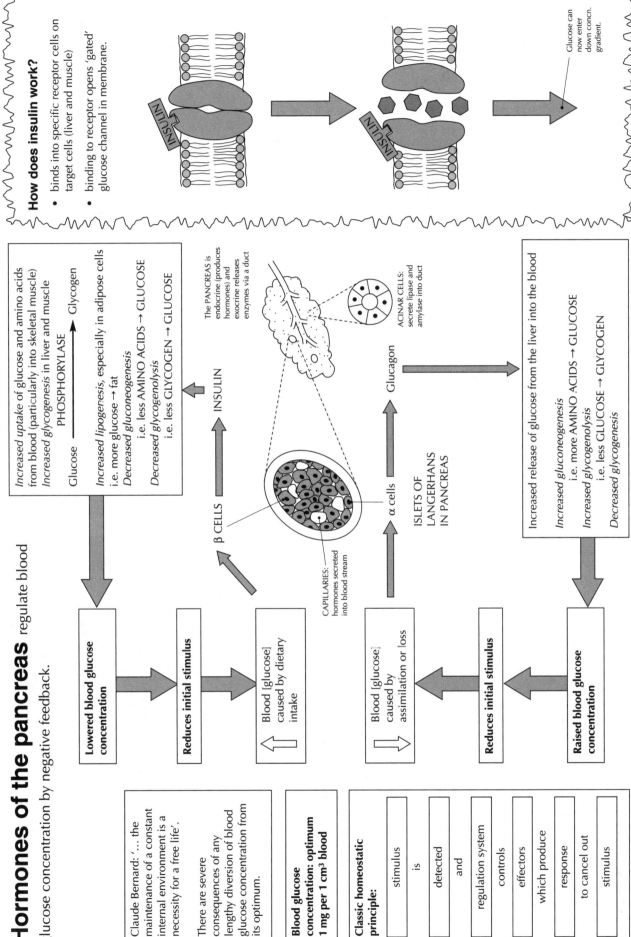

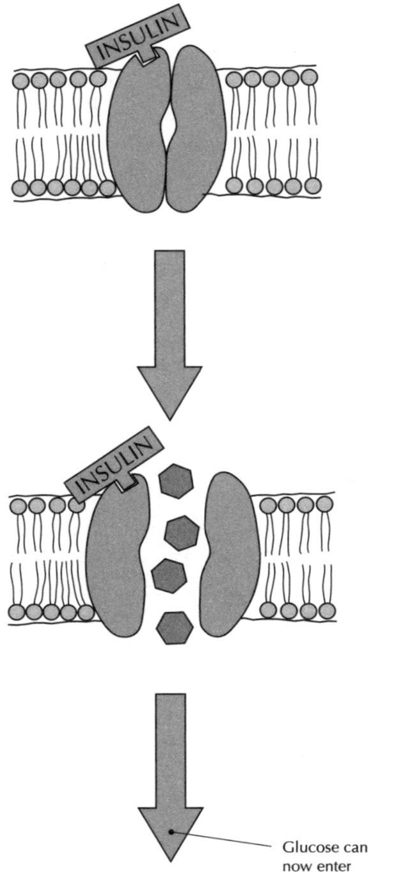

Diabetes mellitus results from failure to control blood glucose concentration.

What is blood glucose concentration?

Glucose is the cells' main source of energy, and it must always be available to them …

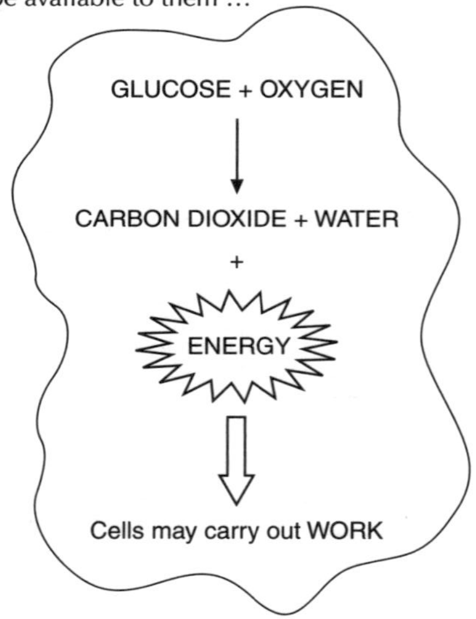
GLUCOSE + OXYGEN
CARBON DIOXIDE + WATER
+
ENERGY
Cells may carry out WORK

… so that the body keeps a constant amount of glucose in the blood. This is the **blood sugar level** and is usually maintained at about 1 mg of glucose per cm^3 of blood. Glucose is also an important solute in maintaining the optimum **plasma water potential.**

Control of insulin secretion

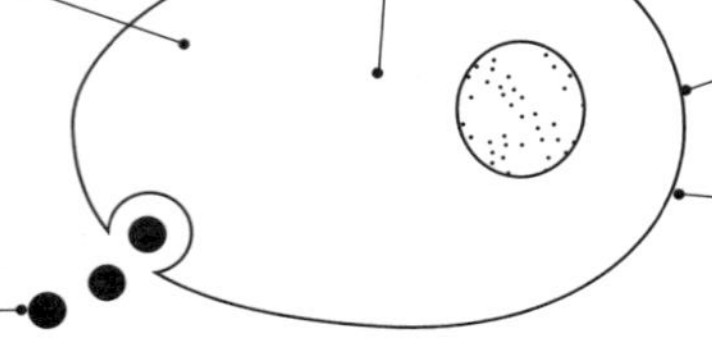
① Glucose enters, β-cells by facilitated diffusion.
② Glucose is respired to ATP.
③ Rising ATP concn. closes K^+ channels in β-cell membrane: **depolarisation**.
④ Voltage-sensitive Ca^{2+} channels open to allow entry of Ca^{2+} ions.
⑤ Rising intracellular Ca^{2+} concn. causes **exocytosis of insulin.**

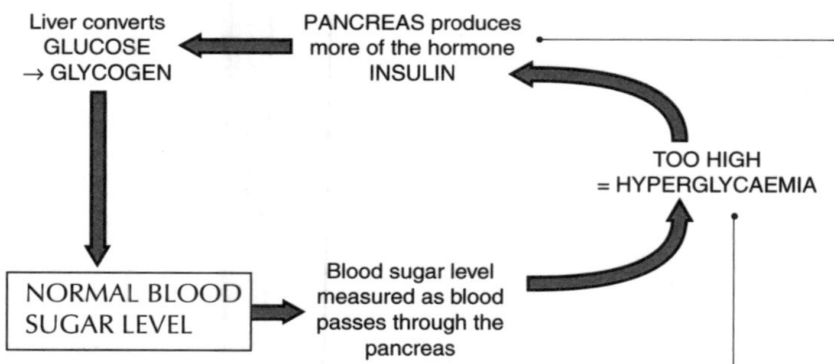
PANCREAS produces more of the hormone INSULIN
Liver converts GLUCOSE → GLYCOGEN
NORMAL BLOOD SUGAR LEVEL
Blood sugar level measured as blood passes through the pancreas
TOO HIGH = HYPERGLYCAEMIA

Glucose tolerance test measures the body's response to high glucose intake (sugar solution).

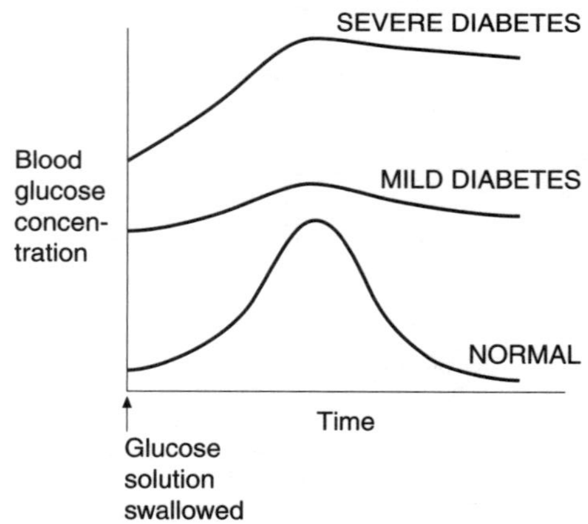
SEVERE DIABETES
MILD DIABETES
NORMAL
Blood glucose concen-tration
Time
Glucose solution swallowed

What is diabetes?

Diabetes is a condition in which there are higher than normal blood glucose concentrations.

Symptoms include:

- excessive thirst, hunger or urine production;
- sweet smelling breath (Ketone bodies);
- high 'overflow' of glucose into urine (test with **Clinistix**).

Damage can be observed in the retina.

Long-term effects if untreated include:

- premature ageing;
- cataract formation;
- hardening of arteries;
- heart disease;
- poor circulation to extremities, sometimes leading to loss of limbs.

In Type I (insulin-dependent) diabetes:

- pancreas does not secrete enough insulin
- may be an autoimmune response, following a viral infection
- treatment is by regular injection of pure insulin – almost all of this is now manufactured by **genetic engineering**
- it may be possible to inject **stem cells** to replace defective or missing ,β-cells.

In Type II (insulin-independent) diabetes:

- liver cells become insensitive to insulin
- blood insulin concentration often higher than normal
- treatment involves adjustment of diet to reduce intake of refined sugars, and moderate exercise regime to increase use of sugars in respiration.

Blood supply of the liver

Hepatic vein – blood is deoxygenated and returns to general circulation.

Hepatic portal vein carries blood from the digestive tract, including branches from stomach, pancreas, spleen, ileum, colon and rectum.

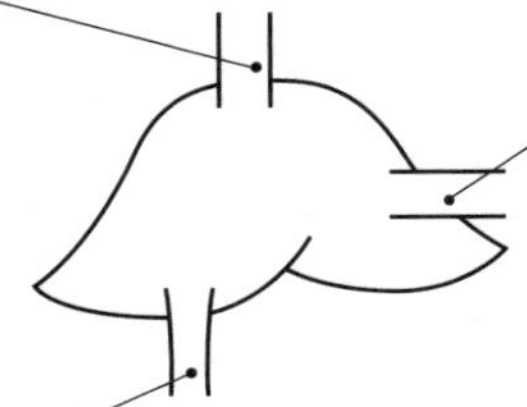

Hepatic artery – blood is fully oxygenated and at normal arterial pressure. Liver has high oxygen demand for aerobic respiration.

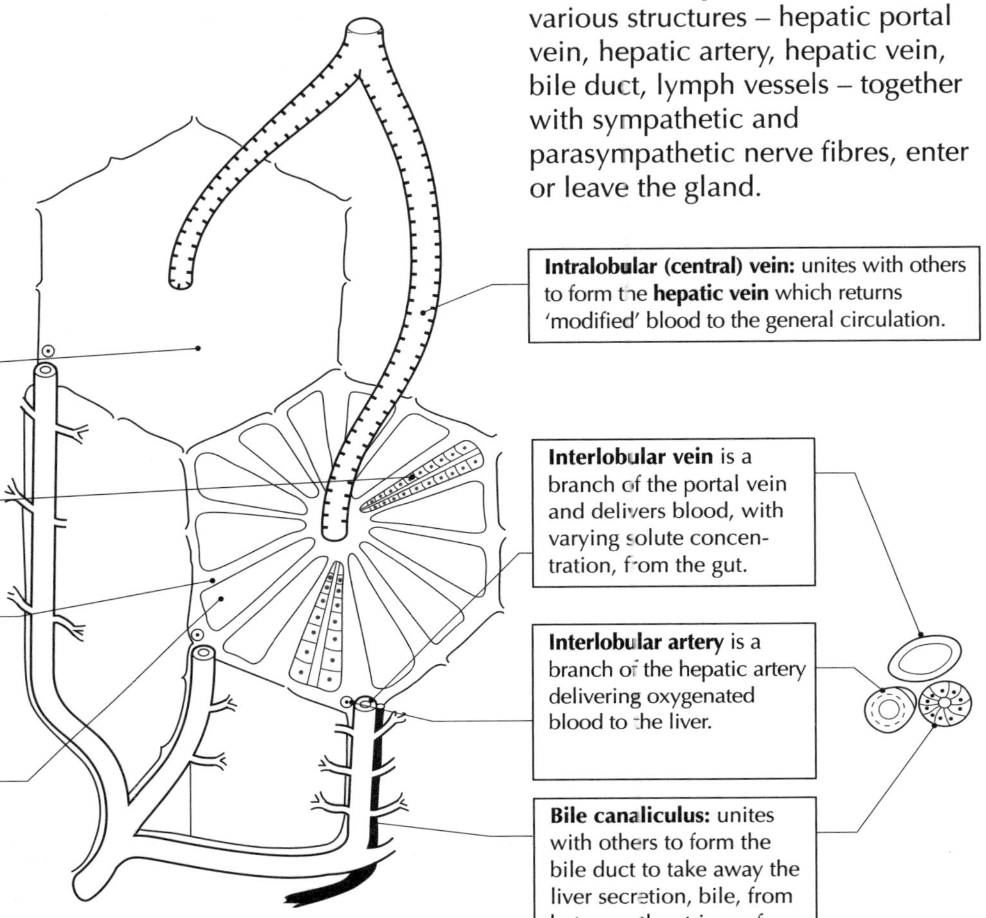

Each lobe is composed of many tiny **lobules** which are the functional units of the liver.

Liver cord or **string of hepatocytes** in contact with blood flowing through sinusoid.

Liver sinusoid is a blood vessel with incomplete walls which allows 'input' blood to come into close sontact with hepatocytes and Küppfer cells.

Each lobule has pairs of columns of **hepatocytes** radiating from a central vein. Between two pairs of columns are **sinusoids** and each sinusoid is serviced by a branch of the hepatic portal vein, a branch of the hepatic artery and a tributary of the bile duct. Each group of cells with its associated blood supply is called a **liver acinus**.

Intralobular (central) vein: unites with others to form the **hepatic vein** which returns 'modified' blood to the general circulation.

Interlobular vein is a branch of the portal vein and delivers blood, with varying solute concentration, from the gut.

Interlobular artery is a branch of the hepatic artery delivering oxygenated blood to the liver.

Bile canaliculus: unites with others to form the bile duct to take away the liver secretion, bile, from between the strings of hepatocytes.

The liver is the largest gland,

weighing between 1.0 and 2.3 kg. It is situated in the upper abdomen, just beneath the diaphragm, and has four lobes. On the posterior surface is the **portal fissure** where various structures – hepatic portal vein, hepatic artery, hepatic vein, bile duct, lymph vessels – together with sympathetic and parasympathetic nerve fibres, enter or leave the gland.

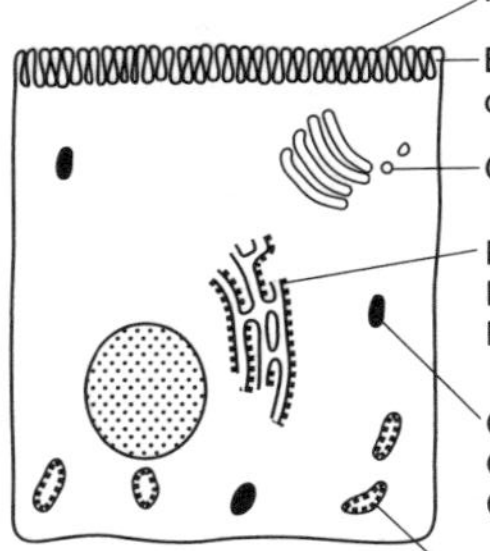

The **hepatocyte** or liver cell has microvilli to increase surface area for absorption from the percolating blood in the sinusoids, prominent Golgi Complex for the preparation of cellular products for secretion, rough endoplasmic reticulum for the synthesis and intracellular transport of numerous proteins, glycogen granules or fat globules to act as energy reserves, numerous mitochondria to generate ATP for the liver's numerous metabolic reactions. The plasma membrane has numerous transporter proteins as well as insulin receptors.

Although all hepatocytes perform all liver functions there is some localisation of function. For example, cells closest to the portal capillaries are most active in gluconeogenesis and glycogenesis and in oxidative ATP formation whilst those closest to the central vein are most active in fat synthesis, glycolysis and drug metabolism. It is these perivenous cells which are most prone to damage following paracetomol overdose.

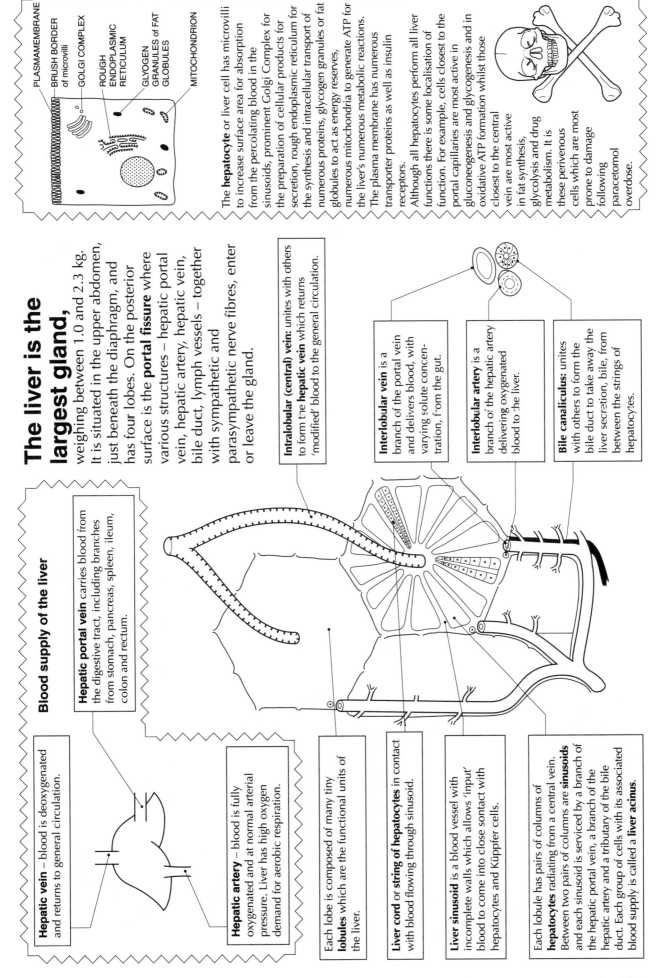

Urea formation from excess amino acids

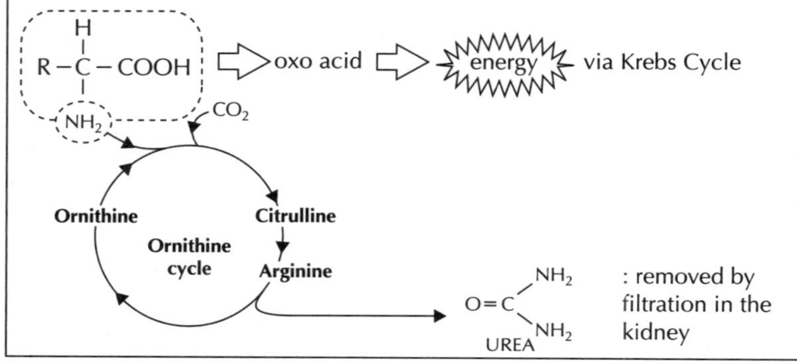

Ammonia is too toxic to be safely transported in the bloodstream, but urea is not.

Küppfer cell
A fixed phagocytic cell which removes old ('effete') red blood cells. Iron is stored as ferritin, globin → amino acid pool and pyrrole rings are excreted as bile pigments.

Storage of minerals and vitamins

principally **iron** (as ferritin), some potassium, copper and trace elements.

mainly the **fat-soluble** vitamins A, D, E. **Small** amounts of B_{12}, C. Synthesis of vitamin A from carotene.

Polar bears store so much vitamin A in their liver that this organ becomes toxic to other animals! Huskies beware!

Branch of the hepatic portal vein (interlobular vein)
Delivers blood from the absorptive regions of the gut carrying a **variable concentration** of the soluble products of digestion.

Blood filled channel

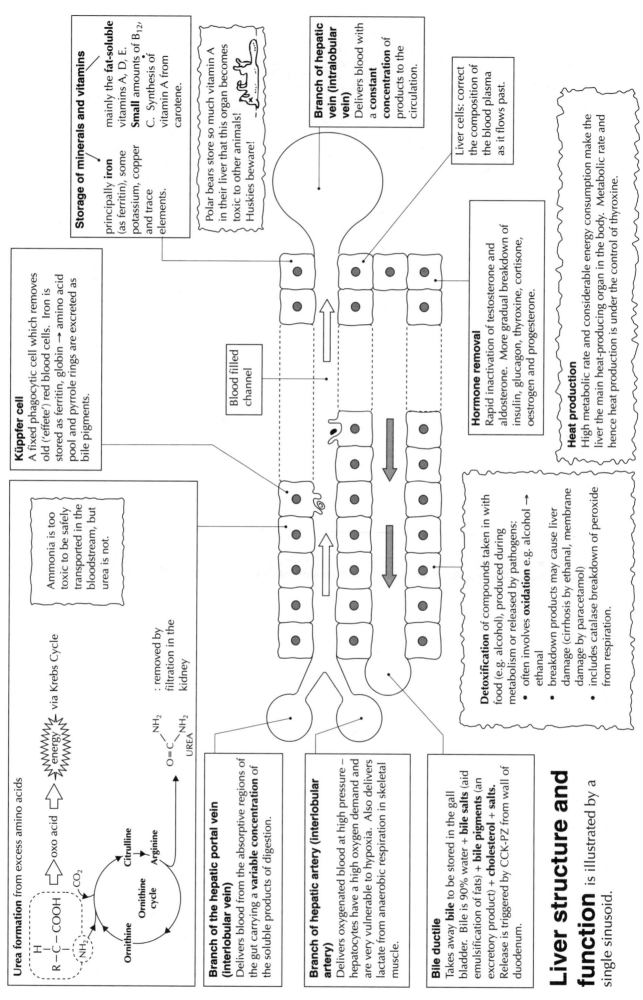

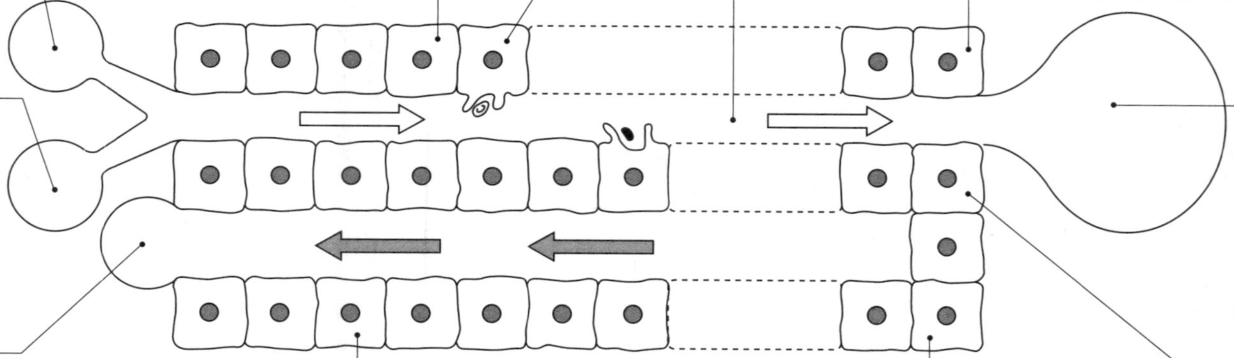

Branch of hepatic artery (interlobular artery)
Delivers oxygenated blood at high pressure – hepatocytes have a high oxygen demand and are very vulnerable to hypoxia. Also delivers lactate from anaerobic respiration in skeletal muscle.

Branch of hepatic vein (intralobular vein)
Delivers blood with a **constant concentration** of products to the circulation.

Bile ductile
Takes away **bile** to be stored in the gall bladder. Bile is 90% water + **bile salts** (aid emulsification of fats) + **bile pigments** (an excretory product) + **cholesterol** + **salts.** Release is triggered by CCK-PZ from wall of duodenum.

Liver cells: correct the composition of the blood plasma as it flows past.

Detoxification of compounds taken in with food (e.g. alcohol), produced during metabolism or released by pathogens:
- often involves **oxidation** e.g. alcohol → ethanal
- breakdown products may cause liver damage (cirrhosis by ethanal, membrane damage by paracetamol)
- includes catalase breakdown of peroxide from respiration.

Hormone removal
Rapid inactivation of testosterone and aldosterone. More gradual breakdown of insulin, glucagon, thyroxine, cortisone, oestrogen and progesterone.

Heat production
High metabolic rate and considerable energy consumption make the liver the main heat-producing organ in the body. Metabolic rate and hence heat production is under the control of thyroxine.

Liver structure and function is illustrated by a single sinusoid.

Excretion involves removal of waste products of metabolism

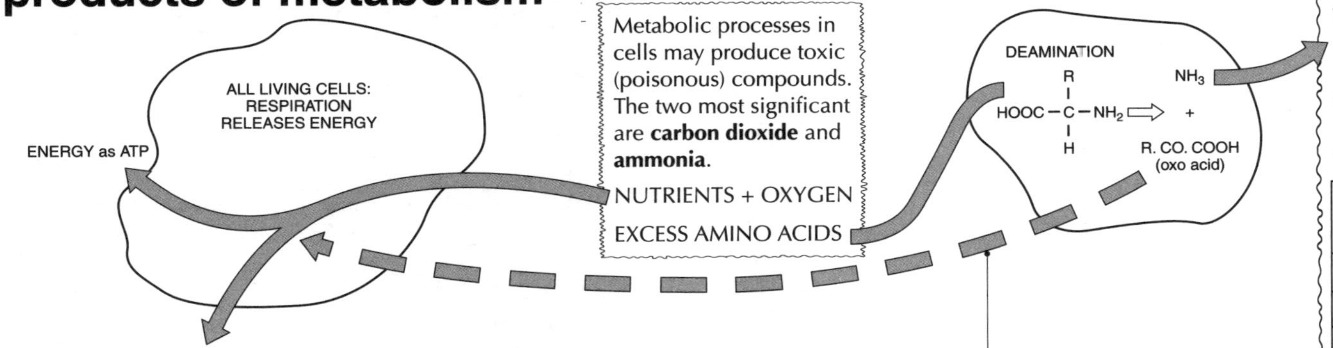

Ammonia is extremely toxic: this compound rapidly raises pH of body fluids.

The ammonia must be:

- removed by dilution in water **if there is no shortage of water**
- converted to a less toxic product **if energy and the appropriate enzymes are available**.

There are **three** common nitrogenous excretory products:

Product	Toxicity	Water required for excretion	Energy required for production
Ammonia	High	High	Low
Urea	Medium	Medium	High
Uric acid	Low	Low	High

Carbon dioxide

- dissolves to form weakly acidic hydrogencarbonate (HCO_3^-) in body fluids such as blood plasma
- removed as CO_2 from respiratory surfaces such as cell membrane (e.g. in *Amoeba*), gill lamellae (fish), or lungs (e.g. birds and mammals)
- may be combined with some other compound for excretion as solid, e.g. earthworm $CaO + CO_2 \rightarrow CaCO_3$.

Using oxo acids for energy

- Carnivorous animals have a high protein diet, so many amino acids are deaminated. These animals make urine with a very high urea content.
- Starving humans (e.g. with eating disorders) break down muscle for energy. Much deamination → NH_3 → high urea concentration in urine.

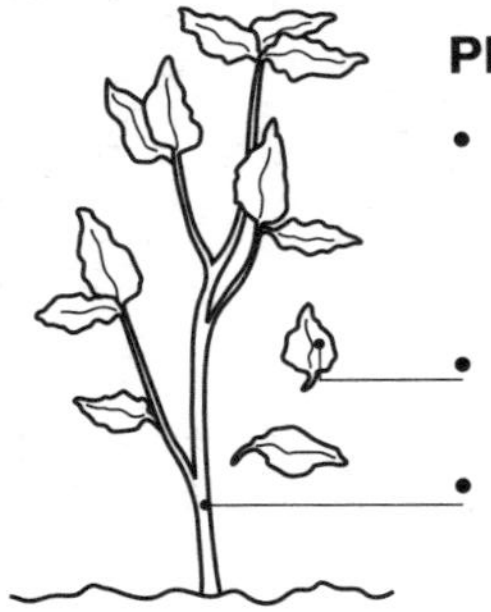

Plants

- autotrophic, so balance intake of nutrients to demand – little need for excretion
- may excrete some wastes in falling leaves
- may store some wastes in old ('heartwood') xylem.

Urine may contain other compounds

Products of metabolism, or substances taken in with the diet, may be filtered by the kidney and excreted in the urine

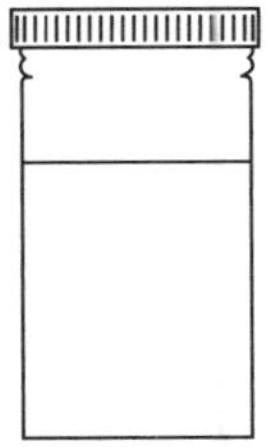

Pregnancy: Human chorionic gonadotrophin (HCG) is released by the developing placenta. It is detected by antibodies on a plastic testing strip dipped in to urine: a dye indicates the result.

Steroid abuse: Testosterone or its products 'overflow' into urine.

Glucose: Blood sugar at high levels in diabetic people 'overflows' into urine, and is detected by a Clinistix (Glucose oxidase strip).

The urinary system

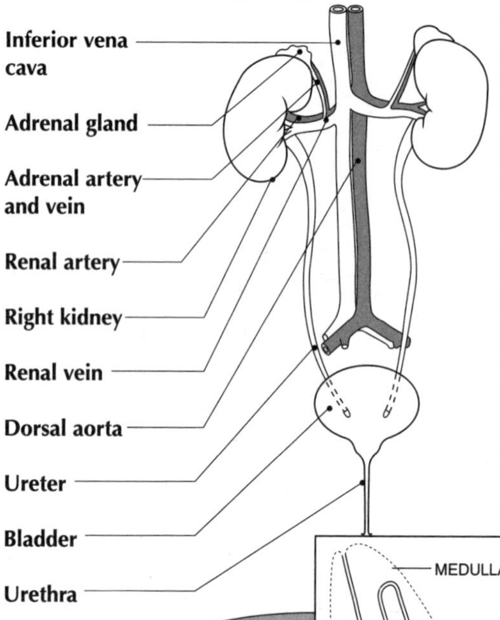

Renal circulation: handles approx. 1200 cm^3 of blood per minute ~ about 25% of cardiac output.

Interlobular artery: delivers blood at high pressure to the glomerular capillaries.

Interlobular vein: drains blood away from the glomerular filtration units and from the Loops of Henle.

Renal papilla: apex of the pyramid from which the ends of the collecting ducts deliver the urine.

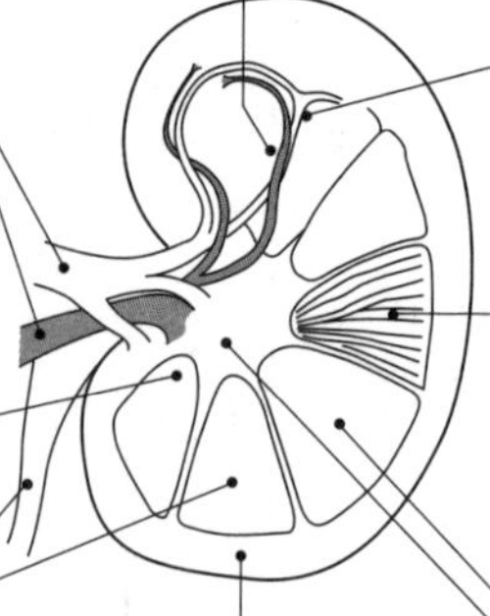

Medulla: has a striated appearance due to the presence of the Loop of Henle, the collecting duct and the vasa rectae.

Ureter: propels urine from the pelvis to the bladder.

Kidney cortex: has a smooth texture and is the site of the Bowman's Capsules and the glomeruli, together with the proximal and distal convoluted tubules and their associated blood supply.

Renal pyramid: one of 5–12 triangular structures which make up the medullary region.

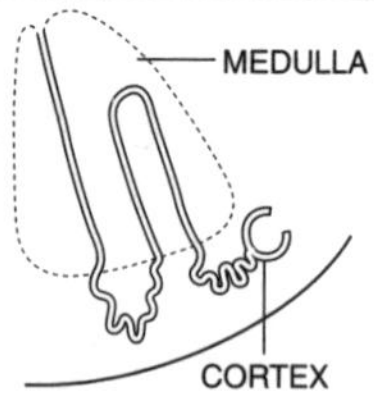

Single nephron: (much enlarged) to show position in medulla and cortex.

Pelvis: urine collects here from tips of renal papillae.

Glomerular filtration occurs in the renal capsule: the selective structure is the **basement membrane of the glomerulus.** Water and solutes of relative molecular mass less than 68 000 form the filtrate.

Glomerular filtrate is formed at above 125 cm^3 per minute in humans (this represents about 20% of the plasma delivered during that time). It contains all the materials present in the blood except blood cells and most proteins, which are too large to cross the basement membrane of the glomerulus.

Active controlled reabsorption of Na^+ occurs in the distal convoluted tubule. This is followed by the osmotic movement of an equivalent volume of water down the water potential gradient.

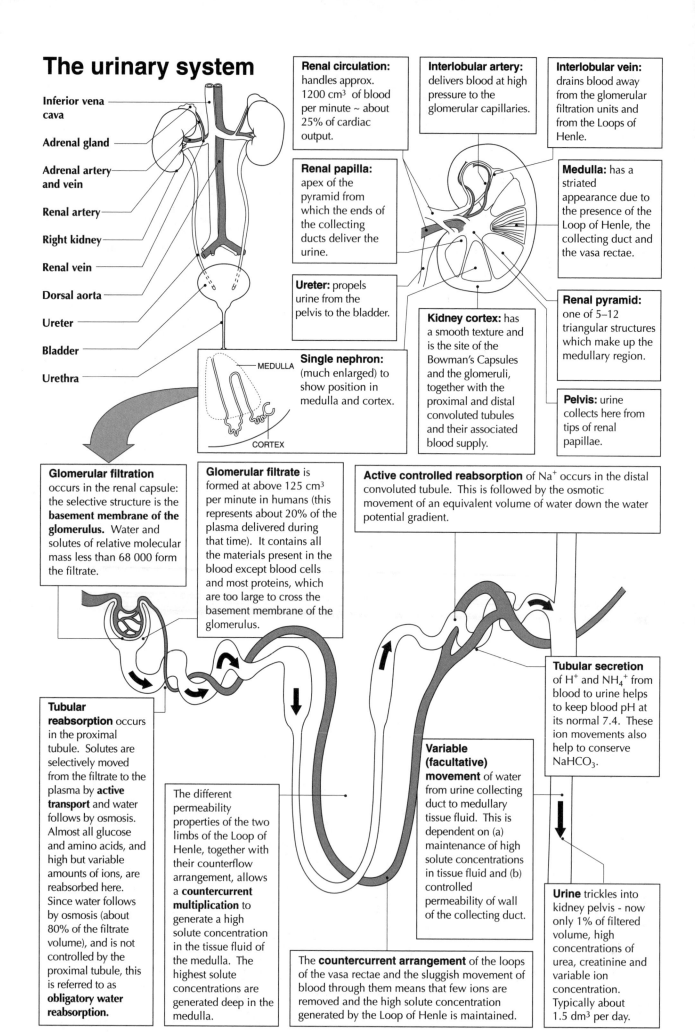

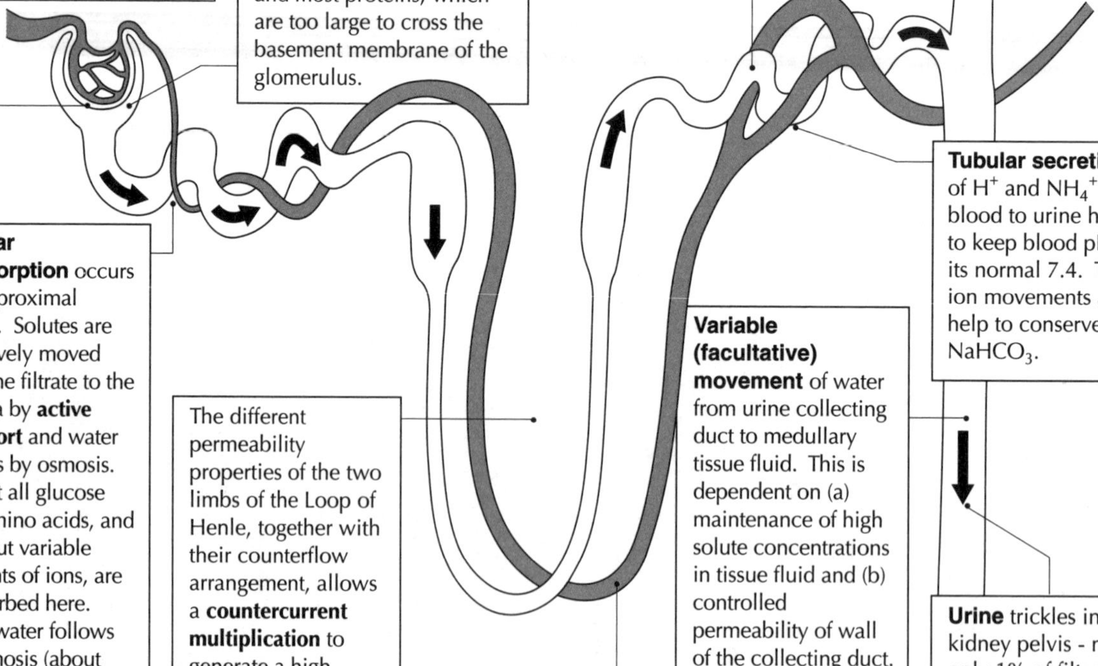

Tubular secretion of H^+ and NH_4^+ from blood to urine helps to keep blood pH at its normal 7.4. These ion movements also help to conserve $NaHCO_3$.

Tubular reabsorption occurs in the proximal tubule. Solutes are selectively moved from the filtrate to the plasma by **active transport** and water follows by osmosis. Almost all glucose and amino acids, and high but variable amounts of ions, are reabsorbed here. Since water follows by osmosis (about 80% of the filtrate volume), and is not controlled by the proximal tubule, this is referred to as **obligatory water reabsorption.**

The different permeability properties of the two limbs of the Loop of Henle, together with their counterflow arrangement, allows a **countercurrent multiplication** to generate a high solute concentration in the tissue fluid of the medulla. The highest solute concentrations are generated deep in the medulla.

Variable (facultative) movement of water from urine collecting duct to medullary tissue fluid. This is dependent on (a) maintenance of high solute concentrations in tissue fluid and (b) controlled permeability of wall of the collecting duct.

The **countercurrent arrangement** of the loops of the vasa rectae and the sluggish movement of blood through them means that few ions are removed and the high solute concentration generated by the Loop of Henle is maintained.

Urine trickles into kidney pelvis - now only 1% of filtered volume, high concentrations of urea, creatinine and variable ion concentration. Typically about 1.5 dm^3 per day.

Osmoregulation involves a negative feedback system which depends on ADH (**A**nti **D**iuretic **H**ormone).

Water intake > Water loss (or decreased **salt intake**)
- water potential rises i.e. becomes **less** negative.

Osmoreceptors in the thirst centre of the **hypothalamus** are **less** stimulated.

Pituitary gland secretes **less** ADH.

Medulla of kidney absorbs **less** water as walls of collecting ducts become **less** permeable.

Large volume of **less** concentrated urine is formed.

Plasma water potential returns to normal.

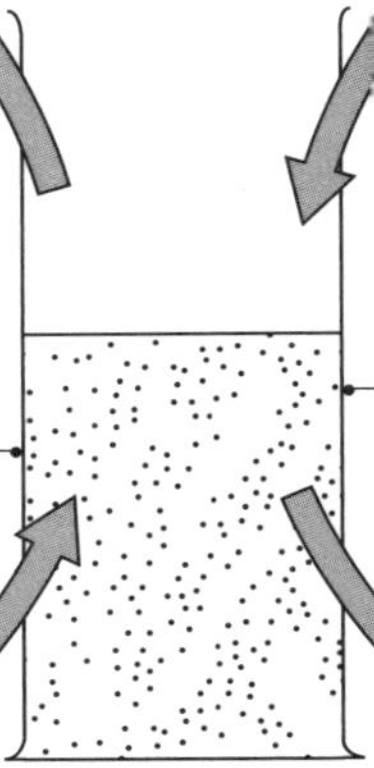

OPTIMUM WATER POTENTIAL IN BLOOD

The optimum blood water potential:
- allows recovery of tissue fluid
- prevents osmotic damage to blood cells (e.g. haemolysis)
- prevents dehydration of tissues.

Water loss > Water intake (or increased **salt intake**)
- water potential falls i.e. becomes **more** negative.

Osmoreceptors in the thirst centre of the **hypothalamus** are **more** stimulated.

Pituitary gland secretes **more** ADH.

Medulla of kidney absorbs **more water** as walls of collecting ducts become **more** permeable.

Small volume of **more** concentrated urine is formed.

Plasma water potential returns to normal.

Hangover! Alcohol reduces the effect of ADH on the collecting ducts. Less water is reabsorbed and the body becomes dehydrated. Low plasma water potential means water leaves brains cells by osmosis: **Headache!**

Sports and water balance
Isotonic sports drinks provide rapid rehydration. They contain:
- glucose and salts, which are quickly absorbed and lower blood water potential, so that
- water follows by osmosis.

Jockeys long to be light!
Jockeys or boxers who want to get down to their ideal weight take **diuretic drugs** that interfere with ADH.
Less water is reabsorbed, more is lost in the urine and the body loses weight!

Intake varies greatly depending on the type of food available. Some desert animals gain 90% of their water from respiration. Camels don't store water in their hump, but fat. Water is released when the fat is respired.

Water intake	
Drinks	1500 cm^3
Water in foods	800 cm^3
From respiration	200 cm^3

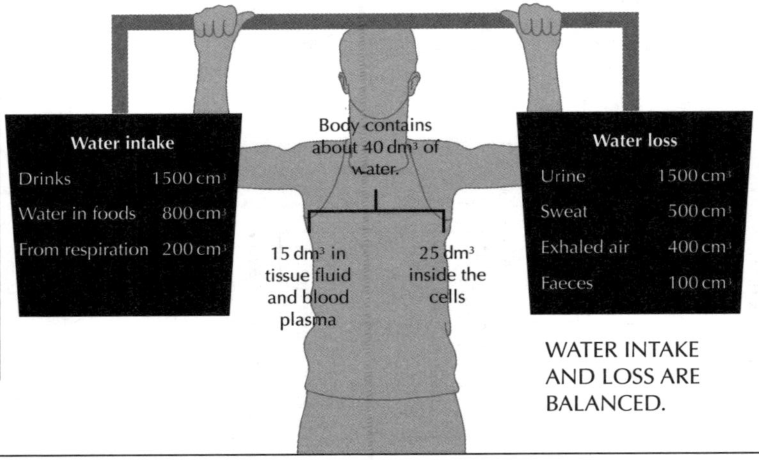
Body contains about 40 dm^3 of water.
15 dm^3 in tissue fluid and blood plasma
25 dm^3 inside the cells

Water loss	
Urine	1500 cm^3
Sweat	500 cm^3
Exhaled air	400 cm^3
Faeces	100 cm^3

WATER INTAKE AND LOSS ARE BALANCED.

The part played by the pituitary gland

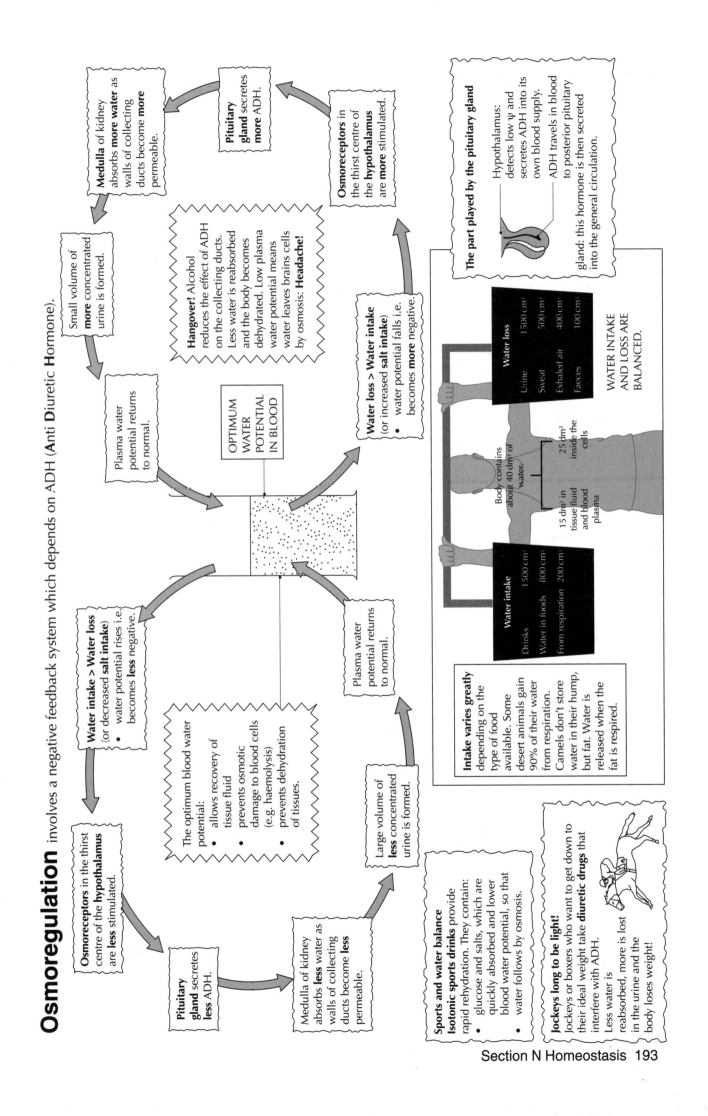

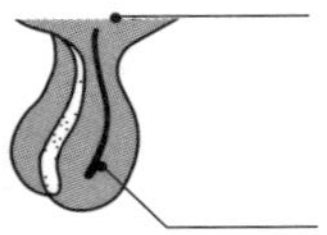
Hypothalamus: detects low ψ and secretes ADH into its own blood supply.
ADH travels in blood to posterior pituitary

gland: this hormone is then secreted into the general circulation.

Control of blood Na^+ concentration and blood pressure

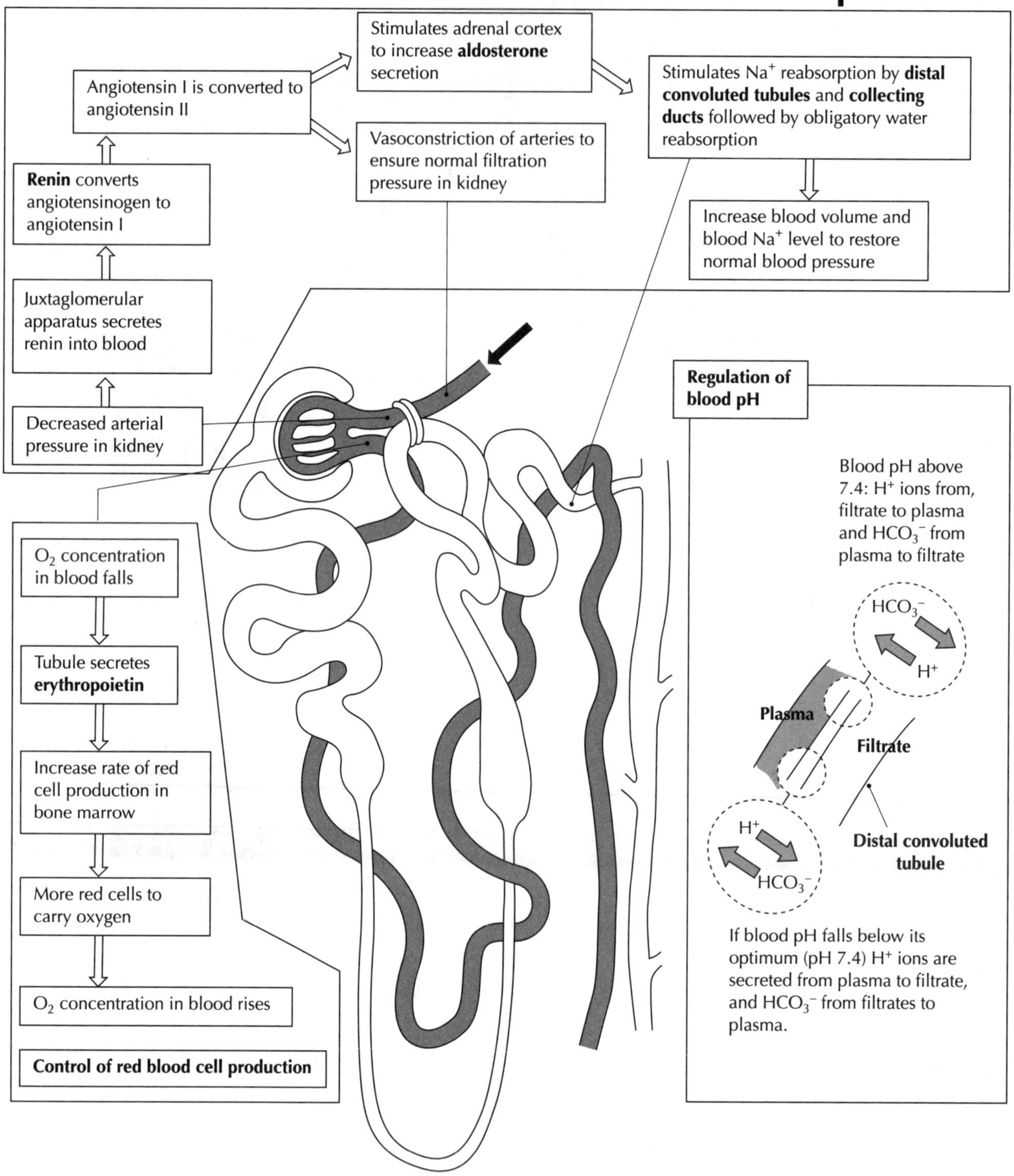

Stimulates adrenal cortex to increase **aldosterone** secretion
Angiotensin I is converted to angiotensin II
Stimulates Na^+ reabsorption by **distal convoluted tubules** and **collecting ducts** followed by obligatory water reabsorption
Vasoconstriction of arteries to ensure normal filtration pressure in kidney
Renin converts angiotensinogen to angiotensin I
Increase blood volume and blood Na^+ level to restore normal blood pressure
Juxtaglomerular apparatus secretes renin into blood
Decreased arterial pressure in kidney
Regulation of blood pH
Blood pH above 7.4: H^+ ions from, filtrate to plasma and HCO_3^- from plasma to filtrate
HCO_3^-
H^+
Plasma
Filtrate
H^+
HCO_3^-
Distal convoluted tubule
If blood pH falls below its optimum (pH 7.4) H^+ ions are secreted from plasma to filtrate, and HCO_3^- from filtrates to plasma.
O_2 concentration in blood falls
Tubule secretes **erythropoietin**
Increase rate of red cell production in bone marrow
More red cells to carry oxygen
O_2 concentration in blood rises
Control of red blood cell production

The basic principle of homeostasis (negative feedback) is illustrated by the action of the nephron

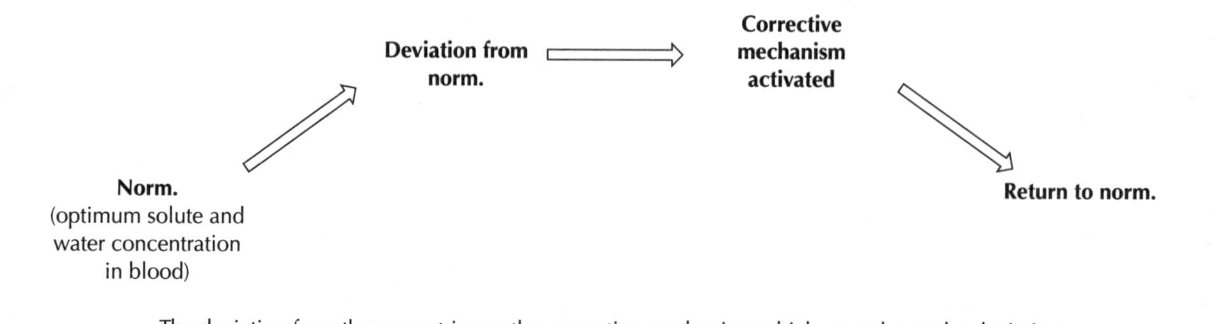

Deviation from norm.
Corrective mechanism activated
Norm.
(optimum solute and water concentration in blood)
Return to norm.

The deviation from the norm. triggers the corrective mechanism which cancels out the deviation.

Kidney failure: if one or both kidneys fail then dialysis is used or a transplant performed to keep urea and solute concentration in the blood constant.

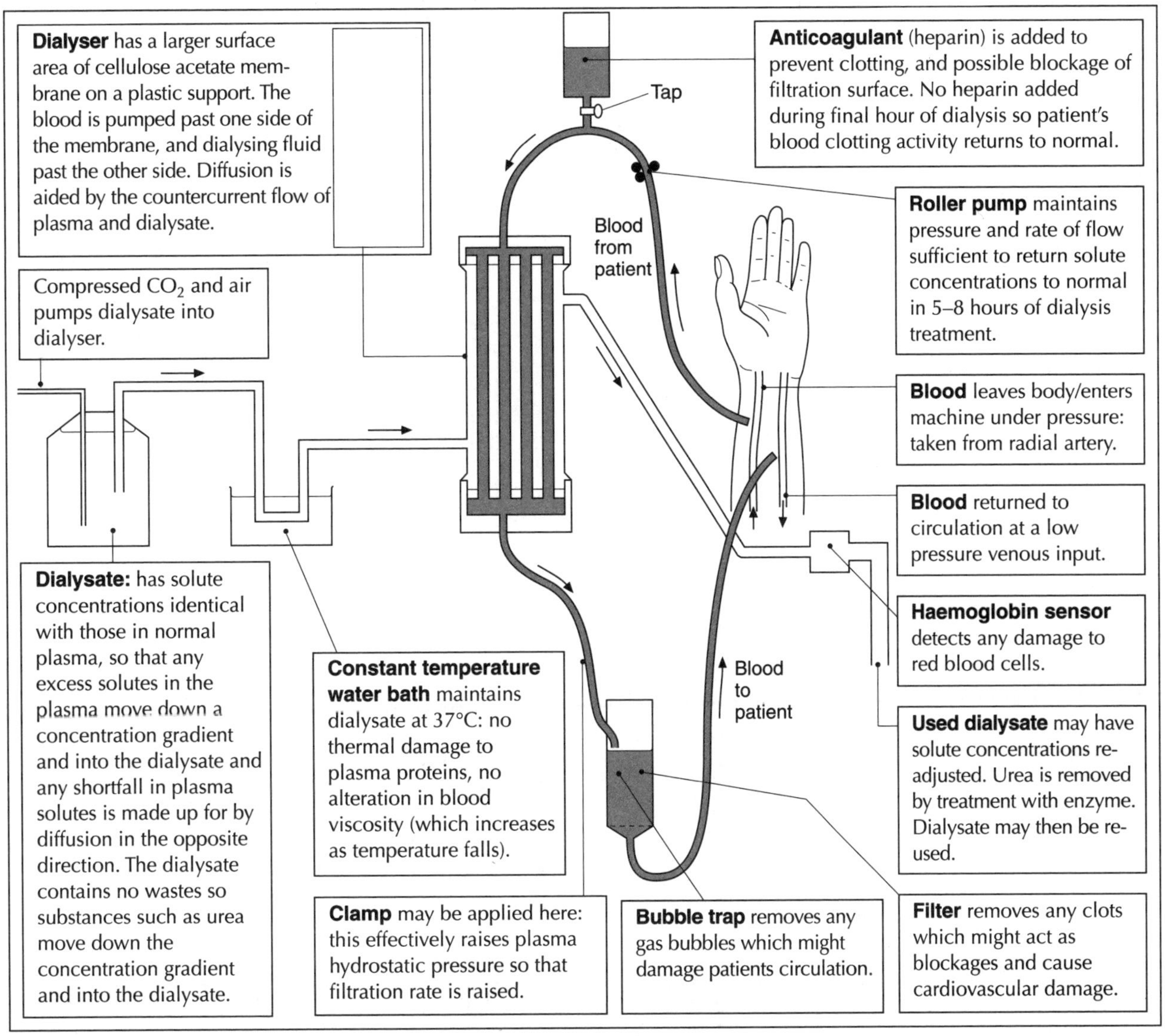
Dialyser has a larger surface area of cellulose acetate membrane on a plastic support. The blood is pumped past one side of the membrane, and dialysing fluid past the other side. Diffusion is aided by the countercurrent flow of plasma and dialysate.
Compressed CO_2 and air pumps dialysate into dialyser.
Tap
Blood from patient
Blood to patient
Anticoagulant (heparin) is added to prevent clotting, and possible blockage of filtration surface. No heparin added during final hour of dialysis so patient's blood clotting activity returns to normal.
Roller pump maintains pressure and rate of flow sufficient to return solute concentrations to normal in 5–8 hours of dialysis treatment.
Blood leaves body/enters machine under pressure: taken from radial artery.
Blood returned to circulation at a low pressure venous input.
Haemoglobin sensor detects any damage to red blood cells.
Used dialysate may have solute concentrations re-adjusted. Urea is removed by treatment with enzyme. Dialysate may then be re-used.
Filter removes any clots which might act as blockages and cause cardiovascular damage.
Dialysate: has solute concentrations identical with those in normal plasma, so that any excess solutes in the plasma move down a concentration gradient and into the dialysate and any shortfall in plasma solutes is made up for by diffusion in the opposite direction. The dialysate contains no wastes so substances such as urea move down the concentration gradient and into the dialysate.
Constant temperature water bath maintains dialysate at 37°C: no thermal damage to plasma proteins, no alteration in blood viscosity (which increases as temperature falls).
Clamp may be applied here: this effectively raises plasma hydrostatic pressure so that filtration rate is raised.
Bubble trap removes any gas bubbles which might damage patients circulation.

Kidney transplantation may be necessary as renal dialysis is inconvenient for the patient and costly.

Kidney transplants have a high success rate because:

1 the vascular connections are simple

2 live donors may be used, so very close blood group matching is possible

3 because of 2 there are fewer immuno-suppression-related problems in which the body's immune system reacts against the new kidney.

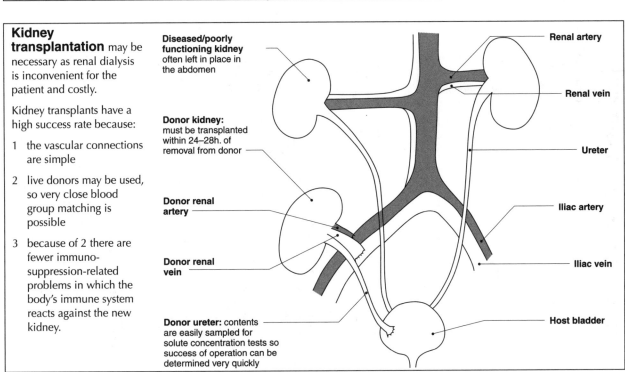

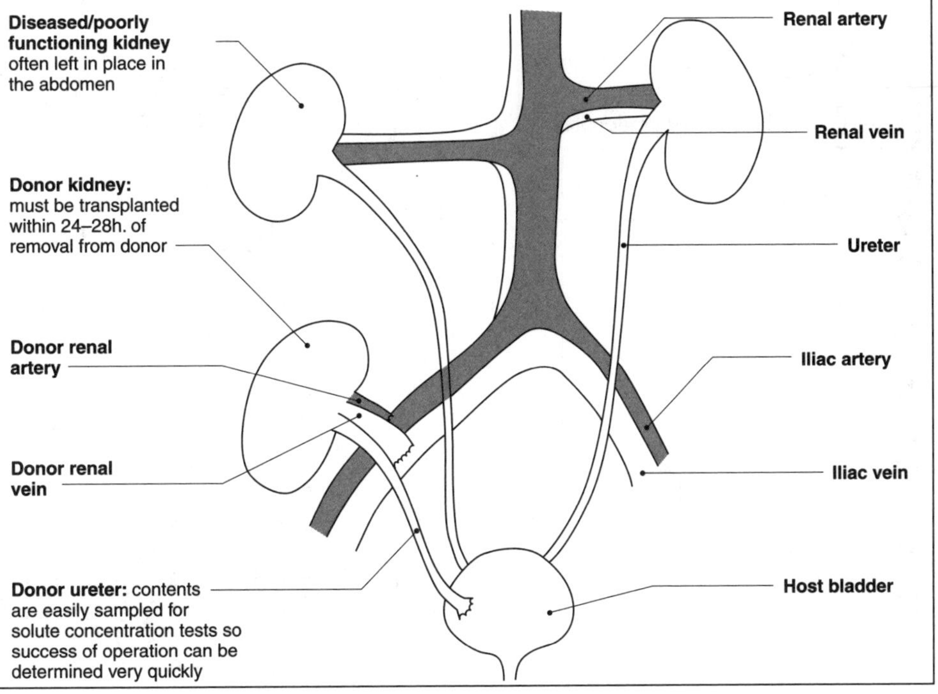
Diseased/poorly functioning kidney often left in place in the abdomen
Donor kidney: must be transplanted within 24–28h. of removal from donor
Donor renal artery
Donor renal vein
Donor ureter: contents are easily sampled for solute concentration tests so success of operation can be determined very quickly
Renal artery
Renal vein
Ureter
Iliac artery
Iliac vein
Host bladder

Control systems in biology

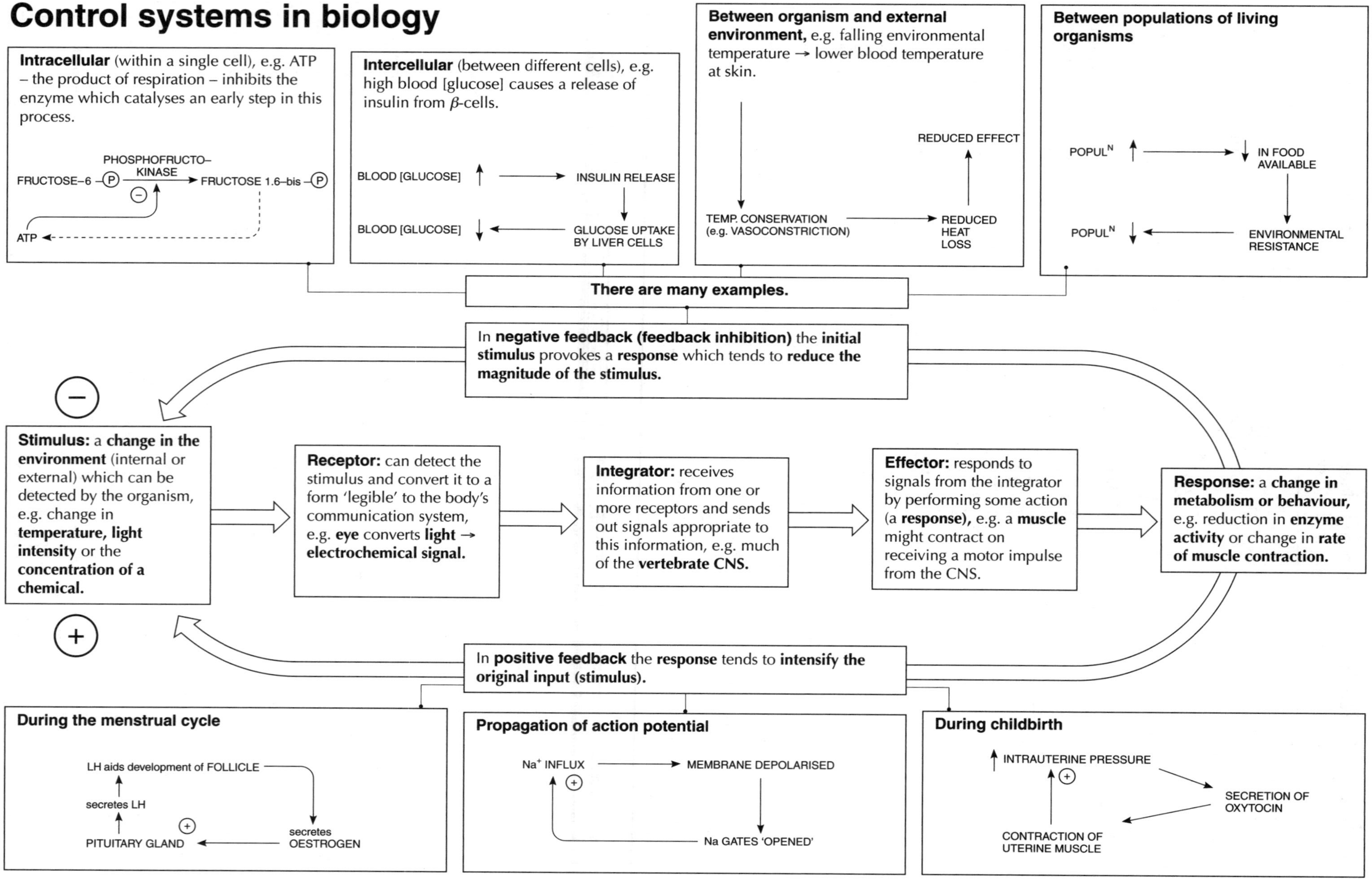
Intracellular (within a single cell), e.g. ATP – the product of respiration – inhibits the enzyme which catalyses an early step in this process.
FRUCTOSE–6 –Ⓟ
PHOSPHOFRUCTO-KINASE
FRUCTOSE 1.6–bis –Ⓟ
⊖
ATP
Intercellular (between different cells), e.g. high blood [glucose] causes a release of insulin from β-cells.
BLOOD [GLUCOSE] ↑
INSULIN RELEASE
GLUCOSE UPTAKE BY LIVER CELLS
BLOOD [GLUCOSE] ↓
Between organism and external environment, e.g. falling environmental temperature → lower blood temperature at skin.
TEMP. CONSERVATION (e.g. VASOCONSTRICTION)
REDUCED HEAT LOSS
REDUCED EFFECT
Between populations of living organisms
POPUL^N ↑
↓ IN FOOD AVAILABLE
ENVIRONMENTAL RESISTANCE
POPUL^N ↓
There are many examples.
In **negative feedback (feedback inhibition)** the **initial stimulus** provokes a **response** which tends to **reduce the magnitude of the stimulus.**
−
+
Stimulus: a **change in the environment** (internal or external) which can be detected by the organism, e.g. change in **temperature, light intensity** or the **concentration of a chemical.**
Receptor: can detect the stimulus and convert it to a form 'legible' to the body's communication system, e.g. **eye** converts **light → electrochemical signal.**
Integrator: receives information from one or more receptors and sends out signals appropriate to this information, e.g. much of the **vertebrate CNS.**
Effector: responds to signals from the integrator by performing some action (a **response),** e.g. a **muscle** might contract on receiving a motor impulse from the CNS.
Response: a **change in metabolism or behaviour,** e.g. reduction in **enzyme activity** or change in **rate of muscle contraction.**
In **positive feedback** the **response** tends to **intensify the original input (stimulus).**
During the menstrual cycle
LH aids development of FOLLICLE
secretes LH
PITUITARY GLAND
⊕
secretes OESTROGEN
Propagation of action potential
Na⁺ INFLUX
⊕
MEMBRANE DEPOLARISED
Na GATES 'OPENED'
During childbirth
↑ INTRAUTERINE PRESSURE
⊕
SECRETION OF OXYTOCIN
CONTRACTION OF UTERINE MUSCLE

Events of the menstrual cycle

Follicle-stimulating hormone (FSH) initiates the development of several primary follicles (each containing a primary oocyte): one follicle continues to develop but the others degenerate by the process of follicular atresia. FSH also increases the activity of the enzymes responsible for formation of oestrogen.

Oestrogen is produced by enzyme modification (in the stroma) of testosterone produced by the thecae of the developing follicle. Oestrogen has several effects:

- It stimulates further growth of the follicle.
- It promotes repair of the endometrium.
- It acts as a feedback inhibitor of the secretion of FSH from the anterior pituitary gland.
- From about day 11 it has a positive feedback action on the secretion of both LH and FSH.

Development of the follicle within the ovary is initiated by FSH but continued by LH. The Graafian follicle (Ⓐ) is mature by day 10–11 and ovulation occurs at day 14 (Ⓑ) following a surge of LH. The remains of the follicle become the corpus luteum (Ⓒ), which secretes steroid hormones. These steroid hormones inhibit LH secretion so that the corpus luteum degenerates and becomes the corpus albicans (Ⓓ).

The **endometrium** begins to thicken and become more vascular under the influence of the ovarian hormone oestrogen. Because of this thickening, to 4–6 mm, the time between menstruation and ovulation is sometimes called the **proliferative phase.**

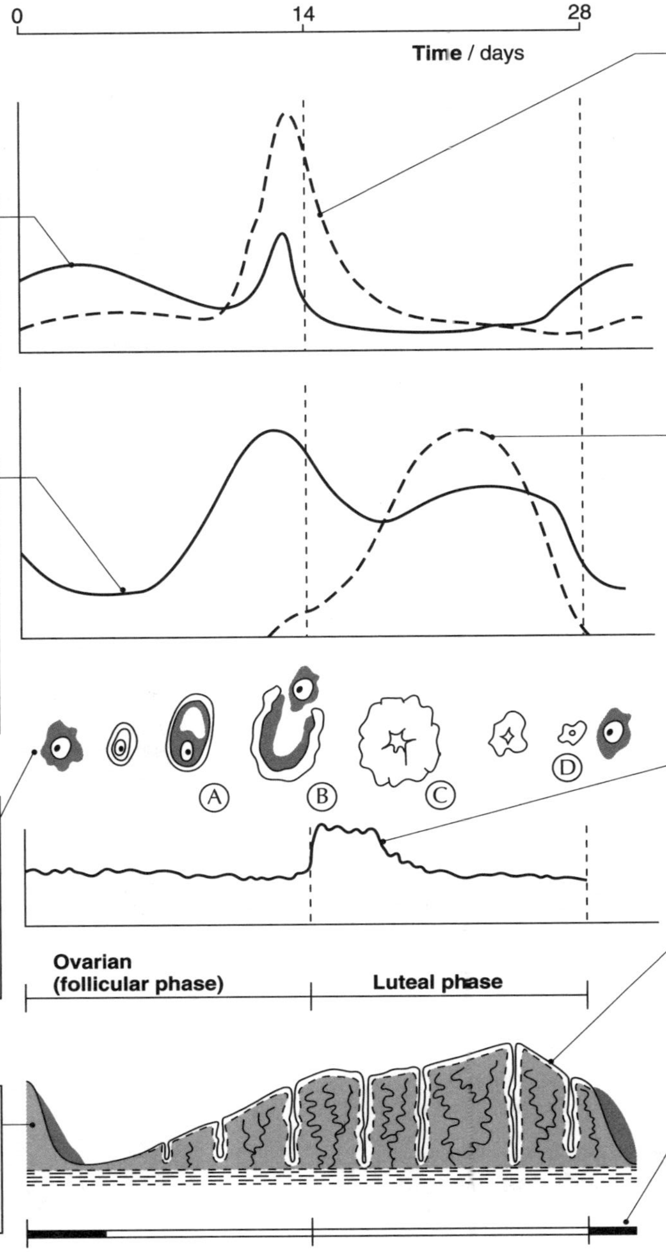

Luteinizing hormone (LH) triggers the secretion of testosterone by the thecae of the follicle, and when its concentration 'surges' it causes release of enzymes which rupture the wall of the ovary, allowing the secondary oocyte to be released at ovulation. After ovulation LH promotes development of the corpus luteum from the remains of the Graafian follicle.

Progesterone is secreted by the corpus luteum. It has several effects:

- It prepares the endometrium for implantation of a fertilised egg by increasing vascularisation, thickening and the storage of glycogen.
- It begins to promote growth of the mammary glands.
- It acts as a feedback inhibitor of FSH secretion, thus arresting development of any further follicles.

This is the basis of the contraceptive 'pill'. The pill contains progesterone at a concentration which feedback-inhibits ovulation.

Body temperature rises by about 1°C at the time of ovulation. This 'heat' is used to determine the 'safe period' for the rhythm method of contraception.

During the post-ovulatory or luteal phase the **endometrium** becomes thicker with more tortuously coiled glands and greater vascularisation of the surface layer, and retains more tissue fluid.

Menstruation is initiated by falling concentrations of oestrogen and progesterone as the corpus luteum degenerates.

At **menstruation** the **stratum functionalis** of the endometrium is shed, leaving the **stratum basilis** to begin proliferation of a new *functionalis.*

Synthetic hormones and control of the oestrous cycle

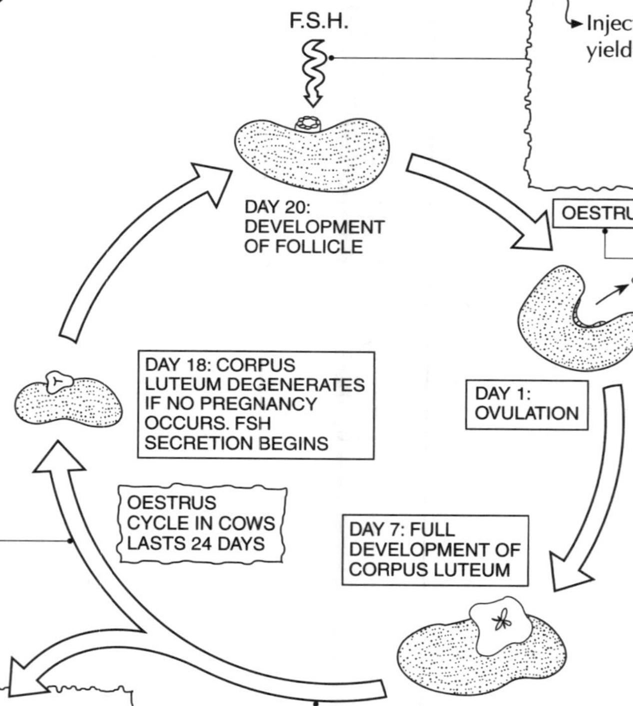

Synchronisation of oestrus is useful in both cows and sheep. For sheep:

- lambing season can be advanced for maximum profit
- cycle can be accelerated → 3 lambs in 2 years
- longer milking season from ewes.

Control

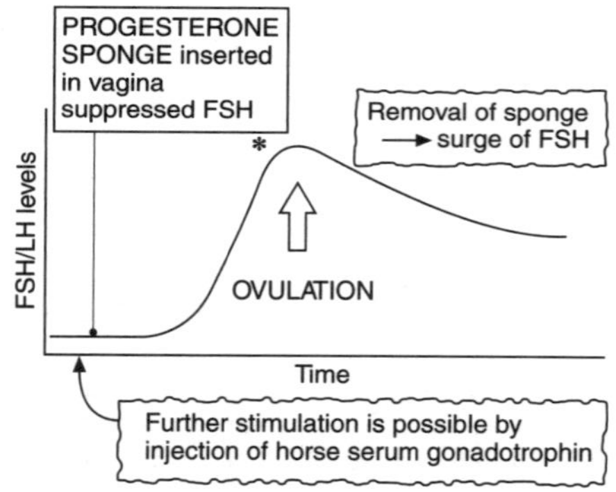

Embryo transplantation benefits from **follicle-stimulating hormone**:

FSH stimulates development of follicles → Inject into valuable (e.g. high milk yielding) cow → releases more eggs

Artificial insemination from bulls with desirable qualities

Cow in oestrus is most receptive to mounting by bull, so is most likely to conceive. Recognise by:

- swollen vulva and watery mucus discharge
- restless behaviour, including licking of other cows
- cow stands with hind legs apart, ready to be mounted

▲ LH levels at peak

▲ optimum time for **artificial insemination**. This technique has several benefits:

- high-quality bull can provide semen for many cows
- very cost effective, and safe (no bulls!)
- reduced transmission of disease.

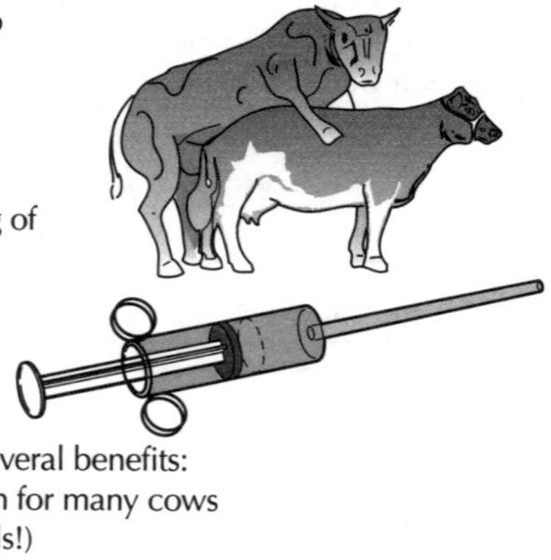

Pregnancy and calving cause lactation

- **Prolactin** stimulates development of mammary glands
- **Oxytocin** stimulates release of milk from udder.

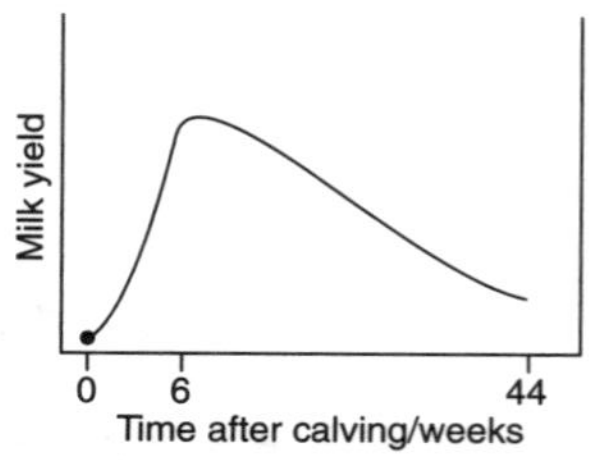

Maximise milk yield by:

- injection with BST (**bovine somatotrophin**) which stimulates cell division and growth of udders
- frequent calving, stimulated by fertility treatment with FSH.

Progesterone inhibits development of more follicles

Corpus luteum

↓

Progesterone

↓

Feedback inhibition

↓

FSH levels fall

Prostaglandin 2 α causes corpus luteum to deteriorate quickly. Next ovulation occurs only 3 days later!

Synthetic hormones work in humans too!

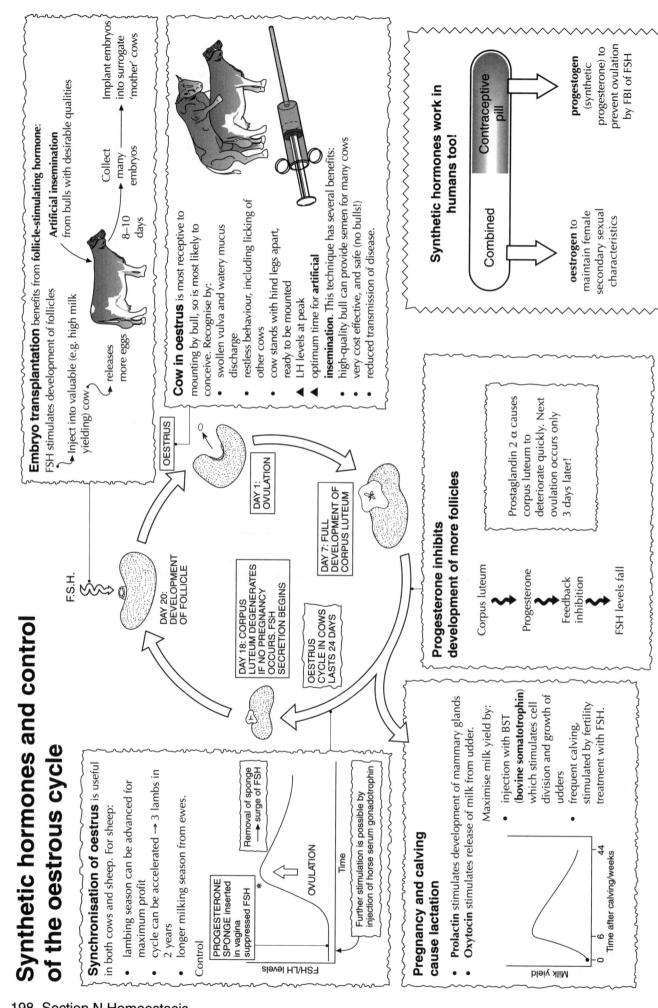

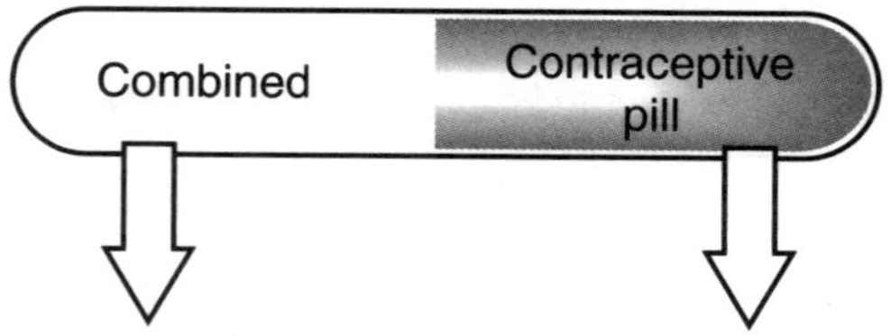

oestrogen to maintain female secondary sexual characteristics

progestogen (synthetic progesterone) to prevent ovulation by FBI of FSH

Oestrogen affects target cell activity by control of transcription.

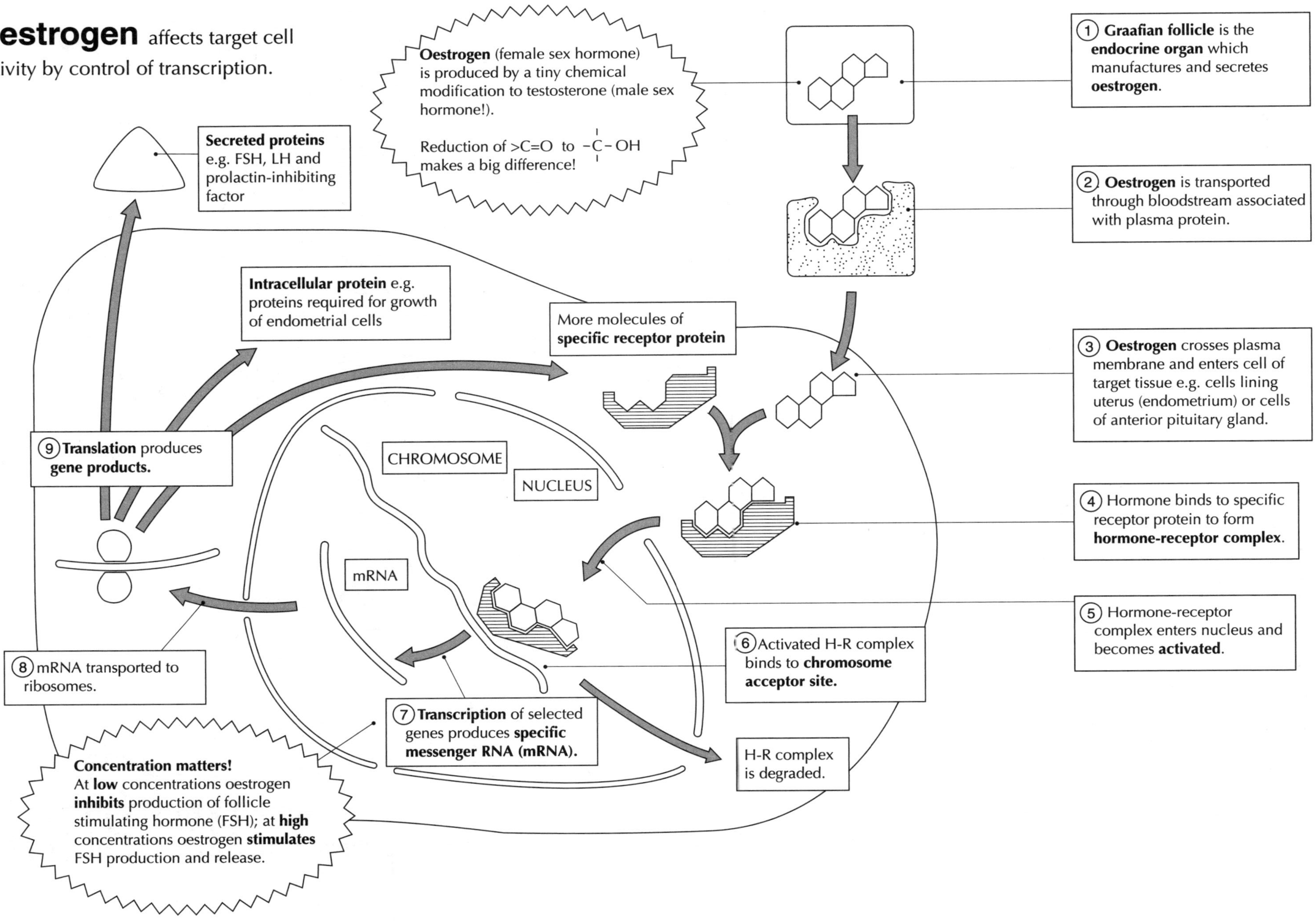

Oestrogen (female sex hormone) is produced by a tiny chemical modification to testosterone (male sex hormone!).
Reduction of >C=O to –C–OH makes a big difference!
1 Graafian follicle is the endocrine organ which manufactures and secretes oestrogen.
2 Oestrogen is transported through bloodstream associated with plasma protein.
3 Oestrogen crosses plasma membrane and enters cell of target tissue e.g. cells lining uterus (endometrium) or cells of anterior pituitary gland.
4 Hormone binds to specific receptor protein to form hormone-receptor complex.
5 Hormone-receptor complex enters nucleus and becomes activated.
6 Activated H-R complex binds to chromosome acceptor site.
7 Transcription of selected genes produces specific messenger RNA (mRNA).
8 mRNA transported to ribosomes.
9 Translation produces gene products.
Secreted proteins e.g. FSH, LH and prolactin-inhibiting factor
Intracellular protein e.g. proteins required for growth of endometrial cells
More molecules of specific receptor protein
CHROMOSOME
NUCLEUS
mRNA
H-R complex is degraded.
Concentration matters!
At low concentrations oestrogen inhibits production of follicle stimulating hormone (FSH); at high concentrations oestrogen stimulates FSH production and release.

Section O Experiments in biology

Design an experiment

An experiment is designed to **test the validity of an hypothesis** and involves the **collection of data.**

e.g. light intensity affects the rate of photosynthesis

using appropriate apparatus and instruments

The volume of oxygen released in a fixed length of time can be used to calculate the rate of photosynthesis: this is the outcome (**dependent) variable**.

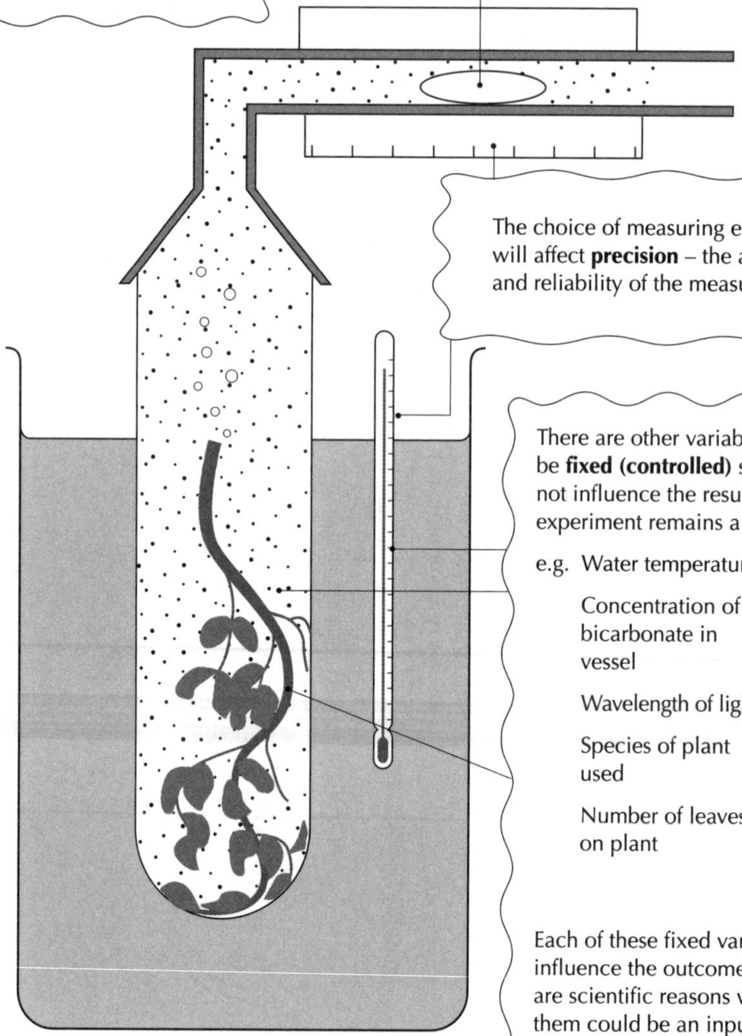

The choice of measuring equipment will affect **precision** – the accuracy and reliability of the measurements.

The light intensity can be varied by the experimenter: this is the input (**independent**) **variable**.

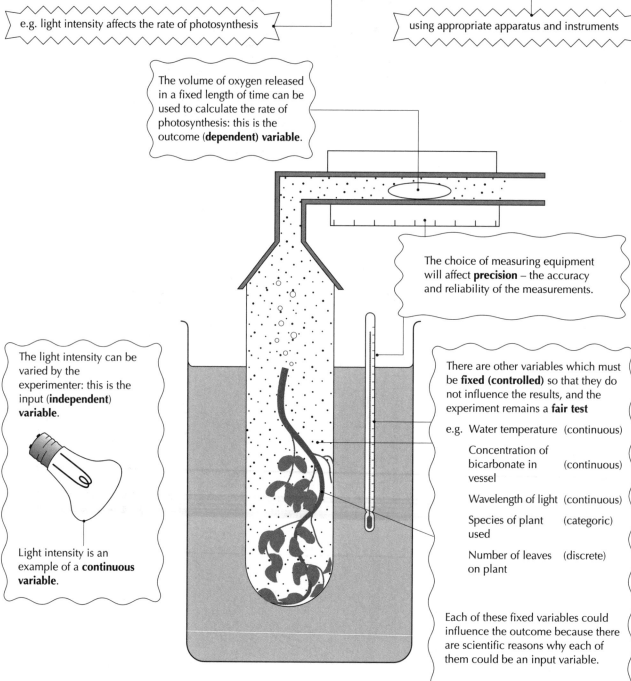

Light intensity is an example of a **continuous variable**.

There are other variables which must be **fixed (controlled)** so that they do not influence the results, and the experiment remains a **fair test**

e.g. Water temperature	(continuous)
Concentration of bicarbonate in vessel	(continuous)
Wavelength of light	(continuous)
Species of plant used	(categoric)
Number of leaves on plant	(discrete)

Each of these fixed variables could influence the outcome because there are scientific reasons why each of them could be an input variable.

A **control** experiment is the same in every respect **except** the manipulated variable is not changed but is kept constant.

A control allows confirmation that **no unknown variable** is responsible for any observed changes in the responding variable: it helps to make the experiment a **fair test**.

A **'repeat'** is performed when the experimenter suspects that misleading data has been obtained through 'operator error'.

'Means' of a series of results minimise the influence of any single result, and therefore reduce the effect of any 'rogue' or anomalous data. The use of means improves the **reliability** of data – how confident you are about the observations or measurements.

Dealing with data

may involve a number of steps.

1. Organisation of the **raw data** (the information which you actually collect during your investigation).
2. Manipulation of the data (converting your measurements into another form).

   and

3. Representation of the data in **graphical** or other form.

Steps 1. and 2. usually involve **preparation of a table of results.**

Raw data: input variable always in LH column

May often involve complex units: take care in presentation e.g. beats per minute or beats/min.

Independent or input variable e.g. temp/°C	Dependent or outcome variable e.g. oxygen given out/cm^3	Manipulated data e.g. rate of oxygen production /cm^3 per min.
20.00		
26.25		
–		
30.75		

Column heading should include **name of variable** and **units** e.g. **temperature/°C** or **temperature in °C**.

Derived from the raw data e.g. if 3 readings of an outcome variable are taken, the **mean** of these readings would be manipulated data OR if you measured heart rates in beats per 15 seconds, the rate in beats per minute would be manipulated data.

Units should **always** be in column heading and **never** alongside the numerical value.

Present **numerical data**

a. in decimal form e.g. 6.25 not $6\frac{1}{4}$
b. to the same number of significant figures e.g. if 6.25 given then use 6.00 <u>not</u> 6
c. as true numbers e.g. 0.7 not .7

There should be an **informative title:** simple **effect of** (variable in LH column) **on** (variable in RH column) e.g. Effect on temperature on pulse rate in humans.

There should be no 'blanks':

a. **missing value** should be shown as –
b. **value of zero** should be shown as 0.

Rule lines around table: looks **neater** and **more professional**.

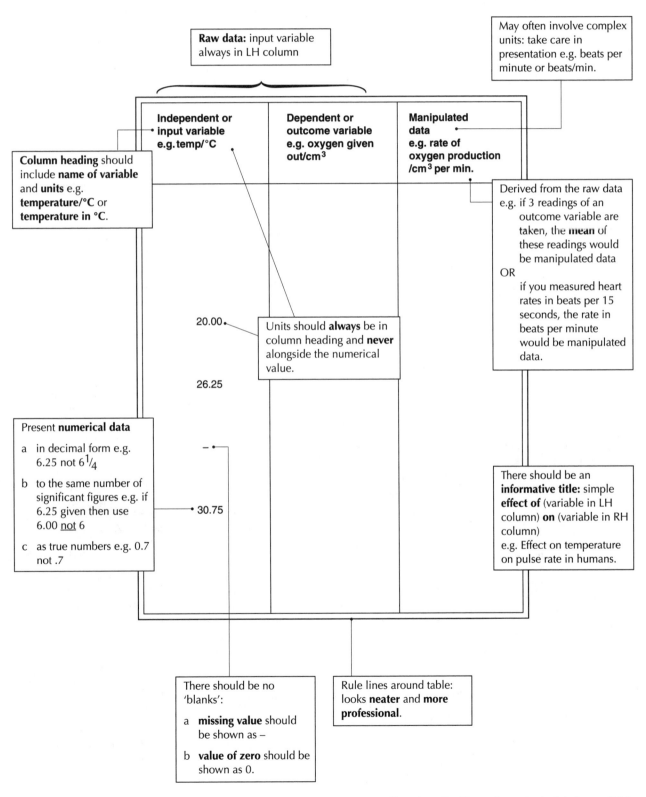

Graphical representation: A graph is a visual presentation of data and may help to make the relationship between variables more obvious, i.e. allow the experimenter to **draw a conclusion**.

For example

Air temperature/°C	**Body temperature of reptile/°C**
20	19.4
25	25.9
30	30.4
35	35.1
40	40.1
45	44.8

isn't as helpful as

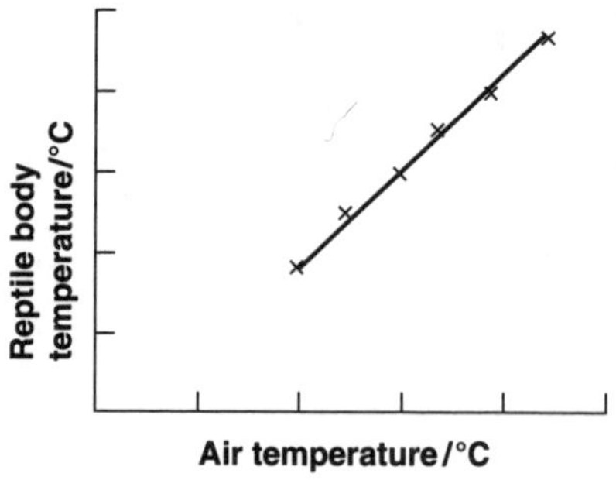

which suggests that **reptile body temperature is directly proportional to air temperature**

Validity – how much confidence can be placed in the conclusion. Can be affected by the **range** and **reliability** of measurement.

A graph may be produced from a table of data **following certain rules**

y axis

outcome (dependent) variable

NB units separated from physical quantity by a solidus

e.g. **reptile temperature /°C**

or in words

e.g. **reptile temperature in °C** or in brackets e.g. reptile temperature (°C)

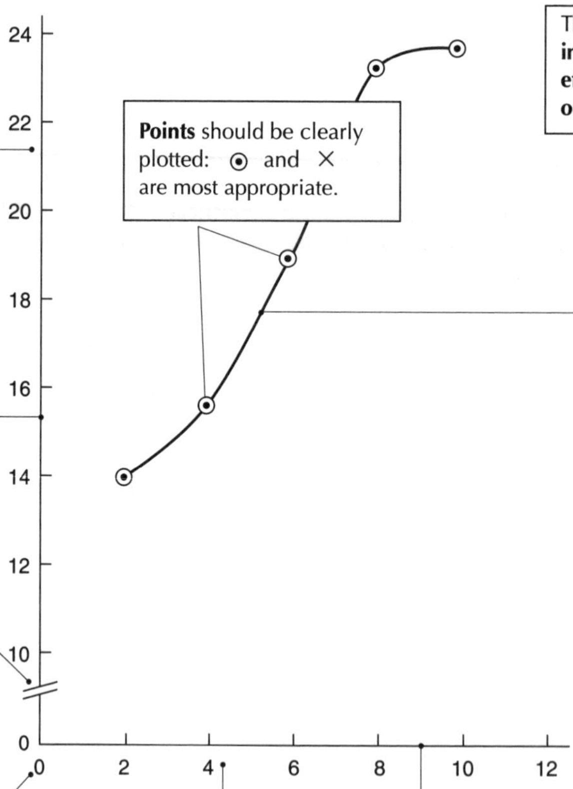

There should be an **informative title:** simply **effect of** (variable on *x* axis) **on** (variable on y axis).

Points should be clearly plotted: ⊙ and × are most appropriate.

The line of the axes should be ruled in: use a **black** line.

Axes should be **linear** and, wherever possible, **unbroken**. If an axis must be broken to make best use of the scale, the break should be shown as ⫲

Origin should be shown on both axes.

Scale on axes should

a be marked **in equal increments**

b extend to cover all plotted points

c make best use of the space available.

x axis

input (independent) variable

physical quantity/units

Once the points are plotted they may be joined by a smooth curve, if theory predicts a smooth, gradual transition...

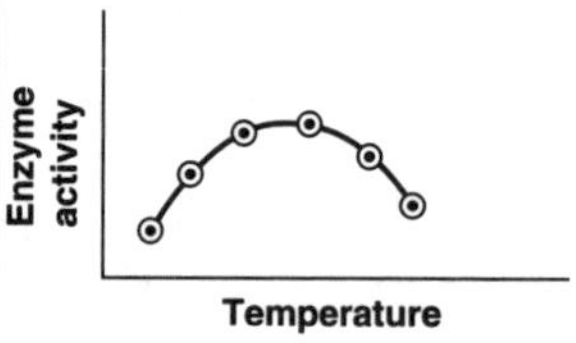

... or by **short, straight lines** if a smooth, gradual change cannot be predicted.

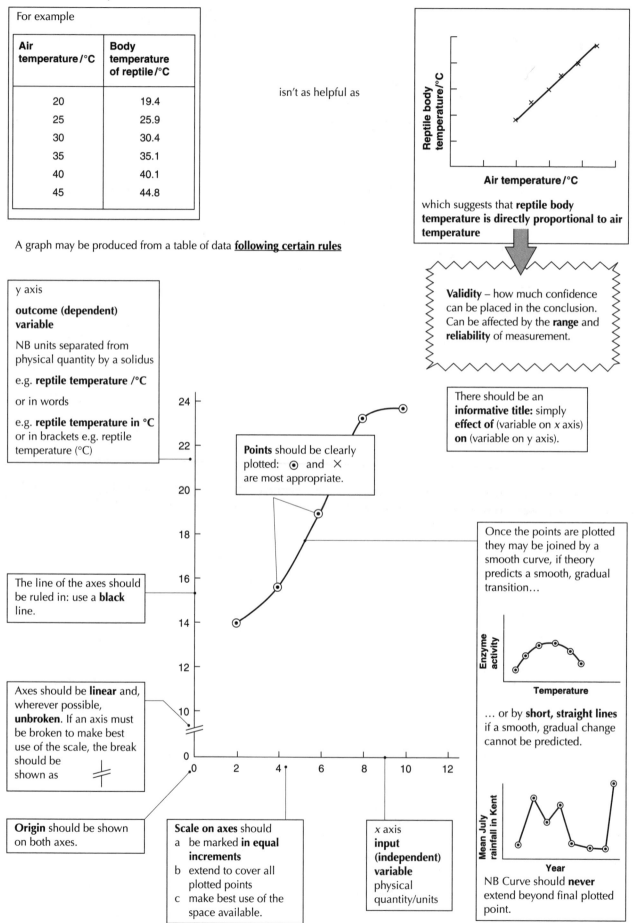

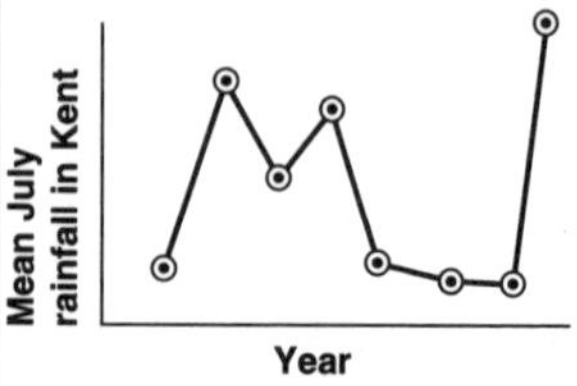

NB Curve should **never** extend beyond final plotted point.

Temperature affects permeability of membranes

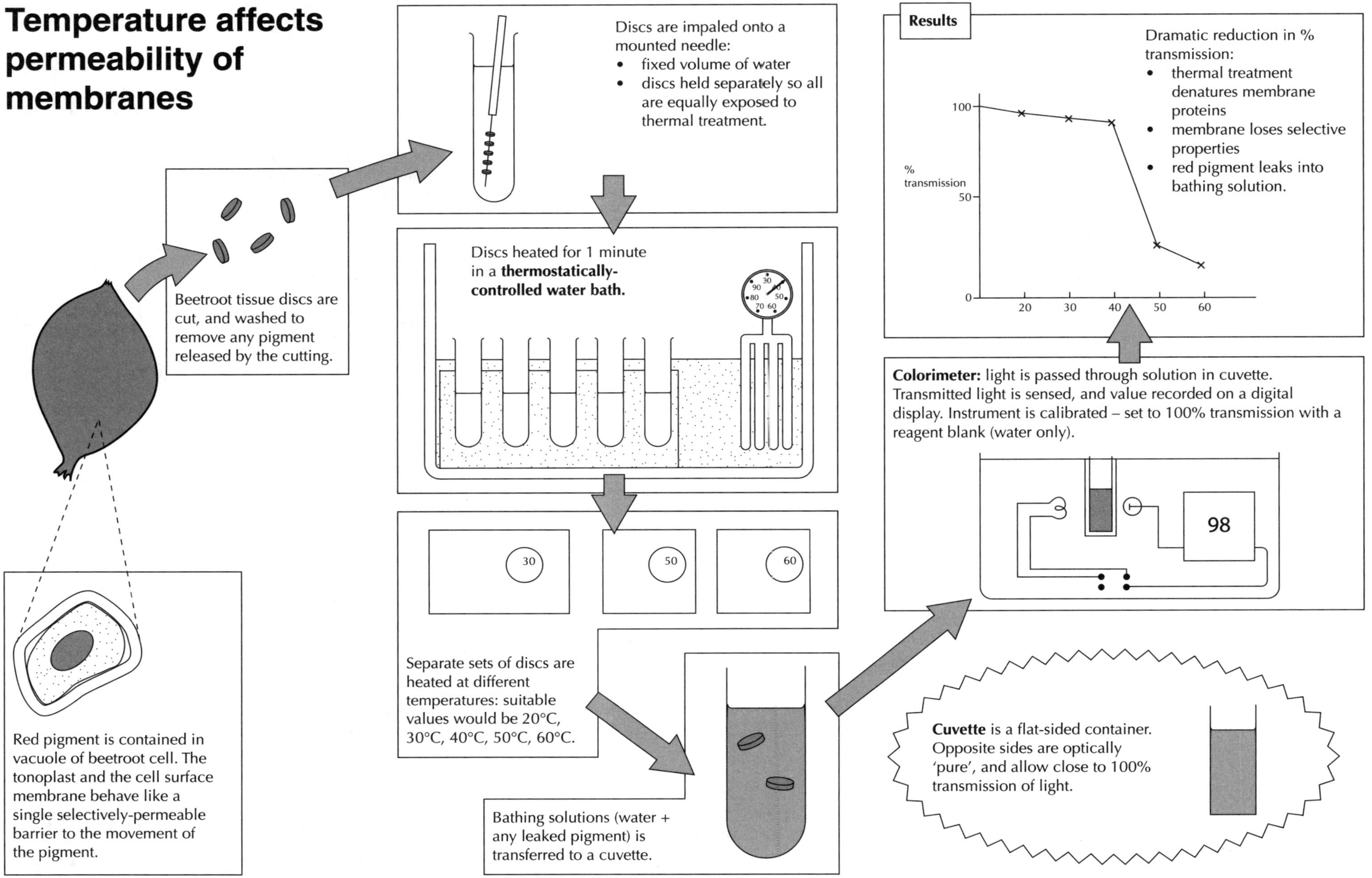
Discs are impaled onto a mounted needle:
• fixed volume of water
• discs held separately so all are equally exposed to thermal treatment.
Beetroot tissue discs are cut, and washed to remove any pigment released by the cutting.
Discs heated for 1 minute in a **thermostatically-controlled water bath.**
90
30
40
80
50
70
60
30
50
60
Separate sets of discs are heated at different temperatures: suitable values would be 20°C, 30°C, 40°C, 50°C, 60°C.
Bathing solutions (water + any leaked pigment) is transferred to a cuvette.
Red pigment is contained in vacuole of beetroot cell. The tonoplast and the cell surface membrane behave like a single selectively-permeable barrier to the movement of the pigment.
Results
100
50
0
% transmission
20
30
40
50
60
Dramatic reduction in % transmission:
• thermal treatment denatures membrane proteins
• membrane loses selective properties
• red pigment leaks into bathing solution.
Colorimeter: light is passed through solution in cuvette. Transmitted light is sensed, and value recorded on a digital display. Instrument is calibrated – set to 100% transmission with a reagent blank (water only).
98
Cuvette is a flat-sided container. Opposite sides are optically 'pure', and allow close to 100% transmission of light.

Determination of water potential of plant tissue

Potato is a suitable tissue: carrot, turnip and apple can be used.

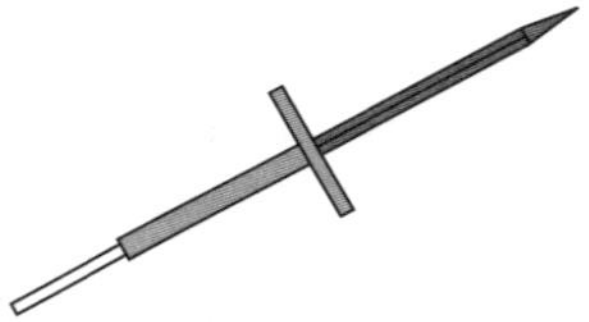

Health and safety! Both cork borer and scalpel are sharp: **Take care!**

Cylinders of potato tuber tissues are cut using a cork borer, on a glazed tile or cutting block.

The cylinders of tissue are pushed out of the cork borer using the blunt end of a pencil.

Quickly, to avoid the tissue drying out …

… the cylinders are cut to shorter cylinders of the same length (30mm is a suitable length).

The mass of each cylinder is measured

accurately, then cylinders are immersed in the sucrose solutions.

A series of solutions is prepared, using distilled water to dilute a 1.0M (1 mol sucrose per dm^3 solution) sucrose solution. (1 mol sucrose = 342g).

Concn. of sucrose solution (mol. dm^{-3})	Vol. 1M sucrose (cm^3)	Vol. distilled water (cm^3)
0.0	0	20
0.1	2	18
0.2	4	16
0.3	6	14
0.4	8	12
0.5	10	10

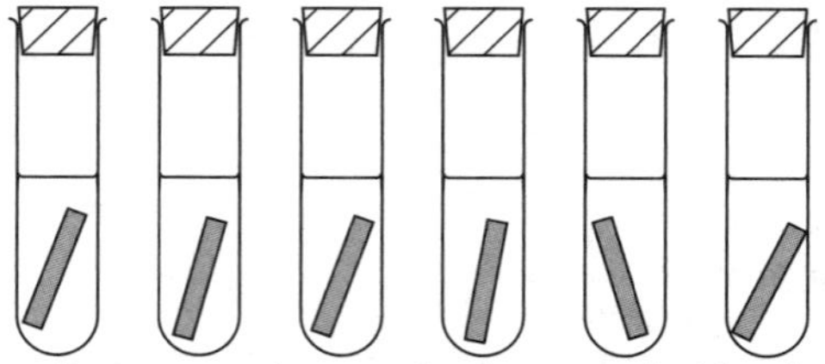

- cylinders must be completely covered with solution: **all then have the same surface area exposed to the sucrose solutions**
- tubes are stoppered: **no evaporation as this would change the sucrose concentration**
- each tube replicated twice: **provides mean of three values.**

Tubes are left
- for the same length of time (up to 24h.)
- at the same temperature

(**control of fixed variables**).

Cylinders are:
- dried to remove surface solution, by rolling on paper towel
- mass is remeasured, using the same top pan balance (**standardisation**).

Results are recorded in a suitable table:

Concn. of soln. (mol. dm^{-3})	Initial mass (g)	Final mass (g)	Change in mass (g)	% change in mass

$$\% \text{ change in mass} = \frac{\text{change in mass}}{\text{initial mass}} \times 100$$

This eliminates any possible effect of variation in initial mass.

… and plotted as a line graph

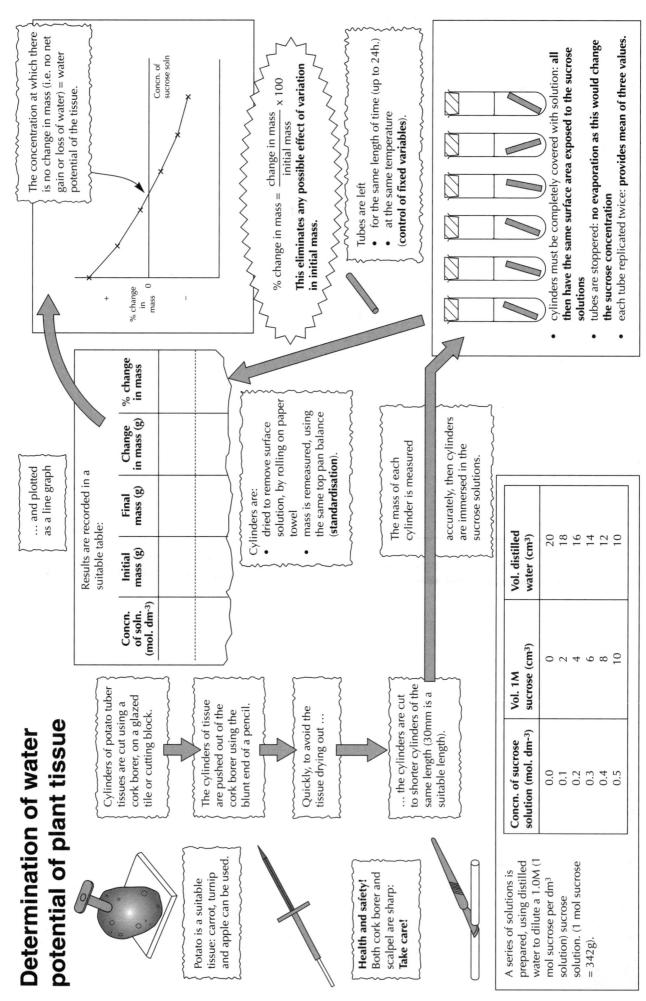

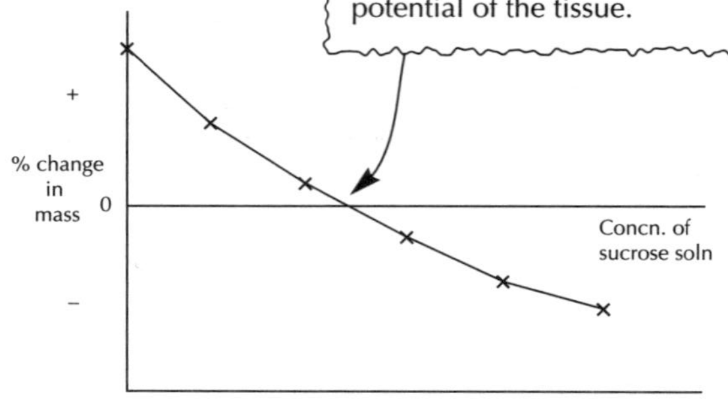

The concentration at which there is no change in mass (i.e. no net gain or loss of water) = water potential of the tissue.

Effect of pH on the activity of the enzyme catalase

$$2H_2O_2 \underset{}{\overset{\text{Catalase}}{\rightleftharpoons}} 2H_2O + O_2$$

This is a very active enzyme! It removes dangerous peroxide ions (produced in respiration) to limit damage to DNA and proteins and slow down cell ageing.

Start reaction by addition of hydrogen peroxide.

Gas syringe: take care with accuracy – can only measure to 0.5 cm^3 (5 mm^3).

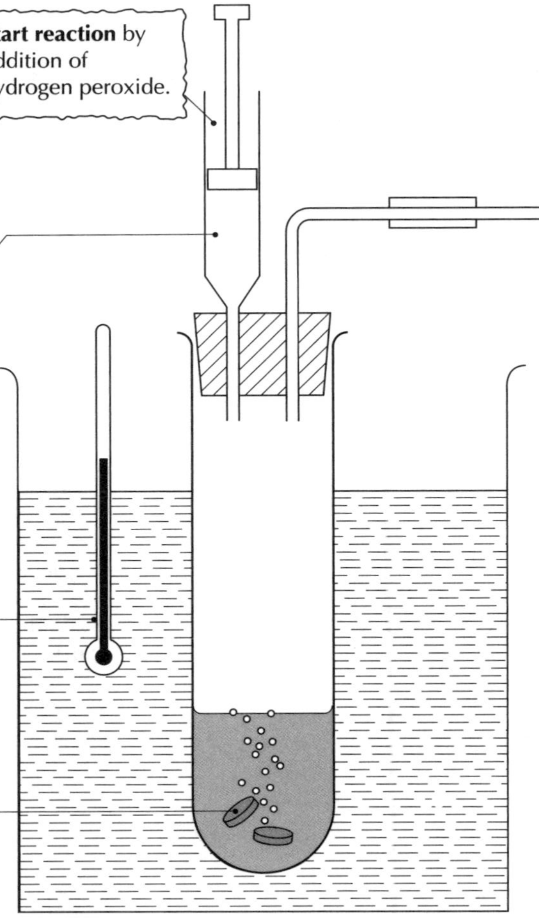

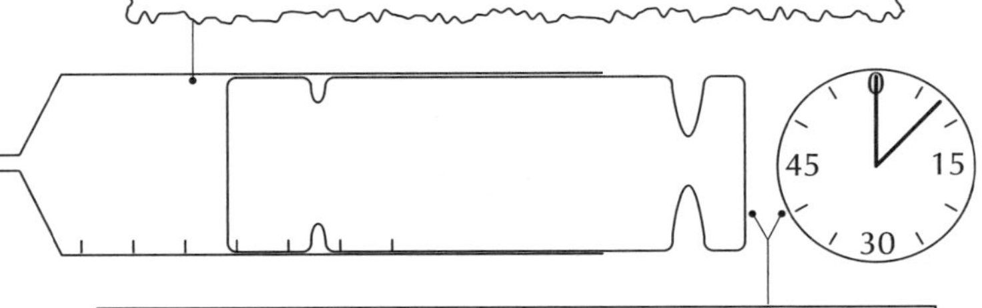

THESE ARE FIXED VARIABLES

Substrate concentration
- fixed volume
- fixed concentration of hydrogen peroxide solution.

Must be fixed as H_2O_2 concentration affects frequency/number of enzyme-substrate collisions.

Temperature
- use thermostatically-controlled water bath
- if no **controlled** water bath available, at least **measure** the temperature to check that it remains constant.

Must be fixed as temperature affects the number of enzyme-substrate collisions which can lead to product.

Enzyme concentration
- fixed **mass** of source (e.g. potato) to provide fixed number of enzyme molecules
- fixed **surface area** of source e.g. fixed number of fixed size potato discs.

Must be fixed as enzyme concentration affects frequency/number of enzyme-substrate collisions.

pH is the input variable
- varied by the use of a range of **buffer solutions**
- affects attraction between substrate and active site of enzyme
- can affect enzyme conformation
- use wide range e.g. pH 4-9 to find 'approximate' optimum pH
- use narrower range e.g. pH 6-8 in increments of 0.5 to obtain more accurate value.

Rate of oxygen production is the outcome variable
- Measure volume produced (mm^3) in measured time (s.).
- Calculate rate = $\frac{\text{volume produced}}{\text{time}}$ in mm^3/s.
- More reliable data uses **initial rate**.

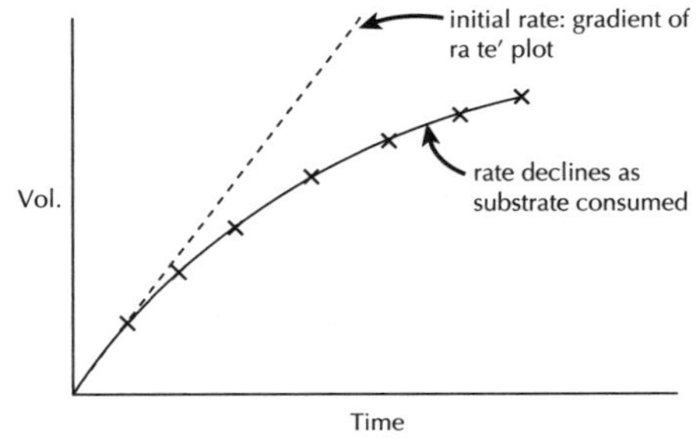

then plot **initial rate** v. pH

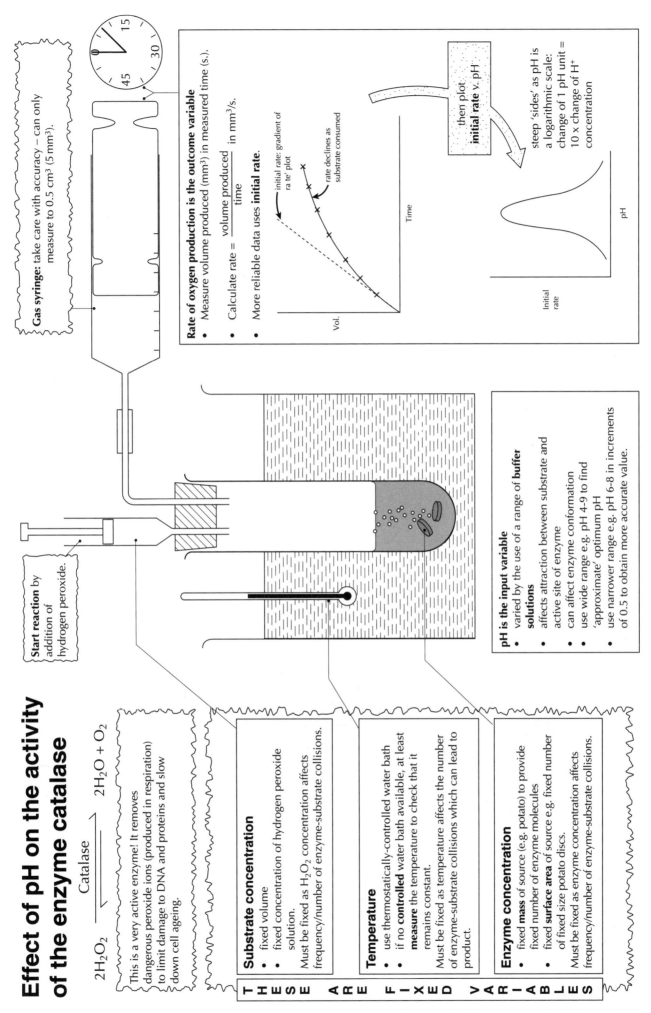

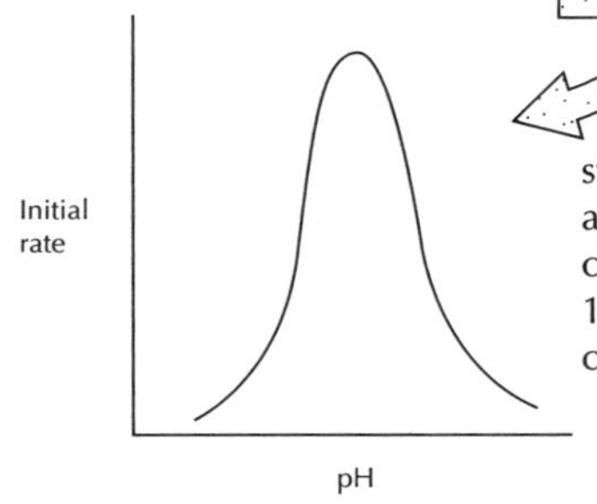

steep 'sides' as pH is a logarithmic scale: change of 1 pH unit = 10 x change of H^+ concentration

Mitosis in root tips

It is easier to observe mitosis in plant cells because:

- mitosis is confined to definite growth regions (meristems);
- cells are more easily 'squashed' without damage.

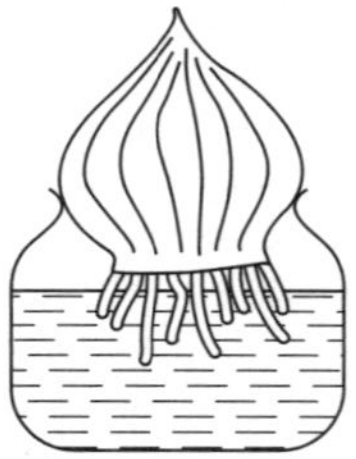

(1) Onion roots develop when onion bulb is allowed to develop above water. Roots increase in length by mitosis.

(2) Root tip (final 5 mm of the visible root) includes the meristems, the region where mitotic cell division is taking place.

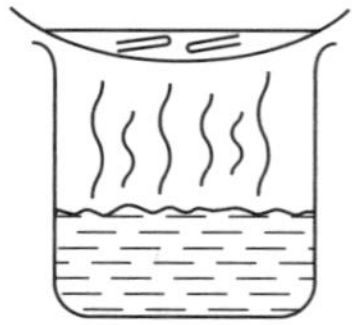

(3) Tips are treated in a mixture of hydrochloric acid (helps to separate cells by dissolving inter-cellular cement) and acetic orcein (which stains chromosomes a red-purple colour). Mixture is heated over a steam bath: **no naked flames as vapour from stain is very flammable.**

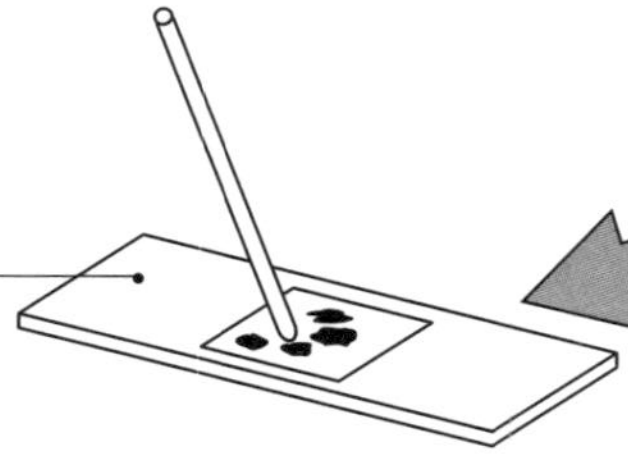

(4) After 3 minutes, tips are placed onto a microscope slide, and covered with a coverslip.

- tap with glass rod
- squash with thumb …

… to separate cells on slide.

(5) View using a light microscope

- x 100 to locate the stained tissue
- x 400 to identify cells at each stage of mitosis.

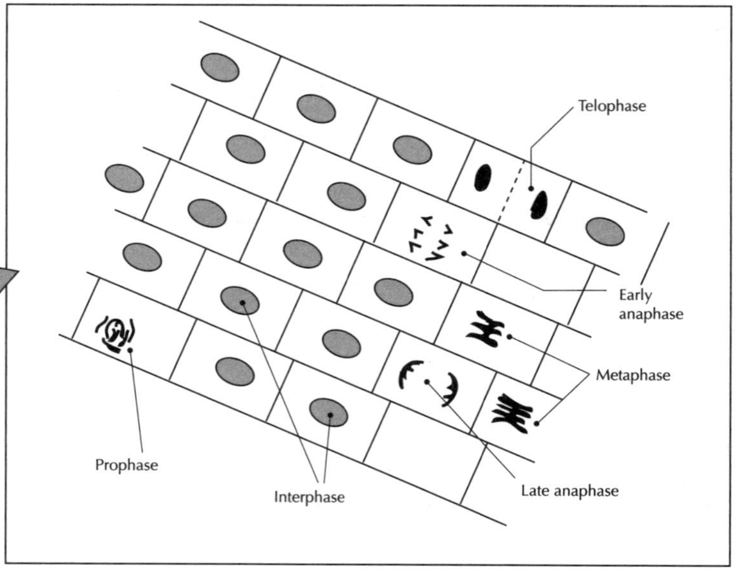

(6) Count number of cells at each stage of mitosis.

Stage	Number of cells	Proportion at each stage	Minutes / 24h.

(7) Present as pie chart.

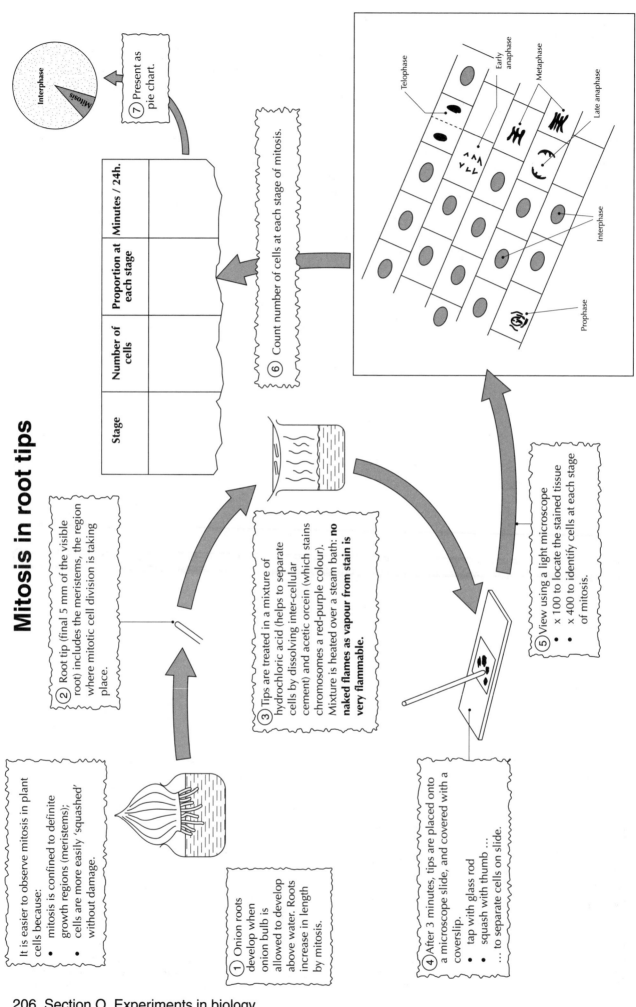

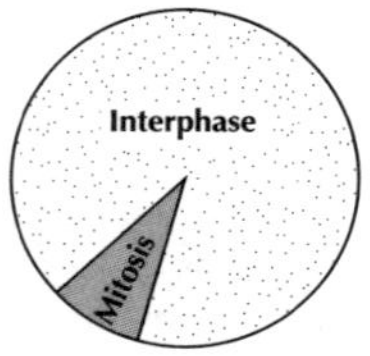

Effect of caffeine on heart rate

The input variable:
- supply as a caffeine solution e.g. coffee or tea, or cola drink
- could vary concentration using decaffeinated version of the drink as a dilutent
- **control** might be same volume of decaffeinated version of the same drink, delivered at the same temperature.

Outcome variable:
- measure using electronic heart rate monitor
- less reliable is to measure radial pulse (in wrist)

(No. of beats in 15s) x 4 = heart rate in beats per minute

- take mean of 3 or 5 values.

The most reliable **fixed variable** is the human subject.

A representative sample would be made up of people (subjects) …

Remember! **Fix** the time interval between drinking and having the heart rate measured!

A single subject would eliminate these issues: perhaps test same individual on several consecutive days.

… of the same **gender** (sex)
- Males usually have a lower heart rate than equivalent females.
- Males usually have more metabolically active tissue than females: this may affect the internal concentration of caffeine.

… of the same **body mass**
- so that the internal concentration of caffeine does not vary too much;
- body mass may have an effect on resting heart rate.

… with the same **fitness level**
- to ensure similar values for resting heart rate.

… of similar **age**
- young, and older than middle age, people often have relatively high resting heart rate.

… and with the **same previous exposure to caffeine**
- heart may become **adapted** to caffeine.

How does caffeine work?

Action potential in neurone of sympathetic nervous system

Post-synaptic cell; effects are due to formation of cyclic AMP, a second messenger. This is removed by hydrolysis.

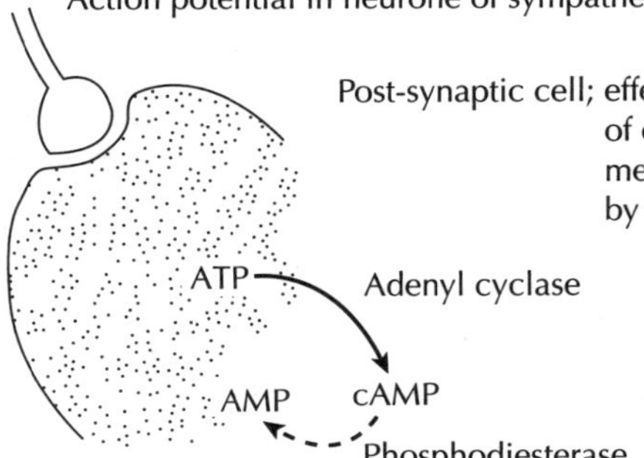

Inhibited by caffeine: continued high concentration of cAMP leads to 'excitation'.

An ideal group might be 30 (6 groups of 5) male members, of similar age, of an athletics club.

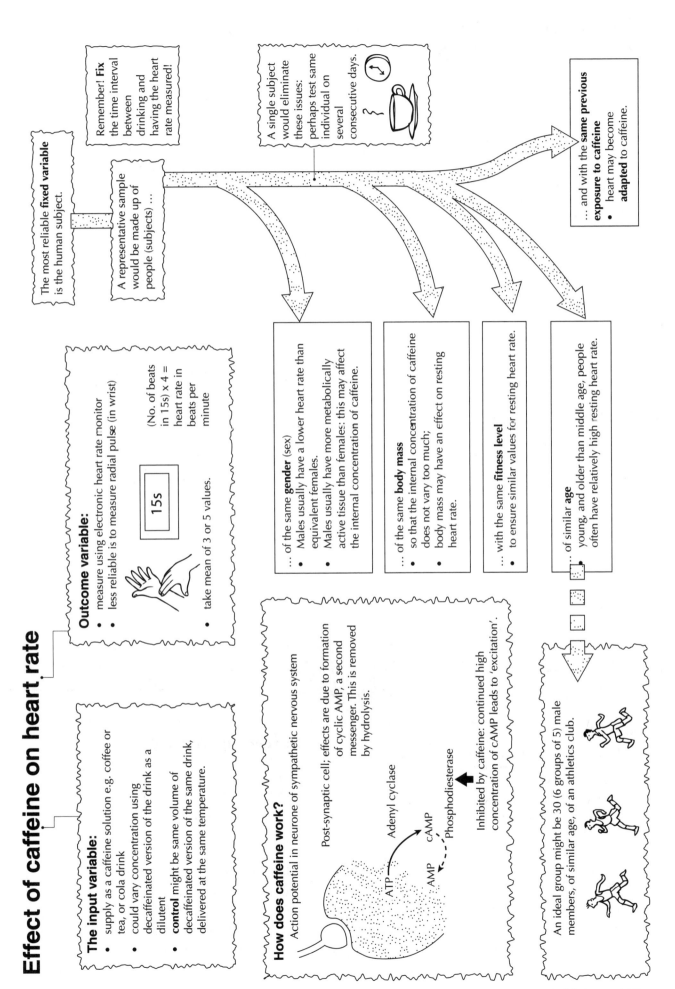

Respirometers and the measurement of respiratory quotient

Principle: Carbon dioxide evolved during respiration is absorbed by potassium hydroxide solution. If the system is closed to the atmosphere a change in volume of the gas within the chamber must be due to the consumption of oxygen. The change in volume, i.e. the oxygen consumption, is measured as the movement of a drop of coloured fluid along a capillary tube.

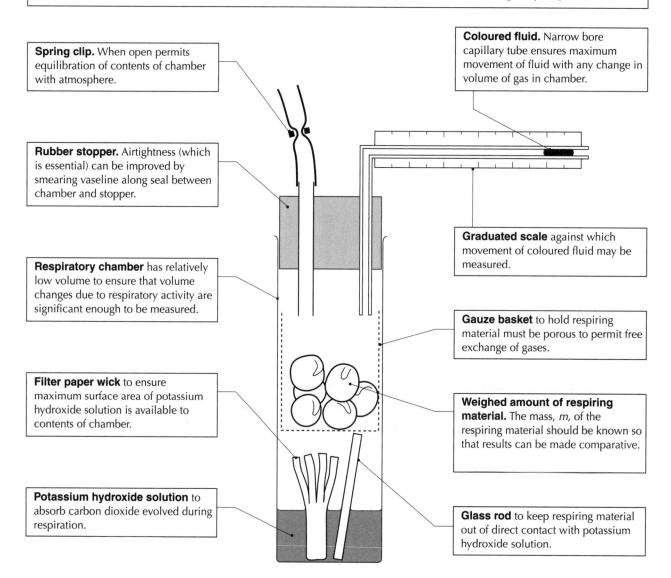

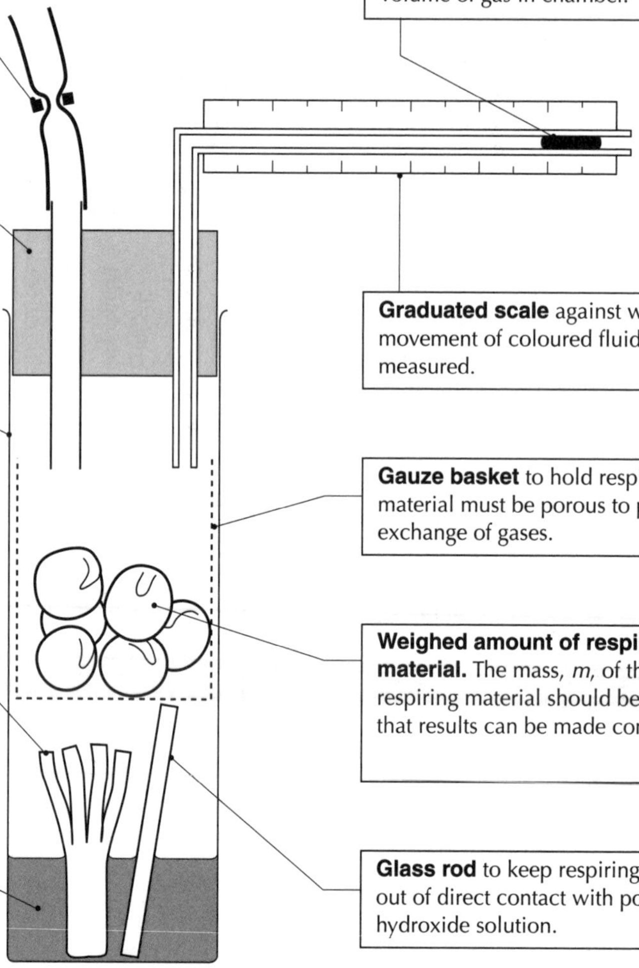

Spring clip. When open permits equilibration of contents of chamber with atmosphere.

Rubber stopper. Airtightness (which is essential) can be improved by smearing vaseline along seal between chamber and stopper.

Respiratory chamber has relatively low volume to ensure that volume changes due to respiratory activity are significant enough to be measured.

Filter paper wick to ensure maximum surface area of potassium hydroxide solution is available to contents of chamber.

Potassium hydroxide solution to absorb carbon dioxide evolved during respiration.

Coloured fluid. Narrow bore capillary tube ensures maximum movement of fluid with any change in volume of gas in chamber.

Graduated scale against which movement of coloured fluid may be measured.

Gauze basket to hold respiring material must be porous to permit free exchange of gases.

Weighed amount of respiring material. The mass, *m*, of the respiring material should be known so that results can be made comparative.

Glass rod to keep respiring material out of direct contact with potassium hydroxide solution.

Method

1 The apparatus, containing a known mass of respiring material, is set up with the spring clip open to permit equilibration.

2 With the spring clip closed the time, *t*, in minutes, for the coloured fluid to move a distance *l* mm. along the capillary tube is noted.

Oxygen Consumed = *l* units

3 Remove KOH solution and note fluid movement in same time, *t*

Oxygen Consumed – Carbon Dioxide Released = m units

Then R.Q. = $\frac{l - m}{l}$

Respiratory quotient (R.Q.)

$$= \frac{CO_2 \text{ evolved}}{O_2 \text{ consumed}} \text{ during respiration of a particular food}$$

Aerobic respiration of carbohydrate

$$C_6H_{12}O_6 + 6O_2 \rightarrow 6CO_2 + 6H_2O$$

Thus R.Q. = $\frac{6}{6}$ = 1.0

For lipid R.Q. $\simeq$ 0.7, for protein R.Q. $\simeq$ 0.95

Thus the nature of the respiratory substrate can be determined by measurement of R.Q.

Mixed diet (ChO/lipid) R.Q. $\simeq$ 0.85

Starvation (body protein reserves) R.Q. $\simeq$ 0.9 – 1.0

Index